T0134395

Applied Multivariate Statistical Analysis

Wolfgang Karl Härdle · Léopold Simar

Applied Multivariate Statistical Analysis

Fifth Edition

 Springer

Wolfgang Karl Härdle
Ladislaus von Bortkiewicz Chair of Statistics
Humboldt-Universität zu Berlin
Berlin, Germany

Léopold Simar
Institute of Statistics, Biostatistics
and Actuarial Sciences
Université Catholique de Louvain
Louvain-la-Neuve, Belgium

ISBN 978-3-030-26005-7   ISBN 978-3-030-26006-4   (eBook)
https://doi.org/10.1007/978-3-030-26006-4

Mathematics Subject Classification (2010): 62H05, 62H10, 62H15, 62H25, 62H30, 62H12

1st–4th editions: © Springer-Verlag Berlin Heidelberg 2003, 2007, 2012, 2015
5th edition: © Springer Nature Switzerland AG 2019

This Springer imprint is published by the registered company Springer Nature Switzerland AG
The registered company address is: Gewerbestrasse 11, 6330 Cham, Switzerland

Preface

The fifth edition of this book on *Applied Multivariate Statistical Analysis* offers an extension of cluster analysis, including unsupervised learning and minimum spanning trees (MST). Examples have been updated as well and light errors have been corrected.

All pictures and numerical examples have been now calculated in the (almost) standard language R and MATLAB. The code of practical examples is indicated with a small **Q** sign. We believe that these publicly available Quantlets on https://github.com/QuantLet/MVA (see also www.quantlet.de) create a valuable contribution to the distribution of knowledge in the statistical science. We welcome all readers of this book to propose changes to the authors' existing codes or add codes in other programming languages. The symbols and notations have also been standardized. Some datasets have been updated in this new edition. In the preparation of the fifth edition, we received valuable inputs from Petra Burdejova, Alla Petukhina, Larisa Adamyan, and Kirill Efimov. Petra Burdejova and Alla Petukhina had very creative and valuable comments on all chapters, in particular, Chaps. 8, 11, and 13. Larisa Adamyan and Kirill Efimov presented their technology on adaptive clustering. Without the input of these young promising researchers, this edition of the text would not have been possible. We would like to thank them.

Berlin, Germany   Wolfgang Karl Härdle
Louvain la Neuve, Belgium   Leopold Simar
January 2019

Contents

Part I   Descriptive Techniques

1   Comparison of Batches   3
    1.1   Boxplots   4
    1.2   Histograms   11
    1.3   Kernel Densities   13
    1.4   Scatterplots   18
    1.5   Chernoff-Flury Faces   21
    1.6   Andrews' Curves   23
    1.7   Parallel Coordinate Plots   27
    1.8   Hexagon Plots   32
    1.9   Boston Housing   35
    1.10   Exercises   42
    References   43

Part II   Multivariate Random Variables

2   A Short Excursion into Matrix Algebra   47
    2.1   Elementary Operations   47
    2.2   Spectral Decompositions   53
    2.3   Quadratic Forms   55
    2.4   Derivatives   58
    2.5   Partitioned Matrices   59
    2.6   Geometrical Aspects   61
    2.7   Exercises   68

3   Moving to Higher Dimensions   71
    3.1   Covariance   72
    3.2   Correlation   76

3.3   Summary Statistics   81
3.4   Linear Model for Two Variables   84
3.5   Simple Analysis of Variance   91
3.6   Multiple Linear Model   95
3.7   Boston Housing   100
3.8   Exercises   103
References   105

4   Multivariate Distributions   107
4.1   Distribution and Density Function   108
4.2   Moments and Characteristic Functions   113
4.3   Transformations   123
4.4   The Multinormal Distribution   125
4.5   Sampling Distributions and Limit Theorems   129
4.6   Heavy-Tailed Distributions   135
4.7   Copulae   151
4.8   Bootstrap   161
4.9   Exercises   164
References   166

5   Theory of the Multinormal   167
5.1   Elementary Properties of the Multinormal   167
5.2   The Wishart Distribution   174
5.3   Hotelling's T^2-Distribution   176
5.4   Spherical and Elliptical Distributions   178
5.5   Exercises   180
References   182

6   Theory of Estimation   183
6.1   The Likelihood Function   184
6.2   The Cramer–Rao Lower Bound   188
6.3   Exercises   192
Reference   193

7   Hypothesis Testing   195
7.1   Likelihood Ratio Test   196
7.2   Linear Hypothesis   205
7.3   Boston Housing   222
7.4   Exercises   225
References   229

Part III Multivariate Techniques

8 Regression Models 233
 8.1 General ANOVA and ANCOVA Models 235
    8.1.1 ANOVA Models 235
    8.1.2 ANCOVA Models 240
    8.1.3 Boston Housing 242
 8.2 Categorical Responses 243
    8.2.1 Multinomial Sampling and Contingency Tables 243
    8.2.2 Log-Linear Models for Contingency Tables 244
    8.2.3 Testing Issues with Count Data 248
    8.2.4 Logit Models 251
 8.3 Exercises 258
 Reference.. 259

9 Variable Selection 261
 9.1 Lasso 262
    9.1.1 Lasso in the Linear Regression Model 262
    9.1.2 Lasso in High Dimensions 271
    9.1.3 Lasso in Logit Model 272
 9.2 Elastic Net 276
    9.2.1 Elastic Net in Linear Regression Model 277
    9.2.2 Elastic Net in Logit Model 278
 9.3 Group Lasso 279
 9.4 Exercises 282
 References... 283

10 Decomposition of Data Matrices by Factors 285
 10.1 The Geometric Point of View 286
 10.2 Fitting the p-Dimensional Point Cloud 287
 10.3 Fitting the n-Dimensional Point Cloud 290
 10.4 Relations Between Subspaces 292
 10.5 Practical Computation 293
 10.6 Exercises 296

11 Principal Components Analysis 299
 11.1 Standardized Linear Combination 300
 11.2 Principal Components in Practice 303
 11.3 Interpretation of the PCs 307
 11.4 Asymptotic Properties of the PCs..................... 310
 11.5 Normalized Principal Components Analysis 313
 11.6 Principal Components as a Factorial Method 315
 11.7 Common Principal Components 320

11.8	Boston Housing	323
11.9	More Examples	326
11.10	Exercises	335
	References	336
12	**Factor Analysis**	337
12.1	The Orthogonal Factor Model	338
12.2	Estimation of the Factor Model	345
12.3	Factor Scores and Strategies	352
12.4	Boston Housing	355
12.5	Exercises	358
	References	361
13	**Cluster Analysis**	363
13.1	The Problem	364
13.2	The Proximity Between Objects	365
13.3	Cluster Algorithms	370
13.4	Adaptive Weights Clustering	381
13.5	Spectral Clustering	385
13.6	Boston Housing	388
13.7	Exercises	391
	References	393
14	**Discriminant Analysis**	395
14.1	Allocation Rules for Known Distributions	395
14.2	Discrimination Rules in Practice	402
14.3	Boston Housing	408
14.4	Exercises	410
	References	411
15	**Correspondence Analysis**	413
15.1	Motivation	413
15.2	Chi-Square Decomposition	416
15.3	Correspondence Analysis in Practice	420
15.4	Exercises	429
16	**Canonical Correlation Analysis**	431
16.1	Most Interesting Linear Combination	431
16.2	Canonical Correlation in Practice	436
16.3	Exercises	441
	References	442
17	**Multidimensional Scaling**	443
17.1	The Problem	443
17.2	Metric Multidimensional Scaling	447

17.3   Nonmetric Multidimensional Scaling   452
17.4   Exercises   458
References   459

18 Conjoint Measurement Analysis   461
18.1   Introduction   461
18.2   Design of Data Generation   463
18.3   Estimation of Preference Orderings   465
18.4   Exercises   472
References   473

19 Applications in Finance   475
19.1   Portfolio Choice   475
19.2   Efficient Portfolio   476
19.3   Efficient Portfolios in Practice   483
19.4   The Capital Asset Pricing Model (CAPM)   484
19.5   Exercises   486
Reference   486

20 Computationally Intensive Techniques   487
20.1   Simplicial Depth   488
20.2   Projection Pursuit   491
20.3   Sliced Inverse Regression   496
20.4   Support Vector Machines   504
20.5   Classification and Regression Trees   519
20.6   Boston Housing   534
20.7   Exercises   538
References   539

Part IV   Appendix

21 Symbols and Notations   543
21.1   Basics   543
21.2   Mathematical Abbreviations   543
21.3   Samples   544
21.4   Densities and Distribution Functions   544
21.5   Moments   544
21.6   Empirical Moments   545
21.7   Distributions   545

22 Data   547
22.1   Boston Housing Data   547
22.2   Swiss Bank Notes   547
22.3   Car Data   548

22.4   Classic Blue Pullovers Data   548
22.5   U.S. Companies Data   549
22.6   French Food Data   549
22.7   Car Marks   549
22.8   U.S. Crime Data   549
22.9   Bankruptcy Data I   550
22.10   Bankruptcy Data II   551
22.11   Journaux Data   551
22.12   Timebudget Data   551
22.13   Vocabulary Data   552
22.14   French Baccalauréat Frequencies   553
References   553

Index   555

Part I
Descriptive Techniques

Chapter 1
Comparison of Batches

Multivariate statistical analysis is concerned with analyzing and understanding data in high dimensions. We suppose that we are given a set $\{x_i\}_{i=1}^n$ of n observations of a variable vector X in $\mathbb{R}^p$. That is, we suppose that each observation x_i has p dimensions:

$$x_i = (x_{i1}, x_{i2}, ..., x_{ip}),$$

and that it is an observed value of a variable vector $X \in \mathbb{R}^p$. Therefore, X is composed of p random variables:

$$X = (X_1, X_2, ..., X_p)$$

where X_j, for $j = 1, \ldots, p$, is a one-dimensional random variable. How do we begin to analyze this kind of data? Before we investigate questions on what inferences we can reach from the data, we should think about how to look at the data. This involves descriptive techniques. Questions that we could answer by descriptive techniques are the following:

- Are there components of X that are more spread out than others?
- Are there some elements of X that indicate subgroups of the data?
- Are there outliers in the components of X?
- How "normal" is the distribution of the data?
- Are there "low-dimensional" linear combinations of X that show "non-normal" behavior?

One difficulty of descriptive methods for high-dimensional data is the human perceptional system. Point clouds in two dimensions are easy to understand and to interpret. With modern interactive computing techniques, we have the possibility to see real-time 3D rotations and thus to perceive also three-dimensional data. A "sliding technique" as described in Härdle and Scott (1992) may give insight into four-dimensional structures by presenting dynamic 3D density contours as the fourth variable is changed over its range.

© Springer Nature Switzerland AG 2019
W. K. Härdle and L. Simar, *Applied Multivariate Statistical Analysis*,
https://doi.org/10.1007/978-3-030-26006-4_1

A qualitative jump in presentation difficulties occurs for dimensions greater than or equal to 5, unless the high-dimensional structure can be mapped into lower dimensional components Klinke and Polzehl (1995). Features like clustered subgroups or outliers, however, can be detected using a purely graphical analysis.

In this chapter, we investigate the basic descriptive and graphical techniques allowing simple exploratory data analysis. We begin the exploration of a data set using boxplots. A boxplot is a simple univariate device that detects outliers component by component and that can compare distributions of the data among different groups. Next, several multivariate techniques are introduced (Flury faces, Andrews' curves, and parallel coordinate plots) which provide graphical displays addressing the questions formulated above. The advantages and the disadvantages of each of these techniques are stressed.

Two basic techniques for estimating densities are also presented: histograms and kernel densities. A density estimate gives a quick insight into the shape of the distribution of the data. We show that kernel density estimates overcome some of the drawbacks of the histograms.

Finally, scatterplots are shown to be very useful for plotting bivariate or trivariate variables against each other: they help to understand the nature of the relationship among variables in a data set and allow for the detection of groups or clusters of points. Draftsman plots or matrix plots are the visualization of several bivariate scatterplots on the same display. They help detect structures in conditional dependencies by *brushing* across the plots. Outliers and observations that need special attention may be discovered with Andrews curves and Parallel Coordinate Plots. This chapter ends with an explanatory analysis of the Boston Housing data.

1.1 Boxplots

Example 1.1 The Swiss bank data (see Appendix, Sect. B.2) consists of 200 measurements on Swiss banknotes. The first half of these measurements are from genuine banknotes, the other half are from counterfeit banknotes.

The authorities measured, as indicated in Fig. 1.1,[1]

$$X_1 = \text{length of the bill,}$$
$$X_2 = \text{height of the bill (left),}$$
$$X_3 = \text{height of the bill (right),}$$
$$X_4 = \text{distance of the inner frame to the lower border,}$$
$$X_5 = \text{distance of the inner frame to the upper border,}$$
$$X_6 = \text{length of the diagonal of the central picture.}$$

[1]Figure provided by Bernhard Flury.

X2   X5   X3

X1

X4

Fig. 1.1 An old Swiss 1000-franc bank note

These data are taken from Flury and Riedwyl (1988). The aim is to study how these measurements may be used in determining whether a bill is genuine or counterfeit.

The *boxplot* is a graphical technique that displays the distribution of variables. It helps us see the location, skewness, spread, tail length, and outlying points.

It is particularly useful in comparing different batches. The boxplot is a graphical representation of the *Five-Number Summary*. To introduce the Five-Number Summary, let us consider for a moment a smaller, one-dimensional data set: the population of the 15 largest world cities in 2006 (Table 1.1).

In the Five-Number Summary, we calculate the upper quartile F_U, the lower quartile F_L, the median, and the extremes. Recall that order statistics $\{x_{(1)}, x_{(2)}, \ldots$ $\ldots, x_{(n)}\}$ are a set of ordered values x_1, x_2, x_n, where $x_{(1)}$ denotes the minimum and $x_{(n)}$ the maximum. The *median M* typically cuts the set of observations into two equal parts, and is defined as

$$M = \begin{cases} x_{\left(\frac{n+1}{2}\right)} & n \text{ odd} \\ \frac{1}{2}\left\{x_{\left(\frac{n}{2}\right)} + x_{\left(\frac{n}{2}+1\right)}\right\} & n \text{ even} \end{cases}. \tag{1.1}$$

The quartiles cut the set into four equal parts, which are often called *fourths* (that is why we use the letter F). Using a definition that goes back to Hoaglin et al. (1983), the definition of a median can be generalized to fourths, eights, etc. Considering the order statistics we can define the depth of a data value $x_{(i)}$ as $\min\{i, n-i+1\}$. If n is odd, the depth of the median is $\frac{n+1}{2}$. If n is even, $\frac{n+1}{2}$ is a fraction. Thus, the median

Table 1.1 The 15 largest world cities in 2006

City	Country	Pop. (10000)	Order statistics
Tokyo	Japan	3420	$x_{(15)}$
Mexico City	Mexico	2280	$x_{(14)}$
Seoul	South Korea	2230	$x_{(13)}$
New York	USA	2190	$x_{(12)}$
Sao Paulo	Brazil	2020	$x_{(11)}$
Bombay	India	1985	$x_{(10)}$
Delhi	India	1970	$x_{(9)}$
Shanghai	China	1815	$x_{(8)}$
Los Angeles	USA	1800	$x_{(7)}$
Osaka	Japan	1680	$x_{(6)}$
Jakarta	Indonesia	1655	$x_{(5)}$
Calcutta	India	1565	$x_{(4)}$
Cairo	Egypt	1560	$x_{(3)}$
Manila	Philippines	1495	$x_{(2)}$
Karachi	Pakistan	1430	$x_{(1)}$

is determined to be the average between the two data values belonging to the next larger and smaller order statistics, i.e., $M = \frac{1}{2}\left\{x_{(\frac{n}{2})} + x_{(\frac{n}{2}+1)}\right\}$. In our example, we have $n = 15$ hence the median $M = x_{(8)} = 1815$.

We proceed in the same way to get the fourths. Take the depth of the median and calculate

$$\text{depth of fourth} = \frac{[\text{depth of median}] + 1}{2}$$

with $[z]$ denoting the largest integer smaller than or equal to z. In our example this gives 4.5 and thus leads to the two-fourths

$$F_L = \frac{1}{2}\left\{x_{(4)} + x_{(5)}\right\}$$
$$F_U = \frac{1}{2}\left\{x_{(11)} + x_{(12)}\right\}$$

(recalling that a depth which is a fraction corresponds to the average of the two nearest data values).

The F-spread, d_F, is defined as $d_F = F_U - F_L$. The *outside bars*

$$F_U + 1.5d_F \tag{1.2}$$
$$F_L - 1.5d_F \tag{1.3}$$

Table 1.2 Five-number summary

World cities				
# 15	Depth			
M	8		1815	
F	4.5	1610		2105
	1	1430		3420

are the borders beyond which a point is regarded as an outlier. For the number of points outside these bars see Exercise 1.3. For the $n = 15$, data points the fourths are $1610 = \frac{1}{2} \left\{ x_{(4)} + x_{(5)} \right\}$ and $2105 = \frac{1}{2} \left\{ x_{(11)} + x_{(12)} \right\}$. Therefore, the *F-spread* and the upper and lower *outside bars* in the above example are calculated as follows:

$$d_F = F_U - F_L = 2105 - 1610 = 495 \tag{1.4}$$
$$F_L - 1.5d_F = 1610 - 1.5 \cdot 495 = 867.5 \tag{1.5}$$
$$F_U + 1.5d_F = 2105 + 1.5 \cdot 495 = 2847.5. \tag{1.6}$$

Since Tokyo is beyond the outside bars, it is considered to be an outlier. The minimum and the maximum are called the *extremes*. The *mean* is defined as

$$\bar{x} = n^{-1} \sum_{i=1}^{n} x_i,$$

which is 1939.7 in our example. The mean is a measure of location. The median (1815), the fourths (1610; 2105), and the extremes (1430; 3420) constitute basic information about the data. The combination of these five numbers leads to the Five-Number Summary as shown in Table 1.2. The depths of each of the five numbers have been added as an additional column.

Construction of the Boxplot

1. Draw a box with borders (edges) at F_L and F_U (i.e., 50% of the data are in this box).
2. Draw the median as a solid line and the mean as a dotted line.
3. Draw "whiskers" from each end of the box to the most remote point that is <u>NOT</u> an outlier.
4. Show outliers as either "★" or "●" depending on whether they are outside of $F_{UL} \pm 1.5d_F$ or $F_{UL} \pm 3d_F$ respectively (this feather is not contained in some software). Label them if possible.

Fig. 1.2 Boxplot for world cities 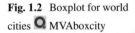 MVAboxcity

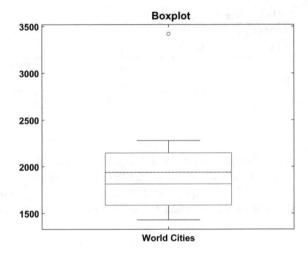

Fig. 1.3 Boxplot for the mileage of American, Japanese and European cars (from left to right) 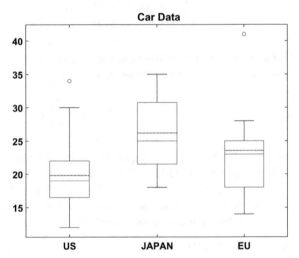 MVAboxcar

In the world cities example, the cut-off points (outside bars) are at 867.5 and 2847.5, hence we can draw whiskers to Karachi and Mexico City. We can see from Fig. 1.2 that the data are very skew: The upper half of the data (above the median) is more spread out than the lower half (below the median), the data contains one outlier marked as a circle and the mean (as a non-robust measure of location) is pulled away from the median.

Boxplots are very useful tools in comparing batches. The relative location of the distribution of different batches tells us a lot about the batches themselves. Before we come back to the Swiss bank data, let us compare the fuel economy of vehicles from different countries, see Fig. 1.3 and Sect. B.3.

Fig. 1.4 The X_6 variable of Swiss bank data (diagonal of bank notes) 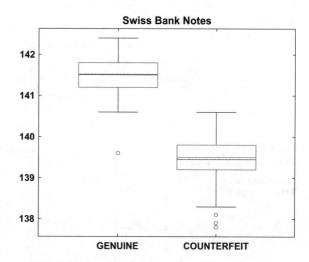 MVAboxbank6

Table 1.3 Five-number summary

Genuine banknotes			
# 100	Depth		
M	50.5		141.5	
F	25.75	141.25		141.8
	1	140.65		142.4

Example 1.2 The data are taken from the second column of Sect. B.3 and show the mileage (miles per gallon) of U.S. American, Japanese and European cars. The five-number summaries for these data sets are {12, 16.8, 18.8, 22, 30}, {18, 22, 25, 30.5, 35}, and {14, 19, 23, 25, 28} for American, Japanese, and European cars, respectively. This reflects the information shown in Fig. 1.3. The following conclusions can be made:

- Japanese cars achieve higher fuel efficiency than U.S. and European cars.
- There is one outlier, a very fuel-efficient car (VW-Rabbit Golf Diesel).
- The main body of the U.S. car data (the box) lies below the Japanese car data.
- The worst Japanese car is more fuel efficient than almost 50% of the U.S. cars.
- The spread of the Japanese and the U.S. cars is almost equal.
- The median of the Japanese data is above that of the European data and the U.S. data.

Now let us apply the boxplot technique to the bank data set. In Fig. 1.4, we show the parallel boxplot of the diagonal variable X_6. On the left is the value of the genuine banknotes, and on the right the value of the counterfeit banknotes. The five-number summary is reported in Tables 1.3 and 1.4.

Table 1.4 Five-number summary

Counterfeit banknotes

# 100	Depth			
M	50.5		139.5	
F	25.75	139.2		139.8
	1	138.3		140.6

Fig. 1.5 The X_1 variable of Swiss bank data (length of banknotes) 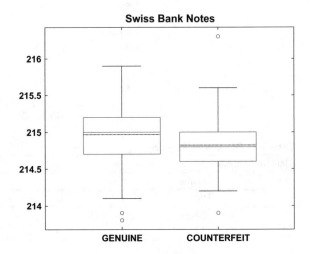 MVAboxbank1

One sees that the diagonals of the genuine banknotes tend to be larger. It is harder to see a clear distinction when comparing the length of the banknotes X_1, see Fig. 1.5. There are a few outliers in both plots. Almost all the observations of the diagonal of the genuine notes are above the ones from the counterfeit notes. There is one observation in Fig. 1.4 of the genuine notes that is almost equal to the median of the counterfeit notes. Can the parallel boxplot technique help us distinguish between the two types of banknotes?

Summary
↪ The median and mean bars are measures of locations.
↪ The relative location of the median (and the mean) in the box is a measure of how skewed it is.
↪ The length of the box and whiskers are a measure of spread.
↪ The length of the whiskers indicate the tail length of the distribution.
↪ The outlying points are indicated with a "⋆" or "•" depending on if they are outside of $F_{UL} \pm 1.5 d_F$ or $F_{UL} \pm 3 d_F$, respectively.
↪ The boxplots do not indicate multimodality or clusters.
↪ If we compare the relative size and location of the boxes, we are comparing distributions.

1.2   Histograms

Histograms are density estimates. A density estimate gives a good impression of
the distribution of the data. In contrast to boxplots, density estimates show possible
multimodality of the data. The idea is to locally represent the data density by counting
the number of observations in a sequence of consecutive intervals (bins) with origin
x_0. Let $B_j(x_0, h)$ denote the *bin* of length h, which is the element of a bin grid starting
at x_0:

$$B_j(x_0, h) = [x_0 + (j - 1)h, x_0 + jh), \quad j \in \mathbb{Z},$$

where $[.,.)$ denotes a left-closed and right-open interval. If $\{x_i\}_{i=1}^n$ is an i.i.d. sample
with density f, the histogram is defined as follows:

$$\widehat{f}_h(x) = n^{-1} h^{-1} \sum_{j \in \mathbb{Z}} \sum_{i=1}^n I\{x_i \in B_j(x_0, h)\} I\{x \in B_j(x_0, h)\}. \qquad (1.7)$$

In sum (1.7) the first indicator function $I\{x_i \in B_j(x_0, h)\}$ counts the number of observations falling into bin $B_j(x_0, h)$. The second indicator function is responsible for
"localizing" the counts around x. The parameter h is a smoothing or localizing parameter and controls the width of the histogram bins. An h that is too large leads to very
big blocks and thus to a very unstructured histogram. On the other hand, an h that is
too small gives a very variable estimate with many unimportant peaks.

The effect of h is given in detail in Fig. 1.6. It contains the histogram (upper
left) for the diagonal of the counterfeit banknotes for $x_0 = 137.8$ (the minimum
of these observations) and $h = 0.1$. Increasing h to $h = 0.2$ and using the same
origin, $x_0 = 137.8$, results in the histogram shown in the lower left of the figure.
This density histogram is somewhat smoother due to the larger h. The bin-width
is next set to $h = 0.3$ (upper right). From this histogram, one has the impression
that the distribution of the diagonal is bimodal with peaks at about 138.5 and 139.9.
The detection of modes requires fine-tuning of the bin-width. Using methods from
smoothing methodology Härdle et al. (2004), one can find an "optimal" bin-width h
for n observations:

$$h_{opt} = \left(\frac{24\sqrt{\pi}}{n}\right)^{1/3}.$$

Unfortunately, the bin-width h is not the only parameter determining the shapes of $\widehat{f}$.

In Fig. 1.7, we show histograms with $x_0 = 137.65$ (upper left), $x_0 = 137.75$ (lower
left), with $x_0 = 137.85$ (upper right), and $x_0 = 137.95$ (lower right). All the graphs
have been scaled equally on the y-axis to allow comparison. One sees that—despite
the fixed bin-width h—the interpretation is not facilitated. The shift of the origin x_0
(to 4 different locations) created 4 different histograms. This property of histograms
strongly contradicts the goal of presenting data features. Obviously, the same data

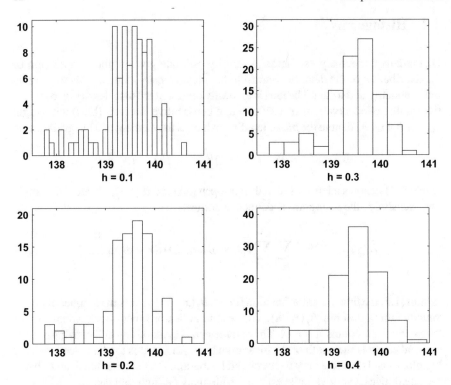

Fig. 1.6 Diagonal of counterfeit banknotes. Histograms with $x_0 = 137.8$ and $h = 0.1$ (upper left), $h = 0.2$ (lower left), $h = 0.3$ (upper right), $h = 0.4$ (lower right) 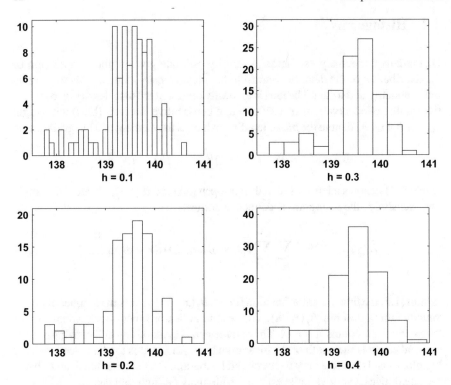 MVAhisbank1

are represented quite differently by the 4 histograms. A remedy has been proposed by Scott (1985): "Average the shifted histograms!". The result is presented in Fig. 1.8.

Here all bank note observations (genuine and counterfeit) have been used. The (so-called) averaged shifted histogram is no longer dependent on the origin and shows a clear bimodality of the diagonals of the Swiss banknotes.

Summary
↪   Modes of the density are detected with a histogram.
↪   Modes correspond to strong peaks in the histogram.
↪   Histograms with the same h need not be identical. They also depend on the origin x_0 of the grid.
↪   The influence of the origin x_0 is drastic. Changing x_0 creates different looking histograms.
↪   The consequence of an h that is too large is an unstructured histogram that is too flat.
↪   A bin-width h that is too small results in an unstable histogram.
↪   There is an "optimal" $h = (24\sqrt{\pi}/n)^{1/3}$.

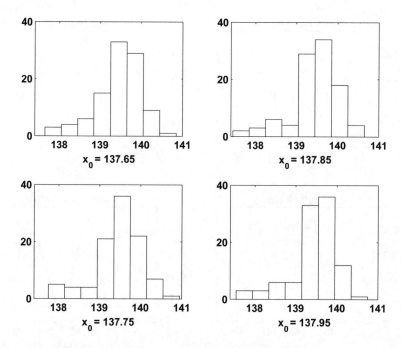

Fig. 1.7 Diagonal of counterfeit banknotes. Histogram with $h = 0.4$ and origins $x_0 = 137.65$ (upper left), $x_0 = 137.75$ (lower left), $x_0 = 137.85$ (upper right), $x_0 = 137.95$ (lower right) **Q** MVAhisbank2

1.3 Kernel Densities

The major difficulties of histogram estimation may be summarized in four critiques:

- determination of the bin-width h, which controls the shape of the histogram,
- choice of the bin origin x_0, which also influences to some extent the shape,
- loss of information since observations are replaced by the central point of the interval in which they fall,
- the underlying density function is often assumed to be smooth, but the histogram is not smooth.

Rosenblatt (1956), Whittle (1958), and Parzen (1962) developed an approach which avoids the last three difficulties. First, a smooth kernel function rather than a box is used as the basic building block. Second, the smooth function is centered directly over each observation. Let us study this refinement by supposing that x is the center value of a bin. The histogram can in fact be rewritten as

$$\widehat{f_h}(x) = n^{-1}h^{-1}\sum_{i=1}^{n}\mathrm{I}\left(|x - x_i| \leq \frac{h}{2}\right). \tag{1.8}$$

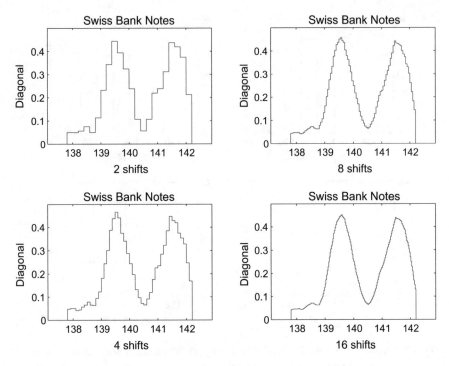

Fig. 1.8 Swiss banknotes: there are 2 shifts (upper left), 4 shifts (lower left), 8 shifts (upper right), and 16 shifts (lower right) 🔍 MVAashbank

If we define $K(u) = I(|u| \leq \frac{1}{2})$, then (1.8) changes to

$$\widehat{f_h}(x) = n^{-1}h^{-1} \sum_{i=1}^{n} K\left(\frac{x - x_i}{h}\right). \tag{1.9}$$

This is the general form of the kernel estimator. Allowing smoother kernel functions like the quartic kernel,

$$K(u) = \frac{15}{16}(1 - u^2)^2 \, I(|u| \leq 1),$$

and computing x not only at bin centers gives us the kernel density estimator. Kernel estimators can also be derived via weighted averaging of rounded points (WARPing) or by averaging histograms with different origins, see Scott (1985). Table 1.5 introduces some commonly used kernels.

Different kernels generate the different shapes of the estimated density. The most important parameter is the bandwidth h, and can be optimized, for example, by cross-validation; see Härdle (1991) for details. The cross-validation method minimizes

Table 1.5 Kernel functions

$K(\bullet)$	Kernel
$K(u) = \frac{1}{2}I(|u| \leq 1)$	Uniform
$K(u) = (1 - |u|)I(|u| \leq 1)$	Triangle
$K(u) = \frac{3}{4}(1 - u^2)I(|u| \leq 1)$	Epanechnikov
$K(u) = \frac{15}{16}(1 - u^2)^2 I(|u| \leq 1)$	Quartic (Biweight)
$K(u) = \frac{1}{\sqrt{2\pi}} \exp(-\frac{u^2}{2}) = \varphi(u)$	Gaussian

the integrated squared error. This measure of discrepancy is based on the squared differences $\left\{\hat{f}_h(x) - f(x)\right\}^2$. Averaging these squared deviations over a grid of points $\{x_l\}_{l=1}^{L}$ leads to

$$L^{-1} \sum_{l=1}^{L} \left\{\hat{f}_h(x_l) - f(x_l)\right\}^2.$$

Asymptotically, if this grid size tends to zero, we obtain the integrated squared error:

$$\int \left\{\hat{f}_h(x) - f(x)\right\}^2 dx.$$

In practice, it turns out that the method consists of selecting a bandwidth that minimizes the cross-validation function

$$\int \hat{f}_h^2 - 2 \sum_{i=1}^{n} \hat{f}_{h,i}(x_i),$$

where $\hat{f}_{h,i}$ is the density estimate obtained by using all datapoints except for the i-th observation. Both terms in the above function involve double sums. Computation may, therefore, be slow. There are many other density bandwidth selection methods. Probably the fastest way to calculate this is to refer to some reasonable reference distribution. The idea of using the Normal distribution as a reference, for example, goes back to Silverman (1986). The resulting choice of h is called the *rule of thumb*.

For the Gaussian kernel from Table 1.5 and a Normal reference distribution, the rule of thumb is to choose

$$h_G = 1.06 \,\hat{\sigma} \, n^{-1/5} \tag{1.10}$$

where $\hat{\sigma} = \sqrt{n^{-1} \sum_{i=1}^{n}(x_i - \overline{x})^2}$ denotes the sample standard deviation. This choice of h_G optimizes the integrated squared distance between the estimator and the true density. For the quartic kernel, we need to transform (1.10). The modified rule of thumb is

$$h_Q = 2.62 \cdot h_G. \tag{1.11}$$

Fig. 1.9 Densities of the
diagonals of genuine and
counterfeit banknotes.
Automatic density estimates

🔍 MVAdenbank

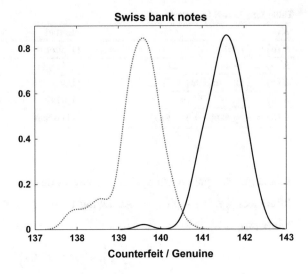

Figure 1.9 shows the automatic density estimates for the diagonals of the counterfeit and genuine banknotes. The density on the left is the density corresponding
to the diagonal of the counterfeit data. The separation is clearly visible, but there is
also an overlap. The problem of distinguishing between the counterfeit and genuine
banknotes is not solved by just looking at the diagonals of the notes. The question
arises whether a better separation could be achieved using not only the diagonals,
but one or two more variables of the data set. The estimation of higher dimensional
densities is analogous to that of one dimensional. We show a two-dimensional density estimate for X_4 and X_5 in Fig. 1.10. The contour lines indicate the height of

Fig. 1.10 Contours of the
density of X_5 and X_6 of
genuine and counterfeit

banknotes 🔍
MVAcontbank2

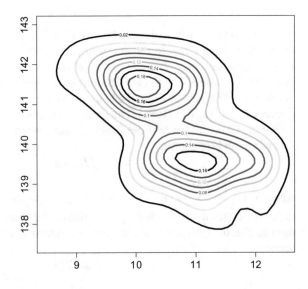

Fig. 1.11 Contours of the density of X_4, X_5, X_6 of genuine and counterfeit banknotes 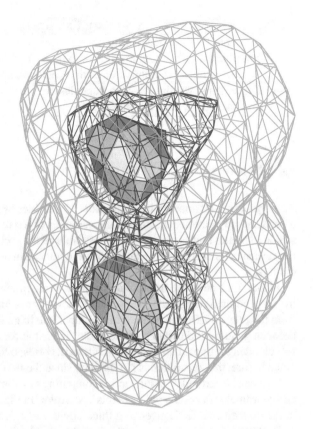 MVAcontbank3

the density. One sees two separate distributions in this higher dimensional space, but they still overlap to some extent.

We can add one more dimension and give a graphical representation of a three-dimensional density estimate, or more precisely an estimate of the joint distribution of X_4, X_5 and X_6. Figure 1.11 shows the contour areas at 3 different levels of the density: 0.2 (green), 0.4 (red), and 0.6 (blue) of this three- dimensional density estimate. One can clearly recognize two "ellipsoids" (at each level), but as before, they overlap. In Chap. 14, we will learn how to separate the two ellipsoids and how to develop a discrimination rule to distinguish between these data points.

Summary

↪ Kernel densities estimate distribution densities by the kernel method.

↪ The bandwidth h determines the degree of smoothness of the estimate $\widehat{f}$.

↪ Kernel densities are smooth functions and they can graphically represent distributions (up to 3 dimensions).

Continued on next page

Summary (continue)
↪   A simple (but not necessarily correct) way to find a good bandwidth is to compute the rule of thumb bandwidth $h_G = 1.06\widehat{\sigma}n^{-1/5}$. This bandwidth is to be used only in combination with a Gaussian kernel φ.
↪   Kernel density estimates are a good descriptive tool for seeing modes, location, skewness, tails, asymmetry, etc.

1.4   Scatterplots

Scatterplots are bivariate or trivariate plots of variables against each other. They help us understand relationships among the variables of a data set. A downward-sloping scatter indicates that as we increase the variable on the horizontal axis, the variable on the vertical axis decreases. An analogous statement can be made for upward-sloping scatters.

Figure 1.12 plots the fifth column (upper inner frame) of the bank data against the sixth column (diagonal). The scatter is downward-sloping. As we already know from the previous section on marginal comparison (e.g., Fig. 1.9) a good separation between genuine and counterfeit banknotes is visible for the diagonal variable. The sub-cloud in the upper half (circles) of Fig. 1.12 corresponds to the true banknotes. As noted before, this separation is not distinct, since the two groups overlap somewhat.

This can be verified in an interactive computing environment by showing the index and coordinates of certain points in this scatterplot. In Fig. 1.12, the 70th observation in the merged data set is given as a thick circle, and it is from a genuine bank note. This observation lies well embedded in the cloud of counterfeit banknotes. One straightforward approach that could be used to tell the counterfeit from the genuine

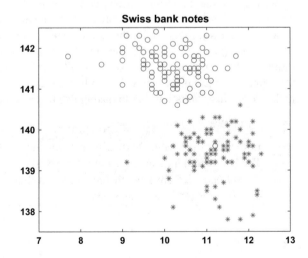

Fig. 1.12 2D scatterplot for X_5 versus X_6 of the banknotes. Genuine notes are circles, counterfeit notes are stars **Q** MVAscabank56

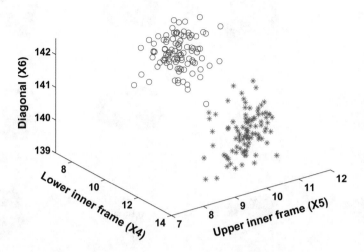

Fig. 1.13 3D Scatterplot of the banknotes for (X_4, X_5, X_6). Genuine notes are circles, counterfeit are stars 🔍 MVAscabank456

banknotes is to draw a straight line and define notes above this value as genuine. We would, of course, misclassify the 70th observation, but can we do better?

If we extend the two-dimensional scatterplot by adding a third variable, e.g., X_4 (lower distance to inner frame), we obtain the scatterplot in three dimensions as shown in Fig. 1.13. It becomes apparent from the location of the point clouds that a better separation is obtained. We have rotated the three-dimensional data until this satisfactory 3D view was obtained. Later, we will see that the rotation is the same as bundling a high-dimensional observation into one or more linear combinations of the elements of the observation vector. In other words, the "separation line" parallel to the horizontal coordinate axis in Fig. 1.12 is, in Fig. 1.13, a plane and no longer parallel to one of the axes. The formula for such a separation plane is a linear combination of the elements of the observation vector:

$$a_1 x_1 + a_2 x_2 + \cdots + a_6 x_6 = \text{const.} \tag{1.12}$$

The algorithm that automatically finds the weights $(a_1, \ldots, a_6)$ will be investigated later on in Chap. 14.

Let us study yet another technique: the scatterplot matrix. If we want to draw all possible two-dimensional scatterplots for the variables, we can create a so-called *draftsman's plot* (named after a draftsman who prepares drafts for parliamentary discussions). Similar to a draftsman's plot the scatterplot matrix helps in creating new ideas and in building knowledge about dependencies and structure.

Figure 1.14 shows a draftsman's plot applied to the last four columns of the full bank data set. For ease of interpretation, we have distinguished between the group of

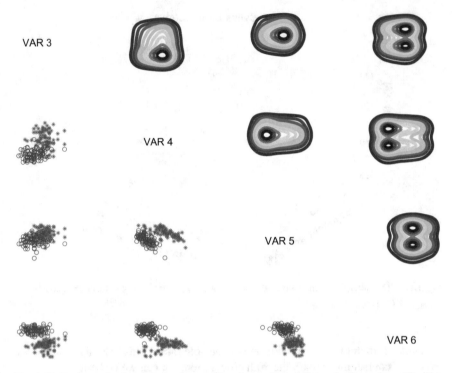

Fig. 1.14 Draftsman's plot of the banknotes. The pictures in the left-hand column show (X_3, X_4), (X_3, X_5) and (X_3, X_6), in the middle we have (X_4, X_5) and (X_4, X_6), and in the lower right (X_5, X_6). The upper right half contains the corresponding density contour plots 🔍 MVAdrafbank4

counterfeit and genuine banknotes by a different color. As discussed several times earlier, the separability of the two types of notes is different for different scatterplots. Not only is it difficult to perform this separation on, say, scatterplot X_3 versus X_4, in addition the "separation line" is no longer parallel to one of the axes. The most obvious separation happens in the scatterplot in the lower right-hand side where indicated, as in Fig. 1.12, X_5 versus X_6. The separation line here would be upward-sloping with an intercept at about $X_6 = 139$. The upper right half of the draftsman's plot shows the density contours that we introduced in Sect. 1.3.

The power of the draftsman's plot lies in its ability to show the internal connections of the scatter diagrams. Define a *brush* as a re-scalable rectangle that we can move via keyboard or mouse over the screen. Inside the brush, we can highlight or color observations. Suppose the technique is installed in such a way that as we move the brush in one scatter, the corresponding observations in the other scatters are also highlighted. By moving the brush, we can study conditional dependence.

If we brush (i.e., highlight or color the observation with the brush) the X_5 versus X_6 plot and move through the upper point cloud, we see that in other plots (e.g., X_3 vs. X_4), the corresponding observations are more embedded in the other sub-cloud.

Summary
$\hookrightarrow$ Scatterplots in two and three dimensions helps in identifying separated points, outliers, or sub-clusters.
$\hookrightarrow$ Scatterplots help us in judging positive or negative dependencies.
$\hookrightarrow$ Draftsman scatterplot matrices help detect structures conditioned on values of other variables.
$\hookrightarrow$ As the brush of a scatterplot matrix moves through a point cloud, we can study conditional dependence.

1.5   Chernoff-Flury Faces

If we are given a data in a numerical form, we tend to also display it numerically. This was done in the preceding sections: an observation $x_1 = (1, 2)$ was plotted as the point $(1, 2)$ in a two-dimensional coordinate system. In multivariate analysis, we want to understand data in low dimensions (e.g., on a 2D computer screen) although the structures are hidden in high dimensions. The numerical display of data structures using coordinates, therefore, ends at dimensions greater than three.

If we are interested in condensing a structure into 2D elements, we have to consider alternative graphical techniques. The Chernoff-Flury faces, for example, provide such a condensation of high-dimensional information into a simple "face". In fact, faces are a simple way of graphically displaying high-dimensional data. The size of the face elements like pupils, eyes, upper and lower hair line, etc., are assigned to certain variables. The idea of using faces goes back to Chernoff (1973) and has been further developed by Bernhard Flury. We follow the design described in Flury and Riedwyl (1988) which uses the following characteristics:

1 right eye size
2 right pupil size
3 position of right pupil
4 right eye slant
5 horizontal position of right eye
6 vertical position of right eye
7 curvature of right eyebrow
8 density of right eyebrow
9 horizontal position of right eyebrow
10 vertical position of right eyebrow
11 right upper hair line
12 right lower hair line
13 right face line

14 darkness of right hair
15 right hair slant
16 right nose line
17 right size of mouth
18 right curvature of mouth
19–36 like 1–18, only for the left side.

First, every variable that is to be coded into a characteristic face element is transformed into a (0, 1) scale, i.e., the minimum of the variable corresponds to 0 and the maximum to 1. The extreme positions of the face elements, therefore, correspond to a certain "grin" or "happy" face element. Dark hair might be coded as 1, and blond hair as 0 and so on.

As an example, consider the observations 91 to 110 of the bank data. Recall that the bank data set consists of 200 observations of dimension 6 where, for example, X_6 is the diagonal of the note. If we assign the six variables to the following face elements:

$$X_1 = 1, 19 \text{ (eye sizes)}$$
$$X_2 = 2, 20 \text{ (pupil sizes)}$$
$$X_3 = 4, 22 \text{ (eye slants)}$$
$$X_4 = 11, 29 \text{ (upper hair lines)}$$
$$X_5 = 12, 30 \text{ (lower hair lines)}$$
$$X_6 = 13, 14, 31, 32 \text{ (face lines and darkness of hair)},$$

we obtain Fig. 1.15.

Also recall that observations 1–100 correspond to the genuine notes, and that observations 101–200 correspond to the counterfeit notes. The counterfeit banknotes then correspond to the upper half of Fig. 1.15. In fact, the faces for these observations look more grim and less happy. The variable X_6 (diagonal) already worked well in the boxplot on Fig. 1.4 in distinguishing between the counterfeit and genuine notes. Here, this variable is assigned to the face line and the darkness of the hair. That is why we clearly see a good separation within these 20 observations.

What happens if we include all 100 genuine and all 100 counterfeit banknotes in the Chernoff-Flury face technique? Figure 1.16 shows the faces of the genuine banknotes with the same assignments as used before and Fig. 1.17 shows the faces of the counterfeit banknotes. Comparing Figs. 1.16 and 1.17 one clearly sees that the diagonal (face line) is longer for genuine banknotes. Equivalently coded is the hair darkness (diagonal) which is lighter (shorter) for the counterfeit banknotes. One sees that the faces of the genuine banknotes have a much darker appearance and have broader face lines. The faces in Fig. 1.16 are obviously different from the ones in Fig. 1.17.

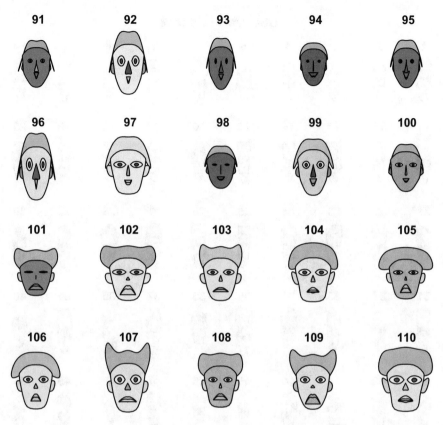

Fig. 1.15 Chernoff-Flury faces for observations 91–110 of the banknotes ⊙ MVAfacebank10

Summary
↪ Faces can be used to detect subgroups in multivariate data.
↪ Subgroups are characterized by similar looking faces.
↪ Outliers are identified by extreme faces, e.g., dark hair, smile or a happy face.
↪ If one element of X is unusual, the corresponding face element significantly changes in shape.

1.6 Andrews' Curves

The basic problem of graphical displays of multivariate data is the dimensionality. Scatterplots work well up to three dimensions (if we use interactive displays). More than three dimensions have to be coded into displayable 2D or 3D structures

Observations 1 to 50

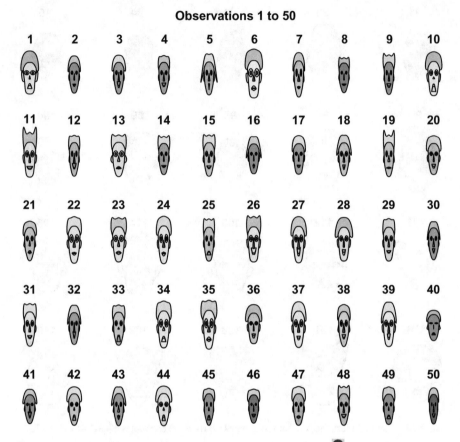

Fig. 1.16 Chernoff-Flury faces for observations 1–50 of the banknotes ▣ MVAfacebank50

(e.g., faces). The idea of coding and representing multivariate data by curves was suggested by Andrews (1972). Each multivariate observation $X_i = (X_{i,1}, .., X_{i,p})$ is transformed into a curve as follows:

$$
f_i(t) = \begin{cases} \frac{X_{i,1}}{\sqrt{2}} + X_{i,2}\sin(t) + X_{i,3}\cos(t) + \cdots \\ \quad + X_{i,p-1}\sin(\frac{p-1}{2}t) + X_{i,p}\cos(\frac{p-1}{2}t) & \text{for } p \text{ odd} \\ \frac{X_{i,1}}{\sqrt{2}} + X_{i,2}\sin(t) + X_{i,3}\cos(t) + \cdots + X_{i,p}\sin(\frac{p}{2}t) & \text{for } p \text{ even} \end{cases} \quad (1.13)
$$

the observation represents the coefficients of a so-called Fourier series ($t \in [-\pi, \pi]$).

Suppose that we have three-dimensional observations: $X_1 = (0, 0, 1)$, $X_2 = (1, 0, 0)$ and $X_3 = (0, 1, 0)$. Here $p = 3$ and the following representations correspond to the Andrews' curves:

Observations 101 to 150

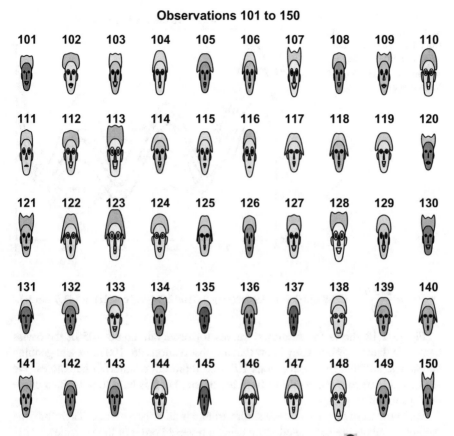

Fig. 1.17 Chernoff-Flury faces for observations 101–150 of the banknotes 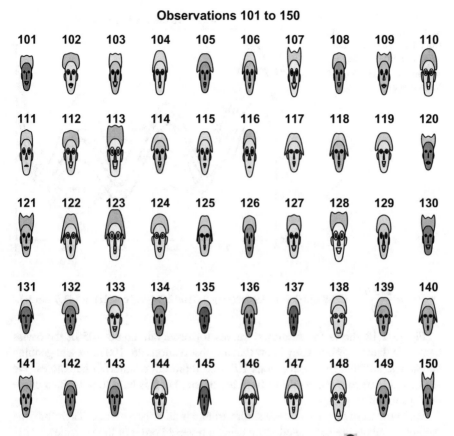 MVAfacebank50

$$f_1(t) = \cos(t)$$
$$f_2(t) = \frac{1}{\sqrt{2}} \quad \text{and}$$
$$f_3(t) = \sin(t).$$

These curves are indeed quite distinct, since the observations X_1, X_2, and X_3 are the 3D unit vectors: each observation has mass only in one of the three dimensions. The order of the variables plays an important role.

Example 1.3 Let us take the 96th observation of the Swiss bank note data set,

$$X_{96} = (215.6, 129.9, 129.9, 9.0, 9.5, 141.7).$$

The Andrews' curve is by (1.13):

Fig. 1.18 Andrews' curves of the observations 96–105 from the Swiss bank note data. The order of the variables is 1,2,3,4,5,6 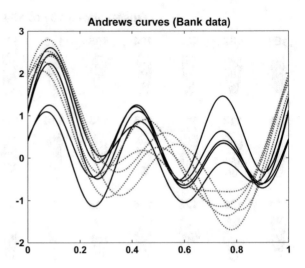 MVAandcur

Andrews curves (Bank data)

$$f_{96}(t) = \frac{215.6}{\sqrt{2}} + 129.9\sin(t) + 129.9\cos(t) + 9.0\sin(2t) + 9.5\cos(2t) + 141.7\sin(3t).$$

Figure 1.18 shows the Andrews' curves for observations 96–105 of the Swiss bank note data set. We already know that the observations 96–100 represent genuine banknotes, and that the observations 101–105 represent counterfeit banknotes. We see that at least four curves differ from the others, but it is hard to tell which curve belongs to which group.

We know from Fig. 1.4 that the sixth variable is an important one. Therefore, the Andrews' curves are calculated again using a reversed order of the variables.

Example 1.4 Let us consider again the 96th observation of the Swiss bank note data set,
$$X_{96} = (215.6, 129.9, 129.9, 9.0, 9.5, 141.7).$$

The Andrews' curve is computed using the reversed order of variables:

$$f_{96}(t) = \frac{141.7}{\sqrt{2}} + 9.5\sin(t) + 9.0\cos(t) + 129.9\sin(2t) + 129.9\cos(2t) + 215.6\sin(3t).$$

In Fig. 1.19, the curves f_{96}–f_{105} for observations 96–105 are plotted. Instead of a difference in high frequency, now we have a difference in the intercept, which makes it more difficult for us to see the differences in observations.

This shows that the order of the variables plays an important role in the interpretation. If X is high dimensional, then the last variables will only have a small visible contribution to the curve: they fall into the high-frequency part of the curve. To overcome this problem, Andrews suggested using an order which is suggested

Fig. 1.19 Andrews' curves
of the observations 96–105
from the Swiss bank note
data. The order of the
variables is 6, 5, 4, 3, 2, 1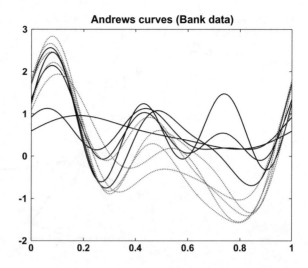
MVAscabank56

by Principal Component Analysis. This technique will be treated in detail in Chap.
11. In fact, the sixth variable will appear there as the most important variable for
discriminating between the two groups. If the number of observations is more than
20, there may be too many curves in one graph. This will result in an overplotting of
curves or a bad "signal-to-ink-ratio", see Tufte (1983). It is, therefore, advisable to
present multivariate observations via Andrews' curves only for a limited number of
observations.

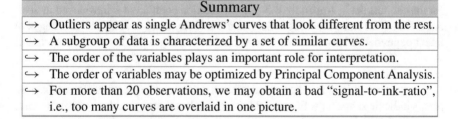

Summary
↪ Outliers appear as single Andrews' curves that look different from the rest.
↪ A subgroup of data is characterized by a set of similar curves.
↪ The order of the variables plays an important role for interpretation.
↪ The order of variables may be optimized by Principal Component Analysis.
↪ For more than 20 observations, we may obtain a bad "signal-to-ink-ratio", i.e., too many curves are overlaid in one picture.

1.7 Parallel Coordinate Plots

Parallel Coordinates Plots (PCP) is a method for representing high-dimensional data,
see Inselberg (1985). Instead of plotting observations in an orthogonal coordinate
system, PCP draws coordinates in parallel axes and connects them with straight lines.
This method helps in representing data with more than four dimensions.

One first scales all variables to $max = 1$ and $min = 0$. The coordinate index j is
drawn onto the horizontal axis, and the scaled value of variable x_{ij} is mapped onto
the vertical axis. This way of representation is very useful for high-dimensional data.
It is, however, also sensitive to the order of the variables, since certain trends in the
data can be shown more clearly in one ordering than in another.

Fig. 1.20 Parallel
coordinates plot of

observations 96–105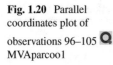
MVAparcoo1

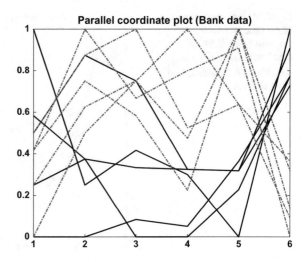

Example 1.5 Take, once again, the observations 96–105 of the Swiss banknotes. These observations are six dimensional, so we can't show them in a six-dimensional Cartesian coordinate system. Using the parallel coordinates plot technique, however, they can be plotted on parallel axes. This is shown in Fig. 1.20.

PCP can also be used for detecting linear dependencies between variables: if all lines are of almost parallel dimensions ($p = 2$), there is a positive linear dependence between them. In Fig. 1.21, we display the two variables, weight and displacement, for the car data set in Appendix B.3. The correlation coefficient ρ introduced in Sect. 3.2 is 0.9. If all lines intersect visibly in the middle, there is evidence of a negative linear dependence between these two variables, see Fig. 1.22. In fact the correlation is $\rho = -0.82$ between two variables mileage and weight: The more the weight the less the mileage.

Another use of PCP is subgroups detection. Lines converging to different discrete points indicate subgroups. Figure 1.23 shows the last three variables—displacement, gear ratio for high gear, and company's headquarters of the car data; we see convergence to the last variable. This last variable is the company's headquarters with three discrete values: U.S., Japan and Europe. PCP can also be used for outlier detection. Figure 1.24 shows the variables headroom, rear seat clearance, and trunk (boot) space in the car data set. There are two outliers visible. The boxplot Fig. 1.25 confirms this.

PCPs have also possible shortcomings: We cannot distinguish observations when two lines cross at one point unless we distinguish them clearly (e.g., by different line style). In Fig. 1.26, observations A and B both have the same value at $j = 2$. Two lines cross at one point here. At the third and fourth dimension, we cannot tell which line belongs to which observation. A dotted line for A and solid line for B could have helped there.

To solve this problem, one uses an interpolation curve instead of straight lines, e.g., cubic curves as in Graham and Kennedy (2003). Figure 1.27 is a variant of Fig. 1.26.

Fig. 1.21 Coordinates Plot indicating strong positive dependence with $\rho = 0.9$, X_1 = weight, X_2 = displacement 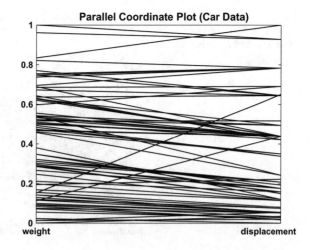 MVApcp2

Fig. 1.22 Coordinates Plot showing strong negative dependence with $\rho = -0.82$, X_1 = mileage, X_2 = weight 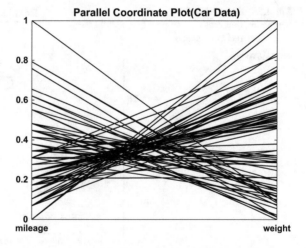 MVApcp3

Fig. 1.23 Parallel Coordinates Plot with subgroups 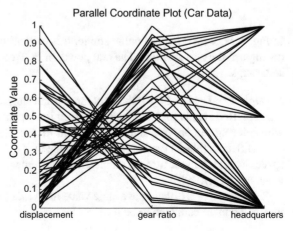 MVApcp4

Fig. 1.24 PCP for $X_1 =$ headroom, $X_2 =$ rear seat clearance, and $X_3 =$ trunk space MVApcp5

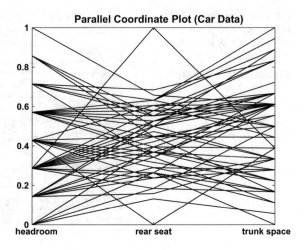

Fig. 1.25 Boxplots for headroom, rear seat clearance, and trunk space MVApcp6

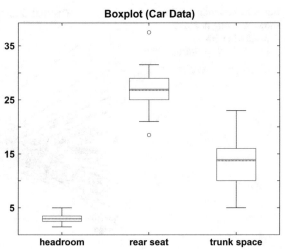

In Fig. 1.27, with a natural cubic spline, it is evident how to follow the curves and distinguish the observations. The real power of PCP comes though through coloring subgroups.

Example 1.6 Data in Fig. 1.28 are colored according to X_{13}—car company's headquarters. Red stands for European car, green for Japan, and black for U.S. This PCP with coloring can provide some information for us:

    1. U.S. cars (black) tend to have large value in X_7, X_8, X_9, X_{10}, X_{11} (trunk (boot) space, weight, length, turning diameter, displacement), which means U.S. cars are generally larger.

    2. Japanese cars (green) have large value in X_3, X_4 (both for repair record), which means Japanese cars tend to be repaired less.

Fig. 1.26 CP with intersection for given data points A = [0, 2, 3, 2] and B = [3, 2, 2, 1] MVApcp7

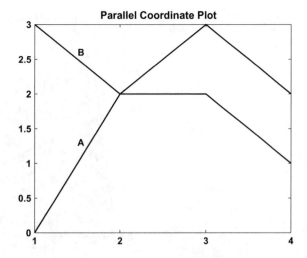

Fig. 1.27 PCP with cubic spline interpolation MVApcp8

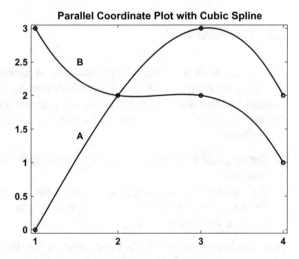

Summary

↪ Parallel coordinates plots overcome the visualization problem of the Cartesian coordinate system for dimensions greater than 4.

↪ Outliers are visible as outlying polygon curves.

↪ The order of variables is important, especially in the detection of subgroups.

↪ Subgroups may be screened by selective coloring.

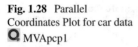

Fig. 1.28 Parallel Coordinates Plot for car data
MVApcp1

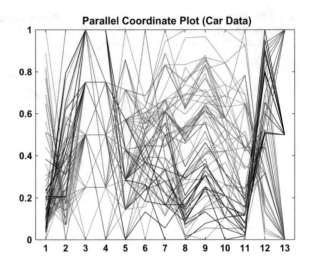

1.8 Hexagon Plots

This section closely follows the presentation of Lewin-Koh (2006). In geometry, a hexagon is a polygon with six edges and six vertices. Hexagon binning is a type of bivariate histogram with hexagon borders. It is useful for visualizing the structure of data sets entailing a large number of observations n. The concept of hexagon binning is as follows:

1. The xy plane over the set (range(x), range(y)) is tessellated by a regular grid of hexagons.
2. The number of points falling in each hexagon is counted.
3. The hexagons with count > 0 are plotted by using a color ramp or varying the radius of the hexagon in proportion to the counts.

This algorithm is extremely fast and effective for displaying the structure of data sets even for $n \geq 10^6$. If the size of the grid and the cuts in the color ramp are chosen in a clever fashion, then the structure inherent in the data should emerge in the binned plot. The same caveats apply to hexagon binning as histograms. Variance and bias vary in opposite directions with bin-width, so we have to settle for finding the value of the bin-width that yields the optimal compromise between variance and bias reduction. Clearly, if we increase the size of the grid, the hexagon plot appears to be smoother, but without some reasonable criterion on hand it remains difficult to say which bin-width provides the "optimal" degree of smoothness. The default number of bins suggested by standard software is 30.

Applications to some data sets are shown as follows. The data is taken from ALL-BUS (2006)[ZA No.3762]. The number of respondents is 2946. The following nine variables have been selected to analyze the relation between each pair of variables:

X_1: Age
X_2: Net income
X_3: Time for television per day in minutes
X_4: Time for work per week in hours
X_5: Time for computer per week in hours
X_6: Days for illness yearly
X_7: Living space (square meters)
X_8: Size
X_9: Weight.

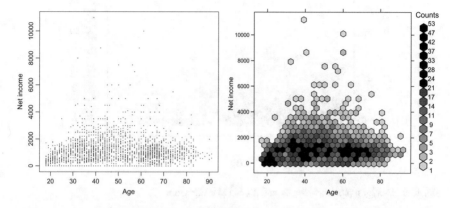

Fig. 1.29 Hexagon plots between X_1 and X_2 ⊙ MVAageIncome

First, we consider two variables X_1 = Age and X_2 = Net income in Fig. 1.29. The top left picture is a scatter plot. The second one is a hexagon plot with borders making it easier to see the separation between hexagons. Looking at these plots one can see that almost all individuals have a net monthly income of less than 2000 EUR. Only two individuals earn more than 10000 EUR per month.

Figure 1.30 shows the relation between X_1 and X_5. About forty percent of respondents from 20 to 80 years old do not use a computer at least once per week. The respondent who deals with a computer 105 h each week was actually not in full-time employment.

Clearly, people who earn modest incomes live in smaller flats. The trend here is relatively clear in Fig. 1.31. The larger the net income, the larger the flat. A few people do however earn high incomes but live in small flats.

Summary
↪ Hexagon binning is a type of bivariate histogram, used for visualizing large data.
↪ Variance and bias vary in opposite directions with bin width.
↪ Hexagons have the property of "symmetry of the nearest neighbours", which lacks in square bins.
↪ Hexagons are visually less biased for displaying densities than other regular tesselations.

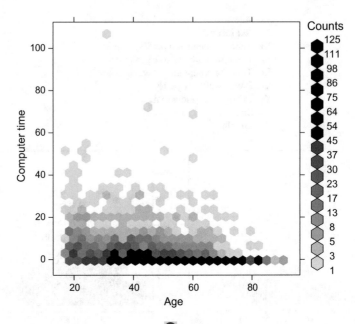

Fig. 1.30 Hexagon plot between X_1 and X_5 🔍 MVAageCom

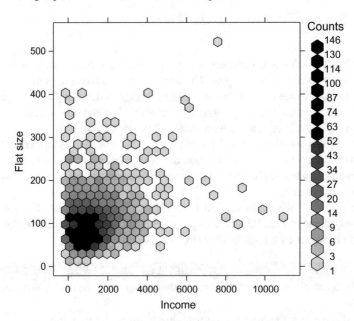

Fig. 1.31 Hexagon plot between X_2 and X_7 🔍 MVAincomeLi

Fig. 1.32 Parallel
coordinates plot for Boston
Housing data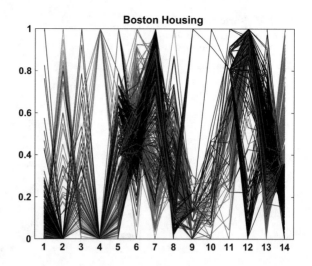
MVApcphousing

1.9 Boston Housing

Aim of the Analysis

The Boston Housing data set was analyzed by Harrison and Rubinfeld (1978), who wanted to find out whether "clean air" had an influence on house prices. We will use this data set in this chapter and in most of the following chapters to illustrate the presented methodology. The data are described in Appendix B.1.

What Can Be Seen from the PCPs

In order to highlight the relations of X_{14} to the remaining 13 variables, we color all of the observations with $X_{14} >$ median(X_{14}) as red lines in Fig. 1.32. Some of the variables seem to be strongly related. The most obvious relation is the negative dependence between X_{13} and X_{14}. It can also be argued that a strong dependence exists between X_{12} and X_{14} since no red lines are drawn in the lower part of X_{12}. The opposite can be said about X_{11}: there are only red lines plotted in the lower part of this variable. Low values of X_{11} induce high values of X_{14}.

For the PCP, the variables have been rescaled over the interval [0, 1] for better graphical representations. The PCP shows that the variables are not distributed in a symmetric manner. It can be clearly seen that the values of X_1 and X_9 are much more concentrated around 0. Therefore, it makes sense to consider transformations of the original data.

Fig. 1.33 Scatterplot matrix
for variables $X_1, \ldots, X_5$ and
X_{14} of the Boston Housing
data 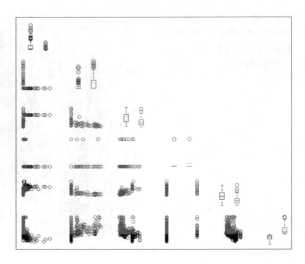 MVAdrafthousing

The Scatterplot Matrix

One characteristic of PCPs is that many lines are drawn on top of each other. This
problem is reduced by depicting the variables in pairs of scatterplots. Including all 14
variables in one large scatterplot matrix is possible, but makes it hard to see anything
from the plots. Therefore, for illustratory purposes we will analyze only one such
matrix from a subset of the variables in Fig. 1.33. On the basis of the PCP and the
scatterplot matrix, we would like to interpret each of the thirteen variables and their
eventual relation to the 14th variable. Included in the figure are images for X_1–X_5
and X_{14}, although each variable is discussed in detail below. All references made to
scatterplots in the following refer to Fig. 1.33.

Per-capita crime rate X_1

Taking the logarithm makes the variable's distribution more symmetric. This can be
seen in the boxplot of $\widetilde{X}_1$ in Fig. 1.35 which shows that the median and the mean
have moved closer to each other than they were for the original X_1. Plotting the
kernel density estimate (KDE) of $\widetilde{X}_1 = \log(X_1)$ would reveal that two subgroups
might exist with different mean values. However, taking a look at the scatterplots in
Fig. 1.34 of the logarithms which include X_1 does not clearly reveal such groups.
Given that the scatterplot of $\log(X_1)$ versus $\log(X_{14})$ shows a relatively strong
negative relation, it might be the case that the two subgroups of X_1 correspond to
houses with two different price levels. This is confirmed by the two boxplots shown

Fig. 1.34 Scatterplot matrix
for variables $\widetilde{X}_1, \ldots, \widetilde{X}_5$ and
$\widetilde{X}_{14}$ of the Boston Housing
data 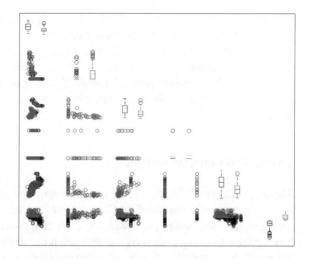 MVAdrafthousingt

to the right of the X_1 versus X_2 scatterplot (in Fig. 1.33): the right boxplot's shape
differs a lot from the black one's, having a much higher median and mean.

Proportion of residential area zoned for large lots X_2

It strikes the eye in Fig. 1.33 that there is a large cluster of observations for which
X_2 is equal to 0. It also strikes the eye that—as the scatterplot of X_1 versus X_2
shows—there is a strong, though nonlinear, negative relation between X_1 and X_2;
almost all observations for which X_2 is high have an X_1-value close to zero, and vice
versa, many observations for which X_2 is zero have quite a high per-capita crime rate
X_1. This could be due to the location of the areas, e.g., urban districts might have a
higher crime rate and at the same time it is unlikely that any residential land would
be zoned in a generous manner.

As far as the house prices are concerned, it can be said that there seems to be no
clear (linear) relation between X_2 and X_{14}, but it is obvious that the more expensive
houses are situated in areas where X_2 is large (this can be seen from the two boxplots on the second position of the diagonal, where the red one has a clearly higher
mean/median than the black one).

Proportion of non-retail business acres X_3

The PCP (in Fig. 1.32) as well as the scatterplot of X_3 versus X_{14} shows an obvious
negative relation between X_3 and X_{14}. The relationship between the logarithms of
both variables seems to be almost linear. This negative relation might be explained
by the fact that non-retail business sometimes causes annoying sounds and other

pollution. Therefore, it seems reasonable to use X_3 as an explanatory variable for the prediction of X_{14} in a linear-regression analysis.

As far as the distribution of X_3 is concerned, it can be said that the kernel density estimate of X_3 clearly has two peaks, which indicates that there are two subgroups. According to the negative relation between X_3 and X_{14}, it could be the case that one subgroup corresponds to the more expensive houses and the other one to the cheaper houses.

Charles River dummy variable X_4

The observation made from the PCP that there are more expensive houses than cheap houses situated on the banks of the Charles River is confirmed by inspecting the scatterplot matrix. Still, we might have some doubt that proximity to the river influences house prices. Looking at the original data set, it becomes clear that the observations for which X_4 equals one are districts that are close to each other. Apparently, the Charles River does not flow through very many different districts. Thus, it may be pure coincidence that the more expensive districts are close to the Charles River— their high values might be caused by many other factors such as the pupil/teacher ratio or the proportion of non-retail business acres.

Nitric oxides concentration X_5

The scatterplot of X_5 versus X_{14} and the separate boxplots of X_5 for more and less expensive houses reveal a clear negative relation between the two variables. As it was the main aim of the authors of the original study to determine whether pollution had an influence on housing prices, it should be considered very carefully whether X_5 can serve as an explanatory variable for price X_{14}. A possible reason against it being an explanatory variable is that people might not like to live in areas where the emissions of nitric oxides are high. Nitric oxides are emitted mainly by automobiles, by factories and from heating private homes. However, as one can imagine there are many good reasons besides nitric oxides not to live in urban or industrial areas. Noise pollution, for example, might be a much better explanatory variable for the price of housing units. As the emission of nitric oxides is usually accompanied by noise pollution, using X_5 as an explanatory variable for X_{14} might lead to the false conclusion that people run away from nitric oxides, whereas in reality it is noise pollution that they are trying to escape.

Average number of rooms per dwelling X_6

The number of rooms per dwelling is a possible measure of the size of the houses. Thus, we expect X_6 to be strongly correlated with X_{14} (the houses' median price). Indeed—apart from some outliers—the scatterplot of X_6 versus X_{14} shows a point cloud which is clearly upward-sloping and which seems to be a realization of a linear dependence of X_{14} on X_6. The two boxplots of X_6 confirm this notion by showing that the quartiles, the mean and the median are all much higher for the red than for the black boxplot.

Proportion of owner-occupied units built prior to 1940 X_7

There is no clear connection visible between X_7 and X_{14}. There could be a weak negative correlation between the two variables, since the (red) boxplot of X_7 for the districts whose price is above the median price indicates a lower mean and median than the (black) boxplot for the district whose price is below the median price. The fact that the correlation is not so clear could be explained by two opposing effects. On the one hand, house prices should decrease if the older houses are not in a good shape. On the other hand, prices could increase, because people often like older houses better than newer houses, preferring their atmosphere of space and tradition. Nevertheless, it seems reasonable that the age of the houses has an influence on their price X_{14}.

Raising X_7 to the power of 2.5 reveals again that the data set might consist of two subgroups. But in this case, it is not obvious that the subgroups correspond to more expensive or cheaper houses. One can furthermore observe a negative relation between X_7 and X_8. This could reflect the way the Boston metropolitan area developed over time; the districts with the newer buildings are further away from employment centers and industrial facilities.

Weighted distance to five Boston employment centers X_8

Since most people like to live close to their place of work, we expect a negative relation between the distances to the employment centers and house prices. The scatterplot hardly reveals any dependence, but the boxplots of X_8 indicate that there might be a slightly positive relation as the red boxplot's median and mean are higher than the black ones. Again, there might be two effects in opposite directions at work here. The first is that living too close to an employment center might not provide enough shelter from the pollution created there. The second, as mentioned above, is that people do not travel very far to their workplace.

Index of accessibility to radial highways X_9

The first obvious thing one can observe from the scatterplots, as well in the histograms and the kernel density estimates, is that there are two subgroups of districts containing X_9 values which are close to the respective group's mean. The scatterplots deliver no hint as to what might explain the occurrence of these two subgroups. The boxplots indicate that for the cheaper and for the more expensive houses the average of X_9 is almost the same.

Full-value property tax X_{10}

X_{10} shows behavior similar to that of X_9: two subgroups exist. A downward-sloping curve seems to underlie the relation of X_{10} and X_{14}. This is confirmed by the two boxplots drawn for X_{10}: the red one has a lower mean and median than the black one.

Pupil/teacher ratio X_{11}

The red and black boxplots of X_{11} indicate a negative relation between X_{11} and X_{14}. This is confirmed by inspection of the scatterplot of X_{11} versus X_{14}: The point cloud is downward-sloping, i.e., the less teachers there are per pupil, the less people pay on median for their dwellings.

Proportion of African American B, $X_{12} = 1000(B - 0.63)^2 I(B < 0.63)$

Interestingly, X_{12} is negatively—though not linearly—correlated with X_3, X_7 and X_{11}, whereas it is positively related with X_{14}. Looking at the data set reveals that for almost all districts X_{12} takes on a value around 390. Since B cannot be larger than 0.63, such values can only be caused by B close to zero. Therefore, the higher X_{12} is, the lower the actual proportion of African-Americans is. Among observations 405 to 470, there are quite a few that have a X_{12} that is much lower than 390. This means that in these districts the proportion of African-Americans is above zero. We can observe two clusters of points in the scatterplots of $\log(X_{12})$: one cluster for which X_{12} is close to 390 and a second one for which X_{12} is between 3 and 100. When X_{12} is positively related with another variable, the actual proportion of African-Americans is negatively correlated with this variable and vice versa. This means that African-Americans live in areas where there is a high proportion of non-retail business land, where there are older houses and where there is a high (i.e., bad) pupil/teacher ratio. It can be observed that districts with housing prices above the median can only be found where the proportion of African-Americans is virtually zero.

Proportion of lower status of the population X_{13}

Of all the variables, X_{13} exhibits the clearest negative relation with X_{14}—hardly any outliers show up. Taking the square root of X_{13} and the logarithm of X_{14} transforms the relation into a linear one.

Transformations

Since most of the variables exhibit an asymmetry with a higher density on the left-hand side, the following transformations are proposed:

$$\widetilde{X_1} = \log(X_1)$$
$$\widetilde{X_2} = X_2/10$$
$$\widetilde{X_3} = \log(X_3)$$
$$\widetilde{X_4} \quad \text{none, since } X_4 \text{ is binary}$$
$$\widetilde{X_5} = \log(X_5)$$
$$\widetilde{X_6} = \log(X_6)$$
$$\widetilde{X_7} = X_7^{2.5}/10000$$
$$\widetilde{X_8} = \log(X_8)$$
$$\widetilde{X_9} = \log(X_9)$$
$$\widetilde{X_{10}} = \log(X_{10})$$
$$\widetilde{X_{11}} = \exp(0.4 \times X_{11})/1000$$
$$\widetilde{X_{12}} = X_{12}/100$$
$$\widetilde{X_{13}} = \sqrt{X_{13}}$$
$$\widetilde{X_{14}} = \log(X_{14})$$

Taking the logarithm or raising the variables to the power of something smaller than one helps to reduce the asymmetry. This is due to the fact that lower values move

Fig. 1.35 Boxplots for all of the variables from the Boston Housing data before and after the proposed transformations MVAboxbhd

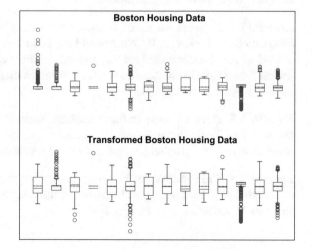

further away from each other, whereas the distance between greater values is reduced by these transformations.

Figure 1.35 displays boxplots for the original mean–variance-scaled variables as well as for the proposed transformed variables. The transformed variables' boxplots are more symmetric and have less outliers than the original variables' boxplots.

1.10   Exercises

Exercise 1.1 Is the upper extreme always an outlier?

Exercise 1.2 Is it possible for the mean or the median to lie outside of the fourths or even outside of the outside bars?

Exercise 1.3 Assume that the data are normally distributed $N(0, 1)$. What percentage of the data do you expect to lie outside the outside bars?

Exercise 1.4 What percentage of the data do you expect to lie outside the outside bars if we assume that the data are normally distributed $N(0, \sigma^2)$ with unknown variance σ^2?

Exercise 1.5 How would the five-number summary of the 15 largest U.S. cities differ from that of the 50 largest U.S. cities? How would the five-number summary of 15 observations of $N(0, 1)$-distributed data differ from that of 50 observations from the same distribution?

Exercise 1.6 Is it possible that all five numbers of the five-number summary could be equal? If so, under what conditions?

Exercise 1.7 Suppose we have 50 observations of $X \sim N(0, 1)$ and another 50 observations of $Y \sim N(2, 1)$. What would the 100 Flury faces look like if you had defined as face elements the face line and the darkness of hair? Do you expect any similar faces? How many faces do you think should look like observations of Y even though they are X observations?

Exercise 1.8 Draw a histogram for the mileage variable of the car data (Sect. B.3). Do the same for the three groups (U.S., Japan, Europe). Do you obtain a similar conclusion as in the parallel boxplot on Fig. 1.3 for these data?

Exercise 1.9 Use some bandwidth selection criterion to calculate the optimally chosen bandwidth h for the diagonal variable of the bank notes. Would it be better to have one bandwidth for the two groups?

References

ALLBUS, Germany General Social Survey 1980–2004 (2006)

D.F. Andrews, Plots of high-dimensional data. Biometrics **28**, 125–136 (1972)

H. Chernoff, Using faces to represent points in k-dimensional space graphically. J. Am. Stat. Assoc. **68**, 361–368 (1973)

B. Flury, H. Riedwyl, *Multivariate Statistics* (Cambridge University Press, A practical Approach, 1988)

M. Graham, J. Kennedy, Using curves to enhance parallel coordinate visualisations, in *Information Visualization, 2003. IV 2003. Proceedings. Seventh International Conference on*, pp. 10–16 (2003)

W. Härdle, M. Müller, S. Sperlich, A. Werwatz, *Non- and Semiparametric Models* (Springer, Heidelberg, 2004)

W. Härdle, *Smoothing Techniques* (With Implementations in S. Springer, New York, 1991)

W. Härdle, D.W. Scott, Smoothing by weighted averaging of rounded points. Comput. Stat. **7**, 97–128 (1992)

D. Harrison, D.L. Rubinfeld, Hedonic prices and the demand for clean air. J. Environ. Econ. Manag. **5**, 81–102 (1978)

W. Hoaglin, F. Mosteller, J.W. Tukey, *Understanding Robust and Exploratory Data Analysis* (Whiley, New York, 1983)

A. Inselberg, A goodness of fit test for binary regression models based on smoohting methods. Vis. Comput. **1**, 69–91 (1985)

S. Klinke, Polzehl, Implementation of kernel based indices in XGobi. Discussion paper 47, SFB 373, Humboldt-University of Berlin (1995)

N. Lewin-Koh, Hexagon binnning. *Technical Report* (2006)

E. Parzen, On estimating of a probability density and mode. Ann. Math. Stat. **35**, 1065–1076 (1962)

M. Rosenblatt, Remarks on some nonparametric estimates of a density function. Ann. Math. Stat. **27**, 832–837 (1956)

D.W. Scott, Averaged shifted histograms: effective nonparametric density estimation in several dimensions. Ann. Stat. **13**, 1024–1040 (1985)

B.W. Silverman, *Density Estimation for Statistics and Data Analysis*, vol. 26, Monographs on Statistics and Applied Probability (Chapman and Hall, London, 1986)

E.R. Tufte. *The Visual Display of Quantitative Information* (Graphics Press, 1983)

P. Whittle, On the smoothing of probability density functions. J. Royal Stat. Soc. Ser. B **55**, 549–557 (1958)

Part II
Multivariate Random Variables

Chapter 2
A Short Excursion into Matrix Algebra

This chapter serves as a reminder of basic concepts of matrix algebra, which are particularly useful in multivariate analysis. It also introduces the notations used in this book for vectors and matrices. Eigenvalues and eigenvectors play an important role in multivariate techniques. In Sects. 2.2 and 2.3, we present the spectral decomposition of matrices and consider the maximization (minimization) of quadratic forms given some constraints.

In analyzing the multivariate normal distribution, partitioned matrices appear naturally. Some of the basic algebraic properties are given in Sect. 2.5. These properties will be heavily used in Chaps. 4 and 5.

The geometry of the multinormal and the geometric interpretation of the multivariate techniques (Part III) intensively uses the notion of angles between two vectors, the projection of a point on a vector, and the distances between two points. These ideas are introduced in Sect. 2.6.

2.1 Elementary Operations

A matrix $\mathcal{A}$ is a system of numbers with n rows and p columns:

$$\mathcal{A} = \begin{pmatrix} a_{11} & a_{12} & \ldots & \ldots & \ldots & a_{1p} \\ \vdots & a_{22} & & & & \vdots \\ \vdots & \vdots & \ddots & & & \vdots \\ \vdots & \vdots & & \ddots & & \vdots \\ \vdots & \vdots & & & \ddots & \vdots \\ a_{n1} & a_{n2} & \ldots & \ldots & \ldots & a_{np} \end{pmatrix}.$$

© Springer Nature Switzerland AG 2019
W. K. Härdle and L. Simar, *Applied Multivariate Statistical Analysis*,
https://doi.org/10.1007/978-3-030-26006-4_2

We also write (a_{ij}) for $\mathcal{A}$ and $\mathcal{A}(n \times p)$ to indicate the numbers of rows and columns. Vectors are matrices with one column and are denoted as x or $x(p \times 1)$. Special matrices and vectors are defined in Table 2.1. Note that we use small letters for scalars as well as for vectors.

Matrix Operations

Elementary operations are summarized below:

$$\mathcal{A}^\top = (a_{ji})$$
$$\mathcal{A} + \mathcal{B} = (a_{ij} + b_{ij})$$
$$\mathcal{A} - \mathcal{B} = (a_{ij} - b_{ij})$$
$$c \cdot \mathcal{A} = (c \cdot a_{ij})$$

$$\mathcal{A} \cdot \mathcal{B} = \mathcal{A}(n \times p)\, \mathcal{B}(p \times m) = \mathcal{C}(n \times m) = (c_{ij}) = \left(\sum_{j=1}^{p} a_{ij} b_{jk} \right).$$

Properties of Matrix Operations

$$\mathcal{A} + \mathcal{B} = \mathcal{B} + \mathcal{A}$$
$$\mathcal{A}(\mathcal{B} + \mathcal{C}) = \mathcal{A}\mathcal{B} + \mathcal{A}\mathcal{C}$$
$$\mathcal{A}(\mathcal{B}\mathcal{C}) = (\mathcal{A}\mathcal{B})\mathcal{C}$$
$$(\mathcal{A}^\top)^\top = \mathcal{A}$$
$$(\mathcal{A}\mathcal{B})^\top = \mathcal{B}^\top \mathcal{A}^\top$$

Matrix Characteristics

Rank

The *rank*, rank$(\mathcal{A})$, of a matrix $\mathcal{A}(n \times p)$ is defined as the maximum number of linearly independent rows (columns). A set of k rows a_j of $\mathcal{A}(n \times p)$ are said to be linearly independent if $\sum_{j=1}^{k} c_j a_j = 0_p$ implies $c_j = 0, \forall j$, where $c_1, \ldots, c_k$ are scalars. In other words, no rows in this set can be expressed as a nontrivial linear combination of the $(k-1)$ remaining rows.

Trace

The *trace* of a matrix $\mathcal{A}(p \times p)$ is the sum of its diagonal elements

$$\mathrm{tr}(\mathcal{A}) = \sum_{i=1}^{p} a_{ii}.$$

Determinant

The *determinant* is an important concept of matrix algebra. For a square matrix $\mathcal{A}$, it is defined as

Table 2.1 Special matrices and vectors

Name	Definition	Notation	Example
Scalar	$p = n = 1$	a	3
Column vector	$p = 1$	a	$\begin{pmatrix} 1 \\ 3 \end{pmatrix}$
Row vector	$n = 1$	$a^\top$	$\begin{pmatrix} 1 & 3 \end{pmatrix}$
Vector of ones	$(\underbrace{1, \ldots, 1}_{n})^\top$	1_n	$\begin{pmatrix} 1 \\ 1 \end{pmatrix}$
Vector of zeros	$(\underbrace{0, \ldots, 0}_{n})^\top$	0_n	$\begin{pmatrix} 0 \\ 0 \end{pmatrix}$
Square matrix	$n = p$	$\mathcal{A}(p \times p)$	$\begin{pmatrix} 2 & 0 \\ 0 & 2 \end{pmatrix}$
Diagonal matrix	$a_{ij} = 0, i \neq j, n = p$	$\mathrm{diag}(a_{ii})$	$\begin{pmatrix} 1 & 0 \\ 0 & 2 \end{pmatrix}$
Identity matrix	$\mathrm{diag}(\underbrace{1, \ldots, 1}_{p})$	$\mathcal{I}_p$	$\begin{pmatrix} 1 & 0 \\ 0 & 1 \end{pmatrix}$
Unit matrix	$a_{ij} = 1, n = p$	$1_n 1_n^\top$	$\begin{pmatrix} 1 & 1 \\ 1 & 1 \end{pmatrix}$
Symmetric matrix	$a_{ij} = a_{ji}$		$\begin{pmatrix} 1 & 2 \\ 2 & 3 \end{pmatrix}$
Null matrix	$a_{ij} = 0$	0	$\begin{pmatrix} 0 & 0 \\ 0 & 0 \end{pmatrix}$
Upper triangular matrix	$a_{ij} = 0, i < j$		$\begin{pmatrix} 1 & 2 & 4 \\ 0 & 1 & 3 \\ 0 & 0 & 1 \end{pmatrix}$
Idempotent matrix	$\mathcal{A}\mathcal{A} = \mathcal{A}$		$\begin{pmatrix} 1 & 0 & 0 \\ 0 & \frac{1}{2} & \frac{1}{2} \\ 0 & \frac{1}{2} & \frac{1}{2} \end{pmatrix}$
Orthogonal matrix	$A^\top \mathcal{A} = \mathcal{I} = \mathcal{A}\mathcal{A}^\top$		$\begin{pmatrix} \frac{1}{\sqrt{2}} & \frac{1}{\sqrt{2}} \\ \frac{1}{\sqrt{2}} & -\frac{1}{\sqrt{2}} \end{pmatrix}$

$$\det(\mathcal{A}) = |\mathcal{A}| = \sum (-1)^{|\tau|} a_{1\tau(1)} \dots a_{p\tau(p)},$$

the summation is over all permutations τ of $\{1, 2, \dots, p\}$, and $|\tau| = 0$ if the permutation can be written as a product of an even number of transpositions and $|\tau| = 1$ otherwise. Some properties of determinant of a matrix are

$$|\mathcal{A}^\top| = |\mathcal{A}|$$
$$|\mathcal{A}\mathcal{B}| = |\mathcal{A}| \cdot |\mathcal{A}|$$
$$|c\mathcal{A}| = c^n |\mathcal{A}|.$$

Example 2.1 In the case of $p = 2$, $\mathcal{A} = \begin{pmatrix} a_{11} & a_{12} \\ a_{21} & a_{22} \end{pmatrix}$ and we can permute the digits "1" and "2" once or not at all. So,

$$|\mathcal{A}| = a_{11}\, a_{22} - a_{12}\, a_{21}.$$

Transpose

For $\mathcal{A}(n \times p)$ and $\mathcal{B}(p \times n)$

$$(\mathcal{A}^\top)^\top = \mathcal{A}, \text{ and } (\mathcal{A}\mathcal{B})^\top = \mathcal{B}^\top \mathcal{A}^\top.$$

Inverse

If $|\mathcal{A}| \neq 0$ and $\mathcal{A}(p \times p)$, then the inverse $\mathcal{A}^{-1}$ exists:

$$\mathcal{A}\,\mathcal{A}^{-1} = \mathcal{A}^{-1}\,\mathcal{A} = \mathcal{I}_p.$$

For small matrices, the inverse of $\mathcal{A} = (a_{ij})$ can be calculated as

$$\mathcal{A}^{-1} = \frac{\mathcal{C}}{|\mathcal{A}|},$$

where $\mathcal{C} = (c_{ij})$ is the adjoint matrix of $\mathcal{A}$. The elements c_{ji} of $\mathcal{C}^\top$ are the co-factors of $\mathcal{A}$:

$$c_{ji} = (-1)^{i+j} \begin{vmatrix} a_{11} & \cdots & a_{1(j-1)} & a_{1(j+1)} & \cdots & a_{1p} \\ \vdots & & & & & \\ a_{(i-1)1} & \cdots & a_{(i-1)(j-1)} & a_{(i-1)(j+1)} & \cdots & a_{(i-1)p} \\ a_{(i+1)1} & \cdots & a_{(i+1)(j-1)} & a_{(i+1)(j+1)} & \cdots & a_{(i+1)p} \\ \vdots & & & & & \\ a_{p1} & \cdots & a_{p(j-1)} & a_{p(j+1)} & \cdots & a_{pp} \end{vmatrix}.$$

The relationship between determinant and inverse of matrix $\mathcal{A}$ is $|\mathcal{A}^{-1}| = |\mathcal{A}|^{-1}$.

G-inverse

A more general concept is the *G-inverse* (Generalized Inverse) $\mathcal{A}^-$ which satisfies the following:

$$\mathcal{A}\,\mathcal{A}^-\mathcal{A} = \mathcal{A}.$$

Later we will see that there may be more than one *G*-inverse.

Example 2.2 The generalized inverse can also be calculated for singular matrices. We have

$$\begin{pmatrix} 1 & 0 \\ 0 & 0 \end{pmatrix} \begin{pmatrix} 1 & 0 \\ 0 & 0 \end{pmatrix} \begin{pmatrix} 1 & 0 \\ 0 & 0 \end{pmatrix} = \begin{pmatrix} 1 & 0 \\ 0 & 0 \end{pmatrix},$$

which means that the generalized inverse of $\mathcal{A} = \begin{pmatrix} 1 & 0 \\ 0 & 0 \end{pmatrix}$ is $\mathcal{A}^- = \begin{pmatrix} 1 & 0 \\ 0 & 0 \end{pmatrix}$ even though the inverse matrix of $\mathcal{A}$ does not exist in this case.

Eigenvalues, Eigenvectors

Consider a $(p \times p)$ matrix $\mathcal{A}$. If there a scalar λ and a vector γ exists such as

$$\mathcal{A}\gamma = \lambda\gamma, \tag{2.1}$$

then we call

$$\begin{array}{ll} \lambda & \text{an eigenvalue} \\ \gamma & \text{an eigenvector.} \end{array}$$

It can be proven that an eigenvalue λ is a root of the p-th order polynomial $|\mathcal{A} - \lambda I_p| = 0$. Therefore, there are up to p eigenvalues $\lambda_1, \lambda_2, \ldots, \lambda_p$ of $\mathcal{A}$. For each eigenvalue λ_j, a corresponding eigenvector γ_j exists given by Eq. (2.1). Suppose the matrix $\mathcal{A}$ has the eigenvalues $\lambda_1, \ldots, \lambda_p$. Let $\Lambda = \text{diag}(\lambda_1, \ldots, \lambda_p)$.

The determinant $|\mathcal{A}|$ and the trace $\text{tr}(\mathcal{A})$ can be rewritten in terms of the eigenvalues:

$$|\mathcal{A}| = |\Lambda| = \prod_{j=1}^{p} \lambda_j \tag{2.2}$$

$$\text{tr}(\mathcal{A}) = \text{tr}(\Lambda) = \sum_{j=1}^{p} \lambda_j. \tag{2.3}$$

An idempotent matrix $\mathcal{A}$ (see the definition in Table 2.1) can only have eigenvalues in $\{0, 1\}$; therefore, $\text{tr}(\mathcal{A}) = \text{rank}(\mathcal{A}) = $ number of eigenvalues $\neq 0$.

Example 2.3 Let us consider the matrix $\mathcal{A} = \begin{pmatrix} 1 & 0 & 0 \\ 0 & \frac{1}{2} & \frac{1}{2} \\ 0 & \frac{1}{2} & \frac{1}{2} \end{pmatrix}$. It is easy to verify that $\mathcal{A}\mathcal{A} = \mathcal{A}$ which implies that the matrix $\mathcal{A}$ is idempotent.

We know that the eigenvalues of an idempotent matrix are equal to 0 or 1. In this case, the eigenvalues of $\mathcal{A}$ are $\lambda_1 = 1$, $\lambda_2 = 1$, and $\lambda_3 = 0$ since $\begin{pmatrix} 1 & 0 & 0 \\ 0 & \frac{1}{2} & \frac{1}{2} \\ 0 & \frac{1}{2} & \frac{1}{2} \end{pmatrix} \begin{pmatrix} 1 \\ 0 \\ 0 \end{pmatrix} =$

$1 \begin{pmatrix} 1 \\ 0 \\ 0 \end{pmatrix}$, $\begin{pmatrix} 1 & 0 & 0 \\ 0 & \frac{1}{2} & \frac{1}{2} \\ 0 & \frac{1}{2} & \frac{1}{2} \end{pmatrix} \begin{pmatrix} 0 \\ \frac{\sqrt{2}}{2} \\ \frac{\sqrt{2}}{2} \end{pmatrix} = 1 \begin{pmatrix} 0 \\ \frac{\sqrt{2}}{2} \\ \frac{\sqrt{2}}{2} \end{pmatrix}$ and $\begin{pmatrix} 1 & 0 & 0 \\ 0 & \frac{1}{2} & \frac{1}{2} \\ 0 & \frac{1}{2} & \frac{1}{2} \end{pmatrix} \begin{pmatrix} 0 \\ \frac{\sqrt{2}}{2} \\ -\frac{\sqrt{2}}{2} \end{pmatrix} = 0 \begin{pmatrix} 0 \\ \frac{\sqrt{2}}{2} \\ -\frac{\sqrt{2}}{2} \end{pmatrix}.$

Using formulas (2.2) and (2.3), we can calculate the trace and the determinant of $\mathcal{A}$ from the eigenvalues: $\mathrm{tr}(\mathcal{A}) = \lambda_1 + \lambda_2 + \lambda_3 = 2$, $|\mathcal{A}| = \lambda_1 \lambda_2 \lambda_3 = 0$, and $\mathrm{rank}(\mathcal{A}) = 2$.

Properties of Matrix Characteristics

$\mathcal{A}(n \times n)$, $\mathcal{B}(n \times n)$, $c \in \mathbb{R}$

$$\mathrm{tr}(\mathcal{A} + \mathcal{B}) = \mathrm{tr}\,\mathcal{A} + \mathrm{tr}\,\mathcal{B} \tag{2.4}$$

$$\mathrm{tr}(c\mathcal{A}) = c\,\mathrm{tr}\,\mathcal{A} \tag{2.5}$$

$$|c\mathcal{A}| = c^n |\mathcal{A}| \tag{2.6}$$

$$|\mathcal{A}\mathcal{B}| = |\mathcal{B}\mathcal{A}| = |\mathcal{A}||\mathcal{B}| \tag{2.7}$$

$\mathcal{A}(n \times p)$, $\mathcal{B}(p \times n)$

$$\mathrm{tr}(\mathcal{A} \cdot \mathcal{B}) = \mathrm{tr}(\mathcal{B} \cdot \mathcal{A}) \tag{2.8}$$

$$\mathrm{rank}(\mathcal{A}) \leq \min(n, p)$$

$$\mathrm{rank}(\mathcal{A}) \geq 0 \tag{2.9}$$

$$\mathrm{rank}(\mathcal{A}) = \mathrm{rank}(\mathcal{A}^\top) \tag{2.10}$$

$$\mathrm{rank}(\mathcal{A}^\top \mathcal{A}) = \mathrm{rank}(\mathcal{A}) \tag{2.11}$$

$$\mathrm{rank}(\mathcal{A} + \mathcal{B}) \leq \mathrm{rank}(\mathcal{A}) + \mathrm{rank}(\mathcal{B}) \tag{2.12}$$

$$\mathrm{rank}(\mathcal{A}\mathcal{B}) \leq \min\{\mathrm{rank}(\mathcal{A}), \mathrm{rank}(\mathcal{B})\} \tag{2.13}$$

$\mathcal{A}(n \times p)$, $\mathcal{B}(p \times q)$, $\mathcal{C}(q \times n)$

$$\mathrm{tr}(\mathcal{A}\mathcal{B}\mathcal{C}) = \mathrm{tr}(\mathcal{B}\mathcal{C}\mathcal{A})$$

$$= \mathrm{tr}(\mathcal{C}\mathcal{A}\mathcal{B}) \tag{2.14}$$

$$\mathrm{rank}(\mathcal{A}\mathcal{B}\mathcal{C}) = \mathrm{rank}(\mathcal{B}) \quad \text{for nonsingular } \mathcal{A}, \mathcal{C} \tag{2.15}$$

$\mathcal{A}(p \times p)$

$$|\mathcal{A}^{-1}| = |\mathcal{A}|^{-1} \tag{2.16}$$

$$\mathrm{rank}(\mathcal{A}) = p \quad \text{if and only if } \mathcal{A} \text{ is nonsingular.} \tag{2.17}$$

2.1 Elementary Operations

Summary
↪ The determinant $\|\mathcal{A}\|$ is the product of the eigenvalues of $\mathcal{A}$.
↪ The inverse of a matrix $\mathcal{A}$ exists if $\|\mathcal{A}\| \neq 0$.
↪ The trace $\text{tr}(\mathcal{A})$ is the sum of the eigenvalues of $\mathcal{A}$.
↪ The sum of the traces of two matrices equals the trace of the sum of the two matrices.
↪ The trace $\text{tr}(\mathcal{AB})$ equals $\text{tr}(\mathcal{BA})$.
↪ The rank$(\mathcal{A})$ is the maximal number of linearly independent rows (columns) of $\mathcal{A}$.

2.2 Spectral Decompositions

The computation of eigenvalues and eigenvectors is an important issue in the analysis of matrices. The spectral decomposition or Jordan decomposition links the structure of a matrix to the eigenvalues and the eigenvectors.

Theorem 2.1 (Jordan Decomposition) *Each symmetric matrix $\mathcal{A}(p \times p)$ can be written as*

$$\mathcal{A} = \Gamma \, \Lambda \, \Gamma^\top = \sum_{j=1}^{p} \lambda_j \gamma_j \gamma_j^\top \tag{2.18}$$

where

$$\Lambda = \text{diag}(\lambda_1, \ldots, \lambda_p)$$

and where

$$\Gamma = (\gamma_1, \gamma_2, \ldots, \gamma_p)$$

is an orthogonal matrix consisting of the eigenvectors γ_j of $\mathcal{A}$.

Example 2.4 Suppose that $\mathcal{A} = \begin{pmatrix} 1 & 2 \\ 2 & 3 \end{pmatrix}$. The eigenvalues are found by solving $|\mathcal{A} - \lambda\mathcal{I}| = 0$. This is equivalent to

$$\begin{vmatrix} 1 - \lambda & 2 \\ 2 & 3 - \lambda \end{vmatrix} = (1 - \lambda)(3 - \lambda) - 4 = 0.$$

Hence, the eigenvalues are $\lambda_1 = 2 + \sqrt{5}$ and $\lambda_2 = 2 - \sqrt{5}$. The eigenvectors are $\gamma_1 = (0.5257, 0.8506)^\top$ and $\gamma_2 = (0.8506, -0.5257)^\top$. They are orthogonal since $\gamma_1^\top \gamma_2 = 0$.

Using spectral decomposition, we can define powers of a matrix $\mathcal{A}(p \times p)$. Suppose $\mathcal{A}$ is a symmetric matrix with positive eigenvalues. Then by Theorem 2.1

$$\mathcal{A} = \Gamma \Lambda \Gamma^\top,$$

and we define for some $\alpha \in \mathbb{R}$

$$\mathcal{A}^\alpha = \Gamma \Lambda^\alpha \Gamma^\top, \qquad (2.19)$$

where $\Lambda^\alpha = \text{diag}(\lambda_1^\alpha, \ldots, \lambda_p^\alpha)$. In particular, we can easily calculate the inverse of the matrix $\mathcal{A}$. Suppose that the eigenvalues of $\mathcal{A}$ are positive. Then with $\alpha = -1$, we obtain the inverse of $\mathcal{A}$ from

$$\mathcal{A}^{-1} = \Gamma \Lambda^{-1} \Gamma^\top. \qquad (2.20)$$

Another interesting decomposition which is later used is given in the following theorem.

Theorem 2.2 (Singular Value Decomposition) *Each matrix $\mathcal{A}(n \times p)$ with rank r can be decomposed as*

$$\mathcal{A} = \Gamma \Lambda \Delta^\top,$$

where $\Gamma(n \times r)$ and $\Delta(p \times r)$. Both Γ and Δ are column orthonormal, i.e., $\Gamma^\top \Gamma = \Delta^\top \Delta = \mathcal{I}_r$ and $\Lambda = \text{diag}\left(\lambda_1^{1/2}, \ldots, \lambda_r^{1/2}\right)$, $\lambda_j > 0$. The values $\lambda_1, \ldots, \lambda_r$ are the nonzero eigenvalues of the matrices $\mathcal{A}\mathcal{A}^\top$ and $\mathcal{A}^\top \mathcal{A}$. Γ and Δ consist of the corresponding r eigenvectors of these matrices.

This is obviously a generalization of Theorem 2.1 (Jordan decomposition). With Theorem 2.2, we can find a G-inverse $\mathcal{A}^-$ of $\mathcal{A}$. Indeed, define $\mathcal{A}^- = \Delta \Lambda^{-1} \Gamma^\top$. Then $\mathcal{A} \mathcal{A}^- \mathcal{A} = \Gamma \Lambda \Delta^\top = \mathcal{A}$. Note that the G-inverse is not unique.

Example 2.5 In Example 2.2, we showed that the generalized inverse of $\mathcal{A} = \begin{pmatrix} 1 & 0 \\ 0 & 0 \end{pmatrix}$ is $\mathcal{A}^- \begin{pmatrix} 1 & 0 \\ 0 & 0 \end{pmatrix}$. The following also holds

$$\begin{pmatrix} 1 & 0 \\ 0 & 0 \end{pmatrix} \begin{pmatrix} 1 & 0 \\ 0 & 8 \end{pmatrix} \begin{pmatrix} 1 & 0 \\ 0 & 0 \end{pmatrix} = \begin{pmatrix} 1 & 0 \\ 0 & 0 \end{pmatrix}$$

which means that the matrix $\begin{pmatrix} 1 & 0 \\ 0 & 8 \end{pmatrix}$ is also a generalized inverse of $\mathcal{A}$.

Summary
↪ The Jordan decomposition gives a representation of a symmetric matrix in terms of eigenvalues and eigenvectors.
↪ The eigenvectors belonging to the largest eigenvalues indicate the "main direction" of the data.
↪ The Jordan decomposition allows one to easily compute the power of a symmetric matrix $\mathcal{A}$: $\mathcal{A}^\alpha = \Gamma \Lambda^\alpha \Gamma^\top$.
↪ The singular value decomposition (SVD) is a generalization of the Jordan decomposition to non-quadratic matrices.

2.3 Quadratic Forms

A quadratic form $Q(x)$ is built from a symmetric matrix $\mathcal{A}(p \times p)$ and a vector $x \in \mathbb{R}^p$:

$$Q(x) = x^\top \mathcal{A} x = \sum_{i=1}^p \sum_{j=1}^p a_{ij} x_i x_j. \tag{2.21}$$

Definiteness of Quadratic Forms and Matrices

$$Q(x) > 0 \text{ for all } x \neq 0 \qquad \textit{positive definite}$$
$$Q(x) \geq 0 \text{ for all } x \neq 0 \qquad \textit{positive semidefinite}$$

A matrix $\mathcal{A}$ is called positive definite (semidefinite) if the corresponding quadratic form $Q(.)$ is positive definite (semidefinite). We write $\mathcal{A} > 0 \ (\geq 0)$.

Quadratic forms can always be diagonalized, as the following result shows.

Theorem 2.3 *If $\mathcal{A}$ is symmetric and $Q(x) = x^\top \mathcal{A} x$ is the corresponding quadratic form, then there exists a transformation $x \mapsto \Gamma^\top x = y$ such that*

$$x^\top \mathcal{A} x = \sum_{i=1}^p \lambda_i y_i^2,$$

where λ_i are the eigenvalues of $\mathcal{A}$.

Proof $\mathcal{A} = \Gamma \ \Lambda \ \Gamma^\top$. By Theorem 2.1 and $y = \Gamma^\top \alpha$ we have that $x^\top \mathcal{A} x = x^\top \Gamma \Lambda \Gamma^\top x = y^\top \Lambda y = \sum_{i=1}^p \lambda_i y_i^2$.

Positive definiteness of quadratic forms can be deduced from positive eigenvalues.

Theorem 2.4 $\mathcal{A} > 0$ *if and only if all $\lambda_i > 0$, $i = 1, \ldots, p$.*

Proof $0 < \lambda_1 y_1^2 + \cdots + \lambda_p y_p^2 = x^\top \mathcal{A} x$ for all $x \neq 0$ by Theorem 2.3.

Corollary 2.1 *If $\mathcal{A} > 0$, then $\mathcal{A}^{-1}$ exists and $|\mathcal{A}| > 0$.*

Example 2.6 The quadratic form $Q(x) = x_1^2 + x_2^2$ corresponds to the matrix $\mathcal{A} = \begin{pmatrix} 1 & 0 \\ 0 & 1 \end{pmatrix}$ with eigenvalues $\lambda_1 = \lambda_2 = 1$ and is thus positive definite. The quadratic form $Q(x) = (x_1 - x_2)^2$ corresponds to the matrix $\mathcal{A} = \begin{pmatrix} 1 & -1 \\ -1 & 1 \end{pmatrix}$ with eigenvalues $\lambda_1 = 2$, $\lambda_2 = 0$ and is positive semidefinite. The quadratic form $Q(x) = x_1^2 - x_2^2$ with eigenvalues $\lambda_1 = 1$, $\lambda_2 = -1$ is indefinite.

In the statistical analysis of multivariate data, we are interested in maximizing quadratic forms given some constraints.

Theorem 2.5 *If $\mathcal{A}$ and $\mathcal{B}$ are symmetric and $\mathcal{B} > 0$, then the maximum of $\frac{x^\top \mathcal{A} x}{x^\top \mathcal{B} x}$ is given by the largest eigenvalue of $\mathcal{B}^{-1} \mathcal{A}$. More generally,*

$$\max_x \frac{x^\top \mathcal{A} x}{x^\top \mathcal{B} x} = \lambda_1 \geq \lambda_2 \geq \cdots \geq \lambda_p = \min_x \frac{x^\top \mathcal{A} x}{x^\top \mathcal{B} x},$$

where $\lambda_1, \ldots, \lambda_p$ denote the eigenvalues of $\mathcal{B}^{-1} \mathcal{A}$. The vector which maximizes (minimizes) $\frac{x^\top \mathcal{A} x}{x^\top \mathcal{B} x}$ is the eigenvector of $\mathcal{B}^{-1} \mathcal{A}$ which corresponds to the largest (smallest) eigenvalue of $\mathcal{B}^{-1} \mathcal{A}$. If $x^\top \mathcal{B} x = 1$, we get

$$\max_x x^\top \mathcal{A} x = \lambda_1 \geq \lambda_2 \geq \cdots \geq \lambda_p = \min_x x^\top \mathcal{A} x$$

Proof Denote norm of vector x as $\|x\| = \sqrt{x^\top x}$. By definition, $\mathcal{B}^{1/2} = \Gamma_\mathcal{B} \Lambda_\mathcal{B}^{1/2} \Gamma_\mathcal{B}^\top$ is symmetric. Then $x^\top \mathcal{B} x = \left\| x^\top \mathcal{B}^{1/2} \right\|^2 = \left\| \mathcal{B}^{1/2} x \right\|^2$. Set $y = \frac{\mathcal{B}^{1/2} x}{\|\mathcal{B}^{1/2} x\|}$, then

$$\max_x \frac{x^\top \mathcal{A} x}{x^\top \mathcal{B} x} = \max_{\{y: y^\top y = 1\}} y^\top \mathcal{B}^{-1/2} \mathcal{A} \mathcal{B}^{-1/2} y. \tag{2.22}$$

From Theorem 2.1, let

$$\mathcal{B}^{-1/2} \mathcal{A} \mathcal{B}^{-1/2} = \Gamma \Lambda \Gamma^\top$$

be the spectral decomposition of $\mathcal{B}^{-1/2} \mathcal{A} \mathcal{B}^{-1/2}$. Set

$$z = \Gamma^\top y, \text{ then } z^\top z = y^\top \Gamma \Gamma^\top y = y^\top y.$$

Thus (2.22) is equivalent to

$$\max_{\{z: z^\top z = 1\}} z^\top \Lambda z = \max_{\{z: z^\top z = 1\}} \sum_{i=1}^{p} \lambda_i z_i^2.$$

But

$$\max_{z} \sum \lambda_i z_i^2 \leq \lambda_1 \underbrace{\max_{z} \sum z_i^2}_{=1} = \lambda_1.$$

The maximum is thus obtained by $z = (1, 0, \ldots, 0)^\top$, i.e.,

$$y = \gamma_1, \text{ hence } x = B^{-1/2}\gamma_1.$$

Since $B^{-1}A$ and $B^{-1/2} A B^{-1/2}$ have the same eigenvalues, the proof is complete. To maximize (minimize) $x^\top Ax$ under $x^\top Bx = 1$, below is another proof using the Lagrange method.

$$\max_{x} x^\top Ax = \max_{x}[x^\top Ax - \lambda(x^\top Bx - 1)].$$

The first derivative of it in respect to x, is equal to 0:

$$2Ax - 2\lambda Bx = 0.$$

so

$$B^{-1}Ax = \lambda x$$

By the definition of eigenvector and eigenvalue, our maximizer x^* is $B^{-1}A's$ eigenvector corresponding to eigenvalue λ. So

$$\max_{\{x : x^\top Bx=1\}} x^\top Ax = \max_{\{x : x^\top Bx=1\}} x^\top BB^{-1}Ax = \max_{\{x : x^\top Bx=1\}} x^\top B\lambda x = \max \lambda$$

which is just the maximum eigenvalue of $B^{-1}A$, and we choose the corresponding eigenvector as our maximizer x^*.

Example 2.7 Consider the matrices $A = \begin{pmatrix} 1 & 2 \\ 2 & 3 \end{pmatrix}$ and $B = \begin{pmatrix} 1 & 0 \\ 0 & 1 \end{pmatrix}$,

we calculate $B^{-1}A = \begin{pmatrix} 1 & 2 \\ 2 & 3 \end{pmatrix}$. The biggest eigenvalue of the matrix $B^{-1}A$ is $2 + \sqrt{5}$.

This means that the maximum of $x^\top Ax$ under the constraint $x^\top Bx = 1$ is $2 + \sqrt{5}$. Notice that the constraint $x^\top Bx = 1$ corresponds to our choice of B, to the points which lie on the unit circle $x_1^2 + x_2^2 = 1$.

Summary
↪   A quadratic form can be described by a symmetric matrix $\mathcal{A}$.
↪   Quadratic forms can always be diagonalized.
↪   Positive definiteness of a quadratic form is equivalent to positiveness of the eigenvalues of the matrix $\mathcal{A}$.
↪   The maximum and minimum of a quadratic form given some constraints can be expressed in terms of eigenvalues.

2.4   Derivatives

For later sections of this book, it will be useful to introduce matrix notation for derivatives of a scalar function of a vector x, i.e., $f(x)$, with respect to x. Consider $f : \mathbb{R}^p \to \mathbb{R}$ and a $(p \times 1)$ vector x, then $\frac{\partial f(x)}{\partial x}$ is the column vector of partial derivatives $\left\{ \frac{\partial f(x)}{\partial x_j} \right\}$, $j = 1, \ldots, p$ and $\frac{\partial f(x)}{\partial x^\top}$ is the row vector of the same derivative ($\frac{\partial f(x)}{\partial x}$ is called the *gradient* of f).

We can also introduce second-order derivatives: $\frac{\partial^2 f(x)}{\partial x \partial x^\top}$ is the $(p \times p)$ matrix of elements $\frac{\partial^2 f(x)}{\partial x_i \partial x_j}$, $i = 1, \ldots, p$ and $j = 1, \ldots, p$. ($\frac{\partial^2 f(x)}{\partial x \partial x^\top}$ is called the *Hessian* of f).

Suppose that a is a $(p \times 1)$ vector and that $\mathcal{A} = \mathcal{A}^\top$ is a $(p \times p)$ matrix. Then

$$\frac{\partial a^\top x}{\partial x} = \frac{\partial x^\top a}{\partial x} = a, \tag{2.23}$$

$$\frac{\partial x^\top \mathcal{A} x}{\partial x} = 2\mathcal{A}x. \tag{2.24}$$

The Hessian of the quadratic form $Q(x) = x^\top \mathcal{A} x$ is

$$\frac{\partial^2 x^\top \mathcal{A} x}{\partial x \partial x^\top} = 2\mathcal{A}. \tag{2.25}$$

Example 2.8   Consider the matrix

$$\mathcal{A} = \begin{pmatrix} 1 & 2 \\ 2 & 3 \end{pmatrix}.$$

From formulas (2.24) and (2.25), it immediately follows that the gradient of $Q(x) = x^\top \mathcal{A} x$ is

$$\frac{\partial x^\top \mathcal{A} x}{\partial x} = 2\mathcal{A}x = 2 \begin{pmatrix} 1 & 2 \\ 2 & 3 \end{pmatrix} x = \begin{pmatrix} 2x & 4x \\ 4x & 6x \end{pmatrix}$$

and the Hessian is

$$\frac{\partial^2 x^\top A x}{\partial x \partial x^\top} = 2A = 2 \begin{pmatrix} 1 & 2 \\ 2 & 3 \end{pmatrix} = \begin{pmatrix} 2 & 4 \\ 4 & 6 \end{pmatrix}.$$

2.5 Partitioned Matrices

Very often we will have to consider certain groups of rows and columns of a matrix $A(n \times p)$. In the case of two groups, we have

$$A = \begin{pmatrix} A_{11} & A_{12} \\ A_{21} & A_{22} \end{pmatrix},$$

where $A_{ij}(n_i \times p_j)$, $i, j = 1, 2$, $n_1 + n_2 = n$ and $p_1 + p_2 = p$.

If $B(n \times p)$ is partitioned accordingly, we have

$$A + B = \begin{pmatrix} A_{11} + B_{11} & A_{12} + B_{12} \\ A_{21} + B_{21} & A_{22} + B_{22} \end{pmatrix}$$

$$B^\top = \begin{pmatrix} B_{11}^\top & B_{21}^\top \\ B_{12}^\top & B_{22}^\top \end{pmatrix}$$

$$AB^\top = \begin{pmatrix} A_{11}B_{11}^\top + A_{12}B_{12}^\top & A_{11}B_{21}^\top + A_{12}B_{22}^\top \\ A_{21}B_{11}^\top + A_{22}B_{12}^\top & A_{21}B_{21}^\top + A_{22}B_{22}^\top \end{pmatrix}.$$

An important particular case is the square matrix $A(p \times p)$, partitioned in such a way that A_{11} and A_{22} are both square matrices (i.e., $n_j = p_j$, $j = 1, 2$). It can be verified that when A is non-singular ($AA^{-1} = I_p$):

$$A^{-1} = \begin{pmatrix} A^{11} & A^{12} \\ A^{21} & A^{22} \end{pmatrix} \tag{2.26}$$

where

$$\begin{cases} A^{11} = (A_{11} - A_{12}A_{22}^{-1}A_{21})^{-1} \stackrel{\text{def}}{=} (A_{11 \cdot 2})^{-1} \\ A^{12} = -(A_{11 \cdot 2})^{-1}A_{12}A_{22}^{-1} \\ A^{21} = -A_{22}^{-1}A_{21}(A_{11 \cdot 2})^{-1} \\ A^{22} = A_{22}^{-1} + A_{22}^{-1}A_{21}(A_{11 \cdot 2})^{-1}A_{12}A_{22}^{-1} \end{cases}.$$

An alternative expression can be obtained by reversing the positions of A_{11} and A_{22} in the original matrix.

The following results will be useful if A_{11} is nonsingular:

$$|A| = |A_{11}||A_{22} - A_{21}A_{11}^{-1}A_{12}| = |A_{11}||A_{22 \cdot 1}|. \tag{2.27}$$

If A_{22} is nonsingular, we have that

$$|A| = |A_{22}||A_{11} - A_{12}A_{22}^{-1}A_{21}| = |A_{22}||A_{11\cdot2}|. \tag{2.28}$$

A useful formula is derived from the alternative expressions for the inverse and the determinant. For instance, let

$$B = \begin{pmatrix} 1 & b^\top \\ a & A \end{pmatrix}$$

where a and b are $(p \times 1)$ vectors and A is non-singular. We then have:

$$|B| = |A - ab^\top| = |A||1 - b^\top A^{-1}a| \tag{2.29}$$

and equating the two expressions for B^{22}, we obtain the following:

$$(A - ab^\top)^{-1} = A^{-1} + \frac{A^{-1}ab^\top A^{-1}}{1 - b^\top A^{-1}a}. \tag{2.30}$$

Example 2.9 Let's consider the matrix

$$A = \begin{pmatrix} 1 & 2 \\ 2 & 2 \end{pmatrix}.$$

We can use formula (2.26) to calculate the inverse of a partitioned matrix, i.e., $A^{11} = -1$, $A^{12} = A^{21} = 1$, $A^{22} = -1/2$. The inverse of A is

$$A^{-1} = \begin{pmatrix} -1 & 1 \\ 1 & -0.5 \end{pmatrix}.$$

It is also easy to calculate the determinant of A:

$$|A| = |1||2 - 4| = -2.$$

Let $A(n \times p)$ and $B(p \times n)$ be any two matrices and suppose that $n \geq p$. From (2.27) and (2.28) we can conclude that

$$\begin{vmatrix} -\lambda I_n & -A \\ B & I_p \end{vmatrix} = (-\lambda)^{n-p}|BA - \lambda I_p| = |AB - \lambda I_n|. \tag{2.31}$$

Since both determinants on the right-hand side of (2.31) are polynomials in λ, we find that the n eigenvalues of AB yield the p eigenvalues of BA plus the eigenvalue 0, $n - p$ times.

The relationship between the eigenvectors is described in the next theorem.

Theorem 2.6 *For $\mathcal{A}(n \times p)$ and $\mathcal{B}(p \times n)$, the nonzero eigenvalues of $\mathcal{A}\mathcal{B}$ and $\mathcal{B}\mathcal{A}$ are the same and have the same multiplicity. If x is an eigenvector of $\mathcal{A}\mathcal{B}$ for an eigenvalue $\lambda \neq 0$, then $y = \mathcal{B}x$ is an eigenvector of $\mathcal{B}\mathcal{A}$.*

Corollary 2.2 *For $\mathcal{A}(n \times p)$, $\mathcal{B}(q \times n)$, $a(p \times 1)$, and $b(q \times 1)$ we have*

$$\text{rank}(\mathcal{A}ab^\top \mathcal{B}) \leq 1.$$

The nonzero eigenvalue, if it exists, equals $b^\top \mathcal{B}\mathcal{A}a$ (with eigenvector $\mathcal{A}a$).

Proof Theorem 2.6 asserts that the eigenvalues of $\mathcal{A}ab^\top \mathcal{B}$ are the same as those of $b^\top \mathcal{B}\mathcal{A}a$. Note that the matrix $b^\top \mathcal{B}\mathcal{A}a$ is a scalar and hence it is its own eigenvalue λ_1.
Applying $\mathcal{A}ab^\top \mathcal{B}$ to $\mathcal{A}a$ yields

$$(\mathcal{A}ab^\top \mathcal{B})(\mathcal{A}a) = (\mathcal{A}a)(b^\top \mathcal{B}\mathcal{A}a) = \lambda_1 \mathcal{A}a.$$

2.6 Geometrical Aspects

Distance

Let $x, y \in \mathbb{R}^p$. A distance d is defined as a function

$$d : \mathbb{R}^{2p} \to \mathbb{R}_+ \quad \text{which fulfills} \quad \begin{cases} d(x, y) > 0 & \forall x \neq y \\ d(x, y) = 0 & \text{if and only if } x = y \\ d(x, y) \leq d(x, z) + d(z, y) & \forall x, y, z \end{cases}.$$

A *Euclidean distance* d between two points x and y is defined as

$$d^2(x, y) = (x - y)^T \mathcal{A}(x - y) \tag{2.32}$$

where $\mathcal{A}$ is a positive-definite matrix ($\mathcal{A} > 0$). $\mathcal{A}$ is called a *metric*.

Example 2.10 A particular case is when $\mathcal{A} = \mathcal{I}_p$, i.e.,

$$d^2(x, y) = \sum_{i=1}^{p} (x_i - y_i)^2. \tag{2.33}$$

Figure 2.1 illustrates this definition for $p = 2$.

Note that the sets $E_d = \{x \in \mathbb{R}^p \mid (x - x_0)^\top (x - x_0) = d^2\}$, i.e., the spheres with radius d and center x_0, are the Euclidean $\mathcal{I}_p$ *iso-distance* curves from the point x_0 (see Fig. 2.2).

Fig. 2.1 Distance d

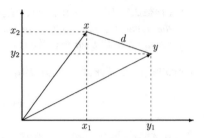

Fig. 2.2 Iso-distance sphere

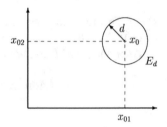

The more general distance (2.32) with a positive-definite matrix $\mathcal{A}$ ($\mathcal{A} > 0$) leads to the iso-distance curves

$$E_d = \{x \in \mathbb{R}^p \mid (x - x_0)^\top \mathcal{A}(x - x_0) = d^2\}, \qquad (2.34)$$

i.e., ellipsoids with center x_0, matrix $\mathcal{A}$ and constant d (see Fig. 2.3).

Let $\gamma_1, \gamma_2, \ldots, \gamma_p$ be the orthonormal eigenvectors of $\mathcal{A}$ corresponding to the eigenvalues $\lambda_1 \geq \lambda_2 \geq \ldots \geq \lambda_p$. The resulting observations are given in the next theorem.

Theorem 2.7 *(i) The principal axes of E_d are in the direction of γ_i; $i = 1, \ldots, p$.*

(ii)   The half-lengths of the axes are $\sqrt{\frac{d^2}{\lambda_i}}$; $i = 1, \ldots, p$.

(iii)   The rectangle surrounding the ellipsoid E_d is defined by the following inequalities:

$$x_{0i} - \sqrt{d^2 a^{ii}} \leq x_i \leq x_{0i} + \sqrt{d^2 a^{ii}}, \quad i = 1, \ldots, p,$$

where a^{ii} is the (i, i) element of $\mathcal{A}^{-1}$. By the rectangle surrounding the ellipsoid E_d we mean the rectangle whose sides are parallel to the coordinate axis.

It is easy to find the coordinates of the tangency points between the ellipsoid and its surrounding rectangle parallel to the coordinate axes. Let us find the coordinates of the tangency point that are in the direction of the j-th coordinate axis (positive direction).

For ease of notation, we suppose the ellipsoid is centered around the origin ($x_0 = 0$). If not, the rectangle will be shifted by the value of x_0.

The coordinate of the tangency point is given by the solution to the following problem:

$$x = \arg \max_{x^\top \mathcal{A}x = d^2} e_j^\top x \qquad (2.35)$$

Fig. 2.3 Iso–distance
ellipsoid

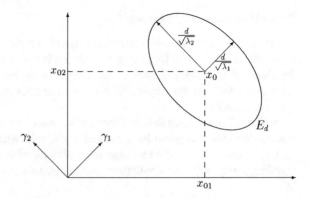

where $e_j^\top$ is the j-th column of the identity matrix $\mathcal{I}_p$. The coordinate of the tangency
point in the negative direction would correspond to the solution of the min problem:
by symmetry, it is the opposite value of the former.

The solution is computed via the Lagrangian $L = e_j^\top x - \lambda(x^\top A x - d^2)$ which
by (2.23) leads to the following system of equations:

$$\frac{\partial L}{\partial x} = e_j - 2\lambda A x = 0 \tag{2.36}$$

$$\frac{\partial L}{\partial \lambda} = x^\top A x - d^2 = 0. \tag{2.37}$$

This gives $x = \frac{1}{2\lambda} A^{-1} e_j$, or componentwise

$$x_i = \frac{1}{2\lambda} a^{ij}, \ i = 1, \ldots, p \tag{2.38}$$

where a^{ij} denotes the (i, j)-th element of A^{-1}.

Premultiplying (2.36) by $x^\top$, we have from (2.37):

$$x_j = 2\lambda d^2.$$

Comparing this to the value obtained by (2.38), for $i = j$ we obtain $2\lambda = \sqrt{\frac{a^{jj}}{d^2}}$.
We choose the positive value of the square root because we are maximizing $e_j^\top x$. A
minimum would correspond to the negative value. Finally, we have the coordinates of
the tangency point between the ellipsoid and its surrounding rectangle in the positive
direction of the j-th axis:

$$x_i = \sqrt{\frac{d^2}{a^{jj}}} \, a^{ij}, \ i = 1, \ldots, p. \tag{2.39}$$

The particular case where $i = j$ provides statement (iii) in Theorem 2.7.

Remark: usefulness of Theorem 2.7

Theorem 2.7 will prove to be particularly useful in many subsequent chapters. First, it provides a helpful tool for graphing an ellipse in two dimensions. Indeed, knowing the slope of the principal axes of the ellipse, their half-lengths and drawing the rectangle inscribing the ellipse, allows one to quickly draw a rough picture of the shape of the ellipse.

In Chap. 7, it is shown that the confidence region for the vector μ of a multivariate normal population is given by a particular ellipsoid whose parameters depend on sample characteristics. The rectangle inscribing the ellipsoid (which is much easier to obtain) will provide the simultaneous confidence intervals for all of the components in μ.

In addition, it will be shown that the contour surfaces of the multivariate normal density are provided by ellipsoids, whose parameters depend on the mean vector and on the covariance matrix. We will see that the tangency points between the contour ellipsoids and the surrounding rectangle are determined by regressing one component on the $(p-1)$ other components. For instance, in the direction of the j-th axis, the tangency points are given by the intersections of the ellipsoid contours with the regression line of the vector of $(p-1)$ variables (all components except the j-th) on the j-th component.

Norm of a Vector

Consider a vector $x \in \mathbb{R}^p$. The norm or length of x (with respect to the metric $\mathcal{I}_p$) is defined as

$$\|x\| = d(0_p, x) = \sqrt{x^\top x}.$$

If $\|x\| = 1$, x is called a *unit vector*. A more general norm can be defined with respect to the metric $\mathcal{A}$:

$$\|x\|_{\mathcal{A}} = \sqrt{x^\top \mathcal{A} x}.$$

Angle between two Vectors

Consider two vectors x and $y \in \mathbb{R}^p$. The angle θ between x and y is defined by the cosine of θ:

$$\cos \theta = \frac{x^\top y}{\|x\| \, \|y\|}, \tag{2.40}$$

see Fig. 2.4. Indeed for $p = 2$, $x = \begin{pmatrix} x_1 \\ x_2 \end{pmatrix}$ and $y = \begin{pmatrix} y_1 \\ y_2 \end{pmatrix}$, we have

$$\begin{aligned} \|x\| \cos \theta_1 = x_1 \; ; \; \|y\| \cos \theta_2 = y_1 \\ \|x\| \sin \theta_1 = x_2 \; ; \; \|y\| \sin \theta_2 = y_2, \end{aligned} \tag{2.41}$$

therefore,

Fig. 2.4 Angle between
vectors

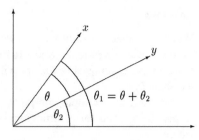

Fig. 2.5 Projection

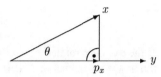

$$\cos\theta = \cos\theta_1\cos\theta_2 + \sin\theta_1\sin\theta_2 = \frac{x_1y_1 + x_2y_2}{\|x\|\,\|y\|} = \frac{x^\top y}{\|x\|\,\|y\|}.$$

Remark 2.1 If $x^\top y = 0$, then the angle θ is equal to $\frac{\pi}{2}$. From trigonometry, we know that the cosine of θ equals the length of the base of a triangle ($\|p_x\|$) divided by the length of the hypotenuse ($\|x\|$). Hence, we have

$$\|p_x\| = \|x\|\,|\cos\theta| = \frac{|x^\top y|}{\|y\|}, \tag{2.42}$$

where p_x is the projection of x on y (which is defined below). It is the coordinate of x on the y vector, see Fig. 2.5.

The angle can also be defined with respect to a general metric $\mathcal{A}$

$$\cos\theta = \frac{x^\top \mathcal{A} y}{\|x\|_\mathcal{A}\,\|y\|_\mathcal{A}}. \tag{2.43}$$

If $\cos\theta = 0$ then x is orthogonal to y with respect to the metric $\mathcal{A}$.

Example 2.11 Assume that there are two centered (i.e., zero mean) data vectors. The cosine of the angle between them is equal to their correlation (defined in (3.8)). Indeed for x and y with $\bar{x} = \bar{y} = 0$ we have

$$r_{XY} = \frac{\sum x_i y_i}{\sqrt{\sum x_i^2 \sum y_i^2}} = \cos\theta$$

according to formula (2.40).

Rotations

When we consider a point $x \in \mathbb{R}^p$, we generally use a p-coordinate system to obtain its geometric representation, like in Fig. 2.1 for instance. There will be situations in multivariate techniques where we will want to rotate this system of coordinates by the angle θ.

Consider, for example, the point P with coordinates $x = (x_1, x_2)^\top$ in $\mathbb{R}^2$ with respect to a given set of orthogonal axes. Let Γ be a (2×2) orthogonal matrix where

$$\Gamma = \begin{pmatrix} \cos\theta & \sin\theta \\ -\sin\theta & \cos\theta \end{pmatrix}. \tag{2.44}$$

If the axes are rotated about the origin through an angle θ in a clockwise direction, the new coordinates of P will be given by the vector y

$$y = \Gamma x, \tag{2.45}$$

and a rotation through the same angle in a anticlockwise direction gives the new coordinates as

$$y = \Gamma^\top x. \tag{2.46}$$

More generally, premultiplying a vector x by an orthogonal matrix Γ geometrically corresponds to a rotation of the system of axes, so that the first new axis is determined by the first row of Γ. This geometric point of view will be exploited in Chaps. 11 and 12.

Column Space and Null Space of a Matrix

Define for $\mathcal{X}(n \times p)$

$$Im(\mathcal{X}) \stackrel{def}{=} C(\mathcal{X}) = \{x \in \mathbb{R}^n \mid \exists a \in \mathbb{R}^p \text{ so that } \mathcal{X}a = x\},$$

the space generated by the columns of $\mathcal{X}$ or the *column space* of $\mathcal{X}$. Note that $C(\mathcal{X}) \subseteq \mathbb{R}^n$ and $\dim\{C(\mathcal{X})\} = \text{rank}(\mathcal{X}) = r \leq \min(n, p)$.

$$Ker(\mathcal{X}) \stackrel{def}{=} N(\mathcal{X}) = \{y \in \mathbb{R}^p \mid \mathcal{X}y = 0\}$$

is the *null space* of $\mathcal{X}$. Note that $N(\mathcal{X}) \subseteq \mathbb{R}^p$ and that $\dim\{N(\mathcal{X})\} = p - r$.

Remark 2.2 $N(\mathcal{X}^\top)$ is the orthogonal complement of $C(\mathcal{X})$ in $\mathbb{R}^n$, i.e., given a vector $b \in \mathbb{R}^n$ it will hold that $x^\top b = 0$ for all $x \in C(\mathcal{X})$, if and only if $b \in N(\mathcal{X}^\top)$.

Example 2.12 Let $\mathcal{X} = \begin{pmatrix} 2 & 3 & 5 \\ 4 & 6 & 7 \\ 6 & 8 & 6 \\ 8 & 2 & 4 \end{pmatrix}$. It is easy to show (e.g., by calculating the determinant of $\mathcal{X}$) that $\text{rank}(\mathcal{X}) = 3$. Hence, the column space of $\mathcal{X}$ is $C(\mathcal{X}) = \mathbb{R}^3$.

The null space of $\mathcal{X}$ contains only the zero vector $(0, 0, 0)^\top$ and its dimension is equal to $\text{rank}(\mathcal{X}) - 3 = 0$.

For $\mathcal{X} = \begin{pmatrix} 2 & 3 & 1 \\ 4 & 6 & 2 \\ 6 & 8 & 3 \\ 8 & 2 & 4 \end{pmatrix}$, the third column is a multiple of the first one and the matrix $\mathcal{X}$ cannot be of full rank. Noticing that the first two columns of $\mathcal{X}$ are independent, we see that $\text{rank}(\mathcal{X}) = 2$. In this case, the dimension of the columns space is 2 and the dimension of the null space is 1.

Projection Matrix

A matrix $\mathcal{P}(n \times n)$ is called an (orthogonal) projection matrix in $\mathbb{R}^n$ if and only if $\mathcal{P} = \mathcal{P}^\top = \mathcal{P}^2$ ($\mathcal{P}$ is idempotent). Let $b \in \mathbb{R}^n$. Then $a = \mathcal{P}b$ is the projection of b on $C(\mathcal{P})$.

Projection on $C(\mathcal{X})$

Consider $\mathcal{X}(n \times p)$ and let

$$\mathcal{P} = \mathcal{X}(\mathcal{X}^\top \mathcal{X})^{-1}\mathcal{X}^\top \tag{2.47}$$

and $\mathcal{Q} = \mathcal{I}_n - \mathcal{P}$. It's easy to check that $\mathcal{P}$ and $\mathcal{Q}$ are idempotent and that

$$\mathcal{P}\mathcal{X} = \mathcal{X} \text{ and } \mathcal{Q}\mathcal{X} = 0. \tag{2.48}$$

Since the columns of $\mathcal{X}$ are projected onto themselves, the projection matrix $\mathcal{P}$ projects any vector $b \in \mathbb{R}^n$ onto $C(\mathcal{X})$. Similarly, the projection matrix $\mathcal{Q}$ projects any vector $b \in \mathbb{R}^n$ onto the orthogonal complement of $C(\mathcal{X})$.

Theorem 2.8 *Let $\mathcal{P}$ be the Projection (2.47) and $\mathcal{Q}$ its orthogonal complement. Then:*

(i) $x = \mathcal{P}b$ entails $x \in C(\mathcal{X})$,
(ii) $y = \mathcal{Q}b$ means that $y^\top x = 0 \; \forall x \in C(\mathcal{X})$.

Proof (i) holds, since $x = \mathcal{X}(\mathcal{X}^\top \mathcal{X})^{-1}\mathcal{X}^\top b = \mathcal{X}a$, where $a = (\mathcal{X}^\top \mathcal{X})^{-1}\mathcal{X}^\top b \in \mathbb{R}^p$.
(ii) follows from $y = b - \mathcal{P}b$ and $x = \mathcal{X}a$.
Hence $y^\top x = b^\top \mathcal{X}a - b^\top \mathcal{X}(\mathcal{X}^\top \mathcal{X})^{-1}\mathcal{X}^\top \mathcal{X}a = 0$.

Remark 2.3 Let $x, y \in \mathbb{R}^n$ and consider $p_x \in \mathbb{R}^n$, the projection of x on y (see Fig. 2.5). With $\mathcal{X} = y$ we have from (2.47)

$$p_x = y(y^\top y)^{-1}y^\top x = \frac{y^\top x}{\|y\|^2} y \tag{2.49}$$

and we can easily verify that

$$\|p_x\| = \sqrt{p_x^\top p_x} = \frac{|y^\top x|}{\|y\|}.$$

See again Remark 2.1.

Summary
↪ A distance between two p-dimensional points x and y is a quadratic form $(x - y)^\top A(x - y)$ in the vectors of differences $(x - y)$. A distance defines the norm of a vector.
↪ Iso-distance curves of a point x_0 are all those points that have the same distance from x_0. Iso-distance curves are ellipsoids whose principal axes are determined by the direction of the eigenvectors of A. The half-length of principal axes is proportional to the inverse of the roots of the eigenvalues of A.
↪ The angle between two vectors x and y is given by $\cos \theta = \frac{x^\top A y}{\|x\|_A \|y\|_A}$ w.r.t. the metric A.
↪ For the Euclidean distance with $A = \mathcal{I}$, the correlation between two centered data vectors x and y is given by the cosine of the angle between them, i.e., $\cos \theta = r_{XY}$.
↪ The projection $\mathcal{P} = \mathcal{X}(\mathcal{X}^\top \mathcal{X})^{-1} \mathcal{X}^\top$ is the projection onto the column space $C(\mathcal{X})$ of $\mathcal{X}$.
↪ The projection of $x \in \mathbb{R}^n$ on $y \in \mathbb{R}^n$ is given by $p_x = \frac{y^\top x}{\|y\|^2} y$.

2.7 Exercises

Exercise 2.1 Compute the determinant for a (3×3) matrix.

Exercise 2.2 Suppose that $|A| = 0$. Is it possible that all eigenvalues of A are positive?

Exercise 2.3 Suppose that all eigenvalues of some (square) matrix A are different from zero. Does the inverse A^{-1} of A exist?

Exercise 2.4 Write a program that calculates the Jordan decomposition of the matrix

$$A = \begin{pmatrix} 1 & 2 & 3 \\ 2 & 1 & 2 \\ 3 & 2 & 1 \end{pmatrix}.$$

Check Theorem 2.1 numerically.

Exercise 2.5 Prove (2.23), (2.24), and (2.25).

Exercise 2.6 Show that a projection matrix only has eigenvalues in $\{0, 1\}$.

Exercise 2.7 Draw some iso-distance ellipsoids for the metric $\mathcal{A} = \Sigma^{-1}$ of Example 3.13.

Exercise 2.8 Find a formula for $|\mathcal{A} + aa^\top|$ and for $(\mathcal{A} + aa^\top)^{-1}$. (Hint: use the inverse partitioned matrix with $\mathcal{B} = \begin{pmatrix} 1 & -a^\top \\ a & \mathcal{A} \end{pmatrix}$.)

Exercise 2.9 Prove the Binomial inverse theorem for two non-singular matrices $\mathcal{A}(p \times p)$ and $\mathcal{B}(p \times p)$: $(\mathcal{A} + \mathcal{B})^{-1} = \mathcal{A}^{-1} - \mathcal{A}^{-1}(\mathcal{A}^{-1} + \mathcal{B}^{-1})^{-1}\mathcal{A}^{-1}$. (Hint: use (2.26) with $\mathcal{C} = \begin{pmatrix} \mathcal{A} & I_p \\ -I_p & \mathcal{B}^{-1} \end{pmatrix}$.)

Chapter 3
Moving to Higher Dimensions

We have seen in the previous chapters how very simple graphical devices can help in understanding the structure and dependency of data. The graphical tools were based on either univariate (bivariate) data representations or on "slick" transformations of multivariate information perceivable by the human eye. Most of the tools are extremely useful in a modeling step, but unfortunately, do not give the full picture of the data set. One reason for this is that the graphical tools presented capture only certain dimensions of the data and do not necessarily concentrate on those dimensions or subparts of the data under analysis that carry the maximum structural information. In Part III of this book, powerful tools for reducing the dimension of a data set will be presented. In this chapter, as a starting point, simple and basic tools are used to describe dependency. They are constructed from elementary facts of probability theory and introductory statistics (for example, the covariance and correlation between two variables).

Sections 3.1 and 3.2 show how to handle these concepts in a multivariate setup and how a simple test on correlation between two variables can be derived. Since linear relationships are involved in these measures, Sect. 3.4 presents the simple linear model for two variables and recalls the basic t-test for the slope. In Sect. 3.5, a simple example of one-factorial analysis of variance introduces the notations for the well-known F-test.

Due to the power of matrix notation, all of this can easily be extended to a more general multivariate setup. Section 3.3 shows how matrix operations can be used to define summary statistics of a data set and for obtaining the empirical moments of linear transformations of the data. These results will prove to be very useful in most of the chapters in Part III.

Finally, matrix notation allows us to introduce the flexible multiple linear model, where more general relationships among variables can be analyzed. In Sect. 3.6, the least squares adjustment of the model and the usual test statistics are presented with their geometric interpretation. Using these notations, the ANOVA model is just a particular case of the multiple linear model.

© Springer Nature Switzerland AG 2019
W. K. Härdle and L. Simar, *Applied Multivariate Statistical Analysis*,
https://doi.org/10.1007/978-3-030-26006-4_3

3.1 Covariance

Covariance is a measure of dependency between random variables. Given two (random) variables X and Y the (theoretical) covariance is defined by

$$\sigma_{XY} = \text{Cov}(X, Y) = E(XY) - (EX)(EY). \tag{3.1}$$

The precise definition of expected values is given in Chap. 4. If X and Y are independent of each other, the covariance $\text{Cov}(X, Y)$ is necessarily equal to zero, see Theorem 3.1. The converse is not true. The covariance of X with itself is the variance:

$$\sigma_{XX} = \text{Var}(X) = \text{Cov}(X, X).$$

If the variable X is p-dimensional multivariate, e.g., $X = \begin{pmatrix} X_1 \\ \vdots \\ X_p \end{pmatrix}$, then the theoretical covariances among all the elements are put into matrix form, i.e., the covariance matrix:

$$\Sigma = \begin{pmatrix} \sigma_{X_1 X_1} & \cdots & \sigma_{X_1 X_p} \\ \vdots & \ddots & \vdots \\ \sigma_{X_p X_1} & \cdots & \sigma_{X_p X_p} \end{pmatrix}.$$

Properties of covariance matrices will be detailed in Chap. 4. Empirical versions of these quantities are

$$s_{XY} = \frac{1}{n} \sum_{i=1}^{n} (x_i - \bar{x})(y_i - \bar{y}) \tag{3.2}$$

$$s_{XX} = \frac{1}{n} \sum_{i=1}^{n} (x_i - \bar{x})^2. \tag{3.3}$$

For small n, say $n \leq 20$, we should replace the factor $\frac{1}{n}$ in (3.2) and (3.3) by $\frac{1}{n-1}$ in order to correct for a small bias. For a p-dimensional random variable, one obtains the empirical covariance matrix (see Sect. 3.3 for properties and details)

$$S = \begin{pmatrix} s_{X_1 X_1} & \cdots & s_{X_1 X_p} \\ \vdots & \ddots & \vdots \\ s_{X_p X_1} & \cdots & s_{X_p X_p} \end{pmatrix}.$$

For a scatterplot of two variables the covariances measure "how close the scatter is to a line." Mathematical details follow but it should already be understood here that in this sense covariance measures only "linear dependence".

Example 3.1 If $\mathcal{X}$ is the entire bank data set, one obtains the covariance matrix $\mathcal{S}$ as indicated below:

$$\mathcal{S} = \begin{pmatrix} 0.14 & 0.03 & 0.02 & -0.10 & -0.01 & 0.08 \\ 0.03 & 0.12 & 0.10 & 0.21 & 0.10 & -0.21 \\ 0.02 & 0.10 & 0.16 & 0.28 & 0.12 & -0.24 \\ -0.10 & 0.21 & 0.28 & 2.07 & 0.16 & -1.03 \\ -0.01 & 0.10 & 0.12 & 0.16 & 0.64 & -0.54 \\ 0.08 & -0.21 & -0.24 & -1.03 & -0.54 & 1.32 \end{pmatrix}. \tag{3.4}$$

The empirical covariance between X_4 and X_5, i.e., $s_{X_4 X_5}$, is found in row 4 and column 5. The value is $s_{X_4 X_5} = 0.16$. Is it obvious that this value is positive? In Exercise 3.1, we will discuss this question further.

If $\mathcal{X}_f$ denotes the counterfeit banknotes, we obtain

$$\mathcal{S}_f = \begin{pmatrix} 0.123 & 0.031 & 0.023 & -0.099 & 0.019 & 0.011 \\ 0.031 & 0.064 & 0.046 & -0.024 & -0.012 & -0.005 \\ 0.024 & 0.046 & 0.088 & -0.018 & 0.000 & 0.034 \\ -0.099 & -0.024 & -0.018 & 1.268 & -0.485 & 0.236 \\ 0.019 & -0.012 & 0.000 & -0.485 & 0.400 & -0.022 \\ 0.011 & -0.005 & 0.034 & 0.236 & -0.022 & 0.308 \end{pmatrix}. \tag{3.5}$$

For the genuine $\mathcal{X}_g$, we have

$$\mathcal{S}_g = \begin{pmatrix} 0.149 & 0.057 & 0.057 & 0.056 & 0.014 & 0.005 \\ 0.057 & 0.131 & 0.085 & 0.056 & 0.048 & -0.043 \\ 0.057 & 0.085 & 0.125 & 0.058 & 0.030 & -0.024 \\ 0.056 & 0.056 & 0.058 & 0.409 & -0.261 & -0.000 \\ 0.014 & 0.049 & 0.030 & -0.261 & 0.417 & -0.074 \\ 0.005 & -0.043 & -0.024 & -0.000 & -0.074 & 0.198 \end{pmatrix}. \tag{3.6}$$

Note that the covariance between X_4 (distance of the frame to the lower border) and X_5 (distance of the frame to the upper border) is negative in both (3.5) and (3.6). Why would this happen? In Exercise 3.2, we will discuss this question in more detail.

At first sight, the matrices $\mathcal{S}_f$ and $\mathcal{S}_g$ look different, but they create almost the same scatterplots (see the discussion in Sect. 1.4). Similarly, the common principal component analysis in Chap. 11 suggests a joint analysis of the covariance structure as in Flury and Riedwyl (1988).

Scatterplots with point clouds that are "upward-sloping", like the one in the upper left of Fig. 1.14, show variables with positive covariance. Scatterplots with "downward-sloping" structure have negative covariance. In Fig. 3.1, we show the

Fig. 3.1 Scatterplot of
variables X_4 versus X_5 of
the entire bank data set
MVAscabank45

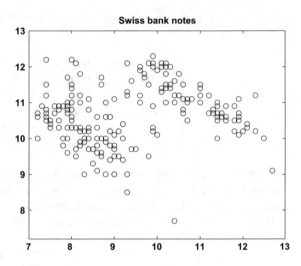

scatterplot of X_4 versus X_5 of the entire bank data set. The point cloud is upward-sloping. However, the two sub-clouds of counterfeit and genuine bank notes are downward-sloping.

Example 3.2 A textile shop manager is studying the sales of "classic blue" pullovers over 10 different periods. He observes the number of pullovers sold (X_1), variation in price (X_2, in EUR), the advertisement costs in local newspapers (X_3, in EUR) and the presence of a sales assistant (X_4, in hours per period). Over the periods, he observes the following data matrix:

$$
\mathcal{X} = \begin{pmatrix}
230 & 125 & 200 & 109 \\
181 & 99 & 55 & 107 \\
165 & 97 & 105 & 98 \\
150 & 115 & 85 & 71 \\
97 & 120 & 0 & 82 \\
192 & 100 & 150 & 103 \\
181 & 80 & 85 & 111 \\
189 & 90 & 120 & 93 \\
172 & 95 & 110 & 86 \\
170 & 125 & 130 & 78
\end{pmatrix}.
$$

He is convinced that the price must have a large influence on the number of pullovers sold. So he makes a scatterplot of X_2 versus X_1, see Fig. 3.2. A rough impression is that the cloud is somewhat downward-sloping. A computation of the empirical covariance yields

Fig. 3.2 Scatterplot of variables X_2 versus X_1 of the pullovers data set

MVAscapull1

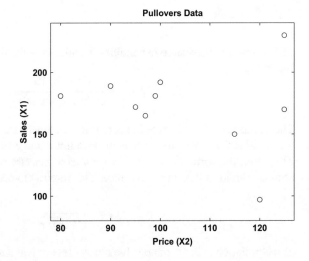

$$s_{X_1 X_2} = \frac{1}{9} \sum_{i=1}^{10} \left(X_{1i} - \bar{X}_1 \right) \left(X_{2i} - \bar{X}_2 \right) = -80.02,$$

a negative value as expected.

Note: The covariance function is scale dependent. Thus, if the prices in this example were in Japanese Yen (JPY), we would obtain a different answer (see Exercise 3.16). A measure of (linear) dependence independent of the scale is the correlation, which we introduce in the next section.

Summary
↪ The covariance is a measure of dependence.
↪ Covariance measures only linear dependence.
↪ Covariance is scale dependent.
↪ There are nonlinear dependencies that have zero covariance.
↪ Zero covariance does not imply independence.
↪ Independence implies zero covariance.
↪ Negative covariance corresponds to downward-sloping scatterplots.
↪ Positive covariance corresponds to upward-sloping scatterplots.
↪ The covariance of a variable with itself is its variance $\mathrm{Cov}(X, X) = \sigma_{XX} = \sigma_X^2$.
↪ For small n, we should replace the factor $\frac{1}{n}$ in the computation of the covariance by $\frac{1}{n-1}$.

3.2 Correlation

The correlation between two variables X and Y is defined from the covariance as the following:

$$\rho_{XY} = \frac{\text{Cov}(X, Y)}{\sqrt{\text{Var}(X)\text{Var}(Y)}}. \tag{3.7}$$

The advantage of the correlation is that it is independent of the scale, i.e., changing the variables' scale of measurement does not change the value of the correlation. Therefore, the correlation is more useful as a measure of association between two random variables than the covariance. The empirical version of ρ_{XY} is as follows:

$$r_{XY} = \frac{s_{XY}}{\sqrt{s_{XX}s_{YY}}}. \tag{3.8}$$

The correlation is in absolute value always less or equal to 1. It is zero if the covariance is zero and vice-versa. For p-dimensional vectors $(X_1, \ldots, X_p)^\top$ we have the theoretical correlation matrix

$$\mathcal{P} = \begin{pmatrix} \rho_{X_1 X_1} & \cdots & \rho_{X_1 X_p} \\ \vdots & \ddots & \vdots \\ \rho_{X_p X_1} & \cdots & \rho_{X_p X_p} \end{pmatrix},$$

and its empirical version, the empirical correlation matrix which can be calculated from the observations,

$$\mathcal{R} = \begin{pmatrix} r_{X_1 X_1} & \cdots & r_{X_1 X_p} \\ \vdots & \ddots & \vdots \\ r_{X_p X_1} & \cdots & r_{X_p X_p} \end{pmatrix}.$$

Example 3.3 We obtain the following correlation matrix for the genuine bank notes:

$$\mathcal{R}_g = \begin{pmatrix} 1.00 & 0.41 & 0.41 & 0.22 & 0.05 & 0.03 \\ 0.41 & 1.00 & 0.66 & 0.24 & 0.20 & -0.25 \\ 0.41 & 0.66 & 1.00 & 0.25 & 0.13 & -0.14 \\ 0.22 & 0.24 & 0.25 & 1.00 & -0.63 & -0.00 \\ 0.05 & 0.20 & 0.13 & -0.63 & 1.00 & -0.25 \\ 0.03 & -0.25 & -0.14 & -0.00 & -0.25 & 1.00 \end{pmatrix}, \tag{3.9}$$

and for the counterfeit banknotes:

$$\mathcal{R}_f = \begin{pmatrix} 1.00 & 0.35 & 0.24 & -0.25 & 0.08 & 0.06 \\ 0.35 & 1.00 & 0.61 & -0.08 & -0.07 & -0.03 \\ 0.24 & 0.61 & 1.00 & -0.05 & 0.00 & 0.20 \\ -0.25 & -0.08 & -0.05 & 1.00 & -0.68 & 0.37 \\ 0.08 & -0.07 & 0.00 & -0.68 & 1.00 & -0.06 \\ 0.06 & -0.03 & 0.20 & 0.37 & -0.06 & 1.00 \end{pmatrix}. \tag{3.10}$$

As noted before for $\mathsf{Cov}(X_4, X_5)$, the correlation between X_4 (distance of the frame to the lower border) and X_5 (distance of the frame to the upper border) is negative. This is natural, since the covariance and correlation always have the same sign (see also Exercise 3.17).

Why is the correlation an interesting statistic to study? It is related to independence of random variables, which we shall define more formally later on. For the moment, we may think of independence as the fact that one variable has no influence on another.

Theorem 3.1 *If X and Y are independent, then $\rho(X, Y) = \mathsf{Cov}(X, Y) = 0$.*

 In general, the converse is not true, as the following example shows.

Example 3.4 Consider a standard normally distributed random variable X and a random variable $Y = X^2$, which is surely not independent of X. Here we have

$$\mathsf{Cov}(X, Y) = \mathsf{E}(XY) - \mathsf{E}(X)\mathsf{E}(Y) = \mathsf{E}(X^3) = 0$$

(because $\mathsf{E}(X) = 0$ and $\mathsf{E}(X^2) = 1$). Therefore $\rho(X, Y) = 0$, as well. This example also shows that correlations and covariances measure only linear dependence. The quadratic dependence of $Y = X^2$ on X is not reflected by these measures of dependence.

Remark 3.1 For two normal random variables, the converse of Theorem 3.1 is true: zero covariance for two normally distributed random variables implies independence. This will be shown later in Corollary 5.2.

Theorem 3.1 enables us to check for independence between the components of a bivariate normal random variable. That is, we can use the correlation and test whether it is zero. The distribution of r_{XY} for an arbitrary (X, Y) is unfortunately complicated. The distribution of r_{XY} will be more accessible if (X, Y) are jointly normal (see Chap. 5). If we transform the correlation by Fisher's Z-transformation,

$$W = \frac{1}{2} \log \left(\frac{1 + r_{XY}}{1 - r_{XY}} \right), \tag{3.11}$$

we obtain a variable that has a more accessible distribution. Under the hypothesis that $\rho = 0$, W has an asymptotic normal distribution. Approximations of the expectation and variance of W are given by the following:

$$E(W) \approx \tfrac{1}{2} \log \left(\tfrac{1+\rho_{XY}}{1-\rho_{XY}} \right)$$
$$\mathrm{Var}(W) \approx \tfrac{1}{(n-3)}. \tag{3.12}$$

The distribution is given in Theorem 3.2.

Theorem 3.2

$$Z = \frac{W - E(W)}{\sqrt{\mathrm{Var}(W)}} \xrightarrow{\mathcal{L}} N(0, 1). \tag{3.13}$$

The symbol "$\xrightarrow{\mathcal{L}}$" denotes convergence in distribution, which will be explained in more detail in Chap. 4.

Theorem 3.2 allows us to test different hypotheses on correlation. We can fix the level of significance α (the probability of rejecting a true hypothesis) and reject the hypothesis if the difference between the hypothetical value and the calculated value of Z is greater than the corresponding critical value of the normal distribution. The following example illustrates the procedure.

Example 3.5 Let's study the correlation between mileage (X_2) and weight (X_8) for the car data set (22.3) where $n = 74$. We have $r_{X_2 X_8} = -0.823$. Our conclusions from the boxplot in Fig. 1.3 ("Japanese cars generally have better mileage than the others") needs to be revised. From Fig. 3.3 and $r_{X_2 X_8}$, we can see that mileage is highly correlated with weight, and that the Japanese cars in the sample are in fact all lighter than the others.

If we want to know whether $\rho_{X_2 X_8}$ is significantly different from $\rho_0 = 0$, we apply Fisher's Z-transform (3.11). This gives us

$$w = \frac{1}{2} \log \left(\frac{1 + r_{X_2 X_8}}{1 - r_{X_2 X_8}} \right) = -1.166 \quad \text{and} \quad z = \frac{-1.166 - 0}{\sqrt{\frac{1}{71}}} = -9.825,$$

i.e., a highly significant value to reject the hypothesis that $\rho = 0$ (the 2.5% and 97.5% quantiles of the normal distribution are -1.96 and 1.96, respectively). If we want to test the hypothesis that, say, $\rho_0 = -0.75$, we obtain

$$z = \frac{-1.166 - (-0.973)}{\sqrt{\frac{1}{71}}} = -1.627.$$

This is a nonsignificant value at the $\alpha = 0.05$ level for z since it is between the critical values at the 5% significance level (i.e., $-1.96 < z < 1.96$).

Example 3.6 Let us consider again the pullovers data set from Example 3.2. Consider the correlation between the presence of the sales assistants (X_4) versus the number of sold pullovers (X_1) (see Fig. 3.4). Here we compute the correlation as

$$r_{X_1 X_4} = 0.633.$$

Fig. 3.3 Mileage (X_2) versus weight (X_8) of U.S. (star), European (plus signs) and Japanese (circle) cars

MVAscacar

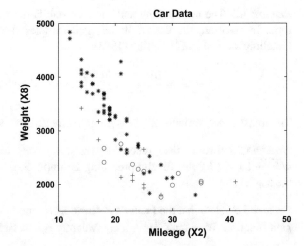

Fig. 3.4 Hours of sales assistants (X_4) versus sales (X_1) of pullovers

MVAscapull2

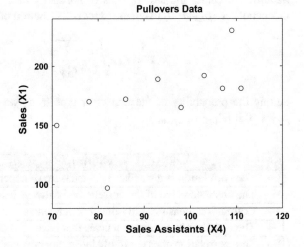

The Z-transform of this value is

$$w = \frac{1}{2} \log \left(\frac{1 + r_{X_1 X_4}}{1 - r_{X_1 X_4}} \right) = 0.746. \tag{3.14}$$

The sample size is $n = 10$, so for the hypothesis $\rho_{X_1 X_4} = 0$, the statistic to consider is

$$z = \sqrt{7}(0.746 - 0) = 1.974 \tag{3.15}$$

which is just statistically significant at the 5% level (i.e., 1.974 is just a little larger than 1.96).

Remark 3.2 The normalizing and variance stabilizing properties of W are asymptotic. In addition, the use of W in small samples (for $n \leq 25$) is improved by Hotelling's transform Hotelling (1953):

$$W^* = W - \frac{3W + \tanh(W)}{4(n-1)} \quad \text{with} \quad \text{Var}(W^*) = \frac{1}{n-1}.$$

The transformed variable W^* is asymptotically distributed as a normal distribution.

Example 3.7 From the preceding remark, we obtain $w^* = 0.6663$ and $\sqrt{10-1}w^* = 1.9989$ for the preceding Example 3.6. This value is significant at the 5% level.

Remark 3.3 Note that the Fisher's Z-transform is the inverse of the hyperbolic tangent function: $W = \tanh^{-1}(r_{XY})$; equivalently $r_{XY} = \tanh(W) = \frac{e^{2W}-1}{e^{2W}+1}$.

Remark 3.4 Under the assumptions of normality of X and Y, we may test their independence ($\rho_{XY} = 0$) using the exact t-distribution of the statistic

$$T = r_{XY}\sqrt{\frac{n-2}{1-r_{XY}^2}} \overset{\rho_{XY}=0}{\sim} t_{n-2}.$$

Setting the probability of the first error type to α, we reject the null hypothesis $\rho_{XY} = 0$ if $|T| \geq t_{1-\alpha/2;n-2}$.

Summary
$\hookrightarrow$ The correlation is a standardized measure of dependence.
$\hookrightarrow$ The absolute value of the correlation is always less or equal to one.
$\hookrightarrow$ Correlation measures only linear dependence.
$\hookrightarrow$ There are nonlinear dependencies that have zero correlation.
$\hookrightarrow$ Zero correlation does not imply independence. For two normal random variables, it does.
$\hookrightarrow$ Independence implies zero correlation.
$\hookrightarrow$ Negative correlation corresponds to downward-sloping scatterplots.
$\hookrightarrow$ Positive correlation corresponds to upward-sloping scatterplots.
$\hookrightarrow$ Fisher's Z-transform helps us in testing hypotheses on correlation.
$\hookrightarrow$ For small samples, Fisher's Z-transform can be improved by the transformation $W^* = W - \frac{3W+\tanh(W)}{4(n-1)}$.

3.3 Summary Statistics

This section focuses on the representation of basic summary statistics (means, covariances and correlations) in matrix notation, since we often apply linear transformations to data. The matrix notation allows us to derive instantaneously the corresponding characteristics of the transformed variables. The Mahalanobis transformation is a prominent example of such linear transformations.

Assume that we have observed n realizations of a p-dimensional random variable; we have a data matrix $\mathcal{X}(n \times p)$:

$$
\mathcal{X} = \begin{pmatrix} x_{11} & \cdots & x_{1p} \\ \vdots & & \vdots \\ \vdots & & \vdots \\ x_{n1} & \cdots & x_{np} \end{pmatrix}.
\tag{3.16}
$$

The rows $x_i = (x_{i1}, \ldots, x_{ip}) \in \mathbb{R}^p$ denote the i-th observation of a p-dimensional random variable $X \in \mathbb{R}^p$.

The statistics that were briefly introduced in Sects. 3.1 and 3.2 can be rewritten in matrix form as follows. The "center of gravity" of the n observations in $\mathbb{R}^p$ is given by the vector $\overline{x}$ of the means $\overline{x}_j$ of the p variables:

$$
\overline{x} = \begin{pmatrix} \overline{x}_1 \\ \vdots \\ \overline{x}_p \end{pmatrix} = n^{-1} \mathcal{X}^\top 1_n.
\tag{3.17}
$$

The dispersion of the n observations can be characterized by the covariance matrix of the p variables. The empirical covariances defined in (3.2) and (3.3) are the elements of the following matrix:

$$
\mathcal{S} = n^{-1} \mathcal{X}^\top \mathcal{X} - \overline{x}\,\overline{x}^\top = n^{-1}(\mathcal{X}^\top \mathcal{X} - n^{-1}\mathcal{X}^\top 1_n 1_n^\top \mathcal{X}).
\tag{3.18}
$$

Note that this matrix is equivalently defined by

$$
\mathcal{S} = \frac{1}{n} \sum_{i=1}^{n} (x_i - \overline{x})(x_i - \overline{x})^\top.
$$

The covariance formula (3.18) can be rewritten as $\mathcal{S} = n^{-1} \mathcal{X}^\top \mathcal{H} \mathcal{X}$ with the *centering matrix*

$$
\mathcal{H} = \mathcal{I}_n - n^{-1} 1_n 1_n^\top.
\tag{3.19}
$$

Note that the centering matrix is symmetric and idempotent. Indeed,

$$\mathcal{H}^2 = (\mathcal{I}_n - n^{-1}1_n1_n^{\top})(\mathcal{I}_n - n^{-1}1_n1_n^{\top})$$
$$= \mathcal{I}_n - n^{-1}1_n1_n^{\top} - n^{-1}1_n1_n^{\top} + (n^{-1}1_n1_n^{\top})(n^{-1}1_n1_n^{\top})$$
$$= \mathcal{I}_n - n^{-1}1_n1_n^{\top} = \mathcal{H}.$$

As a consequence $\mathcal{S}$ is positive semidefinite, i.e.,

$$\mathcal{S} \geq 0. \tag{3.20}$$

Indeed for all $a \in \mathbb{R}^p$,

$$a^{\top}\mathcal{S}a = n^{-1}a^{\top}\mathcal{X}^{\top}\mathcal{H}\mathcal{X}a$$
$$= n^{-1}(a^{\top}\mathcal{X}^{\top}\mathcal{H}^{\top})(\mathcal{H}\mathcal{X}a) \quad \text{since } \mathcal{H}^{\top}\mathcal{H} = \mathcal{H},$$
$$= n^{-1}y^{\top}y = n^{-1}\sum_{j=1}^{p} y_j^2 \geq 0$$

for $y = \mathcal{H}\mathcal{X}a$. It is well known from the one-dimensional case that $n^{-1}\sum_{i=1}^{n}(x_i - \bar{x})^2$ as an estimate of the variance exhibits a bias of the order n^{-1} Breiman (1973). In the multidimensional case, $\mathcal{S}_u = \frac{n}{n-1}\mathcal{S}$ is an unbiased estimate of the true covariance. (This will be shown in Example 4.15.)

The sample correlation coefficient between the i-th and j-th variables is $r_{X_i X_j}$, see (3.8). If $\mathcal{D} = \mathrm{diag}(s_{X_i X_i})$, then the correlation matrix is

$$\mathcal{R} = \mathcal{D}^{-1/2}\mathcal{S}\mathcal{D}^{-1/2}, \tag{3.21}$$

where $\mathcal{D}^{-1/2}$ is a diagonal matrix with elements $(s_{X_i X_i})^{-1/2}$ on its main diagonal.

Example 3.8 The empirical covariances are calculated for the pullover data set.

The vector of the means of the four variables in the dataset is $\bar{x} = (172.7, 104.6, 104.0, 93.8)^{\top}$.

The sample covariance matrix is $\mathcal{S} = \begin{pmatrix} 1037.2 & -80.2 & 1430.7 & 271.4 \\ -80.2 & 219.8 & 92.1 & -91.6 \\ 1430.7 & 92.1 & 2624 & 210.3 \\ 271.4 & -91.6 & 210.3 & 177.4 \end{pmatrix}.$

The unbiased estimate of the variance ($n = 10$) is equal to

$$\mathcal{S}_u = \frac{10}{9}\mathcal{S} = \begin{pmatrix} 1152.5 & -88.9 & 1589.7 & 301.6 \\ -88.9 & 244.3 & 102.3 & -101.8 \\ 1589.7 & 102.3 & 2915.6 & 233.7 \\ 301.6 & -101.8 & 233.7 & 197.1 \end{pmatrix}.$$

$$\text{The sample correlation matrix is } \mathcal{R} = \begin{pmatrix} 1 & -0.17 & 0.87 & 0.63 \\ -0.17 & 1 & 0.12 & -0.46 \\ 0.87 & 0.12 & 1 & 0.31 \\ 0.63 & -0.46 & 0.31 & 1 \end{pmatrix}.$$

Linear Transformation

In many practical applications, we need to study linear transformations of the original data. This motivates the question of how to calculate summary statistics after such linear transformations.

Let $\mathcal{A}$ be a $(q \times p)$ matrix and consider the transformed data matrix

$$\mathcal{Y} = \mathcal{X}\mathcal{A}^\top = (y_1, \ldots, y_n)^\top. \tag{3.22}$$

The row $y_i = (y_{i1}, \ldots, y_{iq}) \in \mathbb{R}^q$ can be viewed as the i-th observation of a q-dimensional random variable $Y = \mathcal{A}X$. In fact we have $y_i = x_i \mathcal{A}^\top$. We immediately obtain the mean and the empirical covariance of the variables (columns) forming the data matrix $\mathcal{Y}$:

$$\overline{y} = \frac{1}{n}\mathcal{Y}^\top 1_n = \frac{1}{n}\mathcal{A}\mathcal{X}^\top 1_n = \mathcal{A}\overline{x} \tag{3.23}$$

$$\mathcal{S}_y = \frac{1}{n}\mathcal{Y}^\top \mathcal{H}\mathcal{Y} = \frac{1}{n}\mathcal{A}\mathcal{X}^\top \mathcal{H}\mathcal{X}\mathcal{A}^\top = \mathcal{A}\mathcal{S}_\mathcal{X}\mathcal{A}^\top. \tag{3.24}$$

Note that if the linear transformation is nonhomogeneous, i.e.,

$$y_i = \mathcal{A}x_i + b \quad \text{where} \quad b(q \times 1),$$

only (3.23) changes: $\overline{y} = \mathcal{A}\overline{x} + b$. Formulas (3.23) and (3.24) are useful in the particular case of $q = 1$, i.e., $y = \mathcal{X}a$, i.e., $y_i = a^\top x_i; i = 1, \ldots, n$:

$$\overline{y} = a^\top \overline{x}$$
$$\mathcal{S}_y = a^\top \mathcal{S}_\mathcal{X} a.$$

Example 3.9 Suppose that $\mathcal{X}$ is the pullover data set. The manager wants to compute his mean expenses for advertisement (X_3) and sales assistant (X_4).

Suppose that the sales assistant charges an hourly wage of 10 EUR. Then the shop manager calculates the expenses Y as $Y = X_3 + 10X_4$. Formula (3.22) says that this is equivalent to defining the matrix $\mathcal{A}(4 \times 1)$ as

$$\mathcal{A} = (0, 0, 1, 10).$$

Using formulas (3.23) and (3.24), it is now computationally very easy to obtain the sample mean $\overline{y}$ and the sample variance $\mathcal{S}_y$ of the overall expenses:

$$\bar{y} = \mathcal{A}\bar{x} = (0, 0, 1, 10) \begin{pmatrix} 172.7 \\ 104.6 \\ 104.0 \\ 93.8 \end{pmatrix} = 1042.0$$

$$\mathcal{S}_y = \mathcal{A}\mathcal{S}_\mathcal{X}\mathcal{A}^\top = (0, 0, 1, 10) \begin{pmatrix} 1152.5 & -88.9 & 1589.7 & 301.6 \\ -88.9 & 244.3 & 102.3 & -101.8 \\ 1589.7 & 102.3 & 2915.6 & 233.7 \\ 301.6 & -101.8 & 233.7 & 197.1 \end{pmatrix} \begin{pmatrix} 0 \\ 0 \\ 1 \\ 10 \end{pmatrix}$$

$$= 2915.6 + 4674 + 19710 = 27299.6.$$

Mahalanobis Transformation

A special case of this linear transformation is

$$z_i = \mathcal{S}^{-1/2}(x_i - \bar{x}), \quad i = 1, \ldots, n. \tag{3.25}$$

Note that for the transformed data matrix $\mathcal{Z} = (z_1, \ldots, z_n)^\top$,

$$\mathcal{S}_\mathcal{Z} = n^{-1}\mathcal{Z}^\top\mathcal{H}\mathcal{Z} = \mathcal{I}_p. \tag{3.26}$$

So the Mahalanobis transformation eliminates the correlation between the variables and standardizes the variance of each variable. If we apply (3.24) using $\mathcal{A} = \mathcal{S}^{-1/2}$, we obtain the identity covariance matrix as indicated in (3.26).

Summary
↪   The center of gravity of a data matrix is given by its mean vector $\bar{x} = n^{-1}\mathcal{X}^\top 1_n$.
↪   The dispersion of the observations in a data matrix is given by the empirical covariance matrix $\mathcal{S} = n^{-1}\mathcal{X}^\top\mathcal{H}\mathcal{X}$.
↪   The empirical correlation matrix is given by $\mathcal{R} = \mathcal{D}^{-1/2}\mathcal{S}\mathcal{D}^{-1/2}$.
↪   A linear transformation $\mathcal{Y} = \mathcal{X}\mathcal{A}^\top$ of a data matrix $\mathcal{X}$ has mean $\mathcal{A}\bar{x}$ and empirical covariance $\mathcal{A}\mathcal{S}_\mathcal{X}\mathcal{A}^\top$.
↪   The Mahalanobis transformation is a linear transformation $z_i = \mathcal{S}^{-1/2}(x_i - \bar{x})$ which gives a standardized, uncorrelated data matrix $\mathcal{Z}$.

3.4   Linear Model for Two Variables

We have looked several times now at downward- and upward-sloping scatterplots. What does the eye define here as a slope? Suppose that we can construct a line corresponding to the general direction of the cloud. The sign of the slope of this line would correspond to the upward and downward directions. Call the variable on the vertical axis Y and the one on the horizontal axis X. A slope line is a linear relationship between X and Y:

$$y_i = \alpha + \beta x_i + \varepsilon_i, \; i = 1, \ldots, n. \tag{3.27}$$

Here, α is the intercept and β is the slope of the line. The errors (or deviations from the line) are denoted as ε_i and are assumed to have zero mean and finite variance σ^2. The task of finding (α, β) in (3.27) is referred to as a linear adjustment.

In Sect. 3.6, we shall derive estimators for α and β more formally, as well as accurately describe what a "good" estimator is. For now, one may try to find a "good" estimator $(\widehat{\alpha}, \widehat{\beta})$ via graphical techniques. A very common numerical and statistical technique is to use those $\widehat{\alpha}$ and $\widehat{\beta}$ that minimize:

$$(\widehat{\alpha}, \widehat{\beta}) = \arg\min_{(\alpha, \beta)} \sum_{i=1}^{n} (y_i - \alpha - \beta x_i)^2. \tag{3.28}$$

The solution to this task are the estimators:

$$\widehat{\beta} = \frac{s_{XY}}{s_{XX}} \tag{3.29}$$

$$\widehat{\alpha} = \bar{y} - \widehat{\beta}\bar{x}. \tag{3.30}$$

The variance of $\widehat{\beta}$ is

$$\mathrm{Var}(\widehat{\beta}) = \frac{\sigma^2}{n \cdot s_{XX}}. \tag{3.31}$$

The standard error (SE) of the estimator is the square root of (3.31),

$$\mathrm{SE}(\widehat{\beta}) = \{\mathrm{Var}(\widehat{\beta})\}^{1/2} = \frac{\sigma}{(n \cdot s_{XX})^{1/2}}. \tag{3.32}$$

We can use this formula to test the hypothesis that $\beta = 0$. In an application the variance σ^2 has to be estimated by an estimator $\widehat{\sigma}^2$ that will be given below. Under a normality assumption of the errors, the t-test for the hypothesis $\beta = 0$ works as follows.

One computes the statistic

$$t = \frac{\widehat{\beta}}{\mathrm{SE}(\widehat{\beta})} \tag{3.33}$$

and rejects the hypothesis at a 5% significance level if $| \, t \, | \geq t_{0.975;n-2}$, where the 97.5% quantile of the Student's t_{n-2} distribution is clearly the 95% critical value for the two-sided test. For $n \geq 30$, this can be replaced by 1.96, the 97.5% quantile of the normal distribution. An estimator $\widehat{\sigma}^2$ of σ^2 will be given in the following.

Fig. 3.5 Regression of sales
(X_1) on price (X_2) of
pullovers 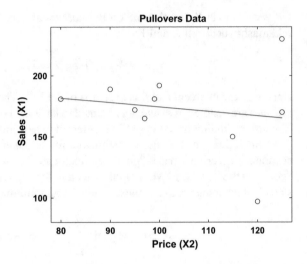MVAregpull

Example 3.10 Let us apply the linear regression model (3.27) to the "classic blue" pullovers. The sales manager believes that there is a strong dependence on the number of sales as a function of price. He computes the regression line as shown in Fig. 3.5.

How good is this fit? This can be judged via goodness-of-fit measures. Define

$$\widehat{y}_i = \widehat{\alpha} + \widehat{\beta} x_i,   \tag{3.34}$$

as the predicted value of y as a function of x. With $\widehat{y}$, the textile shop manager in the above example can predict sales as a function of prices x. The variation in the response variable is

$$ns_{YY} = \sum_{i=1}^{n}(y_i - \overline{y})^2.   \tag{3.35}$$

The variation explained by the linear regression (3.27) with the predicted values (3.34) is

$$\sum_{i=1}^{n}(\widehat{y}_i - \overline{y})^2.   \tag{3.36}$$

The residual sum of squares, the minimum in (3.28), is given by

$$RSS = \sum_{i=1}^{n}(y_i - \widehat{y}_i)^2.   \tag{3.37}$$

An unbiased estimator $\widehat{\sigma}^2$ of σ^2 is given by $RSS/(n-2)$.

The following relation holds between (3.35) and (3.37):

$$\sum_{i=1}^{n}(y_i - \bar{y})^2 = \sum_{i=1}^{n}(\widehat{y}_i - \bar{y})^2 + \sum_{i=1}^{n}(y_i - \widehat{y}_i)^2, \tag{3.38}$$

$$\textit{Total variation} = \textit{Explained variation} + \textit{Unexplained variation.}$$

The *coefficient of determination* is r^2:

$$r^2 = \frac{\sum_{i=1}^{n}(\widehat{y}_i - \bar{y})^2}{\sum_{i=1}^{n}(y_i - \bar{y})^2} = \frac{\textit{explained variation}}{\textit{total variation}}. \tag{3.39}$$

The coefficient of determination increases with the proportion of explained variation by the linear relation (3.27). In the extreme cases where $r^2 = 1$, all of the variation is explained by the linear regression (3.27). The other extreme, $r^2 = 0$, is where the empirical covariance is $s_{XY} = 0$. The coefficient of determination can be rewritten as

$$r^2 = 1 - \frac{\sum_{i=1}^{n}(y_i - \widehat{y}_i)^2}{\sum_{i=1}^{n}(y_i - \bar{y})^2}. \tag{3.40}$$

From (3.39), it can be seen that in the linear regression (3.27), $r^2 = r_{XY}^2$ is the square of the correlation between X and Y.

Example 3.11 For the above pullover example, we estimate

$$\widehat{\alpha} = 210.774 \quad \text{and} \quad \widehat{\beta} = -0.364.$$

The coefficient of determination is

$$r^2 = 0.028.$$

The textile shop manager concludes that sales are not influenced very much by the price (in a linear way).

The geometrical representation of formula (3.38) can be graphically evaluated using Fig. 3.6. This plot shows a section of the linear regression of the "sales" on "price" for the pullovers data. The distance between any point and the overall mean is given by the distance between the point and the regression line and the distance between the regression line and the mean. The sums of these two distances represent the total variance (solid blue lines from the observations to the overall

Fig. 3.6 Regression of sales
(X_1) on price (X_2) of
pullovers. The overall mean
is given by the dashed line

MVAregzoom

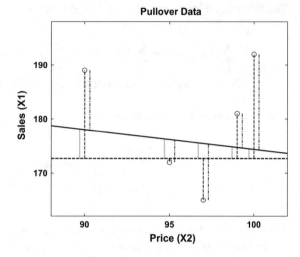

Fig. 3.7 Regression of X_5
(upper inner frame) on X_4
(lower inner frame) for
genuine bank notes

MVAregbank

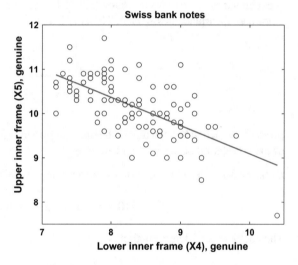

mean), i.e., the explained variance (distance from the regression curve to the mean)
and the unexplained variance (distance from the observation to the regression line),
respectively.

In general, the regression of Y on X is different from that of X on Y. We will
demonstrate this, once again, using the Swiss bank notes data.

Example 3.12 The least squares fit of the variables X_4 (X) and X_5 (Y) from the genuine bank notes are calculated. Figure 3.7 shows the fitted line if X_5 is approximated
by a linear function of X_4. In this case the parameters are

$$\widehat{\alpha} = 15.464 \quad \text{and} \quad \widehat{\beta} = -0.638.$$

If we predict X_4 by a function of X_5 instead, we would arrive at a different intercept and slope

$$\widehat{\alpha} = 14.666 \quad \text{and} \quad \widehat{\beta} = -0.626.$$

The linear regression of Y on X is given by minimizing (3.28), i.e., the vertical errors ε_i. The linear regression of X on Y does the same but here the errors to be minimized in the least squares sense are measured horizontally. As seen in Example 3.12, the two least squares lines are different although both measure (in a certain sense) the slope of the cloud of points.

As shown in the next example, there is still one other way to measure the main direction of a cloud of points: it is related to the spectral decomposition of covariance matrices.

Example 3.13 Suppose that we have the following covariance matrix:

$$\Sigma = \begin{pmatrix} 1 & \rho \\ \rho & 1 \end{pmatrix}.$$

Figure 3.8 shows a scatterplot of a sample of two normal random variables with such a covariance matrix (with $\rho = 0.8$).

The eigenvalues of Σ are, as was shown in Example 2.4, solutions to

$$\begin{vmatrix} 1 - \lambda & \rho \\ \rho & 1 - \lambda \end{vmatrix} = 0.$$

Hence, $\lambda_1 = 1 + \rho$ and $\lambda_2 = 1 - \rho$. Therefore $\Lambda = \text{diag}(1 + \rho, 1 - \rho)$. The eigenvector corresponding to $\lambda_1 = 1 + \rho$ can be computed from the system of linear equations:

$$\begin{pmatrix} 1 & \rho \\ \rho & 1 \end{pmatrix} \begin{pmatrix} x_1 \\ x_2 \end{pmatrix} = (1 + \rho) \begin{pmatrix} x_1 \\ x_2 \end{pmatrix}$$

or

$$x_1 + \rho x_2 = x_1 + \rho x_1$$
$$\rho x_1 + x_2 = x_2 + \rho x_2$$

and thus

$$x_1 = x_2.$$

The first (standardized) eigenvector is

$$\gamma_1 = \begin{pmatrix} 1/\sqrt{2} \\ 1/\sqrt{2} \end{pmatrix}.$$

Fig. 3.8 Scatterplot for
a sample of two correlated
normal random variables
(sample size $n = 150$,
$\rho = 0.8$) 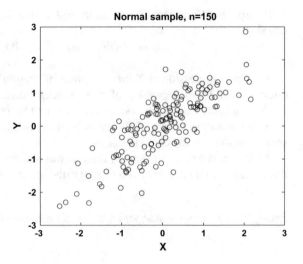MVAcorrnorm

The direction of this eigenvector is the diagonal in Fig. 3.8 and captures the main variation in this direction. We shall come back to this interpretation in Chap. 11. The second eigenvector (orthogonal to γ_1) is

$$\gamma_2 = \begin{pmatrix} 1/\sqrt{2} \\ -1/\sqrt{2} \end{pmatrix}.$$

So finally

$$\Gamma = (\gamma_1, \gamma_2) = \begin{pmatrix} 1/\sqrt{2} & 1/\sqrt{2} \\ 1/\sqrt{2} & -1/\sqrt{2} \end{pmatrix}$$

and we can check our calculation by

$$\Sigma = \Gamma \Lambda \Gamma^\top .$$

The first eigenvector captures the main direction of a point cloud. The linear regression of Y on X and X on Y accomplished, in a sense, the same thing. In general, the direction of the eigenvector and the least squares slope are different. The reason is that the least squares estimator minimizes either vertical or horizontal errors (in 3.28), whereas the first eigenvector corresponds to a minimization that is orthogonal to the eigenvector (see Chap. 11).

Summary
↪ The linear regression $y = \alpha + \beta x + \varepsilon$ models a linear relation between two one-dimensional variables.
↪ The sign of the slope $\widehat{\beta}$ is the same as that of the covariance and the correlation of x and y.
↪ A linear regression predicts values of Y given a possible observation x of X.
↪ The coefficient of determination r^2 measures the amount of variation in Y which is explained by a linear regression on X.
↪ If the coefficient of determination is $r^2 = 1$, then all points lie on one line.
↪ The regression line of X on Y and the regression line of Y on X are in general different.
↪ The t-test for the hypothesis $\beta = 0$ is $t = \frac{\widehat{\beta}}{\text{SE}(\widehat{\beta})}$, where $\text{SE}(\widehat{\beta}) = \frac{\widehat{\sigma}}{(n \cdot s_{XX})^{1/2}}$.
↪ The t-test rejects the null hypothesis $\beta = 0$ at the level of significance α if $\mid t \mid \geq t_{1-\alpha/2;n-2}$ where $t_{1-\alpha;n-2}$ is the $1 - \alpha/2$ quantile of the Student's t-distribution with $(n - 2)$ degrees of freedom.
↪ The standard error $\text{SE}(\widehat{\beta})$ increases/decreases with less/more spread in the X variables.
↪ The direction of the first eigenvector of the covariance matrix of a two-dimensional point cloud is different from the least squares regression line.

3.5 Simple Analysis of Variance

In a simple (i.e., one-factorial) analysis of variance (ANOVA), it is assumed that the average values of the response variable y are induced by one simple factor. Suppose that this factor takes on p values and that for each factor level, we have $m = n/p$ observations. The sample is of the form given in Table 3.1, where all of the observations are independent.

The goal of a simple ANOVA is to analyze the observation structure

$$y_{kl} = \mu_l + \varepsilon_{kl} \text{ for } k = 1, \ldots, m, \text{ and } l = 1, \ldots, p. \tag{3.41}$$

Table 3.1 Observation structure of a simple ANOVA

Sample element	Factor levels l				
1	y_{11}	$\cdots$	y_{1l}	$\cdots$	y_{1p}
2	⋮		⋮		⋮
⋮	⋮		⋮		⋮
k	y_{k1}	$\cdots$	y_{kl}	$\cdots$	y_{kp}
⋮	⋮		⋮		⋮
$m = n/p$	y_{m1}	$\cdots$	y_{ml}	$\cdots$	y_{mp}

Each factor has a mean value μ_l. Each observation y_{kl} is assumed to be a sum of the corresponding factor mean value μ_l and a zero-mean random error ε_{kl}. The linear regression model falls into this scheme with $m = 1$, $p = n$ and $\mu_i = \alpha + \beta x_i$, where x_i is the i-th level value of the factor.

Example 3.14 The "classic blue" pullover company analyzes the effect of three marketing strategies

1. advertisement in local newspaper,
2. presence of sales assistant,
3. luxury presentation in shop windows.

All of these strategies are tried in 10 different shops. The resulting sale observations are given in Table 3.2.

There are $p = 3$ factors and $n = mp = 30$ observations in the data. The "classic blue" pullover company wants to know whether all three marketing strategies have the same mean effect or whether there are differences. Having the same effect means that all μ_l in (3.41) equal one value, μ. The hypothesis to be tested is, therefore,

$$H_0 : \mu_l = \mu \text{ for } l = 1, \ldots, p.$$

The alternative hypothesis, that the marketing strategies have different effects, can be formulated as

$$H_1 : \mu_l \neq \mu_{l'} \text{ for some } l \text{ and } l'.$$

This means that one marketing strategy is better than the others.

The method used to test this problem is to compute as in (3.38) the total variation and to decompose it into the sources of variation. This gives

Table 3.2 Pullover sales as function of marketing strategy

Shop k	Marketing strategy factor l		
	1	2	3
1	9	10	18
2	11	15	14
3	10	11	17
4	12	15	9
5	7	15	14
6	11	13	17
7	12	7	16
8	10	15	14
9	11	13	17
10	13	10	15

$$\sum_{l=1}^{p}\sum_{k=1}^{m}(y_{kl}-\bar{y})^2 = m\sum_{l=1}^{p}(\bar{y}_l-\bar{y})^2 + \sum_{l=1}^{p}\sum_{k=1}^{m}(y_{kl}-\bar{y}_l)^2 \qquad (3.42)$$

The total variation (sum of squares = SS) is

$$SS(\text{reduced}) = \sum_{l=1}^{p}\sum_{k=1}^{m}(y_{kl}-\bar{y})^2 \qquad (3.43)$$

where $\bar{y} = n^{-1}\sum_{l=1}^{p}\sum_{k=1}^{m} y_{kl}$ is the overall mean. Here the total variation is denoted as SS(reduced), since in comparison with the model under the alternative H_1, we have a reduced set of parameters. In fact there is 1 parameter $\mu = \mu_l$ under H_0. Under H_1, the "full" model, we have three parameters, namely the three different means μ_l.

The variation under H_1 is, therefore,

$$SS(\text{full}) = \sum_{l=1}^{p}\sum_{k=1}^{m}(y_{kl}-\bar{y}_l)^2 \qquad (3.44)$$

where $\bar{y}_l = m^{-1}\sum_{k=1}^{m} y_{kl}$ is the mean of each factor l. The hypothetical model H_0 is called reduced, since it has (relative to H_1) fewer parameters.

The F-test of the linear hypothesis is used to compare the difference in the variations under the reduced model H_0 (3.43) and the full model H_1 (3.44) to the variation under the full model H_1:

$$F = \frac{\{SS(\text{reduced}) - SS(\text{full})\}/\{df(r) - df(f)\}}{SS(\text{full})/df(f)}. \qquad (3.45)$$

Here $df(f)$ and $df(r)$ denote the degrees of freedom under the full model and the reduced model, respectively. The degrees of freedom are essential in specifying the shape of the F-distribution. They have a simple interpretation: $df(\cdot)$ is equal to the number of observations minus the number of parameters in the model.

From Example 3.14, $p = 3$ parameters are estimated under the full model, i.e., $df(f) = n - p = 30 - 3 = 27$. Under the reduced model, there is one parameter to estimate, namely the overall mean, i.e., $df(r) = n - 1 = 29$. We can compute

$$SS(\text{reduced}) = 260.3$$

and

$$SS(\text{full}) = 157.7.$$

The F-statistic (3.45) is, therefore,

$$F = \frac{(260.3 - 157.7)/2}{157.7/27} = 8.78.$$

This value needs to be compared to the quantiles of the $F_{2,27}$ distribution. Looking up the critical values in a F-distribution shows that the test statistic above is highly significant. We conclude that the marketing strategies have different effects.

The F-test in a linear regression model

The t-test of a linear regression model can be put into this framework. For a linear regression model (3.27), the reduced model is the one with $\beta = 0$:

$$y_i = \alpha + 0 \cdot x_i + \varepsilon_i.$$

The reduced model has $n - 1$ degrees of freedom and one parameter, the intercept α.
   The full model is given by $\beta \neq 0$,

$$y_i = \alpha + \beta \cdot x_i + \varepsilon_i,$$

and has $n - 2$ degrees of freedom, since there are two parameters (α, β).
   The SS(reduced) equals

$$\text{SS(reduced)} = \sum_{i=1}^{n} (y_i - \bar{y})^2 = \textit{total variation.}$$

The SS(full) equals

$$\text{SS(full)} = \sum_{i=1}^{n} (y_i - \widehat{y}_i)^2 = \text{RSS} = \textit{unexplained variation.}$$

The F-test is, therefore, from (3.45)

$$F = \frac{(\textit{total variation - unexplained variation})/1}{(\textit{unexplained variation})/(n - 2)} \tag{3.46}$$

$$= \frac{\textit{explained variation}}{(\textit{unexplained variation})/(n - 2)}. \tag{3.47}$$

Using the estimators $\widehat{\alpha}$ and $\widehat{\beta}$ the explained variation is

$$\sum_{i=1}^{n} (\widehat{y}_i - \bar{y})^2 = \sum_{i=1}^{n} (\widehat{\alpha} + \widehat{\beta} x_i - \bar{y})^2$$

$$= \sum_{i=1}^{n} \{(\bar{y} - \widehat{\beta}\bar{x}) + \widehat{\beta} x_i - \bar{y}\}^2$$

$$= \sum_{i=1}^{n} \widehat{\beta}^2 (x_i - \bar{x})^2$$

$$= \widehat{\beta}^2 n s_{XX}.$$

From (3.32) the F-ratio (3.46) is, therefore,

$$F = \frac{\widehat{\beta}^2 n s_{XX}}{\text{RSS}/(n-2)} \tag{3.48}$$

$$= \left(\frac{\widehat{\beta}}{\text{SE}(\widehat{\beta})}\right)^2. \tag{3.49}$$

The t-test statistic (3.33) is just the square root of the F- statistic (3.49).

Note, using (3.39) the F-statistic can be rewritten as

$$F = \frac{r^2/1}{(1-r^2)/(n-2)}.$$

In the pullover Example 3.11, we obtain $F = \frac{0.028}{0.972}\frac{8}{1} = 0.2305$, so that the null hypothesis $\beta = 0$ cannot be rejected. We conclude, therefore, that there is only a minor influence of prices on sales.

Summary
↪ Simple ANOVA models an output Y as a function of one factor.
↪ The reduced model is the hypothesis of equal means.
↪ The full model is the alternative hypothesis of different means.
↪ The F-test is based on a comparison of the sum of squares under the full and the reduced models.
↪ The degrees of freedom are calculated as the number of observations minus the number of parameters.
↪ The F-statistic is $$F = \frac{\{\text{SS(reduced)} - \text{SS(full)}\}/\{df(r) - df(f)\}}{\text{SS(full)}/df(f)}.$$
↪ The F-test rejects the null hypothesis if the F-statistic is larger than the 95% quantile of the $F_{df(r)-df(f),df(f)}$ distribution.
↪ The F-test statistic for the slope of the linear regression model $y_i = \alpha + \beta x_i + \varepsilon_i$ is the square of the t-test statistic.

3.6 Multiple Linear Model

The simple linear model and the analysis of variance model can be viewed as a particular case of a more general linear model, where the variations of one variable y are explained by p explanatory variables x, respectively. Let y $(n \times 1)$ and $\mathcal{X}$ $(n \times p)$ be a vector of observations on the response variable and a data matrix on the p

explanatory variables. An important application of the developed theory is the least squares fitting. The idea is to approximate y by a linear combination $\widehat{y}$ of columns of $\mathcal{X}$, i.e., $\widehat{y} \in C(\mathcal{X})$. The problem is to find $\widehat{\beta} \in \mathbb{R}^p$ such that $\widehat{y} = \mathcal{X}\widehat{\beta}$ is the best fit of y in the least squares sense. The linear model can be written as

$$y = \mathcal{X}\beta + \varepsilon, \tag{3.50}$$

where ε are the errors. The least squares solution is given by $\widehat{\beta}$:

$$\widehat{\beta} = \arg\min_{\beta} (y - \mathcal{X}\beta)^\top (y - \mathcal{X}\beta) = \arg\min_{\beta} \varepsilon^\top \varepsilon. \tag{3.51}$$

Suppose that $(\mathcal{X}^\top \mathcal{X})$ is of full rank and thus invertible. Minimizing the expression (3.51) with respect to β yields

$$\widehat{\beta} = (\mathcal{X}^\top \mathcal{X})^{-1} \mathcal{X}^\top y. \tag{3.52}$$

The fitted value $\widehat{y} = \mathcal{X}\widehat{\beta} = \mathcal{X}(\mathcal{X}^\top \mathcal{X})^{-1} \mathcal{X}^\top y = \mathcal{P}y$ is the projection of y onto $C(\mathcal{X})$ as computed in (2.47).

The least squares residuals are

$$e = y - \widehat{y} = y - \mathcal{X}\widehat{\beta} = \mathcal{Q}y = (\mathcal{I}_n - \mathcal{P})y.$$

The vector e is the projection of y onto the orthogonal complement of $C(\mathcal{X})$.

Remark 3.5 A linear model with an intercept α can also be written in this framework. The approximating equation is

$$y_i = \alpha + \beta_1 x_{i1} + \ldots + \beta_p x_{ip} + \varepsilon_i \ ; \ i = 1, \ldots, n.$$

This can be written as

$$y = \mathcal{X}^* \beta^* + \varepsilon$$

where $\mathcal{X}^* = (1_n \ \mathcal{X})$ (we add a column of ones to the data). We have by (3.52)

$$\widehat{\beta}^* = \begin{pmatrix} \widehat{\alpha} \\ \widehat{\beta} \end{pmatrix} = (\mathcal{X}^{*\top} \mathcal{X}^*)^{-1} \mathcal{X}^{*\top} y.$$

Example 3.15 Let us come back to the "classic blue" pullovers example. In Example 3.11, we considered the regression fit of the sales X_1 on the price X_2 and concluded that there was only a small influence of sales by changing the prices. A linear model incorporating all three variables allows us to approximate sales as a linear function of price (X_2), advertisement (X_3) and presence of sales assistants (X_4) simultaneously. Adding a column of ones to the data (in order to estimate the intercept α) leads to

$$\widehat{\alpha} = 65.670 \text{ and } \widehat{\beta}_1 = -0.216, \ \widehat{\beta}_2 = 0.485, \ \widehat{\beta}_3 = 0.844.$$

The coefficient of determination is computed as before in (3.40) and is

$$r^2 = 1 - \frac{e^\top e}{\sum (y_i - \bar{y})^2} = 0.907.$$

We conclude that the variation of X_1 is well approximated by the linear relation.

Remark 3.6 The coefficient of determination is influenced by the number of regressors. For a given sample size n, the r^2 value will increase by adding more regressors into the linear model. The value of r^2 may, therefore, be high even if possibly irrelevant regressors are included. An adjusted coefficient of determination for p regressors and a constant intercept ($p + 1$ parameters) is

$$r^2_{\text{adj}} = r^2 - \frac{p(1 - r^2)}{n - (p + 1)}. \tag{3.53}$$

Example 3.16 The corrected coefficient of determination for Example 3.15 is

$$r^2_{\text{adj}} = 0.907 - \frac{3(1 - 0.907^2)}{10 - 3 - 1}$$
$$= 0.818.$$

This means that 81.8% of the variation of the response variable is explained by the explanatory variables.

Note that the linear model (3.50) is very flexible and can model nonlinear relationships between the response y and the explanatory variables x. For example, a quadratic relation in one variable x could be included. Then $y_i = \alpha + \beta_1 x_i + \beta_2 x_i^2 + \varepsilon_i$ could be written in matrix notation as in (3.50), $y = \mathcal{X}\beta + \varepsilon$ where

$$\mathcal{X} = \begin{pmatrix} 1 & x_1 & x_1^2 \\ 1 & x_2 & x_2^2 \\ \vdots & \vdots & \vdots \\ 1 & x_n & x_n^2 \end{pmatrix}.$$

Properties of $\widehat{\beta}$

When y_i is the i-th observation of a random variable Y, the errors are also random. Under standard assumptions (independence, zero mean, and constant variance σ^2), inference can be conducted on β. Using the properties of Chap. 4, it is easy to prove

$$\mathsf{E}(\widehat{\beta}) = \beta$$
$$\mathsf{Var}(\widehat{\beta}) = \sigma^2 (\mathcal{X}^\top \mathcal{X})^{-1}.$$

The analog of the t-test for the multivariate linear regression situation is

$$t = \frac{\widehat{\beta}_j}{\text{SE}(\widehat{\beta}_j)}.$$

The standard error of each coefficient $\widehat{\beta}_j$ is given by the square root of the diagonal elements of the matrix $\text{Var}(\widehat{\beta})$. In standard situations, the variance σ^2 of the error ε is not known. For linear model with intercept, one may estimate it by

$$\widehat{\sigma}^2 = \frac{1}{n - (p + 1)} (y - \widehat{y})^\top (y - \widehat{y}),$$

where p is the dimension of β. In testing $\beta_j = 0$, we reject the hypothesis at the significance level α if $| t | \geq t_{1-\alpha/2; n-(p+1)}$. More general issues on testing linear models are addressed in Chap. 7.

The ANOVA Model in Matrix Notation

The simple ANOVA problem (Sect. 3.5) may also be rewritten in matrix terms. Recall the definition of a vector of ones from (2.1) and define a vector of zeros as 0_n. Then construct the following $(n \times p)$ matrix, (here $p = 3$),

$$\mathcal{X} = \begin{pmatrix} 1_m & 0_m & 0_m \\ 0_m & 1_m & 0_m \\ 0_m & 0_m & 1_m \end{pmatrix}, \tag{3.54}$$

where $m = 10$. Equation (3.41) then reads as follows.

The parameter vector is $\beta = (\mu_1, \mu_2, \mu_3)^\top$. The data set from Example 3.14 can, therefore, be written as a linear model $y = \mathcal{X}\beta + \varepsilon$ where $y \in \mathbb{R}^n$ with $n = m \cdot p$ is the stacked vector of the columns of Table 3.1. The projection into the column space $C(\mathcal{X})$ of (3.54) yields the least squares estimator $\widehat{\beta} = (\mathcal{X}^\top \mathcal{X})^{-1} \mathcal{X}^\top y$. Note that $(\mathcal{X}^\top \mathcal{X})^{-1} = (1/10) \mathcal{I}_3$ and that $\mathcal{X}^\top y = (106, 124, 151)^\top$ is the sum $\sum_{k=1}^m y_{kj}$ for each factor, i.e., the 3 column sums of Table 3.1. The least squares estimator is, therefore, the vector $\widehat{\beta}_{H_1} = (\widehat{\mu}_1, \widehat{\mu}_2, \widehat{\mu}_3) = (10.6, 12.4, 15.1)^\top$ of sample means for each factor level $j = 1, 2, 3$. Under the null hypothesis of equal mean values $\mu_1 = \mu_2 = \mu_3 = \mu$, we estimate the parameters under the same constraints. This can be put into the form of a linear constraint:

$$-\mu_1 + \mu_2 = 0$$
$$-\mu_1 + \mu_3 = 0.$$

This can be written as $\mathcal{A}\beta = a$, where

$$a = \begin{pmatrix} 0 \\ 0 \end{pmatrix}$$

and

$$A = \begin{pmatrix} -1 & 1 & 0 \\ -1 & 0 & 1 \end{pmatrix}.$$

The constrained least squares solution can be shown (Exercise 3.24) to be given by

$$\widehat{\beta}_{H_0} = \widehat{\beta}_{H_1} - (\mathcal{X}^\top \mathcal{X})^{-1} A^\top \{A(\mathcal{X}^\top \mathcal{X})^{-1} A^\top\}^{-1}(A\widehat{\beta}_{H_1} - a). \tag{3.55}$$

It turns out that (3.55) amounts to simply calculating the overall mean $\bar{y} = 12.7$ of the response variable y: $\widehat{\beta}_{H_0} = (12.7, 12.7, 12.7)^\top$.

The F-test that has already been applied in Example 3.14 can be written as

$$F = \frac{\{||y - \mathcal{X}\widehat{\beta}_{H_0}||^2 - ||y - \mathcal{X}\widehat{\beta}_{H_1}||^2\}/2}{||y - \mathcal{X}\widehat{\beta}_{H_1}||^2/27} \tag{3.56}$$

which gives the same significant value 8.78. Note that again we compare the RSS_{H_0} of the reduced model to the RSS_{H_1} of the full model. It corresponds to comparing the lengths of projections into different column spaces. This general approach in testing linear models is described in detail in Chap. 7.

Summary
$\hookrightarrow$ The relation $y = \mathcal{X}\beta + \varepsilon$ models a linear relation between a one-dimensional variable Y and a p-dimensional variable X. $\mathcal{P}y$ gives the best linear regression fit of the vector y onto $C(\mathcal{X})$. The least squares parameter estimator is $\widehat{\beta} = (\mathcal{X}^\top \mathcal{X})^{-1}\mathcal{X}^\top y$.
$\hookrightarrow$ The simple ANOVA model can be written as a linear model.
$\hookrightarrow$ The ANOVA model can be tested by comparing the length of the projection vectors.
$\hookrightarrow$ The test statistic of the F-test can be written as $$\frac{\{||y - \mathcal{X}\widehat{\beta}_{H_0}||^2 - ||y - \mathcal{X}\widehat{\beta}_{H_1}||^2\}/\{df(r) - df(f)\}}{||y - \mathcal{X}\widehat{\beta}_{H_1}||^2/df(f)}.$$
$\hookrightarrow$ The adjusted coefficient of determination is $$r^2_{\text{adj}} = r^2 - \frac{p(1 - r^2)}{n - (p + 1)}.$$

3.7 Boston Housing

The main statistics presented so far can be computed for the data matrix $\mathcal{X}$(506 × 14) from our Boston Housing data set. The sample means and the sample medians of each variable are displayed in Table 3.3. The table also provides the unbiased estimates of the variance of each variable and the corresponding standard deviations. The comparison of the means and the medians confirms the assymmetry of the components of $\mathcal{X}$ that was pointed out in Sect. 1.9.

The (unbiased) sample covariance matrix is given by the following (14 × 14) matrix $\mathcal{S}_n$:

$$
\begin{pmatrix}
73.99 & -40.22 & 23.99 & -0.12 & 0.42 & -1.33 & 85.41 & -6.88 & 46.85 & 844.82 & 5.40 & -302.38 & 27.99 & -30.72 \\
-40.22 & 543.94 & -85.41 & -0.25 & -1.40 & 5.11 & -373.90 & 32.63 & -63.35 & -1236.45 & -19.78 & 373.72 & -68.78 & 77.32 \\
23.99 & -85.41 & 47.06 & 0.11 & 0.61 & -1.89 & 124.51 & -10.23 & 35.55 & 833.36 & 5.69 & -223.58 & 29.58 & -30.52 \\
-0.12 & -0.25 & 0.11 & 0.06 & 0.00 & 0.02 & 0.62 & -0.05 & -0.02 & -1.52 & -0.07 & 1.13 & -0.10 & 0.41 \\
0.42 & -1.40 & 0.61 & 0.00 & 0.01 & -0.02 & 2.39 & -0.19 & 0.62 & 13.05 & 0.05 & -4.02 & 0.49 & -0.46 \\
-1.33 & 5.11 & -1.89 & 0.02 & -0.02 & 0.49 & -4.75 & 0.30 & -1.28 & -34.58 & -0.54 & 8.22 & -3.08 & 4.49 \\
85.41 & -373.90 & 124.51 & 0.62 & 2.39 & -4.75 & 792.36 & -44.33 & 111.77 & 2402.69 & 15.94 & -702.94 & 121.08 & -97.59 \\
-6.88 & 32.63 & -10.23 & -0.05 & -0.19 & 0.30 & -44.33 & 4.43 & -9.07 & -189.66 & -1.06 & 56.04 & -7.47 & 4.84 \\
46.85 & -63.35 & 35.55 & -0.02 & 0.62 & -1.28 & 111.77 & -9.07 & 75.82 & 1335.76 & 8.76 & -353.28 & 30.39 & -30.56 \\
844.82 & -1236.45 & 833.36 & -1.52 & 13.05 & -34.58 & 2402.69 & -189.66 & 1335.76 & 28404.76 & 168.15 & -6797.91 & 654.71 & -726.26 \\
5.40 & -19.78 & 5.69 & -0.07 & 0.05 & -0.54 & 15.94 & -1.06 & 8.76 & 168.15 & 4.69 & -35.06 & 5.78 & -10.11 \\
-302.38 & 373.72 & -223.58 & 1.13 & -4.02 & 8.22 & -702.94 & 56.04 & -353.28 & -6797.91 & -35.06 & 8334.75 & -238.67 & 279.99 \\
27.99 & -68.78 & 29.58 & -0.10 & 0.49 & -3.08 & 121.08 & -7.47 & 30.39 & 654.71 & 5.78 & -238.67 & 50.99 & -48.45 \\
-30.72 & 77.32 & -30.52 & 0.41 & -0.46 & 4.49 & -97.59 & 4.84 & -30.56 & -726.26 & -10.11 & 279.99 & -48.45 & 84.59
\end{pmatrix}
$$

and the corresponding correlation matrix $\mathcal{R}$(14 × 14) is

Table 3.3 Descriptive statistics for the Boston Housing data set 🔍MVAdescbh

X	$\bar{x}$	Median(X)	Var(X)	Std(X)
X_1	3.61	0.26	73.99	8.60
X_2	11.36	0.00	543.94	23.32
X_3	11.14	9.69	47.06	6.86
X_4	0.07	0.00	0.06	0.25
X_5	0.55	0.54	0.01	0.12
X_6	6.28	6.21	0.49	0.70
X_7	68.57	77.50	792.36	28.15
X_8	3.79	3.21	4.43	2.11
X_9	9.55	5.00	75.82	8.71
X_{10}	408.24	330.00	28405.00	168.54
X_{11}	18.46	19.05	4.69	2.16
X_{12}	356.67	391.44	8334.80	91.29
X_{13}	12.65	11.36	50.99	7.14
X_{14}	22.53	21.20	84.59	9.20

$$\begin{pmatrix}
1.00 & -0.20 & 0.41 & -0.06 & 0.42 & -0.22 & 0.35 & -0.38 & 0.63 & 0.58 & 0.29 & -0.39 & 0.46 & -0.39 \\
-0.20 & 1.00 & -0.53 & -0.04 & -0.52 & 0.31 & -0.57 & 0.66 & -0.31 & -0.31 & -0.39 & 0.18 & -0.41 & 0.36 \\
0.41 & -0.53 & 1.00 & 0.06 & 0.76 & -0.39 & 0.64 & -0.71 & 0.60 & 0.72 & 0.38 & -0.36 & 0.60 & -0.48 \\
-0.06 & -0.04 & 0.06 & 1.00 & 0.09 & 0.09 & 0.09 & -0.10 & -0.01 & -0.04 & -0.12 & 0.05 & -0.05 & 0.18 \\
0.42 & -0.52 & 0.76 & 0.09 & 1.00 & -0.30 & 0.73 & -0.77 & 0.61 & 0.67 & 0.19 & -0.38 & 0.59 & -0.43 \\
-0.22 & 0.31 & -0.39 & 0.09 & -0.30 & 1.00 & -0.24 & 0.21 & -0.21 & -0.29 & -0.36 & 0.13 & -0.61 & 0.70 \\
0.35 & -0.57 & 0.64 & 0.09 & 0.73 & -0.24 & 1.00 & -0.75 & 0.46 & 0.51 & 0.26 & -0.27 & 0.60 & -0.38 \\
-0.38 & 0.66 & -0.71 & -0.10 & -0.77 & 0.21 & -0.75 & 1.00 & -0.49 & -0.53 & -0.23 & 0.29 & -0.50 & 0.25 \\
0.63 & -0.31 & 0.60 & -0.01 & 0.61 & -0.21 & 0.46 & -0.49 & 1.00 & 0.91 & 0.46 & -0.44 & 0.49 & -0.38 \\
0.58 & -0.31 & 0.72 & -0.04 & 0.67 & -0.29 & 0.51 & -0.53 & 0.91 & 1.00 & 0.46 & -0.44 & 0.54 & -0.47 \\
0.29 & -0.39 & 0.38 & -0.12 & 0.19 & -0.36 & 0.26 & -0.23 & 0.46 & 0.46 & 1.00 & -0.18 & 0.37 & -0.51 \\
-0.39 & 0.18 & -0.36 & 0.05 & -0.38 & 0.13 & -0.27 & 0.29 & -0.44 & -0.44 & -0.18 & 1.00 & -0.37 & 0.33 \\
0.46 & -0.41 & 0.60 & -0.05 & 0.59 & -0.61 & 0.60 & -0.50 & 0.49 & 0.54 & 0.37 & -0.37 & 1.00 & -0.74 \\
-0.39 & 0.36 & -0.48 & 0.18 & -0.43 & 0.70 & -0.38 & 0.25 & -0.38 & -0.47 & -0.51 & 0.33 & -0.74 & 1.00
\end{pmatrix}$$

Analyzing $\mathcal{R}$ confirms most of the comments made from examining the scatterplot matrix in Chap. 1. In particular, the correlation between X_{14} (the value of the house) and all the other variables is given by the last row (or column) of $\mathcal{R}$. The highest correlations (in absolute values) are in decreasing order X_{13}, X_6, X_{11}, X_{10}, etc.

Using the Fisher's Z-transform on each of the correlations between X_{14} and the other variables would confirm that all are significantly different from zero, except the correlation between X_{14} and X_4 (the indicator variable for the Charles River). We know, however, that the correlation and Fisher's Z-transform are not appropriate for binary variable.

The same descriptive statistics can be calculated for the transformed variables (transformations were motivated in Sect. 1.9). The results are given in Table 3.4 and as can be seen, most of the variables are now more symmetric.

Note that the covariances and the correlations are sensitive to these nonlinear transformations. For example, the correlation matrix is now

$$\begin{pmatrix}
1.00 & -0.52 & 0.74 & 0.03 & 0.81 & -0.32 & 0.70 & -0.74 & 0.84 & 0.81 & 0.45 & -0.48 & 0.62 & -0.57 \\
-0.52 & 1.00 & -0.66 & -0.04 & -0.57 & 0.31 & -0.53 & 0.59 & -0.35 & -0.31 & -0.35 & 0.18 & -0.45 & 0.36 \\
0.74 & -0.66 & 1.00 & 0.08 & 0.75 & -0.43 & 0.66 & -0.73 & 0.58 & 0.66 & 0.46 & -0.33 & 0.62 & -0.55 \\
0.03 & -0.04 & 0.08 & 1.00 & 0.08 & 0.08 & 0.07 & -0.09 & 0.01 & -0.04 & -0.13 & 0.05 & -0.06 & 0.16 \\
0.81 & -0.57 & 0.75 & 0.08 & 1.00 & -0.32 & 0.78 & -0.86 & 0.61 & 0.67 & 0.34 & -0.38 & 0.61 & -0.52 \\
-0.32 & 0.31 & -0.43 & 0.08 & -0.32 & 1.00 & -0.28 & 0.28 & -0.21 & -0.31 & -0.32 & 0.13 & -0.64 & 0.61 \\
0.70 & -0.53 & 0.66 & 0.07 & 0.78 & -0.28 & 1.00 & -0.80 & 0.47 & 0.54 & 0.38 & -0.29 & 0.64 & -0.48 \\
-0.74 & 0.59 & -0.73 & -0.09 & -0.86 & 0.28 & -0.80 & 1.00 & -0.54 & -0.60 & -0.32 & 0.32 & -0.56 & 0.41 \\
0.84 & -0.35 & 0.58 & 0.01 & 0.61 & -0.21 & 0.47 & -0.54 & 1.00 & 0.82 & 0.40 & -0.41 & 0.46 & -0.43 \\
0.81 & -0.31 & 0.66 & -0.04 & 0.67 & -0.31 & 0.54 & -0.60 & 0.82 & 1.00 & 0.48 & -0.43 & 0.53 & -0.56 \\
0.45 & -0.35 & 0.46 & -0.13 & 0.34 & -0.32 & 0.38 & -0.32 & 0.40 & 0.48 & 1.00 & -0.20 & 0.43 & -0.51 \\
-0.48 & 0.18 & -0.33 & 0.05 & -0.38 & 0.13 & -0.29 & 0.32 & -0.41 & -0.43 & -0.20 & 1.00 & -0.36 & 0.40 \\
0.62 & -0.45 & 0.62 & -0.06 & 0.61 & -0.64 & 0.64 & -0.56 & 0.46 & 0.53 & 0.43 & -0.36 & 1.00 & -0.83 \\
-0.57 & 0.36 & -0.55 & 0.16 & -0.52 & 0.61 & -0.48 & 0.41 & -0.43 & -0.56 & -0.51 & 0.40 & -0.83 & 1.00
\end{pmatrix}$$

Notice that some of the correlations between $\widetilde{X}_{14}$ and the other variables have increased.

Table 3.4 Descriptive statistics for the Boston Housing data set after the transformation 🔍MVAdescbh

$\widetilde{X}$	$\overline{\widetilde{x}}$	Median($\widetilde{X}$)	Var($\widetilde{X}$)	Std($\widetilde{X}$)
$\widetilde{X}_1$	−0.78	−1.36	4.67	2.16
$\widetilde{X}_2$	1.14	0.00	5.44	2.33
$\widetilde{X}_3$	2.16	2.27	0.60	0.78
$\widetilde{X}_4$	0.07	0.00	0.06	0.25
$\widetilde{X}_5$	−0.61	−0.62	0.04	0.20
$\widetilde{X}_6$	1.83	1.83	0.01	0.11
$\widetilde{X}_7$	5.06	5.29	12.72	3.57
$\widetilde{X}_8$	1.19	1.17	0.29	0.54
$\widetilde{X}_9$	1.87	1.61	0.77	0.87
$\widetilde{X}_{10}$	5.93	5.80	0.16	0.40
$\widetilde{X}_{11}$	2.15	2.04	1.86	1.36
$\widetilde{X}_{12}$	3.57	3.91	0.83	0.91
$\widetilde{X}_{13}$	3.42	3.37	0.97	0.99
$\widetilde{X}_{14}$	3.03	3.05	0.17	0.41

Table 3.5 Linear regression results for all variables of Boston Housing data set 🔍MVAlinregbh

Variable	$\widehat{\beta}_j$	$SE(\widehat{\beta}_j)$	t	p-value
constant	4.1769	0.3790	11.020	0.0000
$\widetilde{X}_1$	−0.0146	0.0117	−1.254	0.2105
$\widetilde{X}_2$	0.0014	0.0056	0.247	0.8051
$\widetilde{X}_3$	−0.0127	0.0223	−0.570	0.5692
$\widetilde{X}_4$	0.1100	0.0366	3.002	0.0028
$\widetilde{X}_5$	−0.2831	0.1053	−2.688	0.0074
$\widetilde{X}_6$	0.4211	0.1102	3.822	0.0001
$\widetilde{X}_7$	0.0064	0.0049	1.317	0.1885
$\widetilde{X}_8$	−0.1832	0.0368	−4.977	0.0000
$\widetilde{X}_9$	0.0684	0.0225	3.042	0.0025
$\widetilde{X}_{10}$	−0.2018	0.0484	−4.167	0.0000
$\widetilde{X}_{11}$	−0.0400	0.0081	−4.946	0.0000
$\widetilde{X}_{12}$	0.0445	0.0115	3.882	0.0001
$\widetilde{X}_{13}$	−0.2626	0.0161	−16.320	0.0000

If we want to explain the variations of the price $\widetilde{X}_{14}$ by the variation of all the other variables $\widetilde{X}_1, \ldots, \widetilde{X}_{13}$ we could estimate the linear model

$$\widetilde{X}_{14} = \beta_0 + \sum_{j=1}^{13} \beta_j \widetilde{X}_j + \varepsilon. \tag{3.57}$$

The result is given in Table 3.5.

The value of r^2 (0.765) and r^2_{adj} (0.759) show that most of the variance of X_{14} is explained by the linear model (3.57).

Again we see that the variations of $\widetilde{X}_{14}$ are mostly explained by (in decreasing order of the absolute value of the t-statistic) $\widetilde{X}_{13}, \widetilde{X}_8, \widetilde{X}_{11}, \widetilde{X}_{10}, \widetilde{X}_{12}, \widetilde{X}_6, \widetilde{X}_9, \widetilde{X}_4$, and $\widetilde{X}_5$. The other variables $\widetilde{X}_1, \widetilde{X}_2, \widetilde{X}_3$, and $\widetilde{X}_7$ seem to have little influence on the variations of $\widetilde{X}_{14}$. This will be confirmed by the testing procedures that will be developed in Chap. 7.

3.8 Exercises

Exercise 3.1 The covariance $s_{X_4 X_5}$ between X_4 and X_5 for the entire bank data set is positive. Given the definitions of X_4 and X_5, we would expect a negative covariance. Using Fig. 3.1 can you explain why $s_{X_4 X_5}$ is positive?

Exercise 3.2 Consider the two sub-clouds of counterfeit and genuine banknotes in Fig. 3.1 separately. Do you still expect $s_{X_4 X_5}$ (now calculated separately for each cloud) to be positive?

Exercise 3.3 We remarked that for two normal random variables, zero covariance implies independence. Why does this remark not apply to Example 3.4?

Exercise 3.4 Compute the covariance between the variables

$$X_2 = \text{miles per gallon,}$$
$$X_8 = \text{weight}$$

from the car data set (Sect. 22.3). What sign do you expect the covariance to have?

Exercise 3.5 Compute the correlation matrix of the variables in Example 3.2. Comment on the sign of the correlations and test the hypothesis

$$\rho_{X_1 X_2} = 0.$$

Exercise 3.6 Suppose you have observed a set of observations $\{x_i\}_{i=1}^n$ with $\bar{x} = 0$, $s_{XX} = 1$ and $n^{-1} \sum_{i=1}^n (x_i - \bar{x})^3 = 0$. Define the variable $y_i = x_i^2$. Can you immediately tell whether $r_{XY} \neq 0$?

Exercise 3.7 Find formulas (3.29) and (3.30) for $\widehat{\alpha}$ and $\widehat{\beta}$ by differentiating the objective function in (3.28) w.r.t. α and β.

Exercise 3.8 How many sales does the textile manager expect with a "classic blue" pullover price of $x = 105$?

Exercise 3.9 What does a scatterplot of two random variables look like for $r^2 = 1$ and $r^2 = 0$?

Exercise 3.10 Prove the variance decomposition (3.38) and show that the coefficient of determination is the square of the simple correlation between X and Y.

Exercise 3.11 Make a boxplot for the residuals $e_i = y_i - \widehat{\alpha} - \widehat{\beta} x_i$ for the "classic blue" pullovers data. If there are outliers, identify them and run the linear regression again without them. Do you obtain a stronger influence of price on sales?

Exercise 3.12 Under what circumstances would you obtain the same coefficients from the linear regression lines of Y on X and of X on Y?

Exercise 3.13 Treat the design of Example 3.14 as if there were 30 shops and not 10. Define x_i as the index of the shop, i.e., $x_i = i, i = 1, 2, \ldots, 30$. The null hypothesis is a constant regression line, $\mathsf{E}Y = \mu$. What does the alternative regression curve look like?

Exercise 3.14 Perform the test in Exercise 3.13 for the shop example with a 0.99 significance level. Do you still reject the hypothesis of equal marketing strategies?

Exercise 3.15 Compute an approximate confidence interval for $\rho_{X_1 X_4}$ in Example 3.2. Hint: start from a confidence interval for $\tanh^{-1}(\rho_{X_1 X_4})$ and then apply the inverse transformation.

Exercise 3.16 In Example 3.2, using the exchange rate of 1 EUR = 106 JPY, compute the same empirical covariance using prices in Japanese Yen rather than in Euros. Is there a significant difference? Why?

Exercise 3.17 Why does the correlation have the same sign as the covariance?

Exercise 3.18 Show that $\text{rank}(\mathcal{H}) = \text{tr}(\mathcal{H}) = n - 1$.

Exercise 3.19 Show that $\mathcal{X}_* = \mathcal{H}\mathcal{X}\mathcal{D}^{-1/2}$ is the standardized data matrix, i.e., $\overline{x}_* = 0$ and $\mathcal{S}_{\mathcal{X}_*} = \mathcal{R}_{\mathcal{X}}$.

Exercise 3.20 Compute for the pullovers data the regression of X_1 on X_2, X_3 and of X_1 on X_2, X_4. Which one has better coefficient of determination?

Exercise 3.21 Compare for the pullovers data the coefficient of determination for the regression of X_1 on X_2 (Example 3.11), of X_1 on X_2, X_3 (Exercise 3.20), and of X_1 on X_2, X_3, X_4 (Example 3.15). Observe that this coefficient is increasing with the number of predictor variables. Is this always the case?

...

Exercise 3.22 Consider the ANOVA problem (Sect. 3.5) again. Establish the constraint Matrix $\mathcal{A}$ for testing $\mu_1 = \mu_2$. Test this hypothesis via an analog of (3.55) and (3.56).

Exercise 3.23 Prove (3.52). (Hint, let $f(\beta) = (y - x\beta)^\top (y - x\beta)$ and solve $\frac{\partial f(\beta)}{\partial \beta} = 0$).

Exercise 3.24 Consider the linear model $Y = \mathcal{X}\beta + \varepsilon$ where $\widehat{\beta} = \arg\min_\beta \varepsilon^\top \varepsilon$ is subject to the linear constraints $\mathcal{A}\widehat{\beta} = a$ where $\mathcal{A}(q \times p)$, $(q \leq p)$ is of rank q and a is of dimension $(q \times 1)$. Show that

$$\widehat{\beta} = \widehat{\beta}_{\text{OLS}} - (\mathcal{X}^\top \mathcal{X})^{-1} \mathcal{A}^\top \left(\mathcal{A}(\mathcal{X}^\top \mathcal{X})^{-1} \mathcal{A}^\top \right)^{-1} \left(\mathcal{A}\widehat{\beta}_{\text{OLS}} - a \right),$$

where $\widehat{\beta}_{\text{OLS}} = (\mathcal{X}^\top \mathcal{X})^{-1} \mathcal{X}^\top y$. (Hint, let $f(\beta, \lambda) = (y - x\beta)^\top (y - x\beta) - \lambda^\top (\mathcal{A}\beta - a)$ where $\lambda \in \mathbb{R}^q$ and solve $\frac{\partial f(\beta,\lambda)}{\partial \beta} = 0$ and $\frac{\partial f(\beta,\lambda)}{\partial \lambda} = 0$).

Exercise 3.25 Compute the covariance matrix $S = \text{Cov}(\mathcal{X})$ where $\mathcal{X}$ denotes the matrix of observations on the counterfeit bank notes. Make a Jordan decomposition of S. Why are all of the eigenvalues positive?

Exercise 3.26 Compute the covariance of the counterfeit notes after they are linearly transformed by the vector $a = (1, 1, 1, 1, 1, 1)^\top$.

References

L. Breiman, *Statistics: with a view towards application* (Houghton Mifflin Company, Boston, 1973)

B. Flury, H. Riedwyl, *Multivariate Statistics* (Cambridge University Press, A practical Approach, 1988)

H. Hotelling, New light on the correlation coefficient and its transform. J. Royal Stat. Soc. Ser. B **15**, 193–232 (1953)

Chapter 4
Multivariate Distributions

The preceding chapter showed that by using the two first moments of a multivariate distribution (the mean and the covariance matrix), a lot of information on the relationship between the variables can be made available. Only basic statistical theory was used to derive tests of independence or of linear relationships. In this chapter, we give an introduction to the basic probability tools useful in statistical multivariate analysis.

Means and covariances share many interesting and useful properties, but they represent only part of the information on a multivariate distribution. Section 4.1 presents the basic probability tools used to describe a multivariate random variable including marginal and conditional distributions and the concept of independence. In Sect. 4.2, basic properties on means and covariances (marginal and conditional ones) are derived.

Since many statistical procedures rely on transformations of a multivariate random variable, Sect. 4.3 proposes the basic techniques needed to derive the distribution of transformations with a special emphasis on linear transforms. As an important example of a multivariate random variable, Sect. 4.4 defines the multinormal distribution. It will be analyzed in more detail in Chap. 5 along with most of its "companion" distributions that are useful in making multivariate statistical inferences.

The normal distribution plays a central role in statistics because it can be viewed as an approximation and limit of many other distributions. The basic justification relies on the central limit theorem presented in Sect. 4.5. We present this central theorem in the framework of sampling theory. A useful extension of this theorem is also given: it is an approximate distribution to transformations of asymptotically normal variables. The increasing power of computers today makes it possible to consider alternative approximate sampling distributions. These are based on resampling techniques and are suitable for many general situations. Section 4.8 gives an introduction to the ideas behind bootstrap approximations.

© Springer Nature Switzerland AG 2019   107
W. K. Härdle and L. Simar, *Applied Multivariate Statistical Analysis*,
https://doi.org/10.1007/978-3-030-26006-4_4

4.1   Distribution and Density Function

Let $X = (X_1, X_2, \ldots, X_p)^\top$ be a random vector. The cumulative distribution function (cdf) of X is defined by

$$F(x) = P(X \le x) = P(X_1 \le x_1, X_2 \le x_2, \ldots, X_p \le x_p).$$

For continuous X, a nonnegative probability density function (pdf) f exists, that

$$F(x) = \int_{-\infty}^{x} f(u)du. \tag{4.1}$$

Note that

$$\int_{-\infty}^{\infty} f(u)\, du = 1.$$

Most of the integrals appearing below are multidimensional. For instance, $\int_{-\infty}^{x} f(u)du$ means $\int_{-\infty}^{x_p} \cdots \int_{-\infty}^{x_1} f(u_1, \ldots, u_p)du_1 \cdots du_p$. Note also that the cdf F is differentiable with

$$f(x) = \frac{\partial^p F(x)}{\partial x_1 \cdots \partial x_p}.$$

For discrete X, the values of this random variable are concentrated on a countable or finite set of points $\{c_j\}_{j \in J}$, the probability of events of the form $\{X \in D\}$ can then be computed as

$$P(X \in D) = \sum_{\{j:c_j \in D\}} P(X = c_j).$$

If we partition X as $X = (X_1, X_2)^\top$ with $X_1 \in \mathbb{R}^k$ and $X_2 \in \mathbb{R}^{p-k}$, then the function

$$F_{X_1}(x_1) = P(X_1 \le x_1) = F(x_{11}, \ldots, x_{1k}, \infty, \ldots, \infty) \tag{4.2}$$

is called the *marginal cdf. $F = F(x)$* is called the joint cdf. For continuous X, the marginal pdf can be computed from the joint density by "integrating out" the variable not of interest.

$$f_{X_1}(x_1) = \int_{-\infty}^{\infty} f(x_1, x_2)dx_2. \tag{4.3}$$

The conditional pdf of X_2 given $X_1 = x_1$ is given as

$$f(x_2|x_1) = \frac{f(x_1, x_2)}{f_{X_1}(x_1)}. \tag{4.4}$$

Example 4.1 Consider the pdf

$$f(x_1, x_2) = \begin{cases} \frac{1}{2}x_1 + \frac{3}{2}x_2 & 0 \le x_1, x_2 \le 1, \\ 0 & \text{otherwise.} \end{cases}$$

$f(x_1, x_2)$ is a density since

$$\int f(x_1, x_2)dx_1 dx_2 = \frac{1}{2}\left[\frac{x_1^2}{2}\right]_0^1 + \frac{3}{2}\left[\frac{x_2^2}{2}\right]_0^1 = \frac{1}{4} + \frac{3}{4} = 1.$$

The marginal densities are

$$f_{X_1}(x_1) = \int f(x_1, x_2)dx_2 = \int_0^1 \left(\frac{1}{2}x_1 + \frac{3}{2}x_2\right)dx_2 = \frac{1}{2}x_1 + \frac{3}{4};$$

$$f_{X_2}(x_2) = \int f(x_1, x_2)dx_1 = \int_0^1 \left(\frac{1}{2}x_1 + \frac{3}{2}x_2\right)dx_1 = \frac{3}{2}x_2 + \frac{1}{4}.$$

The conditional densities are, therefore,

$$f(x_2|x_1) = \frac{\frac{1}{2}x_1 + \frac{3}{2}x_2}{\frac{1}{2}x_1 + \frac{3}{4}} \quad \text{and} \quad f(x_1|x_2) = \frac{\frac{1}{2}x_1 + \frac{3}{2}x_2}{\frac{3}{2}x_2 + \frac{1}{4}}.$$

Note that these conditional pdf's are nonlinear in x_1 and x_2 although the joint pdf has a simple (linear) structure.

Independence of two random variables is defined as follows.

Definition 4.1 X_1 and X_2 are independent iff $f(x) = f(x_1, x_2) = f_{X_1}(x_1)f_{X_2}(x_2)$.

That is, X_1 and X_2 are independent if the conditional pdf's are equal to the marginal densities, i.e., $f(x_1|x_2) = f_{X_1}(x_1)$ and $f(x_2|x_1) = f_{X_2}(x_2)$. Independence can be interpreted as follows: knowing $X_2 = x_2$ does not change the probability assessments on X_1, and conversely.

 Different joint pdf's may have the same marginal pdf's.

Example 4.2 Consider the pdf's

$$f(x_1, x_2) = 1, \quad 0 < x_1, x_2 < 1,$$

and

$$f(x_1, x_2) = 1 + \alpha(2x_1 - 1)(2x_2 - 1), \quad 0 < x_1, \; x_2 < 1, \quad -1 \le \alpha \le 1.$$

We compute in both cases the marginal pdf's as

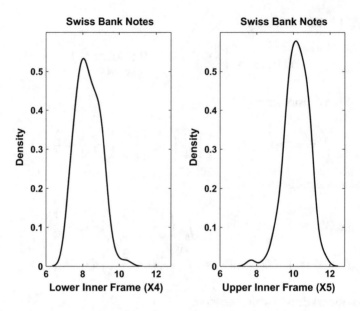

Fig. 4.1 Univariate estimates of the density of X_4 (left) and X_5 (right) of the banknotes
MVAdenbank2

$$f_{X_1}(x_1) = 1, \quad f_{X_2}(x_2) = 1.$$

Indeed

$$\int_0^1 1 + \alpha(2x_1 - 1)(2x_2 - 1)dx_2 = 1 + \alpha(2x_1 - 1)[x_2^2 - x_2]_0^1 = 1.$$

Hence, we obtain identical marginals from different joint distributions.

Let us study the concept of independence using the banknotes example. Consider the variables X_4 (lower inner frame) and X_5 (upper inner frame). From Chap. 3, we already know that they have significant correlation, so they are almost surely not independent. Kernel estimates of the marginal densities, $\widehat{f}_{X_4}$ and $\widehat{f}_{X_5}$, are given in Fig. 4.1. In Fig. 4.2 (left), we show the product of these two densities. The kernel density technique was presented in Sect. 1.3. If X_4 and X_5 are independent, this product $\widehat{f}_{X_4} \cdot \widehat{f}_{X_5}$ should be roughly equal to $\widehat{f}(x_4, x_5)$, the estimate of the joint density of (X_4, X_5). Comparing the two graphs in Fig. 4.2 reveals that the two densities are different. The two variables X_4 and X_5 are, therefore, not independent.

An elegant concept of connecting marginals with joint cdfs is given by *copulae*. Copulae are important in Value-at-Risk calculations and are an essential tool in quantitative finance (Härdle et al. 2009).

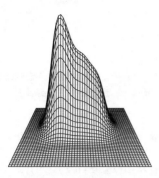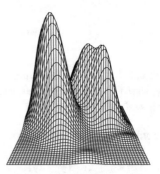

Fig. 4.2 The product of univariate density estimates (left) and the joint density estimate (right) for X_4 (left) and X_5 of the banknotes MVAdenbank3

For simplicity of presentation, we concentrate on the $p = 2$ dimensional case. A two-dimensional copula is a function $C : [0, 1]^2 \rightarrow [0, 1]$ with the following properties:

- For every $u \in [0, 1]$: $C(0, u) = C(u, 0) = 0$.
- For every $u \in [0, 1]$: $C(u, 1) = u$ and $C(1, u) = u$.
- For every $(u_1, u_2), (v_1, v_2) \in [0, 1] \times [0, 1]$ with $u_1 \leq v_1$ and $u_2 \leq v_2$:

$$C(v_1, v_2) - C(v_1, u_2) - C(u_1, v_2) + C(u_1, u_2) \geq 0.$$

The usage of the name "copula" for the function C is explained by the following theorem.

Theorem 4.1 (Sklar's theorem) *Let F be a joint distribution function with marginal distribution functions F_{X_1} and F_{X_2}. Then a copula C exists with*

$$F(x_1, x_2) = C\{F_{X_1}(x_1), F_{X_2}(x_2)\} \tag{4.5}$$

for every $x_1, x_2 \in \mathbb{R}$. If F_{X_1} and F_{X_2} are continuous, then C is unique. On the other hand, if C is a copula and F_{X_1} and F_{X_2} are distribution functions, then the function F defined by (4.5) is a joint distribution function with marginals F_{X_1} and F_{X_2}.

With Sklar's Theorem, the use of the name "copula" becomes obvious. It was chosen to describe "a function that links a multidimensional distribution to its one-dimensional margins" and appeared in the mathematical literature for the first time in Sklar (1959).

Example 4.3 The structure of independence implies that the product of the distribution functions F_{X_1} and F_{X_2} equals their joint distribution function F,

$$F(x_1, x_2) = F_{X_1}(x_1) \cdot F_{X_2}(x_2). \tag{4.6}$$

Thus, we obtain the *independence copula* $C = \Pi$ from

$$\Pi(u_1, \ldots, u_n) = \prod_{i=1}^{n} u_i \ .$$

Theorem 4.2 *Let X_1 and X_2 be random variables with continuous distribution functions F_{X_1} and F_{X_2} and the joint distribution function F. Then X_1 and X_2 are independent if and only if $C_{X_1, X_2} = \Pi$.*

Proof From Sklar's Theorem, we know that there exists a unique copula C with

$$P(X_1 \le x_1, X_2 \le x_2) = F(x_1, x_2) = C\{F_{X_1}(x_1), F_{X_2}(x_2)\} \ . \tag{4.7}$$

Independence can be seen using (4.5) for the joint distribution function F and the definition of Π,

$$F(x_1, x_2) = C\{F_{X_1}(x_1), F_{X_2}(x_2)\} = F_{X_1}(x_1) F_{X_2}(x_2) \ . \tag{4.8}$$

Example 4.4 The *Gumbel–Hougaard* family of copulae Nelsen (1999) is given by the function

$$C_\theta(u, v) = \exp\left[-\left\{(-\log u)^\theta + (-\log v)^\theta\right\}^{1/\theta}\right] \ . \tag{4.9}$$

The parameter θ may take all values in the interval $[1, \infty)$. The Gumbel–Hougaard copulae are suited to describe bivariate extreme value distributions.

For $\theta = 1$, the expression (4.9) reduces to the product copula, i.e., $C_1(u, v) = \Pi(u, v) = u\,v$. For $\theta \to \infty$ one finds for the Gumbel–Hougaard copula:

$$C_\theta(u, v) \longrightarrow \min(u, v) = M(u, v),$$

where the function M is also a copula such that $C(u, v) \le M(u, v)$ for arbitrary copula C. The copula M is called the *Fréchet–Hoeffding upper bound*.

Similarly, we obtain the *Fréchet–Hoeffding lower bound* $W(u, v) = \max(u + v - 1, 0)$ which satisfies $W(u, v) \le C(u, v)$ for any other copula C.

Summary
$\hookrightarrow$ The cumulative distribution function (cdf) is defined as $F(x) = P(X < x)$.
$\hookrightarrow$ If a probability density function (pdf) f exists then $F(x) = \int_{-\infty}^{x} f(u)du$.
$\hookrightarrow$ The pdf integrates to one, i.e., $\int_{-\infty}^{\infty} f(x)dx = 1$.
$\hookrightarrow$ Let $X = (X_1, X_2)^\top$ be partitioned into sub-vectors X_1 and X_2 with joint cdf F. Then $F_{X_1}(x_1) = P(X_1 \le x_1)$ is the marginal cdf of X_1. The marginal pdf of X_1 is obtained by $f_{X_1}(x_1) = \int_{-\infty}^{\infty} f(x_1, x_2)dx_2$. Different joint pdf's may have the same marginal pdf's.
$\hookrightarrow$ The conditional pdf of X_2 given $X_1 = x_1$ is defined as $f(x_2\|x_1) = \dfrac{f(x_1, x_2)}{f_{X_1}(x_1)}$.

Continued on next page

Summary (continue)	
$\hookrightarrow$	Two random variables X_1 and X_2 are called independent iff $f(x_1, x_2) = f_{X_1}(x_1) f_{X_2}(x_2)$. This is equivalent to $f(x_2|x_1) = f_{X_2}(x_2)$.
$\hookrightarrow$	Different joint pdf's may have identical marginal pdf's.
$\hookrightarrow$	Copula is a function which connects marginals to form joint cdf's.

4.2 Moments and Characteristic Functions

Moments—Expectation and Covariance Matrix

If X is a random vector with density $f(x)$ then the expectation of X is

$$\mathsf{E}X = \begin{pmatrix} \mathsf{E}X_1 \\ \vdots \\ \mathsf{E}X_p \end{pmatrix} = \int x f(x) dx = \begin{pmatrix} \int x_1 f(x) dx \\ \vdots \\ \int x_p f(x) dx \end{pmatrix} = \mu. \tag{4.10}$$

Accordingly, the expectation of a matrix of random elements has to be understood component by component. The operation of forming expectations is linear:

$$\mathsf{E}(\alpha X + \beta Y) = \alpha \mathsf{E}X + \beta \mathsf{E}Y. \tag{4.11}$$

If $\mathcal{A}(q \times p)$ is a matrix of real numbers, we have

$$\mathsf{E}(\mathcal{A}X) = \mathcal{A}\mathsf{E}X. \tag{4.12}$$

When X and Y are independent,

$$\mathsf{E}(XY^\top) = \mathsf{E}X\mathsf{E}Y^\top. \tag{4.13}$$

The matrix

$$\mathsf{Var}(X) = \Sigma = \mathsf{E}(X - \mu)(X - \mu)^\top \tag{4.14}$$

is the (theoretical) covariance matrix. We write for a vector X with mean vector μ and covariance matrix Σ,

$$X \sim (\mu, \Sigma). \tag{4.15}$$

The $(p \times q)$ matrix

$$\Sigma_{XY} = \mathsf{Cov}(X, Y) = \mathsf{E}(X - \mu)(Y - \nu)^\top \tag{4.16}$$

is the covariance matrix of $X \sim (\mu, \Sigma_{XX})$ and $Y \sim (\nu, \Sigma_{YY})$. Note that $\Sigma_{XY} = \Sigma_{YX}^\top$ and that $Z = \begin{pmatrix} X \\ Y \end{pmatrix}$ has covariance $\Sigma_{ZZ} = \begin{pmatrix} \Sigma_{XX} & \Sigma_{XY} \\ \Sigma_{YX} & \Sigma_{YY} \end{pmatrix}$. From

$$\mathsf{Cov}(X, Y) = \mathsf{E}(XY^\top) - \mu v^\top = \mathsf{E}(XY^\top) - \mathsf{E}X\mathsf{E}Y^\top \qquad (4.17)$$

it follows that $\mathsf{Cov}(X, Y) = 0$ in the case where X and Y are independent. We often say that $\mu = \mathsf{E}(X)$ is the first-order moment of X and that $\mathsf{E}(XX^\top)$ provides the second-order moments of X:

$$\mathsf{E}(XX^\top) = \{\mathsf{E}(X_iX_j)\}, \text{ for } i = 1, \ldots, p \text{ and } j = 1, \ldots, p. \qquad (4.18)$$

Properties of the Covariance Matrix $\Sigma = \mathsf{Var}(X)$

$$\Sigma = (\sigma_{X_iX_j}), \quad \sigma_{X_iX_j} = \mathsf{Cov}(X_i, X_j), \quad \sigma_{X_iX_i} = \mathsf{Var}(X_i) \qquad (4.19)$$

$$\Sigma = \mathsf{E}(XX^\top) - \mu\mu^\top \qquad (4.20)$$

$$\Sigma \geq 0 \qquad (4.21)$$

Properties of Variances and Covariances

$$\mathsf{Var}(a^\top X) = a^\top \mathsf{Var}(X)a = \sum_{i,j} a_i a_j \sigma_{X_iX_j} \qquad (4.22)$$

$$\mathsf{Var}(\mathcal{A}X + b) = \mathcal{A}\mathsf{Var}(X)\mathcal{A}^\top \qquad (4.23)$$

$$\mathsf{Cov}(X + Y, Z) = \mathsf{Cov}(X, Z) + \mathsf{Cov}(Y, Z) \qquad (4.24)$$

$$\mathsf{Var}(X + Y) = \mathsf{Var}(X) + \mathsf{Cov}(X, Y) + \mathsf{Cov}(Y, X) + \mathsf{Var}(Y) \qquad (4.25)$$

$$\mathsf{Cov}(\mathcal{A}X, \mathcal{B}Y) = \mathcal{A}\mathsf{Cov}(X, Y)\mathcal{B}^\top. \qquad (4.26)$$

Let us compute these quantities for a specific joint density.

Example 4.5 Consider the pdf of Example 4.1. The mean vector $\mu = \binom{\mu_1}{\mu_2}$ is

$$\mu_1 = \int\int x_1 f(x_1, x_2)dx_1dx_2 = \int_0^1\int_0^1 x_1\left(\frac{1}{2}x_1 + \frac{3}{2}x_2\right)dx_1dx_2$$

$$= \int_0^1 x_1\left(\frac{1}{2}x_1 + \frac{3}{4}\right)dx_1 = \frac{1}{2}\left[\frac{x_1^3}{3}\right]_0^1 + \frac{3}{4}\left[\frac{x_1^2}{2}\right]_0^1$$

$$= \frac{1}{6} + \frac{3}{8} = \frac{4+9}{24} = \frac{13}{24},$$

$$\mu_2 = \int\int x_2 f(x_1, x_2)dx_1dx_2 = \int_0^1\int_0^1 x_2\left(\frac{1}{2}x_1 + \frac{3}{2}x_2\right)dx_1dx_2$$

$$= \int_0^1 x_2\left(\frac{1}{4} + \frac{3}{2}x_2\right)dx_2 = \frac{1}{4}\left[\frac{x_2^2}{2}\right]_0^1 + \frac{3}{2}\left[\frac{x_2^3}{3}\right]_0^1$$

$$= \frac{1}{8} + \frac{1}{2} = \frac{1+4}{8} = \frac{5}{8}.$$

The elements of the covariance matrix are

$$\sigma_{X_1 X_1} = \mathsf{E}X_1^2 - \mu_1^2 \quad \text{with}$$

$$\mathsf{E}X_1^2 = \int_0^1 \int_0^1 x_1^2 \left(\frac{1}{2}x_1 + \frac{3}{2}x_2\right) dx_1 dx_2 = \frac{1}{2}\left[\frac{x_1^4}{4}\right]_0^1 + \frac{3}{4}\left[\frac{x_1^3}{3}\right]_0^1 = \frac{3}{8}$$

$$\sigma_{X_2 X_2} = \mathsf{E}X_2^2 - \mu_2^2 \quad \text{with}$$

$$\mathsf{E}X_2^2 = \int_0^1 \int_0^1 x_2^2 \left(\frac{1}{2}x_1 + \frac{3}{2}x_2\right) dx_1 dx_2 = \frac{1}{4}\left[\frac{x_2^3}{3}\right]_0^1 + \frac{3}{2}\left[\frac{x_2^4}{4}\right]_0^1 = \frac{11}{24}$$

$$\sigma_{X_1 X_2} = \mathsf{E}(X_1 X_2) - \mu_1 \mu_2 \quad \text{with}$$

$$\mathsf{E}(X_1 X_2) = \int_0^1 \int_0^1 x_1 x_2 \left(\frac{1}{2}x_1 + \frac{3}{2}x_2\right) dx_1 dx_2 = \int_0^1 \left(\frac{1}{6}x_2 + \frac{3}{4}x_2^2\right) dx_2$$

$$= \frac{1}{6}\left[\frac{x_2^2}{2}\right]_0^1 + \frac{3}{4}\left[\frac{x_2^3}{3}\right]_0^1 = \frac{1}{3}.$$

Hence the covariance matrix is

$$\Sigma = \begin{pmatrix} 0.0815 & 0.0052 \\ 0.0052 & 0.0677 \end{pmatrix}.$$

Conditional Expectations

The conditional expectations are

$$\mathsf{E}(X_2|x_1) = \int x_2 f(x_2|x_1)\, dx_2 \quad \text{and} \quad \mathsf{E}(X_1|x_2) = \int x_1 f(x_1|x_2)\, dx_1. \quad (4.27)$$

$\mathsf{E}(X_2|x_1)$ represents the location parameter of the conditional pdf of X_2 given that $X_1 = x_1$. In the same way, we can define $\mathsf{Var}(X_2|X_1 = x_1)$ as a measure of the dispersion of X_2 given that $X_1 = x_1$. We have from (4.20) that

$$\mathsf{Var}(X_2|X_1 = x_1) = \mathsf{E}(X_2 X_2^\top | X_1 = x_1) - \mathsf{E}(X_2|X_1 = x_1)\,\mathsf{E}(X_2^\top | X_1 = x_1).$$

Using the conditional covariance matrix, the conditional correlations may be defined as

$$\rho_{X_2 X_3|X_1=x_1} = \frac{\mathsf{Cov}(X_2, X_3|X_1 = x_1)}{\sqrt{\mathsf{Var}(X_2|X_1 = x_1)\,\mathsf{Var}(X_3|X_1 = x_1)}}.$$

These conditional correlations are known as partial correlations between X_2 and X_3, conditioned on X_1 being equal to x_1.

Example 4.6 Consider the following pdf

$$f(x_1, x_2, x_3) = \frac{2}{3}(x_1 + x_2 + x_3), \text{ where } 0 < x_1, x_2, x_3 < 1.$$

Note that the pdf is symmetric in x_1, x_2 and x_3 which facilitates the computations. For instance,

$$\begin{aligned} f(x_1, x_2) &= \tfrac{2}{3}(x_1 + x_2 + \tfrac{1}{2}) & 0 < x_1, x_2 < 1 \\ f(x_1) &= \tfrac{2}{3}(x_1 + 1) & 0 < x_1 < 1 \end{aligned}$$

and the other marginals are similar. We also have

$$f(x_1, x_2 | x_3) = \frac{x_1 + x_2 + x_3}{x_3 + 1}, \quad 0 < x_1, x_2 < 1$$

$$f(x_1 | x_3) = \frac{x_1 + x_3 + \frac{1}{2}}{x_3 + 1}, \quad 0 < x_1 < 1.$$

It is easy to compute the following moments:
$E(X_i) = \frac{5}{9}$; $E(X_i^2) = \frac{7}{18}$; $E(X_i X_j) = \frac{11}{36}$ ($i \neq j$ and $i, j = 1, 2, 3$)
$E(X_1 | X_3 = x_3) = E(X_2 | X_3 = x_3) = \frac{1}{12}\left(\frac{6x_3+7}{x_3+1}\right)$;
$E(X_1^2 | X_3 = x_3) = E(X_2^2 | X_3 = x_3) = \frac{1}{12}\left(\frac{4x_3+5}{x_3+1}\right)$
and
$E(X_1 X_2 | X_3 = x_3) = \frac{1}{12}\left(\frac{3x_3+4}{x_3+1}\right).$

Note that the conditional means of X_1 and of X_2, given $X_3 = x_3$, are not linear in x_3. From these moments, we obtain

$$\Sigma = \begin{pmatrix} \frac{13}{162} & -\frac{1}{324} & -\frac{1}{324} \\ -\frac{1}{324} & \frac{13}{162} & -\frac{1}{324} \\ -\frac{1}{324} & -\frac{1}{324} & \frac{13}{162} \end{pmatrix} \text{ in particular } \rho_{X_1 X_2} = -\frac{1}{26} \approx -0.0385.$$

The conditional covariance matrix of X_1 and X_2, given $X_3 = x_3$ is

$$\text{Var}\left(\begin{pmatrix} X_1 \\ X_2 \end{pmatrix} \Big| X_3 = x_3 \right) = \begin{pmatrix} \frac{12x_3^2+24x_3+11}{144(x_3+1)^2} & \frac{-1}{144(x_3+1)^2} \\ \frac{-1}{144(x_3+1)^2} & \frac{12x_3^2+24x_3+11}{144(x_3+1)^2} \end{pmatrix}.$$

In particular, the partial correlation between X_1 and X_2, given that X_3 is fixed at x_3, is given by $\rho_{X_1 X_2 | X_3 = x_3} = -\frac{1}{12x_3^2+24x_3+11}$ which ranges from -0.0909 to -0.0213 when x_3 goes from 0 to 1. Therefore, in this example, the partial correlation may be larger or smaller than the simple correlation, depending on the value of the condition $X_3 = x_3$.

Example 4.7 Consider the following joint pdf

$$f(x_1, x_2, x_3) = 2x_2(x_1 + x_3); \ 0 < x_1, x_2, x_3 < 1.$$

Note the symmetry of x_1 and x_3 in the pdf and that X_2 is independent of (X_1, X_3). It immediately follows that

$$f(x_1, x_3) = (x_1 + x_3) \quad 0 < x_1, x_3 < 1$$

$$f(x_1) = x_1 + \frac{1}{2};$$
$$f(x_2) = 2x_2;$$
$$f(x_3) = x_3 + \frac{1}{2}.$$

Simple computations lead to

$$\mathsf{E}(X) = \begin{pmatrix} \frac{7}{12} \\ \frac{2}{3} \\ \frac{7}{12} \end{pmatrix} \text{ and } \Sigma = \begin{pmatrix} \frac{11}{144} & 0 & -\frac{1}{144} \\ 0 & \frac{1}{18} & 0 \\ -\frac{1}{144} & 0 & \frac{11}{144} \end{pmatrix}.$$

Let us analyze the conditional distribution of (X_1, X_2) given $X_3 = x_3$. We have

$$f(x_1, x_2|x_3) = \frac{4(x_1 + x_3)x_2}{2x_3 + 1} \quad 0 < x_1, x_2 < 1$$

$$f(x_1|x_3) = 2 \left(\frac{x_1 + x_3}{2x_3 + 1} \right) \quad 0 < x_1 < 1$$

$$f(x_2|x_3) = f(x_2) = 2x_2 \quad 0 < x_2 < 1$$

so that again X_1 and X_2 are independent conditionals on $X_3 = x_3$. In this case

$$\mathsf{E}\left(\begin{pmatrix} X_1 \\ X_2 \end{pmatrix} \bigg| X_3 = x_3 \right) = \begin{pmatrix} \frac{1}{3} \left(\frac{2+3x_3}{1+2x_3} \right) \\ \frac{2}{3} \end{pmatrix}$$

$$\mathsf{Var}\left(\begin{pmatrix} X_1 \\ X_2 \end{pmatrix} \bigg| X_3 = x_3 \right) = \begin{pmatrix} \frac{1}{18} \left(\frac{6x_3^2+6x_3+1}{(2x_3+1)^2} \right) & 0 \\ 0 & \frac{1}{18} \end{pmatrix}.$$

Properties of Conditional Expectations

Since $\mathsf{E}(X_2|X_1 = x_1)$ is a function of x_1, say $h(x_1)$, we can define the random variable $h(X_1) = \mathsf{E}(X_2|X_1)$. The same can be done when defining the random variable $\mathsf{Var}(X_2|X_1)$. These two random variables share some interesting properties:

$$\mathsf{E}(X_2) = \mathsf{E}\{\mathsf{E}(X_2|X_1)\} \tag{4.28}$$
$$\mathsf{Var}(X_2) = \mathsf{E}\{\mathsf{Var}(X_2|X_1)\} + \mathsf{Var}\{\mathsf{E}(X_2|X_1)\}. \tag{4.29}$$

Example 4.8 Consider the following pdf

$$f(x_1, x_2) = 2e^{-\frac{x_2}{x_1}};\ 0 < x_1 < 1,\ x_2 > 0.$$

It is easy to show that

$$f(x_1) = 2x_1 \text{ for } 0 < x_1 < 1;\quad \mathsf{E}(X_1) = \frac{2}{3} \text{ and } \mathsf{Var}(X_1) = \frac{1}{18}$$

$$f(x_2|x_1) = \frac{1}{x_1}e^{-\frac{x_2}{x_1}} \text{ for } x_2 > 0;\quad \mathsf{E}(X_2|X_1) = X_1 \text{ and } \mathsf{Var}(X_2|X_1) = X_1^2.$$

Without explicitly computing $f(x_2)$, we can obtain

$$\mathsf{E}(X_2) = \mathsf{E}\{\mathsf{E}(X_2|X_1)\} = \mathsf{E}(X_1) = \frac{2}{3}$$
$$\mathsf{Var}(X_2) = \mathsf{E}\{\mathsf{Var}(X_2|X_1)\} + \mathsf{Var}\{\mathsf{E}(X_2|X_1)\} = \mathsf{E}(X_1^2) + \mathsf{Var}(X_1)$$
$$= \frac{2}{4} + \frac{1}{18} = \frac{10}{18}.$$

The conditional expectation $\mathsf{E}(X_2|X_1)$ viewed as a function $h(X_1)$ of X_1 (known as the regression function of X_2 on X_1), can be interpreted as a conditional approximation of X_2 by a function of X_1. The error term of the approximation is then given by

$$U = X_2 - \mathsf{E}(X_2|X_1).$$

Theorem 4.3 *Let $X_1 \in \mathbb{R}^k$ and $X_2 \in \mathbb{R}^{p-k}$ and $U = X_2 - \mathsf{E}(X_2|X_1)$. Then we have:*

1. $\mathsf{E}(U) = 0$.
2. $\mathsf{E}(X_2|X_1)$ *is the best approximation of X_2 by a function $h(X_1)$ of X_1 where $h : \mathbb{R}^k \longrightarrow \mathbb{R}^{p-k}$. "Best" is the minimum mean squared error (MSE) sense, where*
$$MSE(h) = \mathsf{E}[\{X_2 - h(X_1)\}^\top \{X_2 - h(X_1)\}].$$

Characteristic Functions

The characteristic function (cf) of a random vector $X \in \mathbb{R}^p$ (respectively its density $f(x)$) is defined as

$$\varphi_X(t) = \mathsf{E}(e^{it^\top X}) = \int e^{it^\top x} f(x)\, dx, \quad t \in \mathbb{R}^p,$$

where $\mathbf{i}$ is the complex unit: $\mathbf{i}^2 = -1$. The cf has the following properties:

$$\varphi_X(0) = 1 \text{ and } |\varphi_X(t)| \le 1. \tag{4.30}$$

If φ is absolutely integrable, i.e., the integral $\int_{-\infty}^{\infty} |\varphi(x)| dx$ exists and is finite, then

$$f(x) = \frac{1}{(2\pi)^p} \int_{-\infty}^{\infty} e^{-i t^\top x} \varphi_X(t) \, dt. \tag{4.31}$$

If $X = (X_1, X_2, \ldots, X_p)^\top$, then for $t = (t_1, t_2, \ldots, t_p)^\top$

$$\varphi_{X_1}(t_1) = \varphi_X(t_1, 0, \ldots, 0), \ldots, \varphi_{X_p}(t_p) = \varphi_X(0, \ldots, 0, t_p). \tag{4.32}$$

If $X_1, \ldots, X_p$ are independent random variables, then for $t = (t_1, t_2, \ldots, t_p)^\top$

$$\varphi_X(t) = \varphi_{X_1}(t_1) \cdot \ldots \cdot \varphi_{X_p}(t_p). \tag{4.33}$$

If $X_1, \ldots, X_p$ are independent random variables, then for $t \in \mathbb{R}$

$$\varphi_{X_1 + \ldots + X_p}(t) = \varphi_{X_1}(t) \cdot \ldots \cdot \varphi_{X_p}(t). \tag{4.34}$$

The characteristic function can recover all the cross-product moments of any order: $\forall j_k \geq 0, k = 1, \ldots, p$ and for $t = (t_1, \ldots, t_p)^\top$ we have

$$\mathsf{E}\left(X_1^{j_1} \cdot \ldots \cdot X_p^{j_p}\right) = \frac{1}{i^{j_1 + \cdots + j_p}} \left[\frac{\partial \varphi_X(t)}{\partial t_1^{j_1} \ldots \partial t_p^{j_p}} \right]_{t=0}. \tag{4.35}$$

Example 4.9 The cf of the density in Example 4.5 is given by

$$\begin{aligned}
\varphi_X(t) &= \int_0^1 \int_0^1 e^{i t^\top x} f(x) dx \\
&= \int_0^1 \int_0^1 \{\cos(t_1 x_1 + t_2 x_2) + i \sin(t_1 x_1 + t_2 x_2)\} \left(\frac{1}{2} x_1 + \frac{3}{2} x_2\right) dx_1 dx_2, \\
&= \frac{0.5 \, e^{i t_1} \left(3 i t_1 - 3 i e^{i t_2} t_1 + i t_2 - i e^{i t_2} t_2 + t_1 t_2 - 4 e^{i t_2} t_1 t_2\right)}{t_1^2 t_2^2} \\
&\quad - \frac{0.5 \left(3 i t_1 - 3 i e^{i t_2} t_1 + i t_2 - i e^{i t_2} t_2 - 3 e^{i t_2} t_1 t_2\right)}{t_1^2 t_2^2}.
\end{aligned}$$

Example 4.10 Suppose $X \in \mathbb{R}^1$ follows the density of the standard normal distribution

$$f_X(x) = \frac{1}{\sqrt{2\pi}} \exp\left(-\frac{x^2}{2}\right)$$

(see Sect. 4.4) then the cf can be computed via

Table 4.1 Characteristic functions for some common distributions

	pdf	cf
Uniform	$f(x) = I(x \in [a, b])/(b - a)$	$\varphi_X(t) = (e^{\mathbf{i}bt} - e^{\mathbf{i}at})/(b - a)\mathbf{i}t$
$N_1(\mu, \sigma^2)$	$f(x) = (2\pi\sigma^2)^{-1/2}\exp\{-(x - \mu)^2/2\sigma^2\}$	$\varphi_X(t) = e^{\mathbf{i}\mu t - \sigma^2 t^2/2}$
$\chi^2(n)$	$f(x) = I(x > 0)x^{n/2-1}e^{-x/2}/\{\Gamma(n/2)2^{n/2}\}$	$\varphi_X(t) = (1 - 2\mathbf{i}t)^{-n/2}$
$N_p(\mu, \Sigma)$	$f(x) = |2\pi\Sigma|^{-1/2}\exp\{-(x - \mu)^\top \Sigma(x - \mu)/2\}$	$\varphi_X(t) = e^{\mathbf{i}t^\top \mu - t^\top \Sigma t/2}$

$$
\begin{aligned}
\varphi_X(t) &= \frac{1}{\sqrt{2\pi}} \int_{-\infty}^{\infty} e^{\mathbf{i}tx} \exp\left(-\frac{x^2}{2}\right) dx \\
&= \frac{1}{\sqrt{2\pi}} \int_{-\infty}^{\infty} \exp\left\{-\frac{1}{2}(x^2 - 2\mathbf{i}tx + \mathbf{i}^2 t^2)\right\} \exp\left\{\frac{1}{2}\mathbf{i}^2 t^2\right\} dx \\
&= \exp\left(-\frac{t^2}{2}\right) \int_{-\infty}^{\infty} \frac{1}{\sqrt{2\pi}} \exp\left\{-\frac{(x - \mathbf{i}t)^2}{2}\right\} dx \\
&= \exp\left(-\frac{t^2}{2}\right),
\end{aligned}
$$

since $\mathbf{i}^2 = -1$ and $\int \frac{1}{\sqrt{2\pi}} \exp\left\{-\frac{(x-\mathbf{i}t)^2}{2}\right\} dx = 1$.

A variety of distributional characteristics can be computed from $\varphi_X(t)$. The standard normal distribution has a very simple cf, as was seen in Example 4.10. Deviations from normal covariance structures can be measured by the deviations from the cf (or characteristics of it). In Table 4.1, we give an overview of the cf's for a variety of distributions.

Theorem 4.4 (Cramer–Wold) *The distribution of $X \in \mathbb{R}^p$ is completely determined by the set of all (one-dimensional) distributions of $t^\top X$ where $t \in \mathbb{R}^p$.*

This theorem says that we can determine the distribution of X in $\mathbb{R}^p$ by specifying all of the one-dimensional distributions of the linear combinations

$$
\sum_{j=1}^{p} t_j X_j = t^\top X, \quad t = (t_1, t_2, \ldots, t_p)^\top.
$$

Cumulant functions

Moments $m_k = \int x^k f(x)dx$ often help in describing distributional characteristics. The normal distribution in $d = 1$ dimension is completely characterized by its standard normal density $f = \varphi$ and the moment parameters are $\mu = m_1$ and $\sigma^2 = m_2 - m_1^2$. Another helpful class of parameters are the cumulants or semi-invariants of a distribution. In order to simplify notation, we concentrate here on the one-dimensional ($d = 1$) case.

For a given one-dimensional random variable X with density f and finite moments of order k the characteristic function $\varphi_X(t) = \mathsf{E}(e^{itX})$ has the derivative

$$\frac{1}{\mathbf{i}^j} \left[\frac{\partial^j \log\{\varphi_X(t)\}}{\partial t^j} \right]_{t=0} = \kappa_j, \qquad j = 1, \ldots, k.$$

The values κ_j are called cumulants or semi-invariants since κ_j does not change (for $j > 1$) under a shift transformation $X \mapsto X + a$. The cumulants are natural parameters for dimension reduction methods, in particular the Projection Pursuit method (see Sect. 20.2).

The relationship between the first k moments $m_1, \ldots, m_k$ and the cumulants is given by

$$\kappa_k = (-1)^{k-1} \begin{vmatrix} m_1 & 1 & \cdots & 0 \\ m_2 & \binom{1}{0} m_1 & \cdots & \\ \vdots & \vdots & \ddots & \vdots \\ m_k & \binom{k-1}{0} m_{k-1} & \cdots & \binom{k-1}{k-2} m_1 \end{vmatrix}. \tag{4.36}$$

Example 4.11 Suppose that $k = 1$, then formula (4.36) above yields

$$\kappa_1 = m_1.$$

For $k = 2$, we obtain

$$\kappa_2 = - \begin{vmatrix} m_1 & 1 \\ m_2 & \binom{1}{0} m_1 \end{vmatrix} = m_2 - m_1^2.$$

For $k = 3$, we have to calculate

$$\kappa_3 = \begin{vmatrix} m_1 & 1 & 0 \\ m_2 & m_1 & 1 \\ m_3 & m_2 & 2m_1 \end{vmatrix}.$$

Calculating the determinant we have

$$\begin{aligned} \kappa_3 &= m_1 \begin{vmatrix} m_1 & 1 \\ m_2 & 2m_1 \end{vmatrix} - m_2 \begin{vmatrix} 1 & 0 \\ m_2 & 2m_1 \end{vmatrix} + m_3 \begin{vmatrix} 1 & 0 \\ m_1 & 1 \end{vmatrix} \\ &= m_1(2m_1^2 - m_2) - m_2(2m_1) + m_3 \\ &= m_3 - 3m_1 m_2 + 2m_1^3. \end{aligned} \tag{4.37}$$

Similarly one calculates

$$\kappa_4 = m_4 - 4m_3 m_1 - 3m_2^2 + 12m_2 m_1^2 - 6m_1^4. \tag{4.38}$$

The same type of process is used to find the moments from the cumulants:

$$
\begin{aligned}
m_1 &= \kappa_1 \\
m_2 &= \kappa_2 + \kappa_1^2 \\
m_3 &= \kappa_3 + 3\kappa_2\kappa_1 + \kappa_1^3 \\
m_4 &= \kappa_4 + 4\kappa_3\kappa_1 + 3\kappa_2^2 + 6\kappa_2\kappa_1^2 + \kappa_1^4.
\end{aligned}
\tag{4.39}
$$

A very simple relationship can be observed between the semi-invariants and the central moments $\mu_k = \mathsf{E}(X - \mu)^k$, where $\mu = m_1$ as defined before. In fact, $\kappa_2 = \mu_2$, $\kappa_3 = \mu_3$ and $\kappa_4 = \mu_4 - 3\mu_2^2$.

Skewness γ_3 and kurtosis γ_4 are defined as

$$
\begin{aligned}
\gamma_3 &= \mathsf{E}(X - \mu)^3 / \sigma^3 \\
\gamma_4 &= \mathsf{E}(X - \mu)^4 / \sigma^4.
\end{aligned}
\tag{4.40}
$$

The skewness and kurtosis determine the shape of one-dimensional distributions. The skewness of a normal distribution is 0 and the kurtosis equals 3. The relation of these parameters to the cumulants is given by

$$
\gamma_3 = \frac{\kappa_3}{\kappa_2^{3/2}}
\tag{4.41}
$$

From (4.39) and Example 4.11

$$
\gamma_4 = \frac{\kappa_4 + 3\kappa_2^2 + \kappa_1^4 - m_1^4}{\sigma^4} = \frac{\kappa_4 + 3\kappa_2^2}{\kappa_2^2} = \frac{\kappa_4}{\kappa_2^2} + 3.
\tag{4.42}
$$

These relations will be used later in Sect. 20.2 on Projection Pursuit to determine deviations from normality.

Summary
$\hookrightarrow$ The expectation of a random vector X is $\mu = \int x f(x)\, dx$, the covariance matrix $\Sigma = \mathsf{Var}(X) = \mathsf{E}(X - \mu)(X - \mu)^\top$. We denote $X \sim (\mu, \Sigma)$.
$\hookrightarrow$ Expectations are linear, i.e., $\mathsf{E}(\alpha X + \beta Y) = \alpha \mathsf{E} X + \beta \mathsf{E} Y$. If X and Y are independent, then $\mathsf{E}(XY^\top) = \mathsf{E} X \mathsf{E} Y^\top$.
$\hookrightarrow$ The covariance between two random vectors X and Y is $\Sigma_{XY} = \mathsf{Cov}(X, Y) = \mathsf{E}(X - \mathsf{E} X)(Y - \mathsf{E} Y)^\top = \mathsf{E}(XY^\top) - \mathsf{E} X \mathsf{E} Y^\top$. If X and Y are independent, then $\mathsf{Cov}(X, Y) = 0$.
$\hookrightarrow$ The characteristic function (cf) of a random vector X is $\varphi_X(t) = \mathsf{E}(e^{\mathbf{i} t^\top X})$.
$\hookrightarrow$ The distribution of a p-dimensional random variable X is completely determined by all one-dimensional distributions of $t^\top X$ where $t \in \mathbb{R}^p$ (Theorem of Cramer–Wold).
$\hookrightarrow$ The conditional expectation $\mathsf{E}(X_2 \mid X_1)$ is the MSE best approximation of X_2 by a function of X_1.

4.3 Transformations

Suppose that X has pdf $f_X(x)$. What is the pdf of $Y = 3X$? Or if $X = (X_1, X_2, X_3)^\top$, what is the pdf of

$$Y = \begin{pmatrix} 3X_1 \\ X_1 - 4X_2 \\ X_3 \end{pmatrix}?$$

This is a special case of asking for the pdf of Y when

$$X = u(Y) \tag{4.43}$$

for a one-to-one transformation $u: \mathbb{R}^p \to \mathbb{R}^p$. Define the Jacobian of u as

$$\mathcal{J} = \left(\frac{\partial x_i}{\partial y_j}\right) = \left(\frac{\partial u_i(y)}{\partial y_j}\right)$$

and let $\mathrm{abs}(|\mathcal{J}|)$ be the absolute value of the determinant of this Jacobian. The pdf of Y is given by

$$f_Y(y) = \mathrm{abs}(|\mathcal{J}|) \cdot f_X\{u(y)\}. \tag{4.44}$$

Using this, we can answer the introductory questions, namely,

$$(x_1, \ldots, x_p)^\top = u(y_1, \ldots, y_p) = \frac{1}{3}(y_1, \ldots, y_p)^\top$$

with

$$\mathcal{J} = \begin{pmatrix} \frac{1}{3} & & 0 \\ & \ddots & \\ 0 & & \frac{1}{3} \end{pmatrix}$$

and hence $\mathrm{abs}(|\mathcal{J}|) = \left(\frac{1}{3}\right)^p$. So the pdf of Y is $\frac{1}{3^p} f_X\left(\frac{y}{3}\right)$.

This introductory example is a special case of

$$Y = \mathcal{A}X + b, \text{ where } \mathcal{A} \text{ is nonsingular.}$$

The inverse transformation is

$$X = \mathcal{A}^{-1}(Y - b).$$

Therefore

$$\mathcal{J} = \mathcal{A}^{-1},$$

and hence

$$f_Y(y) = \text{abs}(|\mathcal{A}|^{-1}) f_X\{\mathcal{A}^{-1}(y - b)\}. \tag{4.45}$$

Example 4.12 Consider $X = (X_1, X_2) \in \mathbb{R}^2$ with density $f_X(x) = f_X(x_1, x_2)$,

$$\mathcal{A} = \begin{pmatrix} 1 & 1 \\ 1 & -1 \end{pmatrix}, \quad b = \begin{pmatrix} 0 \\ 0 \end{pmatrix}.$$

Then

$$Y = \mathcal{A}X + b = \begin{pmatrix} X_1 + X_2 \\ X_1 - X_2 \end{pmatrix}$$

and

$$|\mathcal{A}| = -2, \quad \text{abs}(|\mathcal{A}|^{-1}) = \frac{1}{2}, \quad \mathcal{A}^{-1} = -\frac{1}{2}\begin{pmatrix} -1 & -1 \\ -1 & 1 \end{pmatrix}.$$

Hence

$$\begin{aligned}
f_Y(y) &= \text{abs}(|\mathcal{A}|^{-1}) \cdot f_X(\mathcal{A}^{-1}y) \\
&= \frac{1}{2} f_X\left\{ \frac{1}{2}\begin{pmatrix} 1 & 1 \\ 1 & -1 \end{pmatrix}\begin{pmatrix} y_1 \\ y_2 \end{pmatrix} \right\} \\
&= \frac{1}{2} f_X\left\{ \frac{1}{2}(y_1 + y_2), \frac{1}{2}(y_1 - y_2) \right\}. \tag{4.46}
\end{aligned}$$

Example 4.13 Consider $X \in \mathbb{R}^1$ with density $f_X(x)$ and $Y = \exp(X)$. According to (4.43) $x = u(y) = \log(y)$ and hence the Jacobian is

$$\mathcal{J} = \frac{dx}{dy} = \frac{1}{y}.$$

The pdf of Y is, therefore,

$$f_Y(y) = \frac{1}{y} f_X\{\log(y)\}.$$

Summary

↪ If X has pdf $f_X(x)$, then a transformed random vector Y, i.e., $X = u(Y)$, has pdf $f_Y(y) = \text{abs}(|\mathcal{J}|) \cdot f_X\{u(y)\}$, where $\mathcal{J}$ denotes the Jacobian $\mathcal{J} = \left(\frac{\partial u(y_i)}{\partial y_j}\right)$.

↪ In the case of a linear relation $Y = \mathcal{A}X + b$ the pdf's of X and Y are related via $f_Y(y) = \text{abs}(|\mathcal{A}|^{-1}) f_X\{\mathcal{A}^{-1}(y - b)\}$.

4.4   The Multinormal Distribution

The multinormal distribution with mean μ and covariance $\Sigma > 0$ has the density

$$f(x) = |2\pi \Sigma|^{-1/2} \exp\left\{-\frac{1}{2}(x - \mu)^\top \Sigma^{-1}(x - \mu)\right\}. \qquad (4.47)$$

We write $X \sim N_p(\mu, \Sigma)$.

How is this multinormal distribution with mean μ and covariance Σ related to the multivariate standard normal $N_p(0, \mathcal{I}_p)$? Through a linear transformation using the results of Sect. 4.3, as shown in the next theorem.

Theorem 4.5 *Let $X \sim N_p(\mu, \Sigma)$ and $Y = \Sigma^{-1/2}(X - \mu)$ (Mahalanobis transformation). Then*

$$Y \sim N_p(0, \mathcal{I}_p),$$

i.e., the elements $Y_j \in \mathbb{R}$ are independent, one-dimensional $N(0, 1)$ variables.

Proof Note that $(X - \mu)^\top \Sigma^{-1}(X - \mu) = Y^\top Y$. Application of (4.45) gives $\mathcal{J} = \Sigma^{1/2}$, hence

$$f_Y(y) = (2\pi)^{-p/2} \exp\left(-\frac{1}{2}y^\top y\right) \qquad (4.48)$$

which is by (4.47) the pdf of a $N_p(0, \mathcal{I}_p)$.

Note that the above Mahalanobis transformation yields in fact a random variable $Y = (Y_1, \ldots, Y_p)^\top$ composed of independent one-dimensional $Y_j \sim N_1(0, 1)$ since

$$\begin{aligned}
f_Y(y) &= \frac{1}{(2\pi)^{p/2}} \exp\left(-\frac{1}{2}y^\top y\right) \\
&= \prod_{j=1}^{p} \frac{1}{\sqrt{2\pi}} \exp\left(-\frac{1}{2}y_j^2\right) \\
&= \prod_{j=1}^{p} f_{Y_j}(y_j).
\end{aligned}$$

Here each $f_{Y_j}(y)$ is a standard normal density $\frac{1}{\sqrt{2\pi}} \exp\left(-\frac{y^2}{2}\right)$. From this, it is clear that $\mathsf{E}(Y) = 0$ and $\mathsf{Var}(Y) = \mathcal{I}_p$.

How can we create $N_p(\mu, \Sigma)$ variables on the basis of $N_p(0, \mathcal{I}_p)$ variables? We use the inverse linear transformation

$$X = \Sigma^{1/2}Y + \mu. \qquad (4.49)$$

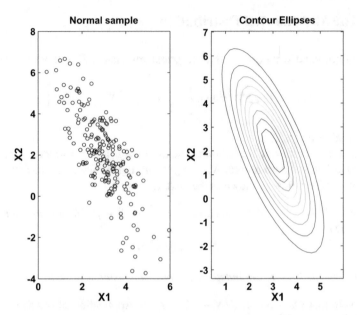

Fig. 4.3 Scatterplot of a normal sample and contour ellipses for $\mu = \begin{pmatrix} 3 \\ 2 \end{pmatrix}$ and $\Sigma = \begin{pmatrix} 1 & -1.5 \\ -1.5 & 4 \end{pmatrix}$

MVAcontnorm

Using (4.11) and (4.23), we can also check that $\mathsf{E}(X) = \mu$ and $\mathsf{Var}(X) = \Sigma$. The following theorem is useful because it presents the distribution of a variable after it has been linearly transformed. The proof is left as an exercise.

Theorem 4.6 *Let* $X \sim N_p(\mu, \Sigma)$ *and* $\mathcal{A}(p \times p)$, $c \in \mathbb{R}^p$, *where* $\mathcal{A}$ *is nonsingular. Then* $Y = \mathcal{A}X + c$ *is again a p-variate Normal, i.e.,*

$$Y \sim N_p(\mathcal{A}\mu + c, \mathcal{A}\Sigma\mathcal{A}^\top). \tag{4.50}$$

Geometry of the $N_p(\mu, \Sigma)$ Distribution

From (4.47), we see that the density of the $N_p(\mu, \Sigma)$ distribution is constant on ellipsoids of the form

$$(x - \mu)^\top \Sigma^{-1}(x - \mu) = d^2. \tag{4.51}$$

Example 4.14 Figure 4.3 shows the contour ellipses of a two-dimensional normal distribution. Note that these contour ellipses are the iso-distance curves (2.34) from the mean of this normal distribution corresponding to the metric Σ^{-1}.

According to Theorem 2.7 in Sect. 2.6, the half-lengths of the axes in the contour ellipsoid are $\sqrt{d^2\lambda_i}$ where λ_i are the eigenvalues of Σ. If Σ is a diagonal matrix, the rectangle circumscribing the contour ellipse has sides with length $2d\sigma_i$ and is thus naturally proportional to the standard deviations of X_i $(i = 1, 2)$.

The distribution of the quadratic form in (4.51) is given in the next theorem.

Theorem 4.7 *If $X \sim N_p(\mu, \Sigma)$, then the variable $U = (X - \mu)^\top \Sigma^{-1}(X - \mu)$ has a χ_p^2 distribution.*

Theorem 4.8 *The characteristic function (cf) of a multinormal $N_p(\mu, \Sigma)$ is given by*

$$\varphi_X(t) = \exp(i\, t^\top \mu - \frac{1}{2} t^\top \Sigma t). \tag{4.52}$$

We can check Theorem 4.8 by transforming the cf back:

$$f(x) = \frac{1}{(2\pi)^p} \int \exp\left(-it^\top x + it^\top \mu - \frac{1}{2} t^\top \Sigma t\right) dt$$

$$= \frac{1}{|2\pi\, \Sigma^{-1}|^{1/2} |2\pi\, \Sigma|^{1/2}} \int \exp\left[-\frac{1}{2}\{t^\top \Sigma t + 2it^\top (x-\mu) - (x-\mu)^\top \Sigma^{-1}(x-\mu)\}\right]$$

$$\cdot \exp\left[-\frac{1}{2}\{(x-\mu)^\top \Sigma^{-1}(x-\mu)\}\right] dt$$

$$= \frac{1}{|2\pi\, \Sigma|^{1/2}} \exp\left[-\frac{1}{2}\{(x-\mu)^\top \Sigma(x-\mu)\}\right]$$

since

$$\int \frac{1}{|2\pi\, \Sigma^{-1}|^{1/2}} \exp\left[-\frac{1}{2}\{t^\top \Sigma t + 2it^\top (x-\mu) - (x-\mu)^\top \Sigma^{-1}(x-\mu)\}\right] dt$$

$$= \int \frac{1}{|2\pi\, \Sigma^{-1}|^{1/2}} \exp\left[-\frac{1}{2}\{(t + i\Sigma^{-1}(x-\mu))^\top \Sigma(t + i\Sigma^{-1}(x-\mu))\}\right] dt$$

$$= 1.$$

Note that if $Y \sim N_p(0, \mathcal{I}_p)$, then

$$\varphi_Y(t) = \exp\left(-\frac{1}{2} t^\top \mathcal{I}_p t\right) = \exp\left(-\frac{1}{2} \sum_{i=1}^p t_i^2\right)$$

$$= \varphi_{Y_1}(t_1) \cdot \ldots \cdot \varphi_{Y_p}(t_p)$$

which is consistent with (4.33).

Singular Normal Distribution

Suppose that we have $\mathrm{rank}(\Sigma) = k < p$, where p is the dimension of X. We define the (singular) density of X with the aid of the G-Inverse Σ^- of Σ,

$$f(x) = \frac{(2\pi)^{-k/2}}{(\lambda_1 \cdots \lambda_k)^{1/2}} \exp\left\{-\frac{1}{2}(x-\mu)^\top \Sigma^-(x-\mu)\right\} \tag{4.53}$$

where

1. x lies on the hyperplane $\mathcal{N}^\top(x - \mu) = 0$ with $\mathcal{N}(p \times (p - k)) : \mathcal{N}^\top \Sigma = 0$ and $\mathcal{N}^\top \mathcal{N} = \mathcal{I}_k$.
2. Σ^- is the G-Inverse of Σ, and $\lambda_1, \ldots, \lambda_k$ are the nonzero eigenvalues of Σ.

What is the connection to a multinormal with k-dimensions? If

$$Y \sim N_k(0, \Lambda_1) \quad \text{and} \quad \Lambda_1 = \text{diag}(\lambda_1, \ldots, \lambda_k), \tag{4.54}$$

then an orthogonal matrix $\mathcal{B}(p \times k)$ with $\mathcal{B}^\top \mathcal{B} = \mathcal{I}_k$ exists that means $X = \mathcal{B}Y + \mu$ where X has a singular pdf of the form (4.53).

Gaussian Copula

In Examples 4.3 and 4.4, we have introduced copulae. Another important copula is the *Gaussian* or *normal copula*,

$$C_\rho(u, v) = \int_{-\infty}^{\Phi_1^{-1}(u)} \int_{-\infty}^{\Phi_2^{-1}(v)} f_\rho(x_1, x_2) dx_2 dx_1 \,, \tag{4.55}$$

see Embrechts et al. (1999). In (4.55), f_ρ denotes the bivariate normal density function with correlation ρ for $n = 2$. The functions Φ_1 and Φ_2 in (4.55) refer to the corresponding one-dimensional standard normal cdfs of the marginals.

In the case of vanishing correlation, $\rho = 0$, the Gaussian copula becomes

$$C_0(u, v) = \int_{-\infty}^{\Phi_1^{-1}(u)} f_{X_1}(x_1) dx_1 \int_{-\infty}^{\Phi_2^{-1}(v)} f_{X_2}(x_2) dx_2$$
$$= u \, v$$
$$= \Pi(u, v) \,.$$

Summary
$\hookrightarrow$ The pdf of a p-dimensional multinormal $X \sim N_p(\mu, \Sigma)$ is $$f(x) = \lvert 2\pi \Sigma \rvert^{-1/2} \exp\left\{-\frac{1}{2}(x - \mu)^\top \Sigma^{-1}(x - \mu)\right\}.$$ The contour curves of a multinormal are ellipsoids with half-lengths proportional to $\sqrt{\lambda_i}$, where λ_i denotes the eigenvalues of Σ ($i = 1, \ldots, p$).
$\hookrightarrow$ The Mahalanobis transformation transforms $X \sim N_p(\mu, \Sigma)$ to $Y = \Sigma^{-1/2}(X - \mu) \sim N_p(0, \mathcal{I}_p)$. Going in the other direction, one can create a $X \sim N_p(\mu, \Sigma)$ from $Y \sim N_p(0, \mathcal{I}_p)$ via $X = \Sigma^{1/2}Y + \mu$.
$\hookrightarrow$ If the covariance matrix Σ is singular (i.e., rank$(\Sigma) < p$), then it defines a singular normal distribution.

Continued on next page

Summary (continue)
$\hookrightarrow$ The Gaussian copula is given by $$C_\rho(u, v) = \int_{-\infty}^{\Phi_1^{-1}(u)} \int_{-\infty}^{\Phi_2^{-1}(v)} f_\rho(x_1, x_2) dx_2 dx_1 \ .$$
$\hookrightarrow$ The density of a singular normal distribution is given by $$\frac{(2\pi)^{-k/2}}{(\lambda_1 \cdots \lambda_k)^{1/2}} \exp\left\{ -\frac{1}{2}(x - \mu)^\top \Sigma^-(x - \mu) \right\} \ .$$

4.5   Sampling Distributions and Limit Theorems

In multivariate statistics, we observe the values of a multivariate random variable X and obtain a sample $\{x_i\}_{i=1}^n$, as described in Chap. 3. Under random sampling, these observations are considered to be realizations of a sequence of i.i.d. random variables $X_1, \ldots, X_n$, where each X_i is a p-variate random variable which replicates the *parent* or *population* random variable X. Some notational confusion is hard to avoid: X_i is not the i-th component of X, but rather the i-th replicate of the p-variate random variable X which provides the i-th observation x_i of our sample.

For a given random sample $X_1, \ldots, X_n$, the idea of statistical inference is to analyze the properties of the population variable X. This is typically done by analyzing some characteristic θ of its distribution like the mean, covariance matrix, etc. Statistical inference in a multivariate setup is considered in more detail in Chaps. 6 and 7.

Inference can often be performed using some observable function of the sample $X_1, \ldots, X_n$, i.e., a *statistics*. Examples of such statistics were given in Chap. 3: the sample mean $\bar{x}$, the sample covariance matrix S. To get an idea of the relationship between a statistics and the corresponding population characteristic, one has to derive the sampling distribution of the statistic. The next example gives some insight into the relation of $(\bar{x}, S)$ to (μ, Σ).

Example 4.15   Consider an i.i.d. sample of n random vectors $X_i \in \mathbb{R}^p$ where $\mathsf{E}(X_i) = \mu$ and $\mathsf{Var}(X_i) = \Sigma$. The sample mean $\bar{x}$ and the covariance matrix S have already been defined in Sect. 3.3. It is easy to prove the following results:

$$\mathsf{E}(\bar{x}) \ = n^{-1} \sum_{i=1}^n \mathsf{E}(X_i) = \mu$$
$$\mathsf{Var}(\bar{x}) = n^{-2} \sum_{i=1}^n \mathsf{Var}(X_i) = n^{-1}\Sigma = \mathsf{E}(\bar{x}\,\bar{x}^\top) - \mu\mu^\top$$

$$E(S) = n^{-1}E\left\{ \sum_{i=1}^{n}(X_i - \overline{x})(X_i - \overline{x})^{\top} \right\}$$

$$= n^{-1}E\left\{ \sum_{i=1}^{n} X_i X_i^{\top} - n\,\overline{x}\,\overline{x}^{\top} \right\}$$

$$= n^{-1}\left\{ n\left(\Sigma + \mu\mu^{\top}\right) - n\left(n^{-1}\Sigma + \mu\mu^{\top}\right) \right\}$$

$$= \frac{n-1}{n}\Sigma.$$

This shows in particular that S is a biased estimator of Σ. By contrast, $S_u = \frac{n}{n-1}S$ is an unbiased estimator of Σ.

Statistical inference often requires more than just the mean and/or the variance of a statistic. We need the sampling distribution of the statistics to derive confidence intervals or to define rejection regions in hypothesis testing for a given significance level. Theorem 4.9 gives the distribution of the sample mean for a multinormal population.

Theorem 4.9 *Let* $X_1, \ldots, X_n$ *be i.i.d. with* $X_i \sim N_p(\mu, \Sigma)$. *Then* $\overline{x} \sim N_p$ $(\mu, n^{-1}\Sigma)$.

Proof $\overline{x} = n^{-1}\sum_{i=1}^{n} X_i$ is a linear combination of independent normal variables, so it has a normal distribution (see Chap. 5). The mean and the covariance matrix were given in the preceding example.

With multivariate statistics, the sampling distributions of the statistics are often more difficult to derive than in the preceding Theorem. In addition, they might be so complicated that approximations have to be used. These approximations are provided by limit theorems. Since they are based on asymptotic limits, the approximations are only valid when the sample size is large enough. In spite of this restriction, they make complicated situations rather simple. The following central limit theorem shows that even if the parent distribution is not normal, when the sample size n is large, the sample mean $\overline{x}$ has an approximate normal distribution.

Theorem 4.10 (Central Limit Theorem (CLT)) *Let* $X_1, X_2, \ldots, X_n$ *be i.i.d. with* $X_i \sim (\mu, \Sigma)$. *Then the distribution of* $\sqrt{n}(\overline{x} - \mu)$ *is asymptotically* $N_p(0, \Sigma)$, *i.e.,*

$$\sqrt{n}(\overline{x} - \mu) \xrightarrow{\mathcal{L}} N_p(0, \Sigma) \qquad as \quad n \longrightarrow \infty.$$

The symbol "$\xrightarrow{\mathcal{L}}$" denotes *convergence in distribution*, which means that the distribution function of the random vector $\sqrt{n}(\overline{x} - \mu)$ converges to the distribution function of $N_p(0, \Sigma)$.

Example 4.16 Assume that $X_1, \ldots, X_n$ are i.i.d. and that they have Bernoulli distributions, where $p = \frac{1}{2}$ (this means that $P(X_i = 1) = \frac{1}{2}$, $P(X_i = 0) = \frac{1}{2}$). Then $\mu = p = \frac{1}{2}$ and $\Sigma = p(1 - p) = \frac{1}{4}$. Hence,

Fig. 4.4 The CLT for
Bernoulli-distributed random
variables. Sample size $n = 5$
(up) and $n = 35$ (down)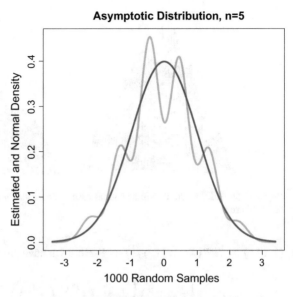
MVAcltbern

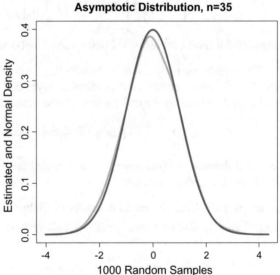

$$\sqrt{n}\left(\overline{x} - \frac{1}{2}\right) \xrightarrow{\mathcal{L}} N_1\left(0, \frac{1}{4}\right) \qquad \text{as} \quad n \longrightarrow \infty.$$

The results are shown in Fig. 4.4 for varying sample sizes.

Example 4.17 Now consider a two-dimensional random sample $X_1, \ldots, X_n$ that is
i.i.d. and created from two independent Bernoulli distributions with $p = 0.5$. The
joint distribution is given by $P(X_i = (0, 0)^\top) = \frac{1}{4}$, $P(X_i = (0, 1)^\top) = \frac{1}{4}$, $P(X_i =$

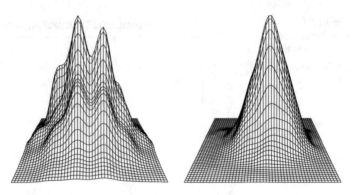

Fig. 4.5 The CLT in the two-dimensional case. Sample size $n = 5$ (left) and $n = 85$ (right)
MVAcltbern2

$(1, 0)^\top) = \frac{1}{4}$, $P(X_i = (1, 1)^\top) = \frac{1}{4}$. Here we have

$$\sqrt{n} \left\{ \overline{x} - \begin{pmatrix} \frac{1}{2} \\ \frac{1}{2} \end{pmatrix} \right\} = N_2 \left(\begin{pmatrix} 0 \\ 0 \end{pmatrix}, \begin{pmatrix} \frac{1}{4} & 0 \\ 0 & \frac{1}{4} \end{pmatrix} \right) \quad \text{as} \quad n \longrightarrow \infty.$$

Figure 4.5 displays the estimated two-dimensional density for different sample sizes.

The asymptotic normal distribution is often used to construct confidence intervals for the unknown parameters. A confidence interval at the level $1 - \alpha$, $\alpha \in (0, 1)$, is an interval that covers the true parameter with probability $1 - \alpha$:

$$P(\theta \in [\widehat{\theta}_l, \widehat{\theta}_u]) = 1 - \alpha,$$

where θ denotes the (unknown) parameter and $\widehat{\theta}_l$ and $\widehat{\theta}_u$ are the lower and upper confidence bounds, respectively.

Example 4.18 Consider the i.i.d. random variables $X_1, \ldots, X_n$ with $X_i \sim (\mu, \sigma^2)$ and σ^2 known. Since we have $\sqrt{n}(\overline{x} - \mu) \overset{\mathcal{L}}{\to} N(0, \sigma^2)$ from the CLT, it follows that

$$P(-u_{1-\alpha/2} \leq \sqrt{n} \frac{(\overline{x} - \mu)}{\sigma} \leq u_{1-\alpha/2}) \longrightarrow 1 - \alpha, \quad \text{as} \quad n \longrightarrow \infty$$

where $u_{1-\alpha/2}$ denotes the $(1 - \alpha/2)$-quantile of the standard normal distribution. Hence the interval

$$\left[\overline{x} - \frac{\sigma}{\sqrt{n}} u_{1-\alpha/2}, \overline{x} + \frac{\sigma}{\sqrt{n}} u_{1-\alpha/2} \right]$$

is an approximate $(1 - \alpha)$-confidence interval for μ.

But what can we do if we do not know the variance σ^2? The following corollary gives the answer.

Corollary 4.1 *If $\widehat{\Sigma}$ is a consistent estimate for Σ, then the CLT still holds, namely,*

$$\sqrt{n}\ \widehat{\Sigma}^{-1/2}(\overline{x} - \mu) \xrightarrow{\mathcal{L}} N_p(0, \mathcal{I}) \quad \text{as} \quad n \longrightarrow \infty.$$

Example 4.19 Consider the i.i.d. random variables $X_1, \ldots, X_n$ with $X_i \sim (\mu, \sigma^2)$, and now with an unknown variance σ^2. From Corollary 4.1 using $\widehat{\sigma}^2 = \frac{1}{n}\sum_{i=1}^{n}(x_i - \overline{x})^2$ we obtain

$$\sqrt{n}\left(\frac{\overline{x} - \mu}{\widehat{\sigma}}\right) \xrightarrow{\mathcal{L}} N(0, 1) \quad \text{as} \quad n \longrightarrow \infty.$$

Hence we can construct an approximate $(1 - \alpha)$-confidence interval for μ using the variance estimate $\widehat{\sigma}^2$:

$$C_{1-\alpha} = \left[\overline{x} - \frac{\widehat{\sigma}}{\sqrt{n}}u_{1-\alpha/2},\ \overline{x} + \frac{\widehat{\sigma}}{\sqrt{n}}u_{1-\alpha/2}\right].$$

Note that by the CLT

$$P(\mu \in C_{1-\alpha}) \longrightarrow 1 - \alpha \quad \text{as} \quad n \longrightarrow \infty.$$

Remark 4.1 One may wonder how large should n be in practice to provide reasonable approximations. There is no definite answer to this question: it mainly depends on the problem at hand (the shape of the distribution of the X_i and the dimension of X_i). If the X_i are normally distributed, the normality of $\overline{x}$ is achieved from $n = 1$. In most situations, however, the approximation is valid in one-dimensional problems for n larger than, say, 50.

Transformation of Statistics

Often in practical problems, one is interested in a function of parameters for which one has an asymptotically normal statistic. Suppose, for instance, that we are interested in a cost function depending on the mean μ of the process: $f(\mu) = \mu^\top A\mu$ where $A > 0$ is given. To estimate μ, we use the asymptotically normal statistic $\overline{x}$. The question is: how does $f(\overline{x})$ behave? More generally, what happens to a statistic t that is asymptotically normal when we transform it by a function $f(t)$? The answer is given by the following theorem.

Theorem 4.11 *If $\sqrt{n}(t - \mu) \xrightarrow{\mathcal{L}} N_p(0, \Sigma)$ and if $f = (f_1, \ldots, f_q)^\top : \mathbb{R}^p \to \mathbb{R}^q$ are real-valued functions which are differentiable at $\mu \in \mathbb{R}^p$, then $f(t)$ is asymptotically normal with mean $f(\mu)$ and covariance $\mathcal{D}^\top \Sigma \mathcal{D}$, i.e.,*

$$\sqrt{n}\{f(t) - f(\mu)\} \xrightarrow{\mathcal{L}} N_q(0, \mathcal{D}^\top \Sigma \mathcal{D}) \quad \text{for} \quad n \longrightarrow \infty, \quad (4.56)$$

where

$$D = \left(\frac{\partial f_j}{\partial t_i}\right)(t)\Big|_{t=\mu}$$

is the $(p \times q)$ matrix of all partial derivatives.

Example 4.20 We are interested in seeing how $f(\overline{x}) = \overline{x}^{\top} \mathcal{A} \overline{x}$ behaves asymptotically with respect to the quadratic cost function of μ, $f(\mu) = \mu^{\top} \mathcal{A} \mu$, where $\mathcal{A} > 0$.

$$D = \frac{\partial f(\overline{x})}{\partial \overline{x}}\Big|_{\overline{x}=\mu} = 2\mathcal{A}\mu.$$

By Theorem 4.11, we have

$$\sqrt{n}(\overline{x}^{\top} \mathcal{A} \overline{x} - \mu^{\top} \mathcal{A} \mu) \xrightarrow{\mathcal{L}} N_1 (0, 4\mu^{\top} \mathcal{A} \Sigma \mathcal{A} \mu).$$

Example 4.21 Suppose

$$X_i \sim (\mu, \Sigma); \quad \mu = \begin{pmatrix} 0 \\ 0 \end{pmatrix}, \quad \Sigma = \begin{pmatrix} 1 & 0.5 \\ 0.5 & 1 \end{pmatrix}, \quad p = 2.$$

We have by the CLT (Theorem 4.10) for $n \to \infty$ that

$$\sqrt{n}(\overline{x} - \mu) \xrightarrow{\mathcal{L}} N(0, \Sigma).$$

Suppose that we would like to compute the distribution of $\begin{pmatrix} \overline{x}_1^2 - \overline{x}_2 \\ \overline{x}_1 + 3\overline{x}_2 \end{pmatrix}$. According to Theorem 4.11 we have to consider $f = (f_1, f_2)^{\top}$ with

$$f_1(x_1, x_2) = x_1^2 - x_2, \quad f_2(x_1, x_2) = x_1 + 3x_2, \quad q = 2.$$

Given this $f(\mu) = \begin{pmatrix} 0 \\ 0 \end{pmatrix}$ and

$$D = (d_{ij}), \quad d_{ij} = \left(\frac{\partial f_j}{\partial x_i}\right)\Big|_{x=\mu} = \begin{pmatrix} 2x_1 & 1 \\ -1 & 3 \end{pmatrix}\Big|_{x=0}.$$

Thus

$$D = \begin{pmatrix} 0 & 1 \\ -1 & 3 \end{pmatrix}.$$

The covariance is

$$\underbrace{\begin{pmatrix} 0 & -1 \\ 1 & 3 \end{pmatrix}}_{D^{\top}} \underbrace{\begin{pmatrix} 1 & \frac{1}{2} \\ \frac{1}{2} & 1 \end{pmatrix}}_{\Sigma} \underbrace{\begin{pmatrix} 0 & 1 \\ -1 & 3 \end{pmatrix}}_{D} = \underbrace{\begin{pmatrix} 0 & -1 \\ 1 & 3 \end{pmatrix}}_{D^{\top}} \underbrace{\begin{pmatrix} -\frac{1}{2} & \frac{5}{2} \\ -1 & \frac{7}{2} \end{pmatrix}}_{\Sigma D} = \underbrace{\begin{pmatrix} 1 & -\frac{7}{2} \\ -\frac{7}{2} & 13 \end{pmatrix}}_{D^{\top} \Sigma D},$$

which yields

$$\sqrt{n}\begin{pmatrix} \overline{x}_1^2 - \overline{x}_2 \\ \overline{x}_1 + 3\overline{x}_2 \end{pmatrix} \xrightarrow{\mathcal{L}} N_2\left(\begin{pmatrix} 0 \\ 0 \end{pmatrix}, \begin{pmatrix} 1 & -\frac{7}{2} \\ -\frac{7}{2} & 13 \end{pmatrix}\right).$$

Example 4.22 Let us continue the previous example by adding one more component to the function f. Since $q = 3 > p = 2$, we might expect a singular normal distribution. Consider $f = (f_1, f_2, f_3)^\top$ with

$$f_1(x_1, x_2) = x_1^2 - x_2, \quad f_2(x_1, x_2) = x_1 + 3x_2, \quad f_3 = x_2^3, \quad q = 3.$$

From this we have that

$$\mathcal{D} = \begin{pmatrix} 0 & 1 & 0 \\ -1 & 3 & 0 \end{pmatrix} \quad \text{and thus} \quad \mathcal{D}^\top \Sigma \mathcal{D} = \begin{pmatrix} 1 & -\frac{7}{2} & 0 \\ -\frac{7}{2} & 13 & 0 \\ 0 & 0 & 0 \end{pmatrix}.$$

The limit is in fact a singular normal distribution!

Summary
$\hookrightarrow$ If $X_1, \ldots, X_n$ are i.i.d. random vectors with $X_i \sim N_p(\mu, \Sigma)$, then $\bar{x} \sim N_p(\mu, \frac{1}{n}\Sigma)$.
$\hookrightarrow$ If $X_1, \ldots, X_n$ are i.i.d. random vectors with $X_i \sim (\mu, \Sigma)$, then the distribution of $\sqrt{n}(\bar{x} - \mu)$ is asymptotically $N(0, \Sigma)$ (Central Limit Theorem).
$\hookrightarrow$ If $X_1, \ldots, X_n$ are i.i.d. random variables with $X_i \sim (\mu, \sigma)$, then an asymptotic confidence interval can be constructed by the CLT: $\bar{x} \pm \frac{\hat{\sigma}}{\sqrt{n}} u_{1-\alpha/2}$.
$\hookrightarrow$ If t is a statistic that is asymptotically normal, i.e., $\sqrt{n}(t - \mu) \xrightarrow{\mathcal{L}} N_p(0, \Sigma)$, then this holds also for a function $f(t)$, i.e., $\sqrt{n}\{f(t) - f(\mu)\}$ is asymptotically normal.

4.6 Heavy-Tailed Distributions

Heavy-tailed distributions were first introduced by the Italian-born Swiss economist Pareto and extensively studied by Paul Lévy. Although in the beginning these distributions were mainly studied theoretically, nowadays they have found many applications in areas as diverse as finance, medicine, seismology, structural engineering. More concretely, they have been used to model returns of assets in financial markets, streamflow in hydrology, precipitation and hurricane damage in meteorology, earthquake prediction in seismology, pollution, material strength, teletraffic, and many others.

A distribution is called heavy-tailed if it has higher probability density in its tail area compared with a normal distribution with same mean μ and variance σ^2. Figure

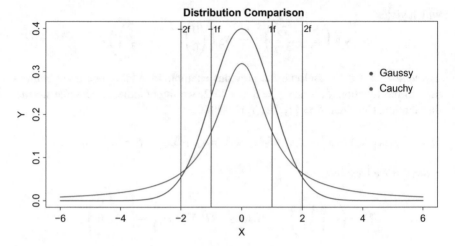

Fig. 4.6 Comparison of the pdf of a standard Gaussian (blue) and a Cauchy distribution (red) with location parameter 0 and scale parameter 1 [Q] MVAgausscauchy

4.6 demonstrates the differences of the pdf curves of a standard Gaussian distribution and a Cauchy distribution with location parameter $\mu = 0$ and scale parameter $\sigma = 1$. The graphic shows that the probability density of the Cauchy distribution is much higher than that of the Gaussian in the tail part, while in the area around the center, the probability density of the Cauchy distribution is much lower.

In terms of kurtosis, a heavy-tailed distribution has kurtosis greater than 3 (see Chap. 4, formula (4.40)), which is called leptokurtic, in contrast to mesokurtic distribution (kurtosis = 3) and platykurtic distribution (kurtosis < 3). Since univariate heavy-tailed distributions serve as basics for their multivariate counterparts and their density properties have been proved useful even in multivariate cases, we will start from introducing some univariate heavy-tailed distributions. Then we will move on to analyze their multivariate counterparts, and their tail-behavior.

Generalized Hyperbolic Distribution

The generalized hyperbolic distribution was introduced by Barndorff–Nielsen and at first applied to model grain size distributions of windblown sands. Today one of its most important uses is in stock price modeling and market risk measurement. The name of the distribution is derived from the fact that its log-density forms a hyperbola, while the log-density of the normal distribution is a parabola (Fig. 4.7).

The density of a one-dimensional generalized hyperbolic (GH) distribution for $x \in \mathbb{R}$ is

$$f_{\text{GH}}(x; \lambda, \alpha, \beta, \delta, \mu)$$
$$= \frac{(\sqrt{\alpha^2 - \beta^2}/\delta)^\lambda}{\sqrt{2\pi} \, K_\lambda(\delta\sqrt{\alpha^2 - \beta^2})} \frac{K_{\lambda-1/2}\{\alpha\sqrt{\delta^2 + (x - \mu)^2}\}}{\sqrt{\delta^2 + (x - \mu)^2}/\alpha)^{1/2-\lambda}} e^{\beta(x-\mu)} \quad (4.57)$$

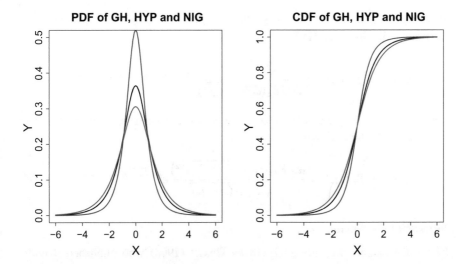

Fig. 4.7 pdf (left) and cdf (right) of GH($\lambda = 0.5$, black), HYP(red), and NIG (blue) with $\alpha = 1, \beta = 0, \delta = 1, \mu = 0$ ⊙ MVAghdis

where K_λ is a modified Bessel function of the third kind with index λ

$$K_\lambda(x) = \frac{1}{2} \int_0^\infty y^{\lambda-1} e^{-\frac{x}{2}(y+y^{-1})} dy \tag{4.58}$$

The domain of variation of the parameters is $\mu \in \mathbb{R}$ and

$$\begin{aligned} \delta \geq 0, |\beta| < \alpha, & \quad \text{if} \quad \lambda > 0 \\ \delta > 0, |\beta| < \alpha, & \quad \text{if} \quad \lambda = 0 \\ \delta > 0, |\beta| \leq \alpha, & \quad \text{if} \quad \lambda < 0 \end{aligned}$$

The generalized hyperbolic distribution has the following mean and variance:

$$\mathsf{E}(X) = \mu + \frac{\delta\beta}{\sqrt{\alpha^2 - \beta^2}} \frac{K_{\lambda+1}(\delta\sqrt{\alpha^2 - \beta^2})}{K_\lambda(\delta\sqrt{\alpha^2 - \beta^2})} \tag{4.59}$$

$$\mathsf{Var}(X) = \delta^2 \left[\frac{K_{\lambda+1}(\delta\sqrt{\alpha^2 - \beta^2})}{\delta\sqrt{\alpha^2 - \beta^2} K_\lambda(\delta\sqrt{\alpha^2 - \beta^2})} + \frac{\beta^2}{\alpha^2 - \beta^2} \left[\frac{K_{\lambda+2}(\delta\sqrt{\alpha^2 - \beta^2})}{K_\lambda(\delta\sqrt{\alpha^2 - \beta^2})} \right. \right.$$
$$\left. \left. - \left\{ \frac{K_{\lambda+1}(\delta\sqrt{\alpha^2 - \beta^2})}{K_\lambda(\delta\sqrt{\alpha^2 - \beta^2})} \right\}^2 \right] \right] \tag{4.60}$$

Where μ and δ play important roles in the density's location and scale, respectively. With specific values of λ, we obtain different subclasses of GH such as hyperbolic

(HYP) or normal-inverse Gaussian (NIG) distribution.   For $\lambda = 1$, we obtain the hyperbolic distributions (HYP)

$$f_{\mathrm{HYP}}(x; \alpha, \beta, \delta, \mu) = \frac{\sqrt{\alpha^2 - \beta^2}}{2\alpha\delta K_1(\delta\sqrt{\alpha^2 - \beta^2})} e^{\{-\alpha\sqrt{\delta^2 + (x-\mu)^2} + \beta(x-\mu)\}} \tag{4.61}$$

where $x, \mu \in \mathbb{R}$, $\delta \geq 0$ and $|\beta| < \alpha$. For $\lambda = -1/2$ we obtain the normal-inverse Gaussian distribution (NIG)

$$f_{\mathrm{NIG}}(x; \alpha, \beta, \delta, \mu) = \frac{\alpha\delta}{\pi} \frac{K_1\left(\alpha\sqrt{\delta^2 + (x-\mu)^2}\right)}{\sqrt{\delta^2 + (x-\mu)^2}} e^{\{\delta\sqrt{\alpha^2 - \beta^2} + \beta(x-\mu)\}} \tag{4.62}$$

Student's t-distribution

The t-distribution was first analyzed by Gosset (1908) who published it under pseudonym "Student" by request of his employer. Let X be a normally distributed random variable with mean μ and variance σ^2, and Y be the random variable such that Y^2/σ^2 has a chi-square distribution with n degrees of freedom. Assume that X and Y are independent, then

$$t \stackrel{\text{def}}{=} \frac{X\sqrt{n}}{Y} \tag{4.63}$$

is distributed as Student's t with n degrees of freedom. The t-distribution has the following density function :

$$f_t(x; n) = \frac{\Gamma\left(\frac{n+1}{2}\right)}{\sqrt{n\pi}\,\Gamma\left(\frac{n}{2}\right)} \left(1 + \frac{x^2}{n}\right)^{-\frac{n+1}{2}} \tag{4.64}$$

where n is the number of degrees of freedom, $-\infty < x < \infty$, and Γ is the gamma function:

$$\Gamma(\alpha) = \int_0^\infty x^{\alpha-1} e^{-x} dx. \tag{4.65}$$

The mean, variance, skewness, and kurtosis of Student's t-distribution ($n > 4$) are

$$\mu = 0$$
$$\sigma^2 = \frac{n}{n-2}$$
$$\text{Skewness} = 0$$
$$\text{Kurtosis} = 3 + \frac{6}{n-4}.$$

The t-distribution is symmetric around 0, which is consistent with the fact that its mean is 0 and skewness is also 0 (Fig. 4.8).

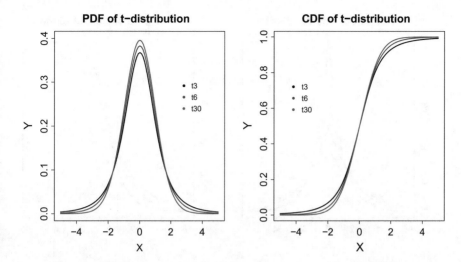

Fig. 4.8 pdf (left) and cdf (right) of t-distribution with different degrees of freedom (t3 stands for t-distribution with degree of freedom 3) 🔍 MVAtdis

Student's t-distribution approaches the normal distribution as n increases, since

$$\lim_{n \to \infty} f_t(x; n) = \frac{1}{\sqrt{2\pi}} e^{-\frac{x^2}{2}}. \tag{4.66}$$

In practice, the t-distribution is widely used, but its flexibility of modeling is restricted because of the integer-valued tail index.

In the tail area of the t-distribution, x is proportional to $|x|^{-(n+1)}$. In Fig. 4.13, we compared the tail-behavior of t-distribution with different degrees of freedom. With higher degree of freedom, the t-distribution decays faster.

Laplace Distribution

The univariate Laplace distribution with mean zero was introduced by Laplace (1774). The Laplace distribution can be defined as the distribution of differences between two independent variates with identical exponential distributions. Therefore, it is also called the double exponential distribution (Fig. 4.9).

The Laplace distribution with mean μ and scale parameter θ has the pdf

$$f_{\text{Laplace}}(x; \mu, \theta) = \frac{1}{2\theta} e^{-\frac{|x-\mu|}{\theta}} \tag{4.67}$$

and the cdf

$$F_{\text{Laplace}}(x; \mu, \theta) = \frac{1}{2}\left\{1 + \text{sign}(x - \mu)(1 - e^{-\frac{|x-\mu|}{\theta}})\right\}, \tag{4.68}$$

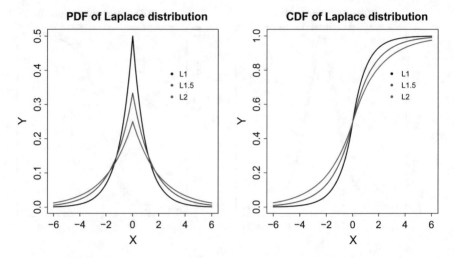

Fig. 4.9 pdf (left) and cdf (right) of Laplace distribution with zero mean and different scale parameters (L1 stands for Laplace distribution with $\theta = 1$) MVAlaplacedis

where sign is sign function. The mean, variance, skewness, and kurtosis of the Laplace distribution are

$$\mu = \mu$$
$$\sigma^2 = 2\theta^2$$
$$\text{Skewness} = 0$$
$$\text{Kurtosis} = 6$$

With mean 0 and $\theta = 1$, we obtain the standard Laplace distribution

$$f(x) = \frac{e^{-|x|}}{2} \tag{4.69}$$

$$F(x) = \begin{cases} \frac{e^x}{2} & \text{for } x < 0 \\ 1 - \frac{e^{-x}}{2} & \text{for } x \geq 0 \end{cases} \tag{4.70}$$

Cauchy Distribution

The Cauchy distribution is motivated by the following example.

Example 4.23 A gangster has just robbed a bank. As he runs to a point s meters away from the wall of the bank, a policeman reaches the crime scene behind the wall of the bank. The robber turns back and starts to shoot but he is such a poor shooter that the angle of his fire (marked in Fig. 4.10 as α) is uniformly distributed. The bullets hit the wall at distance x (from the center). Obviously the distribution of x, the random variable where the bullet hits the wall, is of vital knowledge to the policeman in order

Fig. 4.10 Introduction to Cauchy distribution—robber versus policeman

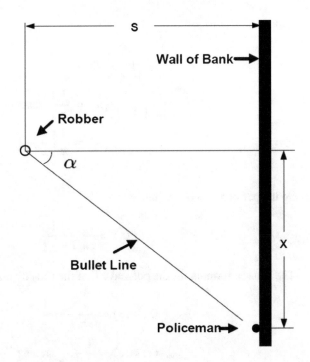

to identify the location of the gangster. (Should the policeman calculate the mean or the median of the observed bullet hits $\{x_i\}_{i=1}^n$ in order to identify the location of the robber?)

Since α is uniformly distributed:

$$f(\alpha) = \frac{1}{\pi} I(\alpha \in [-\pi/2, \pi/2])$$

and

$$\tan \alpha = \frac{x}{s}$$

$$\alpha = \arctan\left(\frac{x}{s}\right)$$

$$d\alpha = \frac{1}{s} \frac{1}{1 + (\frac{x}{s})^2} dx$$

For a small interval $d\alpha$, the probability is given by

$$f(\alpha)d\alpha = \frac{1}{\pi} d\alpha$$

$$= \frac{1}{s\pi} \frac{1}{1 + (\frac{x}{s})^2} dx$$

with

$$\int_{-\frac{\pi}{2}}^{\frac{\pi}{2}} \frac{1}{\pi} d\alpha = 1$$

$$\int_{-\infty}^{\infty} \frac{1}{s\pi} \frac{1}{1 + (\frac{x}{s})^2} dx = \frac{1}{\pi} \left\{ \arctan\left(\frac{x}{s}\right) \right\}_{-\infty}^{\infty}$$

$$= \frac{1}{\pi} \left\{ \frac{\pi}{2} - \left(-\frac{\pi}{2}\right) \right\}$$

$$= 1$$

So the pdf of x can be written as

$$f(x) = \frac{1}{s\pi} \frac{1}{1 + (\frac{x}{s})^2}$$

The general formula for the pdf and cdf of the Cauchy distribution is

$$f_{\text{Cauchy}}(x; m, s) = \frac{1}{s\pi} \frac{1}{1 + (\frac{x-m}{s})^2} \tag{4.71}$$

$$F_{\text{Cauchy}}(x; m, s) = \frac{1}{2} + \frac{1}{\pi} \arctan\left(\frac{x-m}{s}\right) \tag{4.72}$$

where m and s are location and scale parameter, respectively. The case in the above example where $m = 0$ and $s = 1$ is called the standard Cauchy distribution with pdf and cdf as following:

$$f_{\text{Cauchy}}(x) = \frac{1}{\pi(1 + x^2)} \tag{4.73}$$

$$F_{\text{Cauchy}}(x; m, s) = \frac{1}{2} + \frac{\arctan(x)}{\pi} \tag{4.74}$$

The mean, variance, skewness, and kurtosis of Cauchy distribution are all undefined, since its moment generating function diverges. But it has mode and median, both equal to the location parameter m (Fig. 4.11).

Mixture Model

Mixture modeling concerns modeling a statistical distribution by a mixture (or weighted sum) of different distributions. For many choices of component density functions, the mixture model can approximate any continuous density to arbitrary accuracy, provided that the number of component density functions is sufficiently large and the parameters of the model are chosen correctly. The pdf of a mixture distribution consists of L distributions and can be written as

$$f(x) = \sum_{l=1}^{L} w_l p_l(x) \tag{4.75}$$

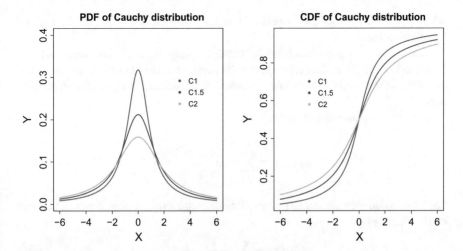

Fig. 4.11 pdf (left) and cdf (right) of Cauchy distribution with $m = 0$ and different scale parameters (C1 stands for Cauchy distribution with $s = 1$) MVAcauchy

under the constraints:

$$0 \leq w_l \leq 1$$

$$\sum_{l=1}^{L} w_l = 1$$

$$\int p_l(x)dx = 1$$

where $p_l(x)$ is the pdf of the l'th component density and w_l is a weight. The mean, variance, skewness, and kurtosis of a mixture are

$$\mu = \sum_{l=1}^{L} w_l \mu_l \qquad (4.76)$$

$$\sigma^2 = \sum_{l=1}^{L} w_l \{\sigma_l^2 + (\mu_l - \mu)^2\} \qquad (4.77)$$

$$\text{Skewness} = \sum_{l=1}^{L} w_l \left\{ \left(\frac{\sigma_l}{\sigma}\right)^3 SK_l + \frac{3\sigma_l^2(\mu_l - \mu)}{\sigma^3} + \left(\frac{\mu_l - \mu}{\sigma}\right)^3 \right\} \qquad (4.78)$$

$$\text{Kurtosis} = \sum_{l=1}^{L} w_l \left\{ \left(\frac{\sigma_l}{\sigma}\right)^4 K_l + \frac{6(\mu_l - \mu)^2\sigma_l^2}{\sigma^4} + \frac{4(\mu_l - \mu)\sigma_l^3}{\sigma^4} SK_l \right.$$

$$\left. + \left(\frac{\mu_l - \mu}{\sigma}\right)^4 \right\}, \qquad (4.79)$$

where μ_l, σ_l, SK_l and K_l are, respectively, mean, variance, skewness, and kurtosis of l'th distribution.

Mixture models are ubiquitous in virtually every facet of statistical analysis, machine learning, and data mining. For data sets comprising continuous variables, the most common approach involves mixture distributions having Gaussian components.

The pdf for a Gaussian mixture is

$$f_{GM}(x) = \sum_{l=1}^{L} \frac{w_l}{\sqrt{2\pi}\sigma_l} e^{-\frac{(x-\mu_l)^2}{2\sigma_l^2}}.$$ (4.80)

For a Gaussian mixture consisting of Gaussian distributions with mean 0, this can be simplified to

$$f_{GM}(x) = \sum_{l=1}^{L} \frac{w_l}{\sqrt{2\pi}\sigma_l} e^{-\frac{x^2}{2\sigma_l^2}},$$ (4.81)

with variance, skewness, and kurtosis

$$\sigma^2 = \sum_{l=1}^{L} w_l \sigma_l^2$$ (4.82)

$$\text{Skewness} = 0$$ (4.83)

$$\text{Kurtosis} = \sum_{l=1}^{L} w_l \left(\frac{\sigma_l}{\sigma}\right)^4 3$$ (4.84)

Example 4.24 Consider a Gaussian Mixture which is 80% N(0,1) and 20% N(0,9). The pdf of N(0,1) and N(0,9) are (Fig 4.12):

$$f_{N(0,1)}(x) = \frac{1}{\sqrt{2\pi}} e^{-\frac{x^2}{2}}$$

$$f_{N(0,9)}(x) = \frac{1}{3\sqrt{2\pi}} e^{-\frac{x^2}{18}}$$

so the pdf of the Gaussian Mixture is

$$f_{GM}(x) = \frac{1}{5\sqrt{2\pi}} \left(4e^{-\frac{x^2}{2}} + \frac{1}{3}e^{-\frac{x^2}{18}}\right)$$

Notice that the Gaussian Mixture is not a Gaussian distribution:

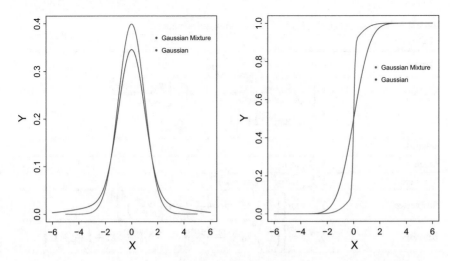

Fig. 4.12 pdf (left) and cdf (right) of a Gaussian mixture (Example 4.24) ◎ MVAmixture

Table 4.2 Basic statistics of t, Laplace, and Cauchy distribution

	t	Laplace	Cauchy
Mean	0	μ	Not defined
Variance	$\frac{n}{n-2}$	$2\theta^2$	Not defined
Skewness	0	0	Not defined
Kurtosis	$3 + \frac{6}{n-4}$	6	Not defined

$$\mu = 0$$
$$\sigma^2 = 0.8 \times 1 + 0.2 \times 9 = 2.6$$
$$\text{Skewness} = 0$$
$$\text{Kurtosis} = 0.8 \times \left(\frac{1}{\sqrt{2.6}}\right)^4 \times 3 + 0.2 \times \left(\frac{\sqrt{9}}{\sqrt{2.6}}\right)^4 \times 3 = 7.54$$

The kurtosis of this Gaussian mixture is higher than 3.

A summary of the basic statistics is given in Tables 4.2 and 4.3.

Multivariate Generalized Hyperbolic Distribution

The multivariate Generalized Hyperbolic Distribution (GH_d) has the following pdf:

Table 4.3 Basic statistics of GH distribution and mixture model

GH	
Mean	$\mu + \dfrac{\delta\beta}{\sqrt{\alpha^2+\beta^2}}\dfrac{K_{\lambda+1}(\delta\sqrt{\alpha^2+\beta^2})}{K_\lambda(\delta\sqrt{\alpha^2+\beta^2})}$
Variance	$\delta^2\left[\dfrac{K_{\lambda+1}(\delta\sqrt{\alpha^2+\beta^2})}{\delta\sqrt{\alpha^2+\beta^2}K_\lambda(\delta\sqrt{\alpha^2+\beta^2})} + \dfrac{\beta^2}{\alpha^2+\beta^2}\left[\dfrac{K_{\lambda+2}(\delta\sqrt{\alpha^2+\beta^2})}{K_\lambda(\delta\sqrt{\alpha^2+\beta^2})} - \left\{\dfrac{K_{\lambda+1}(\delta\sqrt{\alpha^2+\beta^2})}{K_\lambda(\delta\sqrt{\alpha^2+\beta^2})}\right\}^2\right]\right]$
Mixture	
Mean	$\sum_{l=1}^{L} w_l \mu_l$
Variance	$\sum_{l=1}^{L} w_l\{\sigma_l^2 + (\mu_l - \mu)^2\}$
Skewness	$\sum_{l=1}^{L} w_l\left\{\left(\dfrac{\sigma_l}{\sigma}\right)^3 SK_l + \dfrac{3\sigma_l^2(\mu_l-\mu)}{\sigma^3} + \left(\dfrac{\mu_l-\mu}{\sigma}\right)^3\right\}$
Kurtosis	$\sum_{l=1}^{L} w_l\left\{\left(\dfrac{\sigma_l}{\sigma}\right)^4 K_l + \dfrac{6(\mu_l-\mu)^2\sigma_l^2}{\sigma^4} + \dfrac{4(\mu_l-\mu)\sigma_l^3}{\sigma^4} SK_l + \left(\dfrac{\mu_l-\mu}{\sigma}\right)^4\right\}$

$$f_{\mathrm{GH}_d}(x; \lambda, \alpha, \beta, \delta, \Delta, \mu) = a_d \frac{K_{\lambda-\frac{d}{2}}\left\{\alpha\sqrt{\delta^2 + (x-\mu)^\top \Delta^{-1}(x-\mu)}\right\}}{\left\{\alpha^{-1}\sqrt{\delta^2 + (x-\mu)^\top \Delta^{-1}(x-\mu)}\right\}^{\frac{d}{2}-\lambda}} e^{\beta^\top(x-\mu)} \quad (4.85)$$

$$a_d = a_d(\lambda, \alpha, \beta, \delta, \Delta) = \frac{\left(\sqrt{\alpha^2 - \beta^\top \Delta\beta}/\delta\right)^\lambda}{(2\pi)^{\frac{d}{2}} K_\lambda(\delta\sqrt{\alpha^2 - \beta^\top \Delta\beta}}, \quad (4.86)$$

and characteristic function

$$\phi(t) = \left(\frac{\alpha^2 - \beta^\top \Delta\beta}{\alpha^2 - \beta^\top \Delta\beta + \frac{1}{2}t^\top \Delta t - i\beta^\top \Delta t}\right)^{\frac{\lambda}{2}}$$

$$\times \frac{K_\lambda\left(\delta\sqrt{\alpha^2 - \beta^\top \Delta\beta^\top + \frac{1}{2}t^\top \Delta t - i\beta^\top \Delta t}\right)}{K_\lambda\left(\delta\sqrt{\alpha^2 - \beta^\top \Delta\beta^\top}\right)} \quad (4.87)$$

These parameters have the following domain of variation:

$$\begin{aligned} &\lambda \in \mathbb{R}, \qquad \beta, \mu \in \mathbb{R}^d \\ &\delta > 0, \qquad \alpha > \beta^\top \Delta\beta \\ &\Delta \in \mathbb{R}^{d\times d} \text{ positive definite matrix} \\ &|\Delta| = 1 \end{aligned}$$

For $\lambda = \frac{d+1}{2}$, we obtain the multivariate hyperbolic (HYP) distribution; for $\lambda = -\frac{1}{2}$ we get the multivariate normal- inverse Gaussian (NIG) distribution.

Blæsild and Jensen (1981) introduced a second parameterization (ζ, Π, Σ), where

$$\zeta = \delta\sqrt{\alpha^2 - \beta^\top \Delta\beta} \tag{4.88}$$

$$\Pi = \beta\sqrt{\frac{\Delta}{\alpha^2 - \beta^\top \Delta\beta}} \tag{4.89}$$

$$\Sigma = \delta^2 \Delta \tag{4.90}$$

The mean and variance of $X \sim GH_d$

$$\mathsf{E}(X) = \mu + \delta R_\lambda(\zeta)\Pi\Delta^{\frac{1}{2}} \tag{4.91}$$

$$\mathsf{Var}(X) = \delta^2\{\zeta^{-1}R_\lambda(\zeta)\Delta + S_\lambda(\zeta)(\Pi\Delta^{\frac{1}{2}})^\top(\Pi\Delta^{\frac{1}{2}})\} \tag{4.92}$$

where

$$R_\lambda(x) = \frac{K_{\lambda+1}(x)}{K_\lambda(x)} \tag{4.93}$$

$$S_\lambda(x) = \frac{K_{\lambda+2}(x)K_\lambda(x) - K_{\lambda+1}^2(x)}{K_\lambda^2(x)} \tag{4.94}$$

Theorem 4.12 *Suppose that X is a d-dimensional variate distributed according to the generalized hyperbolic distribution GH_d. Let (X_1, X_2) be a partitioning of X, let r and k denote the dimensions of X_1 and X_2, respectively, and let (β_1, β_2) and (μ_1, μ_2) be similar partitions of β and μ, let*

$$\Delta = \begin{pmatrix} \Delta_{11} & \Delta_{12} \\ \Delta_{21} & \Delta_{22} \end{pmatrix} \tag{4.95}$$

be a partition of Δ such that Δ_{11} is a $r \times r$ matrix. Then one has the following:

1. *The distribution of X_1 is the r-dimensional generalized hyperbolic distribution, $GH_r(\lambda^*, \alpha^*, \beta^*, \delta^*, \mu^*, \Delta^*)$, where*

$$\lambda^* = \lambda$$
$$\alpha^* = |\Delta_{11}|^{-\frac{1}{2r}}\{\alpha^2 - \beta_2(\Delta_{22} - \Delta_{21}\Delta_{11}^{-1}\Delta_{12})\beta_2^\top\}^{\frac{1}{2}}$$
$$\beta^* = \beta_1 + \beta_2\Delta_{21}\Delta_{11}^{-1}$$
$$\delta^* = \delta|\Delta_{11}|^{\frac{1}{2r}}$$
$$\mu^* = \mu_1$$
$$\Delta^* = |\Delta|^{-\frac{1}{r}}\Delta_{11}$$

2. *The conditional distribution of X_2 given $X_1 = x_1$ is the k-dimensional generalized hyperbolic distribution $GH_k(\tilde{\lambda}, \tilde{\alpha}, \tilde{\beta}, \tilde{\delta}, \tilde{\mu}, \tilde{\Delta})$, where*

$$\tilde{\lambda} = \lambda - \frac{r}{2}$$

$$\tilde{\alpha} = \alpha |\Delta_{11}|^{\frac{1}{2k}}$$

$$\tilde{\beta} = \beta_2$$

$$\tilde{\delta} = |\Delta_{11}|^{-\frac{1}{2k}} \{\delta^2 + (x_1 - \mu_1)\Delta_{11}^{-1}(x_1 - \mu_1)^\top\}^{\frac{1}{2}}$$

$$\tilde{\mu} = \mu_2 + (x_1 - \mu_1)\Delta_{11}^{-1}\Delta_{12}$$

$$\tilde{\Delta} = |\Delta_{11}|^{\frac{1}{k}}(\Delta_{22} - \Delta_{21}\Delta_{11}^{-1}\Delta_{12})$$

3. Let $Y = XA + B$ be a regular affine transformation of X and let $||A||$ denote the absolute value of the determinant of A. The distribution of Y is the d-dimensional generalized hyperbolic distribution $GH_d(\lambda^+, \alpha^+, \beta^+, \delta^+, \mu^+, \Delta^+)$, where

$$\lambda^+ = \lambda$$

$$\alpha^+ = \alpha ||A||^{-\frac{1}{d}}$$

$$\beta^+ = \beta(A^{-1})^\top$$

$$\delta^+ = ||A||^{\frac{1}{d}}$$

$$\mu^+ = \mu A + B$$

$$\Delta^+ = ||A||^{-\frac{2}{d}} A^\top \Delta A$$

Multivariate t-distribution

If X and Y are independent and distributed as $N_p(\mu, \Sigma)$ and $\mathcal{X}_n^2$, respectively, and $X\sqrt{n/Y} = t - \mu$, then the pdf of t is given by

$$f_t(t; n, \Sigma, \mu) = \frac{\Gamma\{(n+p)/2\}}{\Gamma(n/2)n^{p/2}\pi^{p/2}|\Sigma|^{1/2}\{1 + \frac{1}{n}(t-\mu)^\top\Sigma^{-1}(t-\mu)\}^{(n+p)/2}} \tag{4.96}$$

The distribution of t is the noncentral t-distribution with n degrees of freedom and the noncentrality parameter μ, Giri (1996).

Multivariate Laplace Distribution

Let g and G be the pdf and cdf of a d-dimensional Gaussian distribution $N_d(0, \Sigma)$, the pdf and cdf of a multivariate Laplace distribution can be written as

$$f_{MLaplace_d}(x; m, \Sigma) = \int_0^\infty g(z^{-\frac{1}{2}}x - z^{\frac{1}{2}}m)z^{-\frac{d}{2}}e^{-z}dz \tag{4.97}$$

$$F_{MLaplace_d}(x, m, \Sigma) = \int_0^\infty G(z^{-\frac{1}{2}}x - z^{\frac{1}{2}}m)e^{-z}dz \tag{4.98}$$

the pdf can also be described as

$$f_{M\text{Laplace}_d}(x; m, \Sigma) = \frac{2e^{x^\top \Sigma^{-1} m}}{(2\pi)^{\frac{d}{2}}|\Sigma|^{\frac{1}{2}}}\left(\frac{x^\top \Sigma^{-1} x}{2 + m^\top \Sigma^{-1} m}\right)^{\frac{\lambda}{2}}$$
$$\times K_\lambda\left(\sqrt{(2 + m^\top \Sigma^{-1} m)(x^\top \Sigma^{-1} x)}\right) \qquad (4.99)$$

where $\lambda = \frac{2-d}{2}$ and $K_\lambda(x)$ is the modified Bessel function of the third kind

$$K_\lambda(x) = \frac{1}{2}\left(\frac{x}{2}\right)^\lambda \int_0^\infty t^{-\lambda-1} e^{-t-\frac{x^2}{4t}} dt, \qquad x > 0 \qquad (4.100)$$

Multivariate Laplace distribution has mean and variance

$$\mathsf{E}(X) = m \qquad (4.101)$$
$$\mathsf{Cov}(X) = \Sigma + mm^\top \qquad (4.102)$$

Multivariate Mixture Model

A multivariate mixture model comprises multivariate distributions, e.g., the pdf of a multivariate Gaussian distribution can be written as

$$f(x) = \sum_{l=1}^{L} \frac{w_l}{|2\pi \Sigma_l|^{\frac{1}{2}}} e^{-\frac{1}{2}(x-\mu_l)^\top \Sigma^{-1}(x-\mu_l)} \qquad (4.103)$$

Generalized Hyperbolic Distribution

The GH distribution has an exponential decaying speed

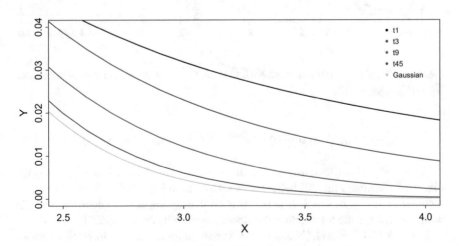

Fig. 4.13 Tail comparison of t-distributions (pdf) ⓠ MVAtdistail

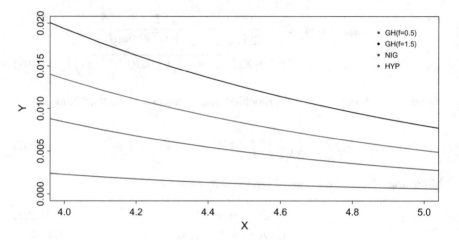

Fig. 4.14 Tail comparison of GH distributions (pdf) 〇 MVAghdistail

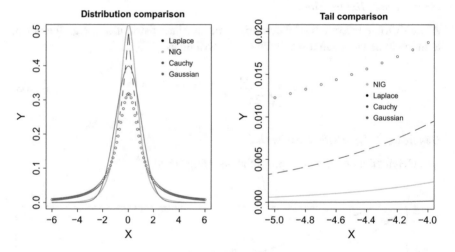

Fig. 4.15 Graphical comparison of the NIG, Laplace, Cauchy, and standard normal distribution
〇 MVAghadatail

$$f_{\text{GH}}(x; \lambda, \alpha, \beta, \delta, \mu = 0) \sim x^{\lambda-1} e^{-(\alpha-\beta)x} \qquad \text{as} \quad x \to \infty, \qquad (4.104)$$

As a comparison to tail-behavior of t-distributions depicted in Fig. 4.13, the
Fig. 4.14 illustrates the tail behavior of GH distributions with different values of
λ with $\alpha = 1$, $\beta = 0$, $\delta = 1$, $\mu = 0$. It is clear that among the four distributions, GH
with $\lambda = 1.5$ has the lowest decaying speed, while NIG decays fastest.

In Fig. 4.15, Chen et al. (2008), four distributions and especially their tail-behavior
are compared. In order to keep the comparability of these distributions, we speci-

fied the means to 0 and standardized the variances to 1. Furthermore, we used one important subclass of the GH distribution: the normal-inverse Gaussian (NIG) distribution with $\lambda = -\frac{1}{2}$ introduced above. On the left panel, the complete forms of these distributions are revealed. The Cauchy (dots) distribution has the lowest peak and the fattest tails. In other words, it has the flattest distribution. The NIG distribution decays second fast in the tails although it has the highest peak, which is more clearly displayed on the right panel.

4.7 Copulae

The cumulative distribution function (cdf) of a two-dimensional vector (X_1, X_2) is given by

$$F(x_1, x_2) = P(X_1 \le x_1, X_2 \le x_2). \tag{4.105}$$

For the case that X_1 and X_2 are independent, their joint cumulative distribution function $F(x_1, x_2)$ can be written as a product of their 1-dimensional marginals:

$$F(x_1, x_2) = F_{X_1}(x_1) F_{X_2}(x_2) = P(X_1 \le x_1) P(X_2 \le x_2). \tag{4.106}$$

But how can we model dependence of X_1 and X_2? Most people would suggest linear correlation. Correlation is though an appropriate measure of dependence only when the random variables have an elliptical or spherical distribution, which include the normal multivariate distribution. Although the terms "correlation" and "dependency" are often used interchangeably, correlation is actually a rather imperfect measure of dependency, and there are many circumstances where correlation should not be used.

Copulae represent an elegant concept of connecting marginals with joint cumulative distribution functions. Copulae are functions that join or "couple" multivariate distribution functions to their 1-dimensional marginal distribution functions. Let us consider a d-dimensional vector $X = (X_1, \ldots, X_d)^\top$. Using copulae, the marginal distribution functions $F_{X_i}(i = 1, \ldots, d)$ can be separately modeled from their dependence structure and then coupled together to form the multivariate distribution F_X. Copula functions have a long history in probability theory and statistics. Their application in finance is very recent. Copulae are important in Value-at-Risk calculations and constitute an essential tool in quantitative finance (Härdle et al. 2009).

First let us concentrate on the two-dimensional case, then we will extend this concept to the d-dimensional case, for a random variable in $\mathbb{R}^d$ with $d \ge 1$. To be able to define a copula function, first we need to represent a concept of the *volume of a rectangle, a 2-increasing function* and *a grounded function*.

Let U_1 and U_2 be two sets in $\overline{\mathbb{R}} = \mathbb{R} \cup \{+\infty\} \cup \{-\infty\}$ and consider the function $F : U_1 \times U_2 \longrightarrow \overline{\mathbb{R}}$.

Definition 4.2 The F-volume of a rectangle $B = [x_1, x_2] \times [y_1, y_2] \subset U_1 \times U_2$ is defined as

$$V_F(B) = F(x_2, y_2) - F(x_1, y_2) - F(x_2, y_1) + F(x_1, y_1) \qquad (4.107)$$

Definition 4.3 F is said to be a 2-increasing function if for every $B = [x_1, x_2] \times [y_1, y_2] \subset U_1 \times U_2$,

$$V_F(B) \geq 0 \qquad (4.108)$$

Remark 4.2 Note, that "to be 2-increasing function" neither implies nor is implied by "to be increasing in each argument".

The following lemmas (Nelsen 1999) will be very useful later for establishing the continuity of copulae.

Lemma 4.1 *Let U_1 and U_2 be nonempty sets in $\overline{\mathbb{R}}$ and let $F : U_1 \times U_2 \longrightarrow \overline{\mathbb{R}}$ be a 2-increasing function. Let x_1, x_2 be in U_1 with $x_1 \leq x_2$, and y_1, y_2 be in U_2 with $y_1 \leq y_2$. Then the function $t \mapsto F(t, y_2) - F(t, y_1)$ is nondecreasing on U_1 and the function $t \mapsto F(x_2, t) - F(x_1, t)$ is nondecreasing on U_2.*

Definition 4.4 If U_1 and U_2 have a smallest element $\min U_1$ and $\min U_2$, respectively, then we say, that a function $F : U_1 \times U_2 \longrightarrow \mathbb{R}$ is grounded if

$$\text{for all } x \in U_1 : \ F(x, \min U_2) = 0 \text{ and} \qquad (4.109)$$
$$\text{for all } y \in U_2 : \ F(\min U_1, y) = 0 \qquad (4.110)$$

In the following, we will refer to this definition of a cdf.

Definition 4.5 A cdf is a function from $\overline{\mathbb{R}}^2 \mapsto [0, 1]$ which

1. is grounded
2. is 2-increasing
3. satisfies $F(\infty, \infty) = 1$

Lemma 4.2 *Let U_1 and U_2 be nonempty sets in $\overline{\mathbb{R}}$ and let $F : U_1 \times U_2 \longrightarrow \overline{\mathbb{R}}$ be a grounded 2-increasing function. Then F is nondecreasing in each argument.*

Definition 4.6 If U_1 and U_2 have a greatest element $\max U_1$ and $\max U_2$, respectively, then we say, that a function $F : U_1 \times U_2 \longrightarrow \mathbb{R}$ has margins and that the margins of F are given by

$$F(x) = F(x, \max U_2) \text{ for all } x \in U_1 \qquad (4.111)$$
$$F(y) = F(\max U_1, y) \text{ for all } y \in U_2 \qquad (4.112)$$

Lemma 4.3 *Let U_1 and U_2 be nonempty sets in $\overline{\mathbb{R}}$ and let $F : U_1 \times U_2 \longrightarrow \overline{\mathbb{R}}$ be a grounded 2-increasing function which has margins. Let $(x_1, y_1), (x_2, y_2) \in S_1 \times S_2$. Then*

$$|F(x_2, y_2) - F(x_1, y_1)| \leq |F(x_2) - F(x_1)| + |F(y_2) - F(y_1)| \quad (4.113)$$

Definition 4.7 A two-dimensional copula is a function C defined on the unit square $I^2 = I \times I$ with $I = [0, 1]$ such that

1. for every $u \in I$ holds: $C(u, 0) = C(0, v) = 0$, i.e., C is grounded.
2. for every $u_1, u_2, v_1, v_2 \in I$ with $u_1 \leq u_2$ and $v_1 \leq v_2$ holds:

$$C(u_2, v_2) - C(u_2, v_1) - C(u_1, v_2) + C(u_1, v_1) \geq 0, \quad (4.114)$$

i.e., C is 2-increasing.
3. for every $u \in I$ holds $C(u, 1) = u$ and $C(1, v) = v$.

Informally, a copula is a joint distribution function defined on the unit square $[0, 1]^2$ which has uniform marginals. That means that if $F_{X_1}(x_1)$ and $F_{X_2}(x_2)$ are univariate distribution functions, then $C\{F_{X_1}(x_1), F_{X_2}(x_2)\}$ is a two-dimensional distribution function with marginals $F_{X_1}(x_1)$ and $F_{X_2}(x_2)$.

Example 4.25 The functions $\max(u + v - 1, 0), uv, \min(u, v)$ can be easily checked to be copula functions. They are called, respectively, the minimum, product, and maximum copula.

Example 4.26 Consider the function

$$C_\rho^{\text{Gauss}}(u, v) = \Phi_\rho\left\{\Phi^{-1}(u), \Phi^{-1}(v)\right\} \quad (4.115)$$

$$= \int_{-\infty}^{\Phi_1^{-1}(u)} \int_{-\infty}^{\Phi_2^{-1}(v)} f_\rho(x_1, x_2) dx_2 dx_1$$

where Φ_ρ is the joint two-dimensional standard normal distribution function with correlation coefficient ρ, while Φ_1 and Φ_2 refer to standard normal cdfs and

$$f_\rho(x_1, x_2) = \frac{1}{2\pi\sqrt{1 - \rho^2}} \exp\left\{-\frac{x_1^2 - 2\rho x_1 x_2 + x_2^2}{2(1 - \rho^2)}\right\} \quad (4.116)$$

denotes the bivariate normal pdf.

It is easy to see, that C^{Gauss} is a copula, the so-called Gaussian or normal copula, since it is 2-increasing and

$$\Phi_\rho\left\{\Phi^{-1}(u), \Phi^{-1}(0)\right\} = \Phi_\rho\left\{\Phi^{-1}(0), \Phi^{-1}(v)\right\} = 0 \quad (4.117)$$

$$\Phi_\rho\left\{\Phi^{-1}(u), \Phi^{-1}(1)\right\} = u \text{ and } \Phi_\rho\left\{\Phi^{-1}(1), \Phi^{-1}(v)\right\} = v \quad (4.118)$$

Fig. 4.16 Surface plot of the
Gumbel–Hougaard copula,
$\theta = 3$ 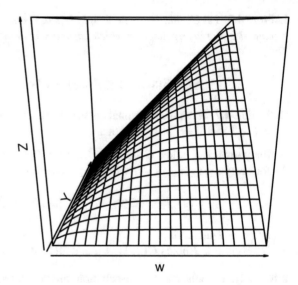 MVAghsurface

A simple and useful way to represent the graph of a copula is the contour diagram
that is, graphs of its level sets—the sets in I^2 given by $C(u, v) =$ a constant. In the
Figs. 4.16 and 4.17, we present the contour diagrams of the Gumbel–Hougard copula
(Example 4.4) for different values of the copula parameter θ.

For $\theta = 1$, the Gumbel–Hougaard copula reduces to the product copula, i.e.,

$$C_1^{\text{GH}}(u, v) = \Pi(u, v) = uv \tag{4.119}$$

For $\theta \to \infty$, one finds for the Gumbel–Hougaard copula:

$$C_\theta^{\text{GH}}(u, v) \longrightarrow \min(u, v) = M(u, v) \tag{4.120}$$

where M is also a copula such that $C(u, v) \le M(u, v)$ for an arbitrary copula C.
The copula M is called the Fréchet–Hoeffding upper bound.
The two-dimensional function $W(u, v) = \max(u + v - 1, 0)$ defines a copula with
$W(u, v) \le C(u, v)$ for any other copula C. W is called the Fréchet–Hoeffding lower
bound.

In Fig. 4.18, we show an example of Gumbel–Hougaard copula sampling for fixed
parameters $\sigma_1 = 1$, $\sigma_2 = 1$ and $\theta = 3$.

One can demonstrate the so-called Fréchet–Hoeffding inequality, which we have
already used in Example 1.3, and which states that each copula function is bounded
by the minimum and maximum one:

$$W(u, v) = \max(u + v - 1, 0) \le C(u, v) \le \min(u, v) = M(u, v) \tag{4.121}$$

The full relationship between copula and joint cdf depends on Sklar's theorem.

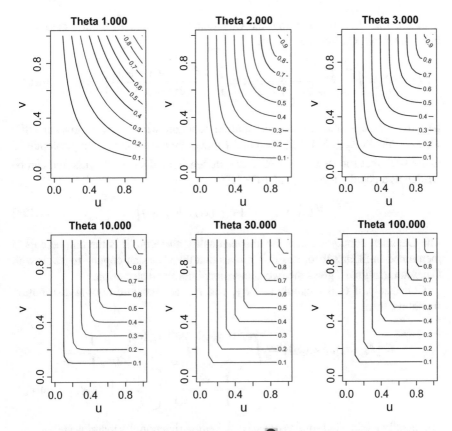

Fig. 4.17 Contour plots of the Gumbel–Hougard copula 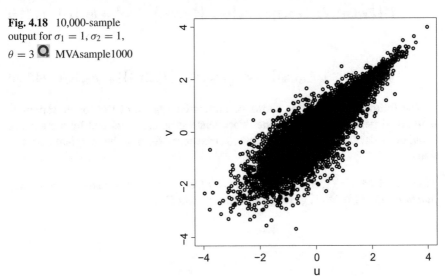 MVAghcontour

Fig. 4.18 10,000-sample output for $\sigma_1 = 1$, $\sigma_2 = 1$, $\theta = 3$ MVAsample1000

Example 4.27 Let us verify that the Gaussian copula satisfies Sklar's theorem in both directions. On the one side, let

$$F(x_1, x_2) = \int_{-\infty}^{x_1} \int_{-\infty}^{x_2} \frac{1}{2\pi \sqrt{1 - \rho^2}} \exp \left\{ -\frac{u_1^2 - 2\rho u_1 u_2 + u_2^2}{2(1 - \rho^2)} \right\} du_2 du_1. \quad (4.122)$$

be a two-dimensional normal distribution function with standard normal cdf's $F_{X_1}(x_1)$ and $F_{X_2}(x_2)$. Since $F_{X_1}(x_1)$ and $F_{X_2}(x_2)$ are continuous, a unique copula C exists such that for all $x_1, x_2 \in \overline{\mathbb{R}}^2$ a two-dimensional distribution function can be written as a copula in $F_{X_1}(x_1)$ and $F_{X_2}(x_2)$:

$$F(x_1, x_2) = C \left\{ \Phi_{X_1}(x_1), \Phi_{X_2}(x_2) \right\} \quad (4.123)$$

The Gaussian copula satisfies the above equality, therefore it is the unique copula mentioned in Sklar's theorem. This proves that the Gaussian copula, together with Gaussian marginals, gives the two-dimensional normal distribution.

Conversely, if C is a copula and F_{X_1} and F_{X_2} are standard normal distribution functions, then

$$C \left\{ F_{X_1}(x_1), F_{X_2}(x_2) \right\} = \int_{-\infty}^{\phi_1^{-1}\{F_{X_1}(x_1)\}} \int_{-\infty}^{\phi_2^{-1}\{F_{X_2}(x_2)\}} \frac{1}{2\pi \sqrt{1 - \rho^2}}$$
$$\times \exp \left\{ -\frac{x_1^2 - 2\rho x_1 x_2 + x_2^2}{2(1 - \rho^2)} \right\} dx_2 dx_1 \quad (4.124)$$

is evidently a joint (two-dimensional) distribution function. Its margins are

$$C \left\{ F_{X_1}(x_1), F_{X_2}(+\infty) \right\} = \Phi_\rho \left[\Phi^{-1} \left\{ F_{X_1}(x_1) \right\}, +\infty \right] = F_{X_1}(x_1) \quad (4.125)$$

$$C \left\{ F_{X_1}(+\infty), F_{X_2}(x_2) \right\} = \Phi_\rho \left[+\infty, \Phi^{-1} \left\{ F_{X_2}(x_2) \right\} \right] = F_{X_2}(x_2) \quad (4.126)$$

The following proposition shows one attractive feature of the copula representation of dependence, i.e., that the dependence structure described by a copula is invariant under increasing and continuous transformations of the marginal distributions.

Theorem 4.13 *If (X_1, X_2) have copula C and set g_1, g_2 two continuously increasing functions, then $\{g_1(X_1), g_2(X_2)\}$ have the copula C, too.*

Example 4.28 Independence implies that the product of the cdf's F_{X_1} and F_{X_2} equals the joint distribution function F, i.e.:

$$F(x_1, x_2) = F_{X_1}(x_1) F_{X_2}(x_2).$$ (4.127)

Thus, we obtain the independence or product copula $C = \Pi(u, v) = uv$.

While it is easily understood how a product copula describes an independence relationship, the converse is also true. Namely, the joint distribution function of two independent random variables can be interpreted as a product copula. This concept is formalized in the following theorem:

Theorem 4.14 *Let X_1 and X_2 be random variables with continuous distribution functions F_{X_1} and F_{X_2} and the joint distribution function F. Then X_1 and X_2 are independent if and only if $C_{X_1, X_2} = \Pi$.*

Example 4.29 Let us consider the Gaussian copula for the case $\rho = 0$, i.e., vanishing correlation. In this case, the Gaussian copula becomes

$$
\begin{aligned}
C_0^{\text{Gauss}}(u, v) &= \int_{-\infty}^{\Phi_1^{-1}(u)} \varphi(x_1) dx_1 \int_{-\infty}^{\Phi_2^{-1}(v)} \varphi(x_2) dx_2 \\
&= uv \\
&= \Pi(u, v).
\end{aligned}
$$ (4.128)

The following theorem, which follows directly from Lemma 4.3, establishes the continuity of copulae .

Theorem 4.15 *Let C be a copula. Then for any $u_1, v_1, u_2, v_2 \in I$ holds*

$$|C(u_2, v_2) - C(u_1, v_1)| \leq |u_2 - u_1| + |v_2 - v_1|$$ (4.129)

From (4.129), it follows that every copula C is uniformly continuous on its domain.

A further important property of copulae concerns the partial derivatives of a copula with respect to its variables:

Theorem 4.16 *Let $C(u, v)$ be a copula. For any $u \in I$, the partial derivative $\frac{\partial C(u,v)}{\partial v}$ exists for almost all $u \in I$. For such u and v one has*

$$\frac{\partial C(u, v)}{\partial v} \in I$$ (4.130)

The analogous statement is true for the partial derivative $\frac{\partial C(u,v)}{\partial u}$:

$$\frac{\partial C(u, v)}{\partial u} \in I$$ (4.131)

Moreover, the functions

$$u \mapsto C_v(u) \stackrel{\text{def}}{=} \partial C(u, v)/\partial v \text{ and}$$

$$v \mapsto C_u(v) \stackrel{\text{def}}{=} \partial C(u, v)/\partial u$$

are defined and nonincreasing almost everywhere on I.

Until now, we have considered copulae only in a two-dimensional setting. Let us now extend this concept to the d-dimensional case, for a random variable in $\mathbb{R}^d$ with $d \geq 1$.

Let $U_1, U_2, \ldots, U_d$ be nonempty sets in $\overline{\mathbb{R}}$ and consider the function $F : U_1 \times U_2 \times \ldots \times U_d \longrightarrow \overline{\mathbb{R}}$. For $a = (a_1, a_2, \ldots, a_d)$ and $b = (b_1, b_2, \ldots, b_d)$ with $a \leq b$ (i.e., $a_k \leq b_k$ for all k) let $B = [a, b] = [a_1, b_1] \times [a_2, b_2] \times \ldots \times [a_n, b_n]$ be the d-box with vertices $c = (c_1, c_2, \ldots, c_d)$. It is obvious, that each c_k is either equal to a_k or to b_k.

Definition 4.8 The F-volume of a d-box $B = [a, b] = [a_1, b_1] \times [a_2, b_2] \times \ldots \times [a_d, b_d] \subset U_1 \times U_2 \times \ldots \times U_d$ is defined as follows:

$$V_F(B) = \sum_{k=1}^{d} \text{sign}(c_k) F(c_k) \tag{4.132}$$

where $\text{sign}(c_k) = 1$, if $c_k = a_k$ for even k and $\text{sign}(c_k) = -1$, if $c_k = a_k$ for odd k.

Example 4.30 For the case $d = 3$, the F-volume of a 3-box $B = [a, b] = [x_1, x_2] \times [y_1, y_2] \times [z_1, z_2]$ is defined as

$$V_F(B) = F(x_2, y_2, z_2) - F(x_2, y_2, z_1) - F(x_2, y_1, z_2) - F(x_1, y_2, z_2)$$
$$+ F(x_2, y_1, z_1) + F(x_1, y_2, z_1) + F(x_1, y_1, z_2) - F(x_1, y_1, z_1)$$

Definition 4.9 F is said to be a d-increasing function if for all d-boxes B with vertices in $U_1 \times U_2 \times \ldots \times U_d$ holds:

$$V_F(B) \geq 0 \tag{4.133}$$

Definition 4.10 If $U_1, U_2, \ldots, U_d$ have a smallest element $\min U_1, \min U_2, \ldots, \min U_d$, respectively, then we say, that a function $F : U_1 \times U_2 \times \ldots \times U_d \longrightarrow \mathbb{R}$ is grounded if :

$$F(x) = 0 \text{ for all } x \in U_1 \times U_2 \times \ldots \times U_d \tag{4.134}$$

such that $x_k = \min U_k$ for at least one k.

The lemmas, which we presented for the two-dimensional case, have analogous multivariate versions, see Nelsen (1999).

Definition 4.11 A d-dimensional copula (or d-copula) is a function C defined on the unit d-cube $I^d = I \times I \times \ldots \times I$ such that

1. for every $u \in I^d$ holds: $C(u) = 0$, if at least one coordinate of u is equal to 0; i.e., C is grounded.
2. for every $a, b \in I^d$ with $a \leq b$ holds:

$$V_C([a, b]) \geq 0; \tag{4.135}$$

i.e., C is 2-increasing.
3. for every $u \in I^d$ holds: $C(u) = u_k$, if all coordinates of u are 1 except u_k.

Analogously to the two-dimensional setting, let us state the Sklar's theorem for the d-dimensional case.

Theorem 4.17 (Sklar's theorem in d-dimensional case) *Let F be a d-dimensional distribution function with marginal distribution functions $F_{X_1}, F_{X_2}, \ldots, F_{X_d}$. Then a d-copula C exists such that for all $x_1, \ldots, x_d \in \overline{\mathbb{R}}^d$:*

$$F(x_1, x_2, \ldots, x_d) = C\left\{F_{X_1}(x_1), F_{X_2}(x_2), \ldots, F_{X_d}(x_d)\right\} \tag{4.136}$$

Moreover, if $F_{X_1}, F_{X_2}, \ldots, F_{X_d}$ are continuous then C is unique. Otherwise C is uniquely determined on the Cartesian product $Im(F_{X_1}) \times Im(F_{X_2}) \times \ldots \times Im(F_{X_d})$.
Conversely, if C is a copula and $F_{X_1}, F_{X_2}, \ldots, F_{X_d}$ are distribution functions then F defined by (4.136) is a d-dimensional distribution function with marginals $F_{X_1}, F_{X_2}, \ldots, F_{X_d}$.

In order to illustrate the d-copulae we present the following examples:

Example 4.31 Let Φ denote the univariate standard normal distribution function and $\Phi_{\Sigma,d}$ the d-dimensional standard normal distribution function with correlation matrix Σ. Then the function

$$
\begin{aligned}
C_\rho^{\text{Gauss}}(u, \Sigma) &= \Phi_{\Sigma,d}\left\{\Phi^{-1}(u_1), \ldots, \Phi^{-1}(u_d)\right\} \\
&= \int_{-\infty}^{\phi_1^{-1}(u_d)} \ldots \int_{-\infty}^{\phi_2^{-1}(u_1)} f_\Sigma(x_1, \ldots, x_n) dx_1 \ldots dx_d
\end{aligned} \tag{4.137}
$$

is the d-dimensional Gaussian or normal copula with correlation matrix Σ. The function

$$
\begin{aligned}
f_\rho(x_1, \ldots, x_d) &= \frac{1}{\sqrt{\det(\Sigma)}} \\
&\times \exp\left\{-\frac{(\Phi^{-1}(u_1), \ldots, \Phi^{-1}(u_d))^\top (\Sigma^{-1} - \mathcal{I}_d)(\Phi^{-1}(u_1), \ldots, \Phi^{-1}(u_d))}{2}\right\}
\end{aligned} \tag{4.138}
$$

is a copula density function. The copula dependence parameter α is the collection of all unknown correlation coefficients in Σ. If $\alpha \neq 0$, then the corresponding normal copula allows to generate joint symmetric dependence. However, it is not possible to model a tail dependence, i.e., joint extreme events have a zero probability.

Example 4.32 Let us consider the following function:

$$C_\theta^{\mathrm{GH}}(u_1, \ldots, u_d) = \exp\left[-\left\{\sum_{j=1}^{d}\left(-\log u_j\right)^\theta\right\}^{1/\theta}\right] \tag{4.139}$$

One recognizes this function as the d-dimensional Gumbel–Hougaard copula function. Unlike the Gaussian copula, the copula (4.139) can generate an upper tail dependence.

Example 4.33 As in the two-dimensional setting, let us consider the d-dimensional Gumbel–Hougaard copula for the case $\theta = 1$. In this case the Gumbel–Hougaard copula reduces to the d-dimensional product copula, i.e.,

$$C_1^{\mathrm{GH}}(u_1, \ldots, u_d) = \prod_{j=1}^{d} u_j = \Pi^d(u) \tag{4.140}$$

The extension of the two-dimensional copula M, which one gets from the d-dimensional Gumbel–Hougaard copula for $\theta \to \infty$ is denoted $M^d(u)$:

$$C_\theta^{\mathrm{GH}}(u_1, \ldots u_d) \longrightarrow \min(u_1, \ldots, u_d) = M^d(u) \tag{4.141}$$

The d-dimensional function

$$W^d(u) = \max(u_1 + u_2 + \ldots + u_d - d + 1, 0) \tag{4.142}$$

defines a copula with $W(u) \leq C(u)$ for any other d-dimensional copula function $C(u)$. $W^d(u)$ is the Fréchet–Hoeffding lower bound in the d-dimensional case.

The functions M^d and Π^d are d-copulae for all $d \geq 2$, whereas the function W^d fails to be a d-copula for any $d > 2$ (Nelsen 1999). However, the d-dimensional version of the Fréchet–Hoeffding inequality can be written as follows:

$$W^d(u) \leq C(u) \leq M^d(u) \tag{4.143}$$

As we have already mentioned, copula functions have been widely applied in empirical finance.

4.8 Bootstrap

Recall that we need large sample sizes in order to sufficiently approximate the critical values computable by the CLT. Here large means $n > 50$ for one-dimensional data. How can we construct confidence intervals in the case of smaller sample sizes? One way is to use a method called the *bootstrap*. The bootstrap algorithm uses the data twice:

1. Estimate the parameter of interest,
2. Simulate from an estimated distribution to approximate the asymptotic distribution of the statistics of interest.

In detail, bootstrap works as follows. Consider the observations $x_1, \ldots, x_n$ of the sample $X_1, \ldots, X_n$ and estimate the empirical distribution function (edf) F_n. In the case of one-dimensional data

$$F_n(x) = \frac{1}{n} \sum_{i=1}^{n} \mathrm{I}(X_i \leq x). \tag{4.144}$$

This is a step function which is constant between neighboring data points.

Example 4.34 Suppose that we have $n = 100$ standard normal $N(0, 1)$ data points $X_i, i = 1, \ldots, n$. The cdf of X is $\Phi(x) = \int_{-\infty}^{x} \varphi(u) du$ and is shown in Fig. 4.19 as the thin, solid line. The empirical distribution function (edf) is displayed as a thick step function line. Figure 4.20 shows the same setup for $n = 1000$ observations.

Fig. 4.19 The standard normal cdf (thick line) and the empirical distribution function (thin line) for $n = 100$ **Q** MVAedfnormal

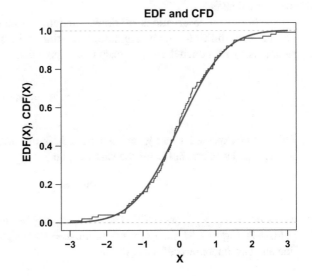

Fig. 4.20 The standard
normal cdf (thick line) and
the empirical distribution
function (thin line) for

$n = 1000$

MVAedfnormal

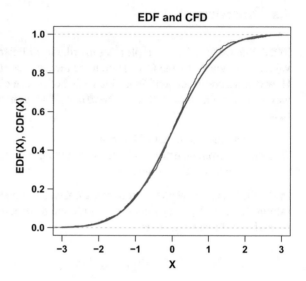

Now draw with replacement a new sample from this empirical distribution. That is
we sample with replacement n^* observations $X_1^*, \ldots, X_{n^*}^*$ from the original sample.
This is called a bootstrap sample. Usually one takes $n^* = n$.

Since we sample with replacement, a single observation from the original sample
may appear several times in the bootstrap sample. For instance, if the original sample
consists of the three observations x_1, x_2, x_3, then a bootstrap sample might look
like $X_1^* = x_3, X_2^* = x_2, X_3^* = x_3$. Computationally, we find the bootstrap sample by
using a uniform random number generator to draw from the indices $1, 2, \ldots, n$ of
the original samples.

The bootstrap observations are drawn randomly from the empirical distribution,
i.e., the probability for each original observation to be selected into the bootstrap
sample is $1/n$ for each draw. It is easy to compute that

$$\mathsf{E}_{F_n}(X_i^*) = \frac{1}{n}\sum_{i=1}^{n} x_i = \bar{x}.$$

This is the expected value given that the cdf is the original mean of the sample
$x_1 \ldots, x_n$. The same holds for the variance, i.e.,

$$\mathsf{Var}_{F_n}(X_i^*) = \widehat{\sigma}^2,$$

where $\widehat{\sigma}^2 = n^{-1}\sum_{i=1}^{n}(x_i - \bar{x})^2$. The cdf of the bootstrap observations is defined as
in (4.144). Figure 4.21 shows the cdf of the $n = 100$ original observations as a thick
line and two bootstrap cdf's as thin lines.

Fig. 4.21 The cdf F_n (thick line) and two bootstrap cdf's F_n^* (thin lines) 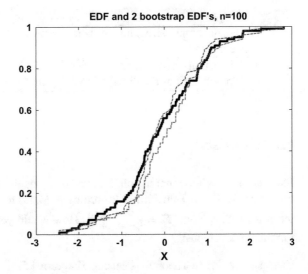 MVAedfbootstrap

The CLT holds for the bootstrap sample. Analogously to Corollary 4.1, we have the following corollary.

Corollary 4.2 *If $X_1^*, \ldots, X_n^*$ is a bootstrap sample from $X_1, \ldots, X_n$, then the distribution of*

$$\sqrt{n}\left(\frac{\overline{x}^* - \overline{x}}{\widehat{\sigma}^*}\right)$$

also becomes $N(0, 1)$ asymptotically, where $\overline{x}^* = n^{-1} \sum_{i=1}^{n} X_i^*$ and $(\widehat{\sigma}^*)^2 = n^{-1} \sum_{i=1}^{n}(X_i^* - \overline{x}^*)^2$.

How do we find a confidence interval for μ using the bootstrap method? Recall that the quantile $u_{1-\alpha/2}$ might be bad for small sample sizes because the true distribution of $\sqrt{n}\left(\frac{\overline{x}-\mu}{\widehat{\sigma}}\right)$ might be far away from the limit distribution $N(0, 1)$. The bootstrap idea enables us to "simulate" this distribution by computing $\sqrt{n}\left(\frac{\overline{x}^*-\overline{x}}{\widehat{\sigma}^*}\right)$ for *many* bootstrap samples . In this way, we can estimate an empirical $(1 - \alpha/2)$-quantile $u_{1-\alpha/2}^*$. The bootstrap improved confidence interval is then

$$C_{1-\alpha}^* = \left[\overline{x} - \frac{\widehat{\sigma}}{\sqrt{n}} u_{1-\alpha/2}^*, \ \overline{x} + \frac{\widehat{\sigma}}{\sqrt{n}} u_{1-\alpha/2}^*\right].$$

By Corollary 4.2 we have

$$P(\mu \in C_{1-\alpha}^*) \longrightarrow 1 - \alpha \quad \text{as } n \to \infty,$$

but with an improved speed of convergence, see Hall (1992).

Summary
$\hookrightarrow$ For small sample sizes, the bootstrap improves the precision of the confidence interval.
$\hookrightarrow$ The bootstrap distribution $\mathcal{L}\left\{\sqrt{n}(\overline{x}^* - \overline{x})/\hat{\sigma}^*\right\}$ converges to the same asymptotic limit as the distribution $\mathcal{L}\left\{\sqrt{n}(\overline{x}^* - \overline{x})/\hat{\sigma}\right\}$.

4.9   Exercises

Exercise 4.1 Assume that the random vector Y has the following normal distribution: $Y \sim N_p(0, \mathcal{I})$. Transform it according to (4.49) to create $X \sim N(\mu, \Sigma)$ with mean $\mu = (3, 2)^\top$ and $\Sigma = \begin{pmatrix} 1 & -1.5 \\ -1.5 & 4 \end{pmatrix}$. How would you implement the resulting formula on a computer?

Exercise 4.2 Prove Theorem 4.7 using Theorem 4.5.

Exercise 4.3 Suppose that X has mean zero and covariance $\Sigma = \begin{pmatrix} 1 & 0 \\ 0 & 2 \end{pmatrix}$. Let $Y = X_1 + X_2$. Write Y as a linear transformation, i.e., find the transformation matrix $\mathcal{A}$. Then compute $\text{Var}(Y)$ via (4.26). Can you obtain the result in another fashion?

Exercise 4.4 Calculate the mean and the variance of the estimate $\hat{\beta}$ in (3.50).

Exercise 4.5 Compute the conditional moments $E(X_2|x_1)$ and $E(X_1|x_2)$ for the pdf of Example 4.5.

Exercise 4.6 Prove the relation (4.28).

Exercise 4.7 Prove the relation (4.29).
Hint: Note that $\text{Var}(E(X_2|X_1)) = E(E(X_2|X_1) E(X_2^\top|X_1)) - E(X_2) E(X_2^\top))$ and that
$E(\text{Var}(X_2|X_1)) = E[E(X_2 X_2^\top|X_1) - E(X_2|X_1) E(X_2^\top|X_1)]$.

Exercise 4.8 Compute (4.46) for the pdf of Example 4.5.

Exercise 4.9

Show that $f_Y(y) = \begin{cases} \frac{1}{2}y_1 - \frac{1}{4}y_2 & 0 \le y_1 \le 2, \ |y_2| \le 1 - |1 - y_1| \\ 0 & \text{otherwise} \end{cases}$ is a pdf.

Exercise 4.10 Compute (4.46) for a two-dimensional standard normal distribution. Show that the transformed random variables Y_1 and Y_2 are independent. Give a geometrical interpretation of this result based on iso-distance curves.

Exercise 4.11 Consider the Cauchy distribution which has no moment, so that the CLT cannot be applied. Simulate the distribution of $\overline{x}$ (for different n's). What can you expect for $n \to \infty$?

Hint: The Cauchy distribution can be simulated by the quotient of two independent standard normally distributed random variables.

Exercise 4.12 A European car company has tested a new model and reports the consumption of petrol (X_1) and oil (X_2). The expected consumption of petrol is 8 liters per 100 km (μ_1) and the expected consumption of oil is 1 liter per 10,000 km (μ_2). The measured consumption of petrol is 8.1 liters per 100 km ($\overline{x}_1$) and the measured consumption of oil is 1.1 liters per 10,000 km ($\overline{x}_2$). The asymptotic distribution of $\sqrt{n} \left\{ \binom{\overline{x}_1}{\overline{x}_2} - \binom{\mu_1}{\mu_2} \right\}$ is $N\left(\binom{0}{0}, \binom{0.1\ \ 0.05}{0.05\ \ 0.1} \right)$.

For the American market, the basic measuring units are miles (1 mile $\approx$ 1.6 km) and gallons (1 gallon $\approx$ 3.8 liter). The consumptions of petrol (Y_1) and oil (Y_2) are usually reported in miles per gallon. Can you express $\overline{y}_1$ and $\overline{y}_2$ in terms of $\overline{x}_1$ and $\overline{x}_2$? Recompute the asymptotic distribution for the American market.

Exercise 4.13 Consider the pdf $f(x_1, x_2) = e^{-(x_1+x_2)}$, $x_1, x_2 > 0$ and let $U_1 = X_1 + X_2$ and $U_2 = X_1 - X_2$. Compute $f(u_1, u_2)$.

Exercise 4.14 Consider the pdf's

$$
\begin{aligned}
f(x_1, x_2) &= 4x_1 x_2 e^{-x_1^2} & x_1, x_2 > 0, \\
f(x_1, x_2) &= 1 & 0 < x_1, x_2 < 1 \text{ and } x_1 + x_2 < 1 \\
f(x_1, x_2) &= \tfrac{1}{2} e^{-x_1} & x_1 > |x_2|.
\end{aligned}
$$

For each of these pdf's compute $E(X)$, $\text{Var}(X)$, $E(X_1|X_2)$, $E(X_2|X_1)$, $\text{Var}(X_1|X_2)$ and $\text{Var}(X_2|X_1)$.

Exercise 4.15 Consider the pdf $f(x_1, x_2) = \tfrac{3}{4} x_1^{-\frac{1}{2}}$, $0 < x_1 < x_2 < 1$. Compute $P(X_1 < 0.25)$, $P(X_2 < 0.25)$ and $P(X_2 < 0.25|X_1 < 0.25)$.

Exercise 4.16 Consider the pdf $f(x_1, x_2) = \tfrac{1}{2\pi}$, $0 < x_1 < 2\pi, 0 < x_2 < 1$. Let $U_1 = \sin X_1 \sqrt{-2 \log X_2}$ and $U_2 = \cos X_1 \sqrt{-2 \log X_2}$. Compute $f(u_1, u_2)$.

Exercise 4.17 Consider $f(x_1, x_2, x_3) = k(x_1 + x_2 x_3)$; $0 < x_1, x_2, x_3 < 1$.

(a) Determine k so that f is a valid pdf of $(X_1, X_2, X_3) = X$.
(b) Compute the (3×3) matrix Σ_X.
(c) Compute the (2×2) matrix of the conditional variance of (X_2, X_3) given $X_1 = x_1$.

Exercise 4.18 Let $X \sim N_2 \left(\binom{1}{2}, \binom{2\ \ a}{a\ \ 2} \right)$.

(a) Represent the contour ellipses for $a = 0$; $-\tfrac{1}{2}$; $+\tfrac{1}{2}$; 1.
(b) For $a = \tfrac{1}{2}$ find the regions of X centered on μ which cover the area of the true parameter with probability 0.90 and 0.95.

Exercise 4.19   Consider the pdf

$$f(x_1, x_2) = \frac{1}{8x_2} e^{-\left(\frac{x_1}{2x_2} + \frac{x_2}{4}\right)} \qquad x_1, x_2 > 0.$$

Compute $f(x_2)$ and $f(x_1|x_2)$. Also give the best approximation of X_1 by a function of X_2. Compute the variance of the error of the approximation.

Exercise 4.20   Prove Theorem 4.6.

References

P. Blæsild, J.L. Jensen, Multivariate distributions of hyperbolic type. Statistical Distributions in Scientific Work - Proceedings of the NATO Advanced Study Institute held at the Università degli studi di Trieste **4**, 45–66 (1981)

Y. Chen, W. Härdle, S.-O. Jeong, Nonparametric risk management with generalized hyperbolic distributions. J. Am. Stat. Assoc. **103**, 910–923 (2008)

P. Embrechts, A. McNeil, D. Straumann. Correlation and Dependence in Risk Management: Properties and Pitfalls. Preprint ETH Zürich (1999)

N.C. Giri, *Multivariate Statistical Analysis* (Marcel Dekker, New York, 1996)

W.S. Gosset, The probable error of a mean. Biometrika **6**, 1–25 (1908)

P. Hall, *The Bootstrap and Edgeworth Expansion Statistical Series* (Springer, New York, 1992)

W. Härdle, N. Hautsch, L. Overbeck, *Applied Quantitative Finance*, 2nd edn. (Springer, Heidelberg, 2009)

R.B. Nelsen, *An Introduction to Copulas* (Springer, New York, 1999)

Pierre-Simon Laplace. Mémoire sur la Probabilité des Causes par les événements. *Savants étranges*, Vol.6:621–656 (1774)

A. Sklar, Fonctions de répartition à *n* dimensions et leurs marges. Publ. Inst. Statist. Univ. Paris **8**, 229–231 (1959)

Chapter 5
Theory of the Multinormal

In the preceding chapter, we saw how the multivariate normal distribution comes into play in many applications. It is useful to know more about this distribution, since it is often a good approximate distribution in many situations. Another reason for considering the multinormal distribution relies on the fact that it has many appealing properties: it is stable under linear transforms, zero correlation corresponds to independence, the marginals and all the conditionals are also multivariate normal variates, etc. The mathematical properties of the multinormal make analyses much simpler.

In this chapter, we will first concentrate on the probabilistic properties of the multinormal, then we will introduce two "companion" distributions of the multinormal which naturally appear when sampling from a multivariate normal population: the Wishart and the Hotelling distributions. The latter is particularly important for most of the testing procedures proposed in Chap. 7.

5.1 Elementary Properties of the Multinormal

Let us first summarize some properties, which were already derived in the previous chapter.

- The pdf of $X \sim N_p(\mu, \Sigma)$ is

$$f(x) = |2\pi \Sigma|^{-1/2} \exp \left\{ -\frac{1}{2}(x - \mu)^\top \Sigma^{-1}(x - \mu) \right\}. \tag{5.1}$$

© Springer Nature Switzerland AG 2019
W. K. Härdle and L. Simar, *Applied Multivariate Statistical Analysis*,
https://doi.org/10.1007/978-3-030-26006-4_5

The expectation is $\mathsf{E}(X) = \mu$, the covariance can be calculated as $\mathsf{Var}(X) = \mathsf{E}(X - \mu)(X - \mu)^\top = \Sigma$.

- Linear transformations turn normal random variables into normal random variables. If $X \sim N_p(\mu, \Sigma)$ and $\mathcal{A}(p \times p), c \in \mathbb{R}^p$, then $Y = \mathcal{A}X + c$ is p-variate Normal, i.e.,

$$Y \sim N_p(\mathcal{A}\mu + c, \mathcal{A}\Sigma\mathcal{A}^\top). \tag{5.2}$$

- If $X \sim N_p(\mu, \Sigma)$, then the Mahalanobis transformation is

$$Y = \Sigma^{-1/2}(X - \mu) \sim N_p(0, \mathcal{I}_p) \tag{5.3}$$

and it holds that

$$Y^\top Y = (X - \mu)^\top \Sigma^{-1}(X - \mu) \sim \chi_p^2. \tag{5.4}$$

Often it is interesting to partition X into sub-vectors X_1 and X_2. The following theorem tells us how to correct X_2 to obtain a vector which is independent of X_1.

Theorem 5.1 *Let* $X = \binom{X_1}{X_2} \sim N_p(\mu, \Sigma)$, $X_1 \in \mathbb{R}^r$, $X_2 \in \mathbb{R}^{p-r}$. *Define* $X_{2.1} = X_2 - \Sigma_{21}\Sigma_{11}^{-1}X_1$ *from the partitioned covariance matrix*

$$\Sigma = \begin{pmatrix} \Sigma_{11} & \Sigma_{12} \\ \Sigma_{21} & \Sigma_{22} \end{pmatrix}.$$

Then

$$X_1 \sim N_r(\mu_1, \Sigma_{11}), \tag{5.5}$$
$$X_{2.1} \sim N_{p-r}(\mu_{2.1}, \Sigma_{22.1}) \tag{5.6}$$

are independent with

$$\mu_{2.1} = \mu_2 - \Sigma_{21}\Sigma_{11}^{-1}\mu_1, \quad \Sigma_{22.1} = \Sigma_{22} - \Sigma_{21}\Sigma_{11}^{-1}\Sigma_{12}. \tag{5.7}$$

Proof

$$X_1 = \mathcal{A}X \quad \text{with} \quad \mathcal{A} = (\mathcal{I}_r, 0)$$
$$X_{2.1} = \mathcal{B}X \quad \text{with} \quad \mathcal{B} = (-\Sigma_{21}\Sigma_{11}^{-1}, \mathcal{I}_{p-r}).$$

Then, by (5.2), X_1 and $X_{2.1}$ are both normal. Note that

$$\mathrm{Cov}(X_1, X_{2.1}) = A\Sigma B^\top = \begin{pmatrix} \begin{bmatrix} 1 & & 0 \\ & \ddots & \\ 0 & & 1 \end{bmatrix} & 0 \end{pmatrix} \begin{pmatrix} \Sigma_{11} & \Sigma_{12} \\ \Sigma_{21} & \Sigma_{22} \end{pmatrix} \begin{pmatrix} (-\Sigma_{21}\Sigma_{11}^{-1})^\top \\ \begin{bmatrix} 1 & & 0 \\ & \ddots & \\ 0 & & 1 \end{bmatrix} \end{pmatrix},$$

$$A\Sigma = (\mathcal{I}_r\ 0) \begin{pmatrix} \Sigma_{11} & \Sigma_{12} \\ \Sigma_{21} & \Sigma_{22} \end{pmatrix} = (\Sigma_{11}\ \ \Sigma_{12}),$$

hence,

$$A\Sigma B^\top = (\Sigma_{11}\ \ \Sigma_{12}) \begin{pmatrix} (-\Sigma_{21}\Sigma_{11}^{-1})^\top \\ \mathcal{I}_{p-r} \end{pmatrix} = \left(-\Sigma_{11} \left(\Sigma_{21}\Sigma_{11}^{-1} \right)^\top + \Sigma_{12} \right).$$

Recall that $\Sigma_{21} = (\Sigma_{12})^\top$. Hence $A\Sigma B^\top = -\Sigma_{11}\Sigma_{11}^{-1}\Sigma_{12} + \Sigma_{12} \equiv 0$.
Using (5.2) again, we also have the joint distribution of $(X_1, X_{2.1})$, namely,

$$\begin{pmatrix} X_1 \\ X_{2.1} \end{pmatrix} = \begin{pmatrix} A \\ B \end{pmatrix} X \sim N_p \left(\begin{pmatrix} \mu_1 \\ \mu_{2.1} \end{pmatrix}, \begin{pmatrix} \Sigma_{11} & 0 \\ 0 & \Sigma_{22.1} \end{pmatrix} \right).$$

With this block diagonal structure of the covariance matrix, the joint pdf of $(X_1, X_{2.1})$ can easily be factorized into

$$f(x_1, x_{2.1}) = |2\pi\Sigma_{11}|^{-\frac{1}{2}} \exp\left\{ -\frac{1}{2}(x_1 - \mu_1)^\top \Sigma_{11}^{-1}(x_1 - \mu_1) \right\} \times$$
$$|2\pi\Sigma_{22.1}|^{-\frac{1}{2}} \exp\left\{ -\frac{1}{2}(x_{2.1} - \mu_{2.1})^\top \Sigma_{22.1}^{-1}(x_{2.1} - \mu_{2.1}) \right\}$$

from which the independence between X_1 and $X_{2.1}$ follows.

The next two corollaries are direct consequences of Theorem 5.1.

Corollary 5.1 Let $X = \begin{pmatrix} X_1 \\ X_2 \end{pmatrix} \sim N_p(\mu, \Sigma)$, $\Sigma = \begin{pmatrix} \Sigma_{11} & \Sigma_{12} \\ \Sigma_{21} & \Sigma_{22} \end{pmatrix}$. $\Sigma_{12} = 0$ if and only if X_1 is independent of X_2.

The independence of two linear transforms of a multinormal X can be shown via the following corollary.

Corollary 5.2 If $X \sim N_p(\mu, \Sigma)$ and given some matrices A and B, then AX and BX are independent if and only if $A\Sigma B^\top = 0$.

The following theorem is also useful. It generalizes Theorem 4.6. The proof is left as an exercise.

Theorem 5.2 *If* $X \sim N_p(\mu, \Sigma)$, $\mathcal{A}(q \times p)$, $c \in \mathbb{R}^q$ *and* $q \leq p$, *then* $Y = \mathcal{A}X + c$
is a q-variate normal, i.e.,

$$Y \sim N_q(\mathcal{A}\mu + c, \mathcal{A}\Sigma\mathcal{A}^\top).$$

The conditional distribution of X_2 given X_1 is given by the next theorem.

Theorem 5.3 *The conditional distribution of* X_2 *given* $X_1 = x_1$ *is normal with mean*
$\mu_2 + \Sigma_{21}\Sigma_{11}^{-1}(x_1 - \mu_1)$ *and covariance* $\Sigma_{22.1}$, *i.e.,*

$$(X_2 \mid X_1 = x_1) \sim N_{p-r}(\mu_2 + \Sigma_{21}\Sigma_{11}^{-1}(x_1 - \mu_1), \Sigma_{22.1}). \tag{5.8}$$

Proof Since $X_2 = X_{2.1} + \Sigma_{21}\Sigma_{11}^{-1}X_1$, for a fixed value of $X_1 = x_1$, X_2 is equivalent
to $X_{2.1}$ plus a constant term:

$$(X_2|X_1 = x_1) = (X_{2.1} + \Sigma_{21}\Sigma_{11}^{-1}x_1),$$

which has the normal distribution $N(\mu_{2.1} + \Sigma_{21}\Sigma_{11}^{-1}x_1, \Sigma_{22.1})$.

Note that the conditional mean of $(X_2 \mid X_1)$ is a linear function of X_1 and that the
conditional variance does not depend on the particular value of X_1. In the following
example, we consider a specific distribution.

Example 5.1 Suppose that $p = 2, r = 1, \mu = \begin{pmatrix} 0 \\ 0 \end{pmatrix}$, and $\Sigma = \begin{pmatrix} 1 & -0.8 \\ -0.8 & 2 \end{pmatrix}$. Then
$\Sigma_{11} = 1$, $\Sigma_{21} = -0.8$ and $\Sigma_{22.1} = \Sigma_{22} - \Sigma_{21}\Sigma_{11}^{-1}\Sigma_{12} = 2 - (0.8)^2 = 1.36$. Hence
the marginal pdf of X_1 is

$$f_{X_1}(x_1) = \frac{1}{\sqrt{2\pi}} \exp\left(-\frac{x_1^2}{2}\right)$$

and the conditional pdf of $(X_2 \mid X_1 = x_1)$ is given by

$$f(x_2 \mid x_1) = \frac{1}{\sqrt{2\pi(1.36)}} \exp\left\{-\frac{(x_2 + 0.8x_1)^2}{2 \times (1.36)}\right\}.$$

As mentioned above, the conditional mean of $(X_2 \mid X_1)$ is linear in X_1. The shift in
the density of $(X_2 \mid X_1)$ can be seen in Fig. 5.1.

Sometimes it will be useful to reconstruct a joint distribution from the marginal
distribution of X_1 and the conditional distribution $(X_2|X_1)$. The following theorem
shows under which conditions this can be easily done in the multinormal framework.

Theorem 5.4 *If* $X_1 \sim N_r(\mu_1, \Sigma_{11})$ *and* $(X_2|X_1 = x_1) \sim N_{p-r}(\mathcal{A}x_1 + b, \Omega)$ *where*
Ω *does not depend on* x_1, *then* $X = \begin{pmatrix} X_1 \\ X_2 \end{pmatrix} \sim N_p(\mu, \Sigma)$, *where*

Fig. 5.1 Shifts in the conditional density 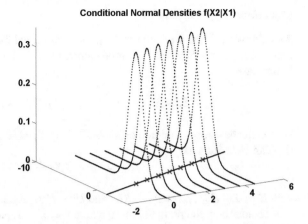 MVAcondnorm

$$\mu = \begin{pmatrix} \mu_1 \\ A\mu_1 + b \end{pmatrix}$$

$$\Sigma = \begin{pmatrix} \Sigma_{11} & \Sigma_{11}A^\top \\ A\Sigma_{11} & \Omega + A\Sigma_{11}A^\top \end{pmatrix}.$$

Example 5.2 Consider the following random variables:

$$X_1 \sim N_1(0, 1),$$

$$X_2 | X_1 = x_1 \sim N_2 \left(\begin{pmatrix} 2x_1 \\ x_1 + 1 \end{pmatrix}, \begin{pmatrix} 1 & 0 \\ 0 & 1 \end{pmatrix} \right).$$

Using Theorem (5.4), where $A = (2 \ 1)^\top$, $b = (0 \ 1)^\top$ and $\Omega = \mathcal{I}_2$, we easily obtain the following result:

$$X = \begin{pmatrix} X_1 \\ X_2 \end{pmatrix} \sim N_3 \left(\begin{pmatrix} 0 \\ 0 \\ 1 \end{pmatrix}, \begin{pmatrix} 1 & 2 & 1 \\ 2 & 5 & 2 \\ 1 & 2 & 2 \end{pmatrix} \right).$$

In particular, the marginal distribution of X_2 is

$$X_2 \sim N_2 \left(\begin{pmatrix} 0 \\ 1 \end{pmatrix}, \begin{pmatrix} 5 & 2 \\ 2 & 2 \end{pmatrix} \right),$$

thus conditional on X_1, the two components of X_2 are independent but marginally they are not.

Note that the marginal mean vector and covariance matrix of X_2 could have also been computed directly by using (4.28)–(4.29). Using the derivation above, however, provides us with useful properties: we have multinormality.

Conditional Approximations

As we saw in Chap. 4 (Theorem 4.3), the conditional expectation $\mathsf{E}(X_2|X_1)$ is the mean squared error (MSE) best approximation of X_2 by a function of X_1. We have in this case

$$X_2 = \mathsf{E}(X_2|X_1) + U = \mu_2 + \Sigma_{21}\Sigma_{11}^{-1}(X_1 - \mu_1) + U. \tag{5.9}$$

Hence, the best approximation of $X_2 \in \mathbb{R}^{p-r}$ by $X_1 \in \mathbb{R}^r$ is the linear approximation that can be written as

$$X_2 = \mathcal{B}_0 + \mathcal{B} X_1 + U \tag{5.10}$$

with $\mathcal{B} = \Sigma_{21}\Sigma_{11}^{-1}$, $\mathcal{B}_0 = \mu_2 - \mathcal{B}\mu_1$ and $U \sim N(0, \Sigma_{22.1})$.

Consider now the particular case where $r = p - 1$. Now $X_2 \in \mathbb{R}$ and $\mathcal{B}$ is a row vector $\beta^{\top}$ of dimension $(1 \times r)$

$$X_2 = \beta_0 + \beta^{\top} X_1 + U. \tag{5.11}$$

This means, geometrically speaking, that the best MSE approximation of X_2 by a function of X_1 is a hyperplane. The marginal variance of X_2 can be decomposed via (5.11):

$$\sigma_{22} = \beta^{\top}\Sigma_{11}\beta + \sigma_{22.1} = \sigma_{21}\Sigma_{11}^{-1}\sigma_{12} + \sigma_{22.1}. \tag{5.12}$$

The ratio

$$\rho_{2.1...r}^2 = \frac{\sigma_{21}\Sigma_{11}^{-1}\sigma_{12}}{\sigma_{22}} \tag{5.13}$$

is known as the square of the multiple correlation between X_2 and the r variables X_1. It is the percentage of the variance of X_2 which is explained by the linear approximation $\beta_0 + \beta^{\top} X_1$. The last term in (5.12) is the residual variance of X_2. The square of the multiple correlation corresponds to the coefficient of determination introduced in Sect. 3.4, see (3.39), but here it is defined in terms of the r.v. X_1 and X_2. It can be shown that $\rho_{2.1...r}$ is also the maximum correlation attainable between X_2 and a linear combination of the elements of X_1, the optimal linear combination being precisely given by $\beta^{\top} X_1$. Note, that when $r = 1$, the multiple correlation $\rho_{2.1}$ coincides with the usual simple correlation $\rho_{X_2 X_1}$ between X_2 and X_1.

Example 5.3 Consider the "classic blue" pullover example (Example 3.15) and suppose that X_1 (sales), X_2 (price), X_3 (advertisement) and X_4 (sales assistants) are normally distributed with

$$\mu = \begin{pmatrix} 172.7 \\ 104.6 \\ 104.0 \\ 93.8 \end{pmatrix} \text{ and } \Sigma = \begin{pmatrix} 1037.21 \\ -80.02 & 219.84 \\ 1430.70 & 92.10 & 2624.00 \\ 271.44 & -91.58 & 210.30 & 177.36 \end{pmatrix}.$$

(These are in fact the sample mean and the sample covariance matrix but in this example we pretend that they are the true parameter values.)

The conditional distribution of X_1 given (X_2, X_3, X_4) is thus a univariate normal with mean

$$\mu_1 + \sigma_{12}\Sigma_{22}^{-1}\begin{pmatrix} X_2 - \mu_2 \\ X_3 - \mu_3 \\ X_4 - \mu_4 \end{pmatrix} = 65.670 - 0.216X_2 + 0.485X_3 + 0.844X_4$$

and variance

$$\sigma_{11.2} = \sigma_{11} - \sigma_{12}\Sigma_{22}^{-1}\sigma_{21} = 96.761$$

The linear approximation of the sales (X_1) by the price (X_2), advertisement (X_3) and sales assistants (X_4) is provided by the conditional mean above.(Note that this coincides with the results of Example 3.15 due to the particular choice of μ and Σ). The quality of the approximation is given by the multiple correlation $\rho_{1.234}^2 = \frac{\sigma_{12}\Sigma_{22}^{-1}\sigma_{21}}{\sigma_{11}} = 0.907$. (Note again that this coincides with the coefficient of determination r^2 found in Example 3.15).

This example also illustrates the concept of partial correlation. The correlation matrix between the 4 variables is given by

$$P = \begin{pmatrix} 1 & -0.168 & 0.867 & 0.633 \\ -0.168 & 1 & 0.121 & -0.464 \\ 0.867 & 0.121 & 1 & 0.308 \\ 0.633 & -0.464 & 0.308 & 1 \end{pmatrix},$$

so that the correlation between X_1 (sales) and X_2 (price) is -0.168. We can compute the conditional distribution of (X_1, X_2) given (X_3, X_4), which is a bivariate normal with mean:

$$\begin{pmatrix} \mu_1 \\ \mu_2 \end{pmatrix} + \begin{pmatrix} \sigma_{13} & \sigma_{14} \\ \sigma_{23} & \sigma_{24} \end{pmatrix}\begin{pmatrix} \sigma_{33} & \sigma_{34} \\ \sigma_{43} & \sigma_{44} \end{pmatrix}^{-1}\begin{pmatrix} X_3 - \mu_3 \\ X_4 - \mu_4 \end{pmatrix} = \begin{pmatrix} 32.516 + 0.467X_3 + 0.977X_4 \\ 153.644 + 0.085X_3 - 0.617X_4 \end{pmatrix}$$

and covariance matrix:

$$\begin{pmatrix} \sigma_{11} & \sigma_{12} \\ \sigma_{21} & \sigma_{22} \end{pmatrix} - \begin{pmatrix} \sigma_{13} & \sigma_{14} \\ \sigma_{23} & \sigma_{24} \end{pmatrix}\begin{pmatrix} \sigma_{33} & \sigma_{34} \\ \sigma_{43} & \sigma_{44} \end{pmatrix}^{-1}\begin{pmatrix} \sigma_{31} & \sigma_{32} \\ \sigma_{41} & \sigma_{42} \end{pmatrix} = \begin{pmatrix} 104.006 & \\ -33.574 & 155.592 \end{pmatrix}.$$

In particular, the last covariance matrix allows the partial correlation between X_1 and X_2 to be computed for a fixed level of X_3 and X_4:

$$\rho_{X_1 X_2|X_3 X_4} = \frac{-33.574}{\sqrt{104.006 \cdot 155.592}} = -0.264,$$

so that in this particular example with a fixed level of advertisement and sales assistance, the negative correlation between price and sales is more important than the marginal one. **Q** MVAbluepullover

Summary
$\hookrightarrow$ If $X \sim N_p(\mu, \Sigma)$, then a linear transformation $\mathcal{A}X + c$, $\mathcal{A}(q \times p)$, where $c \in \mathbb{R}^q$, $q \leq p$, has distribution $N_q(\mathcal{A}\mu + c, \mathcal{A}\Sigma\mathcal{A}^\top)$.
$\hookrightarrow$ Two linear transformations $\mathcal{A}X$ and $\mathcal{B}X$ with $X \sim N_p(\mu, \Sigma)$ are independent if and only if $\mathcal{A}\Sigma\mathcal{B}^\top = 0$.
$\hookrightarrow$ If X_1 and X_2 are partitions of $X \sim N_p(\mu, \Sigma)$, then the conditional distribution of X_2 given $X_1 = x_1$ is again normal.
$\hookrightarrow$ In the multivariate normal case, X_1 is independent of X_2 if and only if $\Sigma_{12} = 0$.
$\hookrightarrow$ The conditional expectation of $(X_2\|X_1)$ is a linear function of X_1 if $\begin{pmatrix} X_1 \\ X_2 \end{pmatrix} \sim N_p(\mu, \Sigma)$.
$\hookrightarrow$ The multiple correlation coefficient is defined as $\rho^2_{2.1...r} = \frac{\sigma_{21}\Sigma_{11}^{-1}\sigma_{12}}{\sigma_{22}}$.
$\hookrightarrow$ The multiple correlation coefficient is the percentage of the variance of X_2 explained by the linear approximation $\beta_0 + \beta^\top X_1$.

5.2   The Wishart Distribution

The Wishart distribution (named after its discoverer) plays a prominent role in the analysis of the estimated covariance matrices. If the mean of $X \sim N_p(\mu, \Sigma)$ is known to be $\mu = 0$, then for a data matrix $\mathcal{X}(n \times p)$ the estimated covariance matrix is proportional to $\mathcal{X}^\top\mathcal{X}$. This is the point where the Wishart distribution comes in, because $\mathcal{M}(p \times p) = \mathcal{X}^\top\mathcal{X} = \sum_{i=1}^n x_i x_i^\top$ has a Wishart distribution $W_p(\Sigma, n)$.

Example 5.4   Set $p = 1$, then for $X \sim N_1(0, \sigma^2)$ the data matrix of the observations

$$\mathcal{X} = (x_1, \ldots, x_n)^\top \quad \text{with} \quad \mathcal{M} = \mathcal{X}^\top\mathcal{X} = \sum_{i=1}^n x_i x_i$$

leads to the Wishart distribution $W_1(\sigma^2, n) = \sigma^2 \chi_n^2$. The one-dimensional Wishart distribution is thus in fact a χ^2 distribution.

When we talk about the distribution of a matrix, we mean of course the joint distribution of all its elements. More exactly: since $\mathcal{M} = \mathcal{X}^\top\mathcal{X}$ is symmetric we only need to consider the elements of the lower triangular matrix

$$
\mathcal{M} = \begin{pmatrix} m_{11} & & & \\ m_{21} & m_{22} & & \\ \vdots & \vdots & \ddots & \\ m_{p1} & m_{p2} & \dots & m_{pp} \end{pmatrix}. \tag{5.14}
$$

Hence the Wishart distribution is defined by the distribution of the vector

$$
(m_{11}, \dots, m_{p1}, m_{22}, \dots, m_{p2}, \dots, m_{pp})^{\top}. \tag{5.15}
$$

Linear transformations of the data matrix $\mathcal{X}$ also lead to Wishart matrices.

Theorem 5.5 *If $\mathcal{M} \sim W_p(\Sigma, n)$ and $\mathcal{B}(p \times q)$, then the distribution of $\mathcal{B}^{\top} \mathcal{M} \mathcal{B}$ is Wishart $W_q(\mathcal{B}^{\top} \Sigma \mathcal{B}, n)$.*

With this theorem, we can standardize Wishart matrices since with $\mathcal{B} = \Sigma^{-1/2}$ the distribution of $\Sigma^{-1/2} \mathcal{M} \Sigma^{-1/2}$ is $W_p(\mathcal{I}, n)$. Another connection to the χ^2-distribution is given by the following theorem.

Theorem 5.6 *If $\mathcal{M} \sim W_p(\Sigma, m)$, and $a \in \mathbb{R}^p$ with $a^{\top} \Sigma a \neq 0$, then the distribution of $\dfrac{a^{\top} \mathcal{M} a}{a^{\top} \Sigma a}$ is χ_m^2.*

This theorem is an immediate consequence of Theorem 5.5 if we apply the linear transformation $x \mapsto a^{\top} x$. Central to the analysis of covariance matrices is the next theorem.

Theorem 5.7 (Cochran) *Let $\mathcal{X}(n \times p)$ be a data matrix from a $N_p(0, \Sigma)$ distribution and let $\mathcal{C}(n \times n)$ be a symmetric matrix.*

(a) *$\mathcal{X}^{\top} \mathcal{C} \mathcal{X}$ has the distribution of weighted Wishart random variables, i.e.,*

$$
\mathcal{X}^{\top} \mathcal{C} \mathcal{X} = \sum_{i=1}^{n} \lambda_i W_p(\Sigma, 1),
$$

   where λ_i, $i = 1, \dots, n$, are the eigenvalues of $\mathcal{C}$.
(b) *$\mathcal{X}^{\top} \mathcal{C} \mathcal{X}$ is Wishart if and only if $\mathcal{C}^2 = \mathcal{C}$. In this case,*

$$
\mathcal{X}^{\top} \mathcal{C} \mathcal{X} \sim W_p(\Sigma, r),
$$

   and $r = \text{rank}(\mathcal{C}) = \text{tr}(\mathcal{C})$.
(c) *$n\mathcal{S} = \mathcal{X}^{\top} \mathcal{H} \mathcal{X}$ is distributed as $W_p(\Sigma, n - 1)$ (note that $\mathcal{S}$ is the sample covariance matrix).*
(d) *$\bar{x}$ and $\mathcal{S}$ are independent.*

The following properties are useful:

1. If $\mathcal{M} \sim W_p(\Sigma, n)$, then $\mathsf{E}(\mathcal{M}) = n\Sigma$.

2. If $\mathcal{M}_i$ are independent Wishart $W_p(\Sigma, n_i)\, i = 1, \cdots, k$, then $\mathcal{M} = \sum_{i=1}^{k} \mathcal{M}_i \sim$ $W_p(\Sigma, n)$ where $n = \sum_{i=1}^{k} n_i$.
3. The density of $W_p(\Sigma, n-1)$ for a positive definite $\mathcal{M}$ is given by

$$f_{\Sigma, n-1}(\mathcal{M}) = \frac{|\mathcal{M}|^{\frac{1}{2}(n-p-2)} e^{-\frac{1}{2} \text{tr}(\mathcal{M}\Sigma^{-1})}}{2^{\frac{1}{2}p(n-1)} \pi^{\frac{1}{4}p(p-1)} |\Sigma|^{\frac{1}{2}(n-1)} \prod_{i=1}^{p} \Gamma\{\frac{n-i}{2}\}}, \tag{5.16}$$

where Γ is the gamma function: $\Gamma(z) = \int_0^\infty t^{z-1} e^{-t} dt$.

For further details on the Wishart distribution, see Mardia et al. (1979).

Summary
$\hookrightarrow$ The Wishart distribution is a generalization of the χ^2-distribution, in particular $W_1(\sigma^2, n) = \sigma^2 \chi_n^2$.
$\hookrightarrow$ The empirical covariance matrix S has a $\frac{1}{n} W_p(\Sigma, n-1)$ distribution.
$\hookrightarrow$ In the normal case, $\bar{x}$ and S are independent.

5.3 Hotelling's T^2-Distribution

Suppose that $Y \in \mathbb{R}^p$ is a standard normal random vector, i.e., $Y \sim N_p(0, \mathcal{I})$, independent of the random matrix $\mathcal{M} \sim W_p(\mathcal{I}, n)$. What is the distribution of $Y^\top \mathcal{M}^{-1} Y$? The answer is provided by the Hotelling T^2-distribution: $n\, Y^\top \mathcal{M}^{-1} Y$ is Hotelling $T^2_{p,n}$ distributed.

The Hotelling T^2-distribution is a generalization of the Student t-distribution. The general multinormal distribution $N(\mu, \Sigma)$ is considered in Theorem 5.8. The Hotelling T^2-distribution will play a central role in hypothesis testing in Chap. 7.

Theorem 5.8 *If $X \sim N_p(\mu, \Sigma)$ is independent of $\mathcal{M} \sim W_p(\Sigma, n)$, then*

$$n(X - \mu)^\top \mathcal{M}^{-1}(X - \mu) \sim T^2_{p,n}.$$

Corollary 5.3 *If $\bar{x}$ is the mean of a sample drawn from a normal population $N_p(\mu, \Sigma)$ and S is the sample covariance matrix, then*

$$(n-1)(\bar{x} - \mu)^\top S^{-1}(\bar{x} - \mu) = n(\bar{x} - \mu)^\top S_u^{-1}(\bar{x} - \mu) \sim T^2_{p,n-1}. \tag{5.17}$$

Recall that $S_u = \frac{n}{n-1} S$ is an unbiased estimator of the covariance matrix. A connection between the Hotelling T^2- and the F-distribution is given by the next theorem.

Theorem 5.9

$$T^2_{p,n} = \frac{np}{n-p+1} F_{p,n-p+1}.$$

Example 5.5 In the univariate case (p = 1), this theorem boils down to the well-known result:

$$\left(\frac{\bar{x} - \mu}{\sqrt{\mathcal{S}_u}/\sqrt{n}}\right)^2 \sim T^2_{1,n-1} = F_{1,n-1} = t^2_{n-1}$$

For further details on Hotelling T^2-distribution see Mardia et al. (1979). The next corollary follows immediately from (3.23), (3.24) and from Theorem 5.8. It will be useful for testing linear restrictions in multinormal populations.

Corollary 5.4 *Consider a linear transform of $X \sim N_p(\mu, \Sigma)$, $Y = \mathcal{A}X$ where $\mathcal{A}(q \times p)$ with $(q \leq p)$. If $\bar{x}$ and $\mathcal{S}_X$ are the sample mean and the sample covariance matrix, we have*

$$\bar{y} = \mathcal{A}\bar{x} \sim N_q(\mathcal{A}\mu, \frac{1}{n}\mathcal{A}\Sigma\mathcal{A}^\top)$$

$$n\mathcal{S}_Y = n\mathcal{A}\mathcal{S}_X\mathcal{A}^\top \sim W_q(\mathcal{A}\Sigma\mathcal{A}^\top, n-1)$$

$$(n-1)(\mathcal{A}\bar{x} - \mathcal{A}\mu)^\top(\mathcal{A}\mathcal{S}_X\mathcal{A}^\top)^{-1}(\mathcal{A}\bar{x} - \mathcal{A}\mu) \sim T^2_{q,n-1}$$

The T^2 distribution is closely connected to the univariate t-statistics. In Example 5.4, we described the manner in which the Wishart distribution generalizes the χ^2-distribution. We can write (5.17) as

$$T^2 = \sqrt{n}(\bar{x} - \mu)^\top \left(\frac{\sum_{j=1}^n (x_j - \bar{x})(x_j - \bar{x})^\top}{n-1}\right)^{-1} \sqrt{n}(\bar{x} - \mu)$$

which is of the form

$$\begin{pmatrix}\text{multivariate normal} \\ \text{random vector}\end{pmatrix}^\top \begin{pmatrix}\dfrac{\text{Wishart random}}{\text{matrix}} \\ \dfrac{}{\text{degrees of freedom}}\end{pmatrix}^{-1} \begin{pmatrix}\text{multivariate normal} \\ \text{random vector}\end{pmatrix}.$$

This is analogous to

$$t^2 = \sqrt{n}(\bar{x} - \mu)(s^2)^{-1}\sqrt{n}(\bar{x} - \mu)$$

or

$$\begin{pmatrix}\text{normal} \\ \text{random variable}\end{pmatrix} \begin{pmatrix}\dfrac{\chi^2\text{-random}}{\text{variable}} \\ \dfrac{}{\text{degrees of freedom}}\end{pmatrix}^{-1} \begin{pmatrix}\text{normal} \\ \text{random variable}\end{pmatrix}$$

for the univariate case. Since the multivariate normal and Wishart random variables are independently distributed, their joint distribution is the product of the marginal normal and Wishart distributions. Using calculus, the distribution of T^2 as given above can be derived from this joint distribution.

Summary
$\hookrightarrow$ Hotelling's T^2-distribution is a generalization of the t-distribution, in particular $T_{1,n}^2 = t_n$.
$\hookrightarrow$ $(n-1)(\overline{x}-\mu)^{\top}S^{-1}(\overline{x}-\mu)$ has a $T_{p,n-1}^2$ distribution.
$\hookrightarrow$ The relation between Hotelling's T^2- and Fisher's F-distribution is given by $T_{p,n}^2 = \frac{np}{n-p+1} F_{p,n-p+1}$.

5.4   Spherical and Elliptical Distributions

The multinormal distribution belongs to the large family of elliptical distributions, which has recently gained a lot of attention in financial mathematics. Elliptical distributions are often used, particularly in risk management.

Definition 5.1 A $(p \times 1)$ random vector Y is said to have a spherical distribution $S_p(\phi)$ if its characteristic function $\psi_Y(t)$ satisfies $\psi_Y(t) = \phi(t^{\top}t)$ for some scalar function $\phi(.)$ which is then called the characteristic generator of the spherical distribution $S_p(\phi)$. We will write $Y \sim S_p(\phi)$.

This is only one of several possible ways to define spherical distributions. We can see spherical distributions as an extension of the standard multinormal distribution $N_p(0, \mathcal{I}_p)$.

Theorem 5.10 *Spherical random variables have the following properties:*

1. *All marginal distributions of a spherically distributed random vector are spherical.*
2. *All the marginal characteristic functions have the same generator.*
3. *Let $X \sim S_p(\phi)$, then X has the same distribution as $r u^{(p)}$ where $u^{(p)}$ is a random vector distributed uniformly on the unit sphere surface in $\mathbb{R}^p$ and $r \geq 0$ is a random variable independent of $u^{(p)}$. If $\mathsf{E}(r^2) < \infty$, then*

$$\mathsf{E}(X) = 0 , \quad \mathsf{Cov}(X) = \frac{\mathsf{E}(r^2)}{p}\mathcal{I}_p.$$

The random radius r is related to the generator ϕ by a relation described in (Fang et al. 1990, p.29). The moments of $X \sim S_p(\phi)$, provided that they exist, can be expressed in terms of one-dimensional integral.

A spherically distributed random vector does not, in general, necessarily possess a density. However, if it does, the marginal densities of dimension smaller than $p-1$ are continuous and the marginal densities of dimension smaller than $p-2$ are differentiable (except possibly at the origin in both cases). Univariate marginal densities for p greater than 2 are nondecreasing on $(-\infty, 0)$ and nonincreasing on $(0, \infty)$.

Definition 5.2 A $(p \times 1)$ random vector X is said to have an elliptical distribution with parameters $\mu(p \times 1)$ and $\Sigma(p \times p)$ if X has the same distribution as $\mu + \mathcal{A}^\top Y$, where $Y \sim S_k(\phi)$ and $\mathcal{A}$ is a $(k \times p)$ matrix such that $\mathcal{A}^\top \mathcal{A} = \Sigma$ with $\mathrm{rank}(\Sigma) = k$. We shall write $X \sim EC_p(\mu, \Sigma, \phi)$.

Remark 5.1 The elliptical distribution can be seen as an extension of $N_p(\mu, \Sigma)$.

Example 5.6 The multivariate t-distribution. Let $Z \sim N_p(0, \mathcal{I}_p)$ and $s \sim \chi_m^2$ be independent. The random vector

$$Y = \sqrt{m}\, \frac{Z}{s}$$

has a multivariate *t*-distribution with m degrees of freedom. Moreover, the *t*-distribution belongs to the family of p-dimensional spherical distributions.

Example 5.7 The multinormal distribution. Let $X \sim N_p(\mu, \Sigma)$.
Then $X \sim EC_p(\mu, \Sigma, \phi)$ and $\phi(u) = \exp(-u/2)$. Figure 4.3 shows a density surface of the multivariate normal distribution: $f(x) = \det(2\pi\Sigma)^{-\frac{1}{2}} \exp\{-\frac{1}{2}(x - \mu)^\top \Sigma^{-1}(x - \mu)\}$ with $\Sigma = \begin{pmatrix} 1 & 0.6 \\ 0.6 & 1 \end{pmatrix}$ and $\mu = \begin{pmatrix} 0 \\ 0 \end{pmatrix}$. Notice that the density is constant on ellipses. This is the reason for calling this family of distributions "elliptical".

Theorem 5.11 *Elliptical random vectors X have the following properties:*

1. *Any linear combination of elliptically distributed variables are elliptical.*
2. *Marginal distributions of elliptically distributed variables are elliptical.*
3. *A scalar function $\phi(.)$ can determine an elliptical distribution $EC_p(\mu, \Sigma, \phi)$ for every $\mu \in \mathbb{R}^p$ and $\Sigma \geq 0$ with $\mathrm{rank}(\Sigma) = k$ iff $\phi(t^\top t)$ is a p-dimensional characteristic function.*
4. *Assume that X is nondegenerate. If $X \sim EC_p(\mu, \Sigma, \phi)$ and $X \sim EC_p(\mu^*, \Sigma^*, \phi^*)$, then a constant $c > 0$ exists that*

$$\mu = \mu^*, \quad \Sigma = c\Sigma^*, \quad \phi^*(.) = \phi(c^{-1}.).$$

   In other words $\Sigma, \phi, \mathcal{A}$ are not unique, unless we impose the condition that $\det(\Sigma) = 1$.
5. *The characteristic function of X, $\psi(t) = \mathsf{E}(e^{\mathbf{i}t^\top X})$ is of the form*

$$\psi(t) = e^{\mathbf{i}t^\top \mu} \phi(t^\top \Sigma t)$$

   for a scalar function ϕ.

6. $X \sim EC_p(\mu, \Sigma, \phi)$ with $\text{rank}(\Sigma) = k$ iff X has the same distribution as:

$$\mu + r\mathcal{A}^\top u^{(k)} \tag{5.18}$$

where $r \geq 0$ is independent of $u^{(k)}$ which is a random vector distributed uniformly on the unit sphere surface in $\mathbb{R}^k$ and $\mathcal{A}$ is a $(k \times p)$ matrix such that $\mathcal{A}^\top \mathcal{A} = \Sigma$.

7. Assume that $X \sim EC_p(\mu, \Sigma, \phi)$ and $\mathsf{E}(r^2) < \infty$. Then

$$\mathsf{E}(X) = \mu \quad \mathsf{Cov}(X) = \frac{\mathsf{E}(r^2)}{\text{rank}(\Sigma)} \Sigma = -2\phi^\top(0)\Sigma.$$

8. Assume that $X \sim EC_p(\mu, \Sigma, \phi)$ with $\text{rank}(\Sigma) = k$. Then

$$Q(X) = (X - \mu)^\top \Sigma^{-1}(X - \mu)$$

has the same distribution as r^2 in Eq. (5.18).

5.5 Exercises

Exercise 5.1 Consider $X \sim N_2(\mu, \Sigma)$ with $\mu = (2, 2)^\top$ and $\Sigma = \begin{pmatrix} 1 & 0 \\ 0 & 1 \end{pmatrix}$ and the matrices $\mathcal{A} = \begin{pmatrix} 1 \\ 1 \end{pmatrix}^\top, \mathcal{B} = \begin{pmatrix} 1 \\ -1 \end{pmatrix}^\top$. Show that $\mathcal{A}X$ and $\mathcal{B}X$ are independent.

Exercise 5.2 Prove Theorem 5.4.

Exercise 5.3 Prove proposition (c) of Theorem 5.7.

Exercise 5.4 Let

$$X \sim N_2\left(\begin{pmatrix} 1 \\ 2 \end{pmatrix}, \begin{pmatrix} 2 & 1 \\ 1 & 2 \end{pmatrix} \right)$$

and

$$Y \mid X \sim N_2\left(\begin{pmatrix} X_1 \\ X_1 + X_2 \end{pmatrix}, \begin{pmatrix} 1 & 0 \\ 0 & 1 \end{pmatrix} \right).$$

(a) Determine the distribution of $Y_2 \mid Y_1$.
(b) Determine the distribution of $W = X - Y$.

Exercise 5.5 Consider $\begin{pmatrix} X \\ Y \\ Z \end{pmatrix} \sim N_3(\mu, \Sigma)$. Compute μ and Σ knowing that

$$Y \mid Z \sim N_1(-Z, 1)$$
$$\mu_{Z \mid Y} = -\frac{1}{3} - \frac{1}{3}Y$$
$$X \mid Y, Z \sim N_1(2 + 2Y + 3Z, 1).$$

Determine the distributions of $X \mid Y$ and of $X \mid Y + Z$.

Exercise 5.6 Knowing that

$$Z \sim N_1(0, 1)$$
$$Y \mid Z \sim N_1(1 + Z, 1)$$
$$X \mid Y, Z \sim N_1(1 - Y, 1)$$

(a) find the distribution of $\begin{pmatrix} X \\ Y \\ Z \end{pmatrix}$ and of $Y \mid X, Z$.

(b) find the distribution of

$$\begin{pmatrix} U \\ V \end{pmatrix} = \begin{pmatrix} 1 + Z \\ 1 - Y \end{pmatrix}.$$

(c) compute $E(Y \mid U = 2)$.

Exercise 5.7 Suppose $\begin{pmatrix} X \\ Y \end{pmatrix} \sim N_2(\mu, \Sigma)$ with Σ positive definite. Is it possible that

(a) $\mu_{X \mid Y} = 3Y^2$,
(b) $\sigma_{XX \mid Y} = 2 + Y^2$,
(c) $\mu_{X \mid Y} = 3 - Y$, and
(d) $\sigma_{XX \mid Y} = 5$?

Exercise 5.8 Let $X \sim N_3 \left(\begin{pmatrix} 1 \\ 2 \\ 3 \end{pmatrix}, \begin{pmatrix} 11 & -6 & 2 \\ -6 & 10 & -4 \\ 2 & -4 & 6 \end{pmatrix} \right)$.

(a) Find the best linear approximation of X_3 by a linear function of X_1 and X_2 and compute the multiple correlation between X_3 and (X_1, X_2).

(b) Let $Z_1 = X_2 - X_3$, $Z_2 = X_2 + X_3$, and $(Z_3 \mid Z_1, Z_2) \sim N_1(Z_1 + Z_2, 10)$. Compute the distribution of $\begin{pmatrix} Z_1 \\ Z_2 \\ Z_3 \end{pmatrix}$.

Exercise 5.9 Let $(X, Y, Z)^\top$ be a trivariate normal r.v. with

$$Y \mid Z \sim N_1(2Z, 24)$$
$$Z \mid X \sim N_1(2X + 3, 14)$$
$$X \sim N_1(1, 4)$$
$$\text{and } \rho_{XY} = 0.5.$$

Find the distribution of $(X, Y, Z)^\top$ and compute the partial correlation between X and Y for fixed Z. Do you think it is reasonable to approximate X by a linear function of Y and Z?

Exercise 5.10 Let $X \sim N_4\left(\begin{pmatrix} 1 \\ 2 \\ 3 \\ 4 \end{pmatrix}, \begin{pmatrix} 4 & 1 & 2 & 4 \\ 1 & 4 & 2 & 1 \\ 2 & 2 & 16 & 1 \\ 4 & 1 & 1 & 9 \end{pmatrix} \right)$.

(a) Give the best linear approximation of X_2 as a function of (X_1, X_4) and evaluate the quality of the approximation.
(b) Give the best linear approximation of X_2 as a function of (X_1, X_3, X_4) and compare your answer with part (a).

Exercise 5.11 Prove Theorem 5.2.
(Hint: complete the linear transformation $Z = \begin{pmatrix} \mathcal{A} \\ \mathcal{I}_{p-q} \end{pmatrix} X + \begin{pmatrix} c \\ 0_{p-q} \end{pmatrix}$ and then use Theorem 5.1 to get the marginal of the first q components of Z.)

Exercise 5.12 Prove Corollaries 5.1 and 5.2.

References

K.T. Fang, S. Kotz, K.W. Ng, *Symmetric Multivariate and Related Distributions* (Chapman and Hall, London, 1990)
K.V. Mardia, J.T. Kent, J.M. Bibby, *Multivariate Analysis* (Academic Press, Duluth, London, 1979)

Chapter 6
Theory of Estimation

We know from our basic knowledge of statistics that one of the objectives in statistics is to better understand and model the underlying process which generates data. This is known as statistical inference: we infer from information contained in sample properties of the population from which the observations are taken. In multivariate statistical inference, we do exactly the same. The basic ideas were introduced in Sect. 4.5 on sampling theory: we observed the values of a multivariate random variable X and obtained a sample $\mathcal{X} = \{x_i\}_{i=1}^n$. Under random sampling, these observations are considered to be realizations of a sequence of i.i.d. random variables $X_1, \ldots, X_n$, where each X_i is a p-variate random variable which replicates the *parent* or *population* random variable X. In this chapter, for notational convenience, we will no longer differentiate between a random variable X_i and an observation of it, x_i, in our notation. We will simply write x_i and it should be clear from the context whether a random variable or an observed value is meant.

Statistical inference infers from the i.i.d. random sample $\mathcal{X}$ the properties of the population: typically, some unknown characteristic θ of its distribution. In parametric statistics, θ is a k-variate vector $\theta \in \mathbb{R}^k$ characterizing the unknown properties of the population pdf $f(x; \theta)$: this could be the mean, the covariance matrix, kurtosis, etc.

The aim will be to estimate θ from the sample $\mathcal{X}$ through estimators $\widehat{\theta}$ which are functions of the sample: $\widehat{\theta} = \widehat{\theta}(\mathcal{X})$. When an estimator $\widehat{\theta}$ is proposed, we must derive its sampling distribution to analyze its properties.

In this chapter, the basic theoretical tools are developed which are needed to derive estimators and to determine their properties in general situations. We will basically rely on the maximum likelihood theory in our presentation. In many situations, the maximum likelihood estimators indeed share asymptotic optimal properties which make their use easy and appealing.

We will illustrate the multivariate normal population and also the linear regression model, where the applications are numerous and the derivations are easy to do. In multivariate setups, the maximum likelihood estimator is at times too complicated to be derived analytically. In such cases, the estimators are obtained using numerical

© Springer Nature Switzerland AG 2019
W. K. Härdle and L. Simar, *Applied Multivariate Statistical Analysis*,
https://doi.org/10.1007/978-3-030-26006-4_6

methods (nonlinear optimization). The general theory and the asymptotic properties of these estimators remain simple and valid. The following Chap. 7 concentrates on hypothesis testing and confidence interval issues.

6.1   The Likelihood Function

Suppose that $\{x_i\}_{i=1}^{n}$ is an i.i.d. sample from a population with pdf $f(x; \theta)$. The aim is to estimate $\theta \in \mathbb{R}^k$, which is a vector of unknown parameters. The *likelihood function* is defined as the joint density $L(\mathcal{X}; \theta)$ of the observations x_i considered as a function of θ:

$$L(\mathcal{X}; \theta) = \prod_{i=1}^{n} f(x_i; \theta), \tag{6.1}$$

where $\mathcal{X}$ denotes the sample of the data matrix with the observations $x_1^\top, \ldots, x_n^\top$ in each row. The *maximum likelihood estimator* (MLE) of θ is defined as

$$\widehat{\theta} = \arg \max_{\theta} L(\mathcal{X}; \theta).$$

Often, it is easier to maximize the *log-likelihood function*

$$\ell(\mathcal{X}; \theta) = \log L(\mathcal{X}; \theta), \tag{6.2}$$

which is equivalent since the logarithm is a monotone one-to-one function. Hence

$$\widehat{\theta} = \arg \max_{\theta} L(\mathcal{X}; \theta) = \arg \max_{\theta} \ell(\mathcal{X}; \theta).$$

The following examples illustrate cases where the maximization process can be performed analytically, i.e., we will obtain an explicit analytical expression for $\widehat{\theta}$. Unfortunately, in other situations, the maximization process can be more intricate, involving nonlinear optimization techniques. In the latter case, given a sample $\mathcal{X}$ and the likelihood function, numerical methods will be used to determine the value of θ maximizing $L(\mathcal{X}; \theta)$ or $\ell(\mathcal{X}; \theta)$. These numerical methods are typically based on Newton–Raphson iterative techniques.

Example 6.1   Consider a sample $\{x_i\}_{i=1}^{n}$ from $N_p(\mu, \mathcal{I})$, i.e., from the pdf

$$f(x; \theta) = (2\pi)^{-p/2} \exp\left\{-\frac{1}{2}(x - \theta)^\top (x - \theta)\right\}$$

where $\theta = \mu \in \mathbb{R}^p$ is the mean vector parameter. The log-likelihood is in this case

$$\ell(\mathcal{X}; \theta) = \sum_{i=1}^{n} \log\{f(x_i; \theta)\} = \log(2\pi)^{-np/2} - \frac{1}{2} \sum_{i=1}^{n} (x_i - \theta)^\top (x_i - \theta). \quad (6.3)$$

The term $(x_i - \theta)^\top (x_i - \theta)$ equals

$$(x_i - \overline{x})^\top (x_i - \overline{x}) + (\overline{x} - \theta)^\top (\overline{x} - \theta) + 2(\overline{x} - \theta)^\top (x_i - \overline{x}).$$

Summing this term over $i = 1, \ldots, n$ we see that

$$\sum_{i=1}^{n} (x_i - \theta)^\top (x_i - \theta) = \sum_{i=1}^{n} (x_i - \overline{x})^\top (x_i - \overline{x}) + n(\overline{x} - \theta)^\top (\overline{x} - \theta).$$

Hence

$$\ell(\mathcal{X}; \theta) = \log(2\pi)^{-np/2} - \frac{1}{2} \sum_{i=1}^{n} (x_i - \overline{x})^\top (x_i - \overline{x}) - \frac{n}{2} (\overline{x} - \theta)^\top (\overline{x} - \theta).$$

Only the last term depends on θ and is obviously maximized for

$$\widehat{\theta} = \widehat{\mu} = \overline{x}.$$

Thus $\overline{x}$ is the MLE of θ for this family of pdf's $f(x, \theta)$.

A more complex example is the following one where we derive the MLE's for μ and Σ.

Example 6.2 Suppose $\{x_i\}_{i=1}^{n}$ is a sample from a normal distribution $N_p(\mu, \Sigma)$. Here $\theta = (\mu, \Sigma)$ with Σ interpreted as a vector. Due to the symmetry of Σ the unknown parameter θ is in fact $\{p + \frac{1}{2} p(p+1)\}$-dimensional. Then

$$L(\mathcal{X}; \theta) = |2\pi \Sigma|^{-n/2} \exp\left\{-\frac{1}{2} \sum_{i=1}^{n} (x_i - \mu)^\top \Sigma^{-1} (x_i - \mu)\right\} \quad (6.4)$$

and

$$\ell(\mathcal{X}; \theta) = -\frac{n}{2} \log|2\pi \Sigma| - \frac{1}{2} \sum_{i=1}^{n} (x_i - \mu)^\top \Sigma^{-1} (x_i - \mu). \quad (6.5)$$

The term $(x_i - \mu)^\top \Sigma^{-1} (x_i - \mu)$ equals

$$(x_i - \overline{x})^\top \Sigma^{-1} (x_i - \overline{x}) + (\overline{x} - \mu)^\top \Sigma^{-1} (\overline{x} - \mu) + 2(\overline{x} - \mu)^\top \Sigma^{-1} (x_i - \overline{x}).$$

Summing this term over $i = 1, \ldots, n$ we see that

$$\sum_{i=1}^{n}(x_i - \mu)^{\top} \Sigma^{-1}(x_i - \mu) = \sum_{i=1}^{n}(x_i - \overline{x})^{\top} \Sigma^{-1}(x_i - \overline{x}) + n(\overline{x} - \mu)^{\top} \Sigma^{-1}(\overline{x} - \mu).$$

Note that from (2.14)

$$(x_i - \overline{x})^{\top} \Sigma^{-1}(x_i - \overline{x}) = \operatorname{tr}\left\{ (x_i - \overline{x})^{\top} \Sigma^{-1}(x_i - \overline{x}) \right\}$$
$$= \operatorname{tr}\left\{ \Sigma^{-1}(x_i - \overline{x})(x_i - \overline{x})^{\top} \right\}.$$

Therefore, by summing over the index i we finally arrive at

$$\sum_{i=1}^{n}(x_i - \mu)^{\top} \Sigma^{-1}(x_i - \mu) = \operatorname{tr}\left\{ \Sigma^{-1} \sum_{i=1}^{n}(x_i - \overline{x})(x_i - \overline{x})^{\top} \right\}$$
$$+ n(\overline{x} - \mu)^{\top} \Sigma^{-1}(\overline{x} - \mu)$$
$$= \operatorname{tr}\{\Sigma^{-1}n\mathcal{S}\} + n(\overline{x} - \mu)^{\top} \Sigma^{-1}(\overline{x} - \mu).$$

Thus the log-likelihood function for $N_p(\mu, \Sigma)$ is

$$\ell(\mathcal{X}; \theta) = -\frac{n}{2} \log |2\pi \Sigma| - \frac{n}{2} \operatorname{tr}\{\Sigma^{-1}\mathcal{S}\} - \frac{n}{2}(\overline{x} - \mu)^{\top} \Sigma^{-1}(\overline{x} - \mu). \qquad (6.6)$$

We can easily see that the third term is maximized by $\mu = \overline{x}$. In fact the MLE's are given by

$$\widehat{\mu} = \overline{x}, \quad \widehat{\Sigma} = \mathcal{S}.$$

The derivation of $\widehat{\Sigma}$ is a lot more complicated. It involves derivatives with respect to matrices with their notational complexities and will not be presented here; for more elaborate proof see Mardia et al. (1979, p. 103–104). Note that the unbiased covariance estimator $\mathcal{S}_u = \frac{n}{n-1}\mathcal{S}$ is not the MLE of Σ!

Example 6.3 Consider the linear regression model $y_i = \beta^{\top} x_i + \varepsilon_i$ for $i = 1, \ldots, n$, where ε_i is i.i.d. and $N(0, \sigma^2)$ and where $x_i \in \mathbb{R}^p$. Here $\theta = (\beta^{\top}, \sigma)$ is a $(p + 1)$-dimensional parameter vector. Denote

$$y = \begin{pmatrix} y_1 \\ \vdots \\ y_n \end{pmatrix}, \quad \mathcal{X} = \begin{pmatrix} x_1^{\top} \\ \vdots \\ x_n^{\top} \end{pmatrix}.$$

Then

$$L(y, \mathcal{X}; \theta) = \prod_{i=1}^{n} \frac{1}{\sqrt{2\pi}\sigma} \exp\left\{ -\frac{1}{2\sigma^2}(y_i - \beta^{\top} x_i)^2 \right\}$$

and

$$\ell(y, \mathcal{X}; \theta) = \log \left\{ \frac{1}{(2\pi)^{n/2}\sigma^n} \right\} - \frac{1}{2\sigma^2} \sum_{i=1}^{n} (y_i - \beta^\top x_i)^2$$

$$= -\frac{n}{2} \log(2\pi) - n \log \sigma - \frac{1}{2\sigma^2} (y - \mathcal{X}\beta)^\top (y - \mathcal{X}\beta)$$

$$= -\frac{n}{2} \log(2\pi) - n \log \sigma - \frac{1}{2\sigma^2} (y^\top y + \beta^\top \mathcal{X}^\top \mathcal{X}\beta - 2\beta^\top \mathcal{X}^\top y).$$

Differentiating w.r.t. the parameters yields

$$\frac{\partial}{\partial \beta} \ell = -\frac{1}{2\sigma^2} (2\mathcal{X}^\top \mathcal{X}\beta - 2\mathcal{X}^\top y)$$

$$\frac{\partial}{\partial \sigma} \ell = -\frac{n}{\sigma} + \frac{1}{\sigma^3} \{ (y - \mathcal{X}\beta)^\top (y - \mathcal{X}\beta) \}.$$

Note that $\frac{\partial}{\partial \beta} \ell$ denotes the vector of the derivatives w.r.t. all components of β (the gradient). Since the first equation only depends on β, we start with deriving $\widehat{\beta}$.

$$\mathcal{X}^\top \mathcal{X}\widehat{\beta} = \mathcal{X}^\top y, \quad \text{hence} \quad \widehat{\beta} = (\mathcal{X}^\top \mathcal{X})^{-1} \mathcal{X}^\top y$$

Plugging $\widehat{\beta}$ into the second equation gives

$$\frac{n}{\sigma} = \frac{1}{\sigma^3} (y - \mathcal{X}\widehat{\beta})^\top (y - \mathcal{X}\widehat{\beta}), \quad \text{hence} \quad \widehat{\sigma}^2 = \frac{1}{n} \|y - \mathcal{X}\widehat{\beta}\|^2,$$

where $\| \cdot \|^2$ denotes the Euclidean vector norm from Sect. 2.6. We see that the MLE $\widehat{\beta}$ is identical with the least squares estimator (3.52). The variance estimator

$$\widehat{\sigma}^2 = \frac{1}{n} \sum_{i=1}^{n} (y_i - \widehat{\beta}^\top x_i)^2$$

is nothing else than the residual sum of squares (RSS) from (3.37) generalized to the case of multivariate x_i.

Note that when the x_i's are considered to be fixed we have

$$\mathsf{E}(y) = \mathcal{X}\beta \quad \text{and} \quad \mathsf{Var}(y) = \sigma^2 \mathcal{I}_n.$$

Then, using the properties of moments from Sect. 4.2 we have

$$\mathsf{E}(\widehat{\beta}) = (\mathcal{X}^\top \mathcal{X})^{-1} \mathcal{X}^\top \mathsf{E}(y) = \beta, \tag{6.7}$$

$$\mathsf{Var}(\widehat{\beta}) = \sigma^2 (\mathcal{X}^\top \mathcal{X})^{-1}. \tag{6.8}$$

Summary
↪   If $\{x_i\}_{i=1}^{n}$ is an i.i.d. sample from a distribution with pdf $f(x; \theta)$, then $L(\mathcal{X}; \theta) = \prod_{i=1}^{n} f(x_i; \theta)$ is the likelihood function. The maximum likelihood estimator (MLE) is that value of θ which maximizes $L(\mathcal{X}; \theta)$. Equivalently one can maximize the log-likelihood $\ell(\mathcal{X}; \theta)$.
↪   The MLE's of μ and Σ from a $N_p(\mu, \Sigma)$ distribution are $\widehat{\mu} = \overline{x}$ and $\widehat{\Sigma} = \mathcal{S}$. Note that the MLE of Σ is not unbiased.
↪   The MLE's of β and σ in the linear model $y = \mathcal{X}\beta + \varepsilon$, $\varepsilon \sim N_n(0, \sigma^2\mathcal{I})$ are given by the least squares estimator $\widehat{\beta} = (\mathcal{X}^\top\mathcal{X})^{-1}\mathcal{X}^\top y$ and $\widehat{\sigma}^2 = \frac{1}{n}\|y - \mathcal{X}\widehat{\beta}\|^2$. $\mathsf{E}(\widehat{\beta}) = \beta$ and $\mathsf{Var}(\widehat{\beta}) = \sigma^2(\mathcal{X}^\top\mathcal{X})^{-1}$.

6.2   The Cramer–Rao Lower Bound

As pointed out above, an important question in estimation theory is whether an estimator $\widehat{\theta}$ has certain desired properties, in particular, if it converges to the unknown parameter θ it is supposed to estimate. One typical property we want for an estimator is unbiasedness, meaning that on the average, the estimator hits its target: $\mathsf{E}(\widehat{\theta}) = \theta$. We have seen, for instance, (see Example 6.2) that $\overline{x}$ is an unbiased estimator of μ and $\mathcal{S}$ is a biased estimator of Σ in finite samples. If we restrict ourselves to unbiased estimation then the natural question is whether the estimator shares some optimality properties in terms of its sampling variance. Since we focus on unbiasedness, we look for an estimator with the smallest possible variance.

In this context, the Cramer–Rao lower bound will give the minimal achievable variance for any unbiased estimator. This result is valid under very general regularity conditions (discussed below). One of the most important applications of the Cramer–Rao lower bound is that it provides the asymptotic optimality property of maximum likelihood estimators. The Cramer–Rao theorem involves the *score function* and its properties which will be derived first.

The score function $s(\mathcal{X}; \theta)$ is the derivative of the log-likelihood function w.r.t. $\theta \in \mathbb{R}^k$

$$s(\mathcal{X}; \theta) = \frac{\partial}{\partial\theta}\ell(\mathcal{X}; \theta) = \frac{1}{L(\mathcal{X}; \theta)}\frac{\partial}{\partial\theta}L(\mathcal{X}; \theta). \qquad (6.9)$$

The covariance matrix $\mathcal{F}_n = \mathsf{Var}\{s(\mathcal{X}; \theta)\}$ is called the *Fisher information matrix*. In what follows, we will give some interesting properties of score functions.

Theorem 6.1 *If $s = s(\mathcal{X}; \theta)$ is the score function and if $\widehat{\theta} = t = t(\mathcal{X}, \theta)$ is any function of $\mathcal{X}$ and θ, then under regularity conditions*

$$\mathsf{E}(st^\top) = \frac{\partial}{\partial\theta}\mathsf{E}(t^\top) - \mathsf{E}\left(\frac{\partial t^\top}{\partial\theta}\right). \qquad (6.10)$$

The proof is left as an exercise (see Exercise 6.9). The regularity conditions required for this theorem are rather technical and ensure that the expressions (expectations and derivations) appearing in (6.10) are well defined. In particular, the support of the density $f(x; \theta)$ should not depend on θ. The next corollary is a direct consequence.

Corollary 6.1 *If $s = s(\mathcal{X}; \theta)$ is the score function, and $\widehat{\theta} = t = t(\mathcal{X})$ is any unbiased estimator of θ (i.e., $\mathsf{E}(t) = \theta$), then*

$$\mathsf{E}(st^\top) = \mathsf{Cov}(s, t) = \mathcal{I}_k. \tag{6.11}$$

Note that the score function has mean zero (see Exercise 6.10).

$$\mathsf{E}\{s(\mathcal{X}; \theta)\} = 0. \tag{6.12}$$

Hence, $\mathsf{E}(ss^\top) = \mathsf{Var}(s) = \mathcal{F}_n$ and by setting $s = t$ in Theorem 6.1 it follows that

$$\mathcal{F}_n = -\mathsf{E}\left\{ \frac{\partial^2}{\partial\theta\partial\theta^\top} \ell(\mathcal{X}; \theta) \right\}.$$

Remark 6.1 If $x_1, \ldots, x_n$ are i.i.d., $\mathcal{F}_n = n\mathcal{F}_1$ where $\mathcal{F}_1$ is the Fisher information matrix for sample size $n = 1$.

Example 6.4 Consider an i.i.d. sample $\{x_i\}_{i=1}^n$ from $N_p(\theta, \mathcal{I})$. In this case, the parameter θ is the mean μ. It follows from (6.3) that

$$
\begin{aligned}
s(\mathcal{X}; \theta) &= \frac{\partial}{\partial\theta} \ell(\mathcal{X}; \theta) \\
&= -\frac{1}{2}\frac{\partial}{\partial\theta} \left\{ \sum_{i=1}^n (x_i - \theta)^\top (x_i - \theta) \right\} \\
&= n(\overline{x} - \theta).
\end{aligned}
$$

Hence, the information matrix is

$$\mathcal{F}_n = \mathsf{Var}\{n(\overline{x} - \theta)\} = n\mathcal{I}_p.$$

How well can we estimate θ? The answer is given in the following theorem which is from Cramer and Rao. As pointed out above, this theorem gives a lower bound for unbiased estimators. Hence, all estimators, which are unbiased <u>and</u> attain this lower bound, are *minimum variance estimators*.

Theorem 6.2 (Cramer–Rao) *If $\widehat{\theta} = t = t(\mathcal{X})$ is any unbiased estimator for θ, then under regularity conditions*

$$\mathsf{Var}(t) \geq \mathcal{F}_n^{-1}, \tag{6.13}$$

where

$$\mathcal{F}_n = \mathsf{E}\{s(\mathcal{X};\theta)s(\mathcal{X};\theta)^\top\} = \mathsf{Var}\{s(\mathcal{X};\theta)\} \qquad (6.14)$$

is the Fisher information matrix.

Proof Consider the correlation $\rho_{Y,Z}$ between Y and Z where $Y = a^\top t$, $Z = c^\top s$. Here s is the score and the vectors $a, c \in \mathbb{R}^p$. By Corollary 6.1 $\mathsf{Cov}(s,t) = \mathcal{I}$ and thus

$$\mathsf{Cov}(Y,Z) = a^\top \mathsf{Cov}(t,s)c = a^\top c$$
$$\mathsf{Var}(Z) = c^\top \mathsf{Var}(s)c = c^\top \mathcal{F}_n c.$$

Hence,

$$\rho_{Y,Z}^2 = \frac{\mathsf{Cov}^2(Y,Z)}{\mathsf{Var}(Y)\mathsf{Var}(Z)} = \frac{(a^\top c)^2}{a^\top \mathsf{Var}(t)a \cdot c^\top \mathcal{F}_n c} \leq 1. \qquad (6.15)$$

In particular, this holds for any $c \neq 0$. Therefore, it holds also for the maximum of the left-hand side of (6.15) with respect to c. Since

$$\max_c \frac{c^\top aa^\top c}{c^\top \mathcal{F}_n c} = \max_{c^\top \mathcal{F}_n c = 1} c^\top aa^\top c$$

and

$$\max_{c^\top \mathcal{F}_n c = 1} c^\top aa^\top c = a^\top \mathcal{F}_n^{-1} a$$

by our maximization Theorem 2.5 we have

$$\frac{a^\top \mathcal{F}_n^{-1} a}{a^\top \mathsf{Var}(t)a} \leq 1 \quad \forall\, a \in \mathbb{R}^p, \quad a \neq 0,$$

i.e.,

$$a^\top \{\mathsf{Var}(t) - \mathcal{F}_n^{-1}\}a \geq 0 \quad \forall\, a \in \mathbb{R}^p, \quad a \neq 0,$$

which is equivalent to $\mathsf{Var}(t) \geq \mathcal{F}_n^{-1}$.

Maximum likelihood estimators (MLE's) attain the lower bound if the sample size n goes to infinity. The next Theorem 6.3 states this and, in addition, gives the asymptotic sampling distribution of the maximum likelihood estimation, which turns out to be multinormal.

Theorem 6.3 *Suppose that the sample $\{x_i\}_{i=1}^n$ is i.i.d. If $\widehat{\theta}$ is the MLE for $\theta \in \mathbb{R}^k$, i.e., $\widehat{\theta} = \arg\max_\theta L(\mathcal{X};\theta)$, then under some regularity conditions, as $n \to \infty$:*

$$\sqrt{n}(\widehat{\theta} - \theta) \xrightarrow{\mathcal{L}} N_k(0, \mathcal{F}_1^{-1}) \qquad (6.16)$$

where $\mathcal{F}_1$ denotes the Fisher information for sample size $n = 1$.

As a consequence of Theorem 6.3, we see that under regularity conditions the MLE is asymptotically unbiased, efficient (minimum variance) and normally distributed. Also it is a consistent estimator of θ.

Note that from Property (5.4) of the multinormal it follows that asymptotically

$$n(\widehat{\theta} - \theta)^{\top} \mathcal{F}_1 (\widehat{\theta} - \theta) \xrightarrow{\mathcal{L}} \chi_p^2. \tag{6.17}$$

If $\widehat{\mathcal{F}}_1$ is a consistent estimator of $\mathcal{F}_1$ (e.g. $\widehat{\mathcal{F}}_1 = \mathcal{F}_1(\widehat{\theta})$), we have equivalently

$$n(\widehat{\theta} - \theta)^{\top} \widehat{\mathcal{F}}_1 (\widehat{\theta} - \theta) \xrightarrow{\mathcal{L}} \chi_p^2. \tag{6.18}$$

This expression is sometimes useful in testing hypotheses about θ and in constructing confidence regions for θ in a very general setup. These issues will be raised in more details in the next chapter but from (6.18) it can be seen, for instance, that when n is large,

$$\P\left\{n(\widehat{\theta} - \theta)^{\top} \widehat{\mathcal{F}}_1 (\widehat{\theta} - \theta) \le \chi_{1-\alpha;p}^2\right\} \approx 1 - \alpha,$$

where $\chi_{v;p}^2$ denotes the v-quantile of a χ_p^2 random variable. So, the ellipsoid $n(\widehat{\theta} - \theta)^{\top} \widehat{\mathcal{F}}_1 (\widehat{\theta} - \theta) \le \chi_{1-\alpha;p}^2$ provides in $\mathbb{R}^p$ an asymptotic $(1 - \alpha)$-confidence region for θ.

Summary
$\hookrightarrow$ The score function is the derivative $s(\mathcal{X}; \theta) = \frac{\partial}{\partial \theta} \ell(\mathcal{X}; \theta)$ of the log-likelihood with respect to θ. The covariance matrix of $s(\mathcal{X}; \theta)$ is the Fisher information matrix.
$\hookrightarrow$ The score function has mean zero: $\mathsf{E}\{s(\mathcal{X}; \theta)\} = 0$.
$\hookrightarrow$ The Cramer–Rao bound says that any unbiased estimator $\widehat{\theta} = t = t(\mathcal{X})$ has a variance that is bounded from below by the inverse of the Fisher information. Thus, an unbiased estimator, which attains this lower bound, is a minimum variance estimator.
$\hookrightarrow$ For i.i.d. data $\{x_i\}_{i=1}^n$ the Fisher information matrix is: $\mathcal{F}_n = n\mathcal{F}_1$.
$\hookrightarrow$ MLE's attain the lower bound in an asymptotic sense, i.e., $$\sqrt{n}(\widehat{\theta} - \theta) \xrightarrow{\mathcal{L}} N_k(0, \mathcal{F}_1^{-1})$$ if $\widehat{\theta}$ is the MLE for $\theta \in \mathbb{R}^k$, i.e., $\widehat{\theta} = \arg\max_{\theta} L(\mathcal{X}; \theta)$.

6.3 Exercises

Exercise 6.1 Consider a uniform distribution on the interval $[0, \theta]$. What is the MLE of θ? (Hint: the maximization here cannot be performed by means of derivatives. Here the support of x depends on θ.)

Exercise 6.2 Consider an i.i.d. sample of size n from the bivariate population with pdf $f(x_1, x_2) = (\theta_1\theta_2)^{-1} \exp(-x_1/\theta_1 - x_2/\theta_2)$, $x_1, x_2 > 0$. Compute the MLE of $\theta = (\theta_1, \theta_2)$. Find the Cramer–Rao lower bound. Is it possible to derive a minimal variance unbiased estimator of θ?

Exercise 6.3 Show that the MLE of Example 6.1, $\widehat{\mu} = \overline{x}$, is a minimal variance estimator for any finite sample size n (i.e., without applying Theorem 6.3).

Exercise 6.4 We know from Example 6.4 that the MLE of Example 6.1 has $\mathcal{F}_1 = \mathcal{I}_p$. This leads to

$$\sqrt{n}(\overline{x} - \mu) \xrightarrow{\mathcal{L}} N_p(0, \mathcal{I})$$

by Theorem 6.3. Can you give an analogous result for the square $\overline{x}^2$ for the case $p = 1$?

Exercise 6.5 Consider an i.i.d. sample of size n from the bivariate population with pdf $f(x_1, x_2) = (\theta_1^2\theta_2 x_2)^{-1} \exp(-x_1/\theta_1 x_2 - x_2/\theta_1\theta_2)$, $x_1, x_2 > 0$. Compute the MLE of $\theta = (\theta_1, \theta_2)$. Find the Cramer–Rao lower bound and the asymptotic variance of $\widehat{\theta}$.

Exercise 6.6 Consider a sample $\{x_i\}_{i=1}^n$ from $N_p(\mu, \Sigma_0)$ where Σ_0 is known. Compute the Cramer–Rao lower bound for μ. Can you derive a minimal unbiased estimator for μ?

Exercise 6.7 Let $X \sim N_p(\mu, \Sigma)$ where Σ is unknown but we know $\Sigma = \text{diag}(\sigma_{11}, \sigma_{22}, \ldots, \sigma_{pp})$. From an i.i.d. sample of size n, find the MLE of μ and of Σ.

Exercise 6.8 Reconsider the setup of the previous exercise. Suppose that

$$\Sigma = \text{diag}(\sigma_{11}, \sigma_{22}, \ldots, \sigma_{pp}).$$

Can you derive in this case the Cramer–Rao lower bound for $\theta^\top = (\mu_1 \ldots \mu_p, \sigma_{11} \ldots \sigma_{pp})$?

Exercise 6.9 Prove Theorem 6.1. (Hint: start from $\frac{\partial}{\partial\theta}\mathsf{E}(t^\top) = \frac{\partial}{\partial\theta}\int t^\top(\mathcal{X}; \theta) L(\mathcal{X}; \theta)d\mathcal{X}$, then permute integral and derivatives and note that $s(\mathcal{X}; \theta) = \frac{1}{L(\mathcal{X};\theta)}\frac{\partial}{\partial\theta}L(\mathcal{X}; \theta)$.)

Exercise 6.10 Prove expression (6.12). (Hint: start from $\mathsf{E}\{s(\mathcal{X};\theta)\} = \int \frac{1}{L(\mathcal{X};\theta)}$ $\frac{\partial}{\partial\theta}L(\mathcal{X};\theta)L(\mathcal{X};\theta)\partial\mathcal{X}$ and then permute integral and derivatives.)

Reference

K.V. Mardia, J.T. Kent, J.M. Bibby, *Multivariate Analysis* (Academic Press, Duluth, London, 1979)

Chapter 7
Hypothesis Testing

In the preceding chapter, the theoretical basis of estimation theory was presented. Now, we turn our interest toward testing issues: we want to test the hypothesis H_0 that the unknown parameter θ belongs to some subspace of $\mathbb{R}^q$. This subspace is called the *null set* and will be denoted by $\Omega_0 \subset \mathbb{R}^q$.

In many cases, this null set corresponds to restrictions which are imposed on the parameter space: H_0 corresponds to a "reduced model". As we have already seen in Chap. 3, the solution to a testing problem is in terms of a *rejection region* R which is a set of values in the sample space which leads to the decision of rejecting the null hypothesis H_0 in favor of an alternative H_1, which is called the "full model".

In general, we want to construct a rejection region R which controls the size of the type I error, i.e., the probability of rejecting the null hypothesis when it is true. More formally, a solution to a testing problem is of predetermined size α if

$$P(\text{Rejecting } H_0 \mid H_0 \text{ is true}) = \alpha.$$

In fact, since H_0 is often a composite hypothesis, it is achieved by finding R such that

$$\sup_{\theta \in \Omega_0} P(\mathcal{X} \in R \mid \theta) = \alpha.$$

In this chapter, we will introduce a tool which allows us to build a rejection region in general situations; it is based on the likelihood ratio principle. This is a very useful technique because it allows us to derive a rejection region with an asymptotically appropriate size α. The technique will be illustrated through various testing problems and examples. We concentrate on multinormal populations and linear models, where the size of the test will often be exact even for finite sample sizes n.

Section 7.1 gives the basic ideas and Sect. 7.2 presents the general problem of testing linear restrictions. This allows us to propose solutions to frequent types of analyses (including comparisons of several means, repeated measurements, and profile

© Springer Nature Switzerland AG 2019
W. K. Härdle and L. Simar, *Applied Multivariate Statistical Analysis*,
https://doi.org/10.1007/978-3-030-26006-4_7

analysis). Each case can be viewed as a simple specific case of testing linear restrictions. Special attention is devoted to confidence intervals and confidence regions for means and for linear restrictions on means in a multinormal setup.

7.1 Likelihood Ratio Test

Suppose that the distribution of $\{x_i\}_{i=1}^n$, $x_i \in \mathbb{R}^p$, depends on a parameter vector θ. We will consider two hypotheses:

$$H_0 : \theta \in \Omega_0$$
$$H_1 : \theta \in \Omega_1.$$

The hypothesis H_0 corresponds to the "reduced model" and H_1 to the "full model". This notation was already used in Chap. 3.

Example 7.1 Consider a multinormal $N_p(\theta, \mathcal{I})$. To test if θ equals a certain fixed value θ_0 we construct the test problem:

$$H_0 : \theta = \theta_0$$
$$H_1 : \text{no constraints on } \theta$$

or, equivalently, $\Omega_0 = \{\theta_0\}$, $\Omega_1 = \mathbb{R}^p$.

Define $L_j^* = \max\limits_{\theta \in \Omega_j} L(\mathcal{X}; \theta)$, the maxima of the likelihood for each of the hypotheses. Consider the *likelihood ratio* (LR)

$$\lambda(\mathcal{X}) = \frac{L_0^*}{L_1^*} \tag{7.1}$$

One tends to favor H_0 if the LR is high and H_1 if the LR is low. The *likelihood ratio test* (LRT) tells us when exactly to favor H_0 over H_1. A likelihood ratio test of size α for testing H_0 against H_1 has the rejection region

$$R = \{\mathcal{X} : \lambda(\mathcal{X}) < c\},$$

where c is determined so that $\sup\limits_{\theta \in \Omega_0} P_\theta(\mathcal{X} \in R) = \alpha$. The difficulty here is to express c as a function of α, because $\lambda(\mathcal{X})$ might be a complicated function of $\mathcal{X}$.

Instead of λ we may equivalently use the log-likelihood

$$-2 \log \lambda = 2(\ell_1^* - \ell_0^*).$$

In this case, the rejection region will be $R = \{\mathcal{X} : -2\log\lambda(\mathcal{X}) > k\}$. What is the distribution of λ or of $-2\log\lambda$ from which we need to compute c or k?

Theorem 7.1 (Wilks Theorem) *If $\Omega_1 \subset \mathbb{R}^q$ is a q-dimensional space and if $\Omega_0 \subset \Omega_1$ is an r-dimensional subspace, then under regularity conditions*

$$\forall \theta \in \Omega_0 : -2\log\lambda \xrightarrow{\mathcal{L}} \chi^2_{q-r} \quad as \ n \to \infty.$$

An asymptotic rejection region can now be given by simply computing the $1 - \alpha$ quantile $k = \chi^2_{1-\alpha;q-r}$. The LRT rejection region is, therefore,

$$R = \{\mathcal{X} : -2\log\lambda(\mathcal{X}) > \chi^2_{1-\alpha;q-r}\}.$$

Theorem 7.1 is thus very helpful: it gives a general way of building rejection regions into many problems. Unfortunately, it is only an asymptotic result, meaning that the size of the test is only approximately equal to α, although the approximation becomes better when the sample size n increases. The question is "how large should n be?" There is no definite rule: we encounter here the same problem that was already discussed with respect to the Central Limit Theorem in Chap. 4.

Fortunately, in many standard circumstances, we can derive exact tests even for finite samples because the test statistic $-2\log\lambda(\mathcal{X})$ or a simple transformation of it turns out to have a simple form. This is the case in most of the following standard testing problems. All of them can be viewed as an illustration of the likelihood ratio principle.

Test Problem 1 is an *amuse-bouche*: in testing the mean of a multinormal population with a known covariance matrix the likelihood ratio statistic has a very simple quadratic form with a known distribution under H_0.

Test Problem 1 Suppose that $X_1, \ldots, X_n$ is an i.i.d. random sample from a $N_p(\mu, \Sigma)$ population.

$$H_0 : \mu = \mu_0, \ \Sigma \ \text{known versus} \ H_1 : \ \text{no constraints.}$$

In this case H_0 is a simple hypothesis, i.e., $\Omega_0 = \{\mu_0\}$ and therefore the dimension r of Ω_0 equals 0. Since we have imposed no constraints in H_1, the space Ω_1 is the whole $\mathbb{R}^p$ which leads to $q = p$. From (6.6) we know that

$$\ell_0^* = \ell(\mu_0, \Sigma) = -\frac{n}{2}\log|2\pi\Sigma| - \frac{1}{2}n\,\mathrm{tr}(\Sigma^{-1}\mathcal{S}) - \frac{1}{2}n(\bar{x} - \mu_0)^\top \Sigma^{-1}(\bar{x} - \mu_0).$$

Under H_1, the maximum of $\ell(\mu, \Sigma)$ is

$$\ell_1^* = \ell(\bar{x}, \Sigma) = -\frac{n}{2}\log|2\pi\Sigma| - \frac{1}{2}n\,\mathrm{tr}(\Sigma^{-1}\mathcal{S}).$$

Therefore,

$$-2 \log \lambda = 2(\ell_1^* - \ell_0^*) = n(\overline{x} - \mu_0)^\top \Sigma^{-1}(\overline{x} - \mu_0) \qquad (7.2)$$

which, by Theorem 4.7, has a χ_p^2-distribution under H_0.

Example 7.2 Consider the bank data again. Let us test whether the population mean of the forged banknotes is equal to

$$\mu_0 = (214.9, 129.9, 129.7, 8.3, 10.1, 141.5)^\top.$$

(This is in fact the sample mean of the genuine banknotes.) The sample mean of the forged banknotes is

$$\overline{x} = (214.8, 130.3, 130.2, 10.5, 11.1, 139.4)^\top.$$

Suppose for the moment that the estimated covariance matrix $\mathcal{S}_f$ given in (3.5) is the true covariance matrix Σ. We construct the likelihood ratio test and obtain

$$\begin{aligned} -2 \log \lambda &= 2(\ell_1^* - \ell_0^*) = n(\overline{x} - \mu_0)^\top \Sigma^{-1}(\overline{x} - \mu_0) \\ &= 7362.32, \end{aligned}$$

the quantile $k = \chi_{0.95;6}^2$ equals 12.592. The rejection consists of all values in the sample space, which lead to values of the likelihood ratio test statistic larger than 12.592. Under H_0, the value of $-2 \log \lambda$ is, therefore, highly significant. Hence, the true mean of the forged banknotes is significantly different from μ_0!

Test Problem 2 is the same as the preceding one but in a more realistic situation, where the covariance matrix is unknown; here the Hotelling's T^2-distribution will be useful to determine an exact test and a confidence region for the unknown μ.

> **Test Problem 2** Suppose that $X_1, \ldots, X_n$ is an i.i.d. random sample from a $N_p(\mu, \Sigma)$ population.
>
> $$H_0 : \mu = \mu_0, \ \Sigma \text{ unknown versus } H_1 : \text{ no constraints.}$$

Under H_0, it can be shown that

$$\begin{aligned} \mathcal{S}_0 &= \frac{1}{n}\left[x - 1_n\mu_0^\top - 1_n\overline{x}^\top + 1_n\overline{x}^\top\right]^\top \left[x - 1_n\mu_0^\top - 1_n\overline{x}^\top + 1_n\overline{x}^\top\right] \\ &= \mathcal{S} + (\overline{x} - \mu_0)(\overline{x} - \mu_0)^\top \\ \ell_0^* &= \ell(\mu_0, \mathcal{S} + dd^\top), \quad d = (\overline{x} - \mu_0) \end{aligned} \qquad (7.3)$$

and under H_1, we have

$$\ell_1^* = \ell(\overline{x}, \mathcal{S}).$$

This leads after some calculation to

$$
\begin{aligned}
-2\log\lambda &= 2(\ell_1^* - \ell_0^*)\\
&= -n\log|\mathcal{S}| - n\operatorname{tr}(\mathcal{S}^{-1}\mathcal{S}) - n\,(\overline{x}-\overline{x})^{\top}\mathcal{S}^{-1}\,(\overline{x}-\overline{x}) + n\log|\mathcal{S}+dd^{\top}|\\
&\quad + n\operatorname{tr}\left[(\mathcal{S}+dd^{\top})^{-1}\mathcal{S}\right] + n\,(\overline{x}-\mu_0)^{\top}\,(\mathcal{S}+dd^{\top})^{-1}\,(\overline{x}-\mu_0)\\
&= n\log\left|\frac{\mathcal{S}+dd^{\top}}{\mathcal{S}}\right| + n\operatorname{tr}\left[(\mathcal{S}+dd^{\top})^{-1}\mathcal{S}\right] + nd^{\top}(\mathcal{S}+dd^{\top})^{-1}d - np\\
&= n\log\left|\frac{\mathcal{S}+dd^{\top}}{\mathcal{S}}\right| + n\operatorname{tr}\left[(\mathcal{S}+dd^{\top})^{-1}(dd^{\top}+\mathcal{S})\right] - np\\
&= n\log\left|\frac{\mathcal{S}+dd^{\top}}{\mathcal{S}}\right|\\
&= n\log|1 + \mathcal{S}^{-1/2}dd^{\top}\mathcal{S}^{-1/2}|
\end{aligned}
$$

By using the result for the determinant of a partitioned matrix, it equals to

$$
n\log\begin{vmatrix} 1 & -d^{\top}\mathcal{S}^{-1/2}\\ \mathcal{S}^{-1/2}d & I \end{vmatrix}
$$

$$
= n\log\begin{vmatrix} 1 & -d^{\top}\mathcal{S}^{-1/2}{}_1 & -d^{\top}\mathcal{S}^{-1/2}{}_2 & \cdots & -d^{\top}\mathcal{S}^{-1/2}{}_p\\ \mathcal{S}^{-1/2}d_1 & 1 & 0 & \cdots & 0\\ \mathcal{S}^{-1/2}d_2 & 0 & 1 & & 0\\ \vdots & \vdots & & \ddots & \\ \mathcal{S}^{-1/2}d_p & 0 & 0 & \cdots & 1 \end{vmatrix}
$$

$$
= n\log 1 + n\log\sum_{i=1}^{p} -d^{\top}\mathcal{S}^{-1/2}{}_i(-1)^{1+(i+1)}\begin{vmatrix} \mathcal{S}^{-1/2}d_1 & 1 & 0 & \cdots & 0\\ \mathcal{S}^{-1/2}d_2 & 0 & 1 & \cdots & 0\\ \vdots & & & \ddots & \\ \mathcal{S}^{-1/2}d_i & 0 & 0 & \cdots & 0\\ \vdots & & & & \\ \mathcal{S}^{-1/2}d_p & 0 & 0 & \cdots & 1 \end{vmatrix}
$$

$$
= n\log 1 + \sum_{i=1}^{p} -d^{\top}\mathcal{S}^{-1/2}{}_i(-1)^{2+i}\mathcal{S}^{-1/2}d_i(-1)^{i+1}
$$

$$
= n\log(1 + d^{\top}\mathcal{S}^{-1}d). \tag{7.4}
$$

This statistic is a monotone function of $(n-1)d^{\top}\mathcal{S}^{-1}d$. This means that $-2\log\lambda > k$ if and only if $(n-1)d^{\top}\mathcal{S}^{-1}d > k'$. The latter statistic has by Corollary 5.3, under H_0, a Hotelling's T^2-distribution. Therefore,

$$
(n-1)(\overline{x}-\mu_0)^{\top}\mathcal{S}^{-1}(\overline{x}-\mu_0) \sim T^2_{p,n-1}, \tag{7.5}
$$

or equivalently

$$\left(\frac{n-p}{p}\right)(\bar{x}-\mu_0)^\top S^{-1}(\bar{x}-\mu_0) \sim F_{p,n-p}. \qquad (7.6)$$

In this case, an exact rejection region may be defined as

$$\left(\frac{n-p}{p}\right)(\bar{x}-\mu_0)^\top S^{-1}(\bar{x}-\mu_0) > F_{1-\alpha;p,n-p}.$$

Alternatively, we have from Theorem 7.1 that under H_0 the asymptotic distribution of the test statistic is

$$-2\log \lambda \xrightarrow{\mathcal{L}} \chi^2_p, \quad \text{as } n \to \infty$$

which leads to the (asymptotically valid) rejection region

$$n \log\{1 + (\bar{x}-\mu_0)^\top S^{-1}(\bar{x}-\mu_0)\} > \chi^2_{1-\alpha;p},$$

but of course, in this case, we would prefer to use the exact F-test provided just above.

Example 7.3 Consider the problem of Example 7.2 again. We know that S_f is the empirical analogue for Σ_f, the covariance matrix for the forged banknotes. The test statistic (7.5) has the value 1153.4 or its equivalent for the F-distribution in (7.6) is 182.5 which is highly significant ($F_{0.95;6,94} = 2.1966$) so that we conclude that $\mu_f \neq \mu_0$.

Confidence Region for μ

When estimating a multidimensional parameter $\theta \in \mathbb{R}^k$ from a sample, we saw in Chap. 6 how to determine the estimator $\widehat{\theta} = \widehat{\theta}(\mathcal{X})$. For the observed data we end up with a point estimate, which is the corresponding observed value of $\widehat{\theta}$. We know $\widehat{\theta}(\mathcal{X})$ is a random variable and we often prefer to determine a *confidence region* for θ. A confidence region (CR) is a random subset of $\mathbb{R}^k$ (determined by appropriate statistics) such that we are "confident", at a certain given level $1 - \alpha$, that this region contains θ:

$$P(\theta \in CR) = 1 - \alpha.$$

This is just a multidimensional generalization of the basic univariate confidence interval. Confidence regions are particularly useful when a hypothesis H_0 on θ is rejected, because they eventually help in identifying which component of θ is responsible for the rejection.

There are only a few cases where confidence regions can be easily assessed, and include most of the testing problems on mean presented in this section.

Corollary 5.3 provides a pivotal quantity, which allows confidence regions for μ to be constructed. Since $\left(\frac{n-p}{p}\right)(\bar{x}-\mu)^\top S^{-1}(\bar{x}-\mu) \sim F_{p,n-p}$, we have

$$P\left\{\left(\frac{n-p}{p}\right)(\mu - \bar{x})^{\top}\mathcal{S}^{-1}(\mu - \bar{x}) < F_{1-\alpha;p,n-p}\right\} = 1 - \alpha.$$

Then,

$$CR = \left\{\mu \in \mathbb{R}^p \mid (\mu - \bar{x})^{\top}\mathcal{S}^{-1}(\mu - \bar{x}) \leq \frac{p}{n-p}F_{1-\alpha;p,n-p}\right\}$$

is a confidence region at level $(1 - \alpha)$ for μ. It is the interior of an iso-distance ellipsoid in $\mathbb{R}^p$ centered at $\bar{x}$, with a scaling matrix $\mathcal{S}^{-1}$ and a distance constant $\left(\frac{p}{n-p}\right)F_{1-\alpha;p,n-p}$. When p is large, ellipsoids are not easy to handle for practical purposes. One is thus interested in finding confidence intervals for $\mu_1, \mu_2, \ldots, \mu_p$ so that simultaneous confidence on all the intervals reaches the desired level of say, $1 - \alpha$.

Below, we consider a more general problem. We construct *simultaneous confidence intervals* for all possible linear combinations $a^{\top}\mu$, $a \in \mathbb{R}^p$ of the elements of μ.

Suppose for a moment that we fix a particular projection vector a. We are back to a standard univariate problem of finding a confidence interval for the mean $a^{\top}\mu$ of a univariate random variable $a^{\top}X$. We can use the t-statistic and an obvious confidence interval for $a^{\top}\mu$ is given by the values $a^{\top}\mu$ such that

$$\left|\frac{\sqrt{n-1}(a^{\top}\mu - a^{\top}\bar{x})}{\sqrt{a^{\top}\mathcal{S}a}}\right| \leq t_{1-\frac{\alpha}{2};n-1}$$

or equivalently

$$t^2(a) = \frac{(n-1)\left\{a^{\top}(\mu - \bar{x})\right\}^2}{a^{\top}\mathcal{S}a} \leq F_{1-\alpha;1,n-1}.$$

This provides the $(1 - \alpha)$ confidence interval for $a^{\top}\mu$:

$$\left(a^{\top}\bar{x} - \sqrt{F_{1-\alpha;1,n-1}\frac{a^{\top}\mathcal{S}a}{n-1}} \leq a^{\top}\mu \leq a^{\top}\bar{x} + \sqrt{F_{1-\alpha;1,n-1}\frac{a^{\top}\mathcal{S}a}{n-1}}\right).$$

Now, it is easy to prove (using Theorem 2.5) that

$$\max_a t^2(a) = (n - 1)(\bar{x} - \mu)^{\top}\mathcal{S}^{-1}(\bar{x} - \mu) \sim T^2_{p,n-1}.$$

Therefore, simultaneously for all $a \in \mathbb{R}^p$, the interval

$$\left(a^{\top}\bar{x} - \sqrt{K_\alpha a^{\top}\mathcal{S}a}, \ a^{\top}\bar{x} + \sqrt{K_\alpha a^{\top}\mathcal{S}a}\right), \tag{7.7}$$

where $K_\alpha = \frac{p}{n-p}F_{1-\alpha;p,n-p}$, will contain $a^{\top}\mu$ with probability $(1 - \alpha)$.

A particular choice of a are the columns of the identity matrix $\mathcal{I}_p$, providing simultaneous confidence intervals for $\mu_1, \ldots, \mu_p$. We, therefore, have with probability $(1 - \alpha)$ for $j = 1, \ldots, p$

$$\bar{x}_j - \sqrt{\frac{p}{n-p} F_{1-\alpha;p,n-p} s_{jj}} \le \mu_j \le \bar{x}_j + \sqrt{\frac{p}{n-p} F_{1-\alpha;p,n-p} s_{jj}}. \qquad (7.8)$$

It should be noted that these intervals define a rectangle inscribing the confidence ellipsoid for μ given above. They are particularly useful when a null hypothesis H_0 of the type described above is rejected and one would like to see which component(s) are mainly responsible for the rejection.

Example 7.4 The 95% confidence region for μ_f, the mean of the forged banknotes, is given by the ellipsoid:

$$\left\{ \mu \in \mathbb{R}^6 \,\middle|\, (\mu - \bar{x}_f)^\top S_f^{-1} (\mu - \bar{x}_f) \le \frac{6}{94} F_{0.95;6,94} \right\}.$$

The 95% simultaneous confidence intervals are given by (we use $F_{0.95;6,94} = 2.1966$)

$$
\begin{aligned}
214.692 &\le \mu_1 \le & 214.954 \\
130.205 &\le \mu_2 \le & 130.395 \\
130.082 &\le \mu_3 \le & 130.304 \\
10.108 &\le \mu_4 \le & 10.952 \\
10.896 &\le \mu_5 \le & 11.370 \\
139.242 &\le \mu_6 \le & 139.658.
\end{aligned}
$$

Comparing the inequalities with $\mu_0 = (214.9, 129.9, 129.7, 8.3, 10.1, 141.5)^\top$
    shows that almost all components (except the first one) are responsible for the rejection of μ_0 in Examples 7.2 and 7.3.
    In addition, the method can provide other confidence intervals. We have at the same level of confidence (choosing $a^\top = (0,\ 0,\ 0,\ 1,\ -1,\ 0)$)

$$-1.211 \le \mu_4 - \mu_5 \le 0.005$$

showing that for the forged bills, the lower border is essentially smaller than the upper border.

Remark 7.1 It should be noted that the confidence region is an ellipsoid, whose characteristics depend on the whole matrix S. In particular, the slope of the axis depends on the eigenvectors of S and therefore on the covariances s_{ij}. However, the rectangle inscribing the confidence ellipsoid provides the simultaneous confidence intervals for μ_j, $j = 1, \ldots, p$. They do not depend on the covariances s_{ij}, but only on the variances s_{jj} (see (7.8)). In particular, it may happen that a tested value μ_0 is covered by the intervals (7.8) but not covered by the confidence ellipsoid. In this case, μ_0 is rejected by a test based on the confidence ellipsoid but not rejected by a test based on the simultaneous confidence intervals. The simultaneous confidence intervals are easier to handle than the full ellipsoid but we have lost some information, namely, the covariance between the components (see Exercise 7.14).

The following problem concerns the covariance matrix in a multinormal population: in this situation, the test statistic has a slightly more complicated distribution. We will, therefore, invoke the approximation of Theorem 7.1 in order to derive a test of approximate size α.

Test Problem 3 Suppose that $X_1, \ldots, X_n$ is an i.i.d. random sample from a $N_p(\mu, \Sigma)$ population.

$$H_0 : \Sigma = \Sigma_0, \ \mu \text{ unknown versus } H_1 : \text{ no constraints.}$$

Under H_0 we have $\widehat{\mu} = \overline{x}$, and $\Sigma = \Sigma_0$, whereas under H_1 we have $\widehat{\mu} = \overline{x}$, and $\widehat{\Sigma} = S$. Hence

$$\ell_0^* = \ell(\overline{x}, \Sigma_0) = -\frac{1}{2}n \log |2\pi \Sigma_0| - \frac{1}{2}n \operatorname{tr}(\Sigma_0^{-1} S)$$

$$\ell_1^* = \ell(\overline{x}, S) = -\frac{1}{2}n \log |2\pi S| - \frac{1}{2}np$$

and thus

$$-2 \log \lambda = 2(\ell_1^* - \ell_0^*)$$
$$= n \operatorname{tr}(\Sigma_0^{-1} S) - n \log |\Sigma_0^{-1} S| - np.$$

Note that this statistic is a function of the eigenvalues of $\Sigma_0^{-1} S$. Unfortunately, the exact finite sample distribution of $-2 \log \lambda$ is very complicated. Asymptotically, we have under H_0

$$-2 \log \lambda \xrightarrow{\mathcal{L}} \chi_m^2 \quad \text{as} \ n \to \infty$$

with $m = \frac{1}{2}\{p(p+1)\}$, since a $(p \times p)$ covariance matrix has only these m parameters as a consequence of its symmetry.

Example 7.5 Consider the US companies data set (Sect. 22.5) and suppose we are interested in the companies of the energy sector, analyzing their assets (X_1) and sales (X_2). The sample is of size 15 and provides the value of $S = 10^7 \times \begin{bmatrix} 1.6635 & 1.2410 \\ 1.2410 & 1.3747 \end{bmatrix}$. We want to test if $\operatorname{Var}\begin{pmatrix} X_1 \\ X_2 \end{pmatrix} = 10^7 \times \begin{bmatrix} 1.2248 & 1.1425 \\ 1.1425 & 1.5112 \end{bmatrix} = \Sigma_0$. ($\Sigma_0$ is in fact the empirical variance matrix for X_1 and X_2 for the manufacturing sector). The test statistic (⬤MVAusenergy) turns out to be $-2 \log \lambda = 5.4046$, which is not significant for χ_3^2 (p-value=0.1445). So we cannot conclude that $\Sigma \neq \Sigma_0$.

In the next testing problem, we address a question that was already stated in Chap. 3, Sect. 3.6: testing a particular value of the coefficients β in a linear model.

The presentation is carried out in general terms so that it can be built on in the next section where we will test linear restrictions on β.

Test Problem 4 Suppose that $Y_1, \ldots, Y_n$ are independent r.v.'s with $Y_i \sim N_1(\beta^\top x_i, \sigma^2)$, $x_i \in \mathbb{R}^p$.

$$H_0 : \beta = \beta_0, \ \sigma^2 \text{ unknown versus } H_1 : \text{ no constraints.}$$

Under H_0, we have $\beta = \beta_0$, $\widehat{\sigma}_0^2 = \frac{1}{n}||y - \mathcal{X}\beta_0||^2$ and under H_1 we have $\widehat{\beta} = (\mathcal{X}^\top \mathcal{X})^{-1}\mathcal{X}^\top y$, $\widehat{\sigma}^2 = \frac{1}{n}||y - \mathcal{X}\widehat{\beta}||^2$ (see Example 6.3). Hence by Theorem 7.1

$$-2\log \lambda = 2(\ell_1^* - \ell_0^*)$$
$$= n \log \left(\frac{||y - \mathcal{X}\beta_0||^2}{||y - \mathcal{X}\widehat{\beta}||^2} \right)$$
$$\overset{\mathcal{L}}{\longrightarrow} \chi_p^2.$$

We draw upon the result (3.45) which gives us

$$F = \frac{(n-p)}{p} \left(\frac{||y - \mathcal{X}\beta_0||^2}{||y - \mathcal{X}\widehat{\beta}||^2} - 1 \right) \sim F_{p,n-p},$$

so that in this case we again have an exact distribution.

Example 7.6 Let us consider our "classic blue" pullovers again. In Example 3.11, we tried to model the dependency of sales on prices. As we have seen in Fig. 3.5 the slope of the regression curve is rather small, hence we might ask if $\binom{\alpha}{\beta} = \binom{211}{0}$. Here

$$y = \begin{pmatrix} y_1 \\ \vdots \\ y_{10} \end{pmatrix} = \begin{pmatrix} x_{1,1} \\ \vdots \\ x_{10,1} \end{pmatrix}, \qquad \mathcal{X} = \begin{pmatrix} 1 & x_{1,2} \\ \vdots & \vdots \\ 1 & x_{10,2} \end{pmatrix}.$$

The test statistic for the LR test is

$$-2\log \lambda = 9.10$$

which under the χ_2^2 distribution is significant. The exact F-test statistic

$$F = 5.93$$

is also significant under the $F_{2,8}$ distribution ($F_{2,8;0.95} = 4.46$).

Summary
$\hookrightarrow$ The hypotheses $H_0 : \theta \in \Omega_0$ against $H_1 : \theta \in \Omega_1$ can be tested using the likelihood ratio test (LRT). The likelihood ratio (LR) is the quotient $\lambda(\mathcal{X}) = L_0^*/L_1^*$, where the L_j^* are the maxima of the likelihood for each of the hypotheses.
$\hookrightarrow$ The test statistic in the LRT is $\lambda(\mathcal{X})$ or equivalently its logarithm $\log \lambda(\mathcal{X})$. If Ω_1 is q-dimensional and $\Omega_0 \subset \Omega_1$ r-dimensional, then the asymptotic distribution of $-2 \log \lambda$ is χ^2_{q-r}. This allows H_0 to be tested against H_1 by calculating the test statistic $-2 \log \lambda = 2(\ell_1^* - \ell_0^*)$ where $\ell_j^* = \log L_j^*$.
$\hookrightarrow$ The hypothesis $H_0 : \mu = \mu_0$ for $X \sim N_p(\mu, \Sigma)$, where Σ is known, leads to $-2 \log \lambda = n(\overline{x} - \mu_0)^{\top} \Sigma^{-1}(\overline{x} - \mu_0) \sim \chi^2_p$.
$\hookrightarrow$ The hypothesis $H_0 : \mu = \mu_0$ for $X \sim N_p(\mu, \Sigma)$, where Σ is unknown, leads to $-2 \log \lambda = n \log\{1 + (\overline{x} - \mu_0)^{\top} S^{-1}(\overline{x} - \mu_0)\} \xrightarrow{\mathcal{L}} \chi^2_p$, and $(n-1)(\overline{x} - \mu_0)^{\top} S^{-1}(\overline{x} - \mu_0) \sim T^2_{p,n-1}$.
$\hookrightarrow$ The hypothesis $H_0 : \Sigma = \Sigma_0$ for $X \sim N_p(\mu, \Sigma)$, where μ is unknown, leads to $-2 \log \lambda = n \operatorname{tr}\left(\Sigma_0^{-1} S\right) - n \log |\Sigma_0^{-1} S| - np \xrightarrow{\mathcal{L}} \chi^2_m$, $m = \frac{1}{2}p(p+1)$.
$\hookrightarrow$ The hypothesis $H_0 : \beta = \beta_0$ for $Y_i \sim N_1(\beta^{\top} x_i, \sigma^2)$, where σ^2 is unknown, leads to $-2 \log \lambda = n \log \left(\frac{\|y - \mathcal{X}\beta_0\|^2}{\|y - \mathcal{X}\beta\|^2}\right) \xrightarrow{\mathcal{L}} \chi^2_p$.

7.2 Linear Hypothesis

In this section, we present a very general procedure which allows a linear hypothesis to be tested, i.e., a linear restriction, either on a vector mean μ or on the coefficient β of a linear model. The presented technique covers many of the practical testing problems on means or regression coefficients.

Linear hypotheses are of the form $\mathcal{A}\mu = a$ with known matrices $\mathcal{A}(q \times p)$ and $a(q \times 1)$ with $q \le p$.

Example 7.7 Let $\mu = (\mu_1, \mu_2)^{\top}$. The hypothesis that $\mu_1 = \mu_2$ can be equivalently written as

$$\mathcal{A}\mu = \begin{pmatrix} 1 & -1 \end{pmatrix} \begin{pmatrix} \mu_1 \\ \mu_2 \end{pmatrix} = 0 = a.$$

The general idea is to test a normal population $H_0 : \mathcal{A}\mu = a$ (restricted model) against the full model H_1, where no restrictions are put on μ. Due to the properties of the multinormal, we can easily adapt the Test Problems 1 and 2 to this new situation. Indeed we know, from Theorem 5.2, that $y_i = \mathcal{A}x_i \sim N_q(\mu_y, \Sigma_y)$, where $\mu_y = \mathcal{A}\mu$ and $\Sigma_y = \mathcal{A}\Sigma\mathcal{A}^{\top}$.

Testing the null $H_0 : \mathcal{A}\mu = a$, is the same as testing $H_0 : \mu_y = a$. The appropriate statistics are $\overline{y}$ and S_y, which can be derived from the original statistics $\overline{x}$ and S

available from $\mathcal{X}$:

$$\bar{y} = \mathcal{A}\bar{x}, \quad \mathcal{S}_y = \mathcal{A}\mathcal{S}\mathcal{A}^\top.$$

Here the difference between the translated sample mean and the tested value is $d = \mathcal{A}\bar{x} - a$. We are now in the situation to proceed to Test Problems 5 and 6.

Test Problem 5 Suppose $X_1, \ldots, X_n$ is an i.i.d. random sample from a $N_p(\mu, \Sigma)$ population.

$$H_0 : \mathcal{A}\mu = a, \ \Sigma \text{ known versus } H_1 : \text{no constraints.}$$

By (7.2) we have that, under H_0:

$$n(\mathcal{A}\bar{x} - a)^\top (\mathcal{A}\Sigma\mathcal{A}^\top)^{-1}(\mathcal{A}\bar{x} - a) \sim \mathcal{X}_q^2,$$

and we reject H_0 if this test statistic is too large at the desired significance level.

Example 7.8 We consider hypotheses on partitioned mean vectors $\mu = \begin{pmatrix} \mu_1 \\ \mu_2 \end{pmatrix}$. Let us first look at

$$H_0 : \mu_1 = \mu_2, \text{ versus } H_1 : \text{no constraints,}$$

for $N_{2p}\left(\begin{pmatrix} \mu_1 \\ \mu_2 \end{pmatrix}, \begin{pmatrix} \Sigma & 0 \\ 0 & \Sigma \end{pmatrix}\right)$ with known Σ. This is equivalent to $\mathcal{A} = (\mathcal{I}, -\mathcal{I})$, $a = (0, \ldots, 0)^\top \in \mathbb{R}^p$ and leads to

$$-2 \log \lambda = n(\bar{x}_1 - \bar{x}_2)(2\Sigma)^{-1}(\bar{x}_1 - \bar{x}_2) \sim \chi_p^2.$$

Another example is the test whether $\mu_1 = 0$, i.e.,

$$H_0 : \mu_1 = 0, \text{ versus } H_1 : \text{no constraints,}$$

for $N_{2p}\left(\begin{pmatrix} \mu_1 \\ \mu_2 \end{pmatrix}, \begin{pmatrix} \Sigma & 0 \\ 0 & \Sigma \end{pmatrix}\right)$ with known Σ. This is equivalent to $\mathcal{A}\mu = a$ with $\mathcal{A} = (\mathcal{I}, 0)$, and $a = (0, \ldots, 0)^\top \in \mathbb{R}^p$. Hence

$$-2 \log \lambda = n\bar{x}_1 \Sigma^{-1}\bar{x}_1 \sim \chi_p^2.$$

Test Problem 6 Suppose $X_1, \ldots, X_n$ is an i.i.d. random sample from a $N_p(\mu, \Sigma)$ population.

$$H_0 : \mathcal{A}\mu = a, \ \Sigma \text{ unknown versus } H_1 : \text{no constraints.}$$

From Corollary (5.4) and under H_0 it follows immediately that

$$(n - 1)(\mathcal{A}\bar{x} - a)^{\top}(\mathcal{A}\mathcal{S}\mathcal{A}^{\top})^{-1}(\mathcal{A}\bar{x} - a) \sim T^2_{q,n-1} \qquad (7.9)$$

since indeed under H_0

$$\mathcal{A}\bar{x} \sim N_q(a, n^{-1}\mathcal{A}\Sigma\mathcal{A}^{\top})$$

is independent of

$$n\mathcal{A}\mathcal{S}\mathcal{A}^{\top} \sim W_q(\mathcal{A}\Sigma\mathcal{A}^{\top}, n - 1).$$

Example 7.9 Let's come back again to the bank data set and suppose that we want to test if $\mu_4 = \mu_5$, i.e., the hypothesis that the lower border mean equals the larger border mean for the forged bills. In this case

$$\mathcal{A} = (0\,0\,0\,1 - 1\,0)$$
$$a = 0.$$

The test statistic is

$$99(\mathcal{A}\bar{x})^{\top}(\mathcal{A}\mathcal{S}_f\mathcal{A}^{\top})^{-1}(\mathcal{A}\bar{x}) \sim T^2_{1,99} = F_{1,99}.$$

The observed value is 13.638, which is significant at the 5% level.

Repeated Measurements

In many situations, n independent sampling units are observed at p different times or under p different experimental conditions (different treatments, ...). So here we repeat p one-dimensional measurements on n different subjects. For instance, we observe the results from n students taking p different exams. We end up with a $(n \times p)$ matrix. We can thus consider the situation, where we have $X_1, \ldots, X_n$ i.i.d. from a normal distribution $N_p(\mu, \Sigma)$ when there are p repeated measurements. The hypothesis of interest in this case is that there are no treatment effects, $H_0 : \mu_1 = \mu_2 = \ldots = \mu_p$. This hypothesis is a direct application of Test Problem 6. Indeed, introducing an appropriate matrix transform on μ we have

$$H_0 : \mathcal{C}\mu = 0 \text{ where } \mathcal{C}((p-1) \times p) = \begin{pmatrix} 1 & -1 & 0 & \cdots & 0 \\ 0 & 1 & -1 & \cdots & 0 \\ \vdots & \vdots & \vdots & \vdots & \vdots \\ 0 & \cdots & 0 & & 1 & -1 \end{pmatrix}. \qquad (7.10)$$

Note that in many cases one of the experimental conditions is the "control" (a placebo, standard drug or reference condition). Suppose it is the first component. In that case, one is interested in studying differences to the control variable. The matrix $\mathcal{C}$ has, therefore, a different form

$$\mathcal{C}((p-1) \times p) = \begin{pmatrix} 1 & -1 & 0 & \cdots & 0 \\ 1 & 0 & -1 & \cdots & 0 \\ \vdots & \vdots & \vdots & \vdots & \vdots \\ 1 & 0 & 0 & \cdots & -1 \end{pmatrix}.$$

By (7.9), the null hypothesis will be rejected if:

$$\frac{(n-p+1)}{p-1} \bar{x}^{\top} \mathcal{C}^{\top} (\mathcal{C}\mathcal{S}\mathcal{C}^{\top})^{-1} \mathcal{C}\bar{x} > F_{1-\alpha; p-1, n-p+1}.$$

As a matter of fact, $\mathcal{C}\mu$ is the mean of the random variable $y_i = \mathcal{C}x_i$

$$y_i \sim N_{p-1}(\mathcal{C}\mu, \mathcal{C}\Sigma\mathcal{C}^{\top}).$$

Simultaneous confidence intervals for linear combinations of the mean of y_i have been derived above in (7.7). For all $a \in \mathbb{R}^{p-1}$, with probability $(1 - \alpha)$ we have

$$a^{\top}\mathcal{C}\mu \in a^{\top}\mathcal{C}\bar{x} \pm \sqrt{\frac{(p-1)}{n-p+1} F_{1-\alpha; p-1, n-p+1} a^{\top}\mathcal{C}\mathcal{S}\mathcal{C}^{\top}a}.$$

Due to the nature of the problem here, the row sums of the elements in $\mathcal{C}$ are zero: $\mathcal{C}1_p = 0$, therefore $a^{\top}\mathcal{C}$ is a vector having sum of elements equals to 0 . This is called a *contrast*. Let $b = \mathcal{C}^{\top}a$. We have $b^{\top}1_p = \sum_{j=1}^{p} b_j = 0$. The result above thus provides for all contrasts of μ, and $b^{\top}\mu$ simultaneous confidence intervals at level $(1 - \alpha)$

$$b^{\top}\mu \in b^{\top}\bar{x} \pm \sqrt{\frac{(p-1)}{n-p+1} F_{1-\alpha; p-1, n-p+1} b^{\top}\mathcal{S}b}.$$

Examples of contrasts for $p = 4$ are $b^{\top} = (1 \ -1 \ 0 \ 0)$ or $(1 \ 0 \ 0 \ -1)$ or even $(1 \ -\frac{1}{3} \ -\frac{1}{3} \ -\frac{1}{3})$ when the control is to be compared with the mean of 3 different treatments.

Example 7.10 Bock (1975) considers the evolution of the vocabulary of children from the eighth through eleventh grade. The data set contains the scores of a vocabulary test of 40 randomly chosen children. This is a repeated measurement situation, ($n = 40$, $p = 4$), since the same children were observed from grades 8 to 11. The statistics of interest are

$$\bar{x} = (1.086, 2.544, 2.851, 3.420)^{\top}$$

$$S = \begin{pmatrix} 2.902 & 2.438 & 2.963 & 2.183 \\ 2.438 & 3.049 & 2.775 & 2.319 \\ 2.963 & 2.775 & 4.281 & 2.939 \\ 2.183 & 2.319 & 2.939 & 3.162 \end{pmatrix}.$$

Suppose we are interested in the yearly evolution of the children. Then the matrix $\mathcal{C}$ providing successive differences of μ_j is

$$\mathcal{C} = \begin{pmatrix} 1 & -1 & 0 & 0 \\ 0 & 1 & -1 & 0 \\ 0 & 0 & 1 & -1 \end{pmatrix}.$$

The value of the test statistic is $F_{\text{obs}} = 53.134$, which is highly significant for $F_{3,37}$. There are significant differences between the successive means. However, the analysis of the contrasts shows the following simultaneous 95% confidence intervals:

$$-1.958 \le \mu_1 - \mu_2 \le -0.959$$
$$-0.949 \le \mu_2 - \mu_3 \le 0.335$$
$$-1.171 \le \mu_3 - \mu_4 \le 0.036.$$

Thus, the rejection of H_0 is mainly due to the difference between the childrens' performances in the first and second year. The confidence intervals for the following contrasts may also be of interest:

$$-2.283 \le \mu_1 - \tfrac{1}{3}(\mu_2 + \mu_3 + \mu_4) \le -1.423$$
$$-1.777 \le \tfrac{1}{3}(\mu_1 + \mu_2 + \mu_3) - \mu_4 \le -0.742$$
$$-1.479 \le \mu_2 - \mu_4 \le -0.272.$$

They show that μ_1 is different from the average of the 3 other years (the same being true for μ_4) and μ_4 turns out to be higher than μ_2 (and of course higher than μ_1).

Test Problem 7 illustrates how the likelihood ratio can be applied to testing a linear restriction on the coefficient β of a linear model. It is also shown how a transformation of the test statistic leads to an exact F-test as presented in Chap. 3.

Test Problem 7 Suppose $Y_1, \ldots, Y_n$, are independent with $Y_i \sim N_1(\beta^\top x_i, \sigma^2)$, and $x_i \in \mathbb{R}^p$.

$H_0 : \mathcal{A}\beta = a$, σ^2 unknown versus $H_1 :$ no constraints.

To get the constrained maximum likelihood estimators under H_0, let $f(\beta, \lambda) = (y - x\beta)^\top(y - x\beta) - \lambda^\top(\mathcal{A}\beta - a)$ where $\lambda \in \mathbb{R}^q$ and solve $\frac{\partial f(\beta,\lambda)}{\partial \beta} = 0$ and $\frac{\partial f(\beta,\lambda)}{\partial \lambda} = 0$ (Exercise 3.24), thus we obtain:

$$\tilde{\beta} = \widehat{\beta} - (\mathcal{X}^\top\mathcal{X})^{-1}\mathcal{A}^\top\{\mathcal{A}(\mathcal{X}^\top\mathcal{X})^{-1}\mathcal{A}^\top\}^{-1}(\mathcal{A}\widehat{\beta} - a)$$

for β and $\tilde{\sigma}^2 = \frac{1}{n}(y - \mathcal{X}\tilde{\beta})^\top(y - \mathcal{X}\tilde{\beta})$. The estimate $\widehat{\beta}$ denotes the unconstrained MLE as before. Hence, the LR statistic is

$$-2 \log \lambda = 2(\ell_1^* - \ell_0^*)$$

$$= n \log \left(\frac{\|y - \mathcal{X}\tilde{\beta}\|^2}{\|y - \mathcal{X}\widehat{\beta}\|^2} \right)$$

$$\xrightarrow{\mathcal{L}} \chi_q^2$$

where q is the number of elements of a. This problem also has an exact F-test since

$$\frac{n - p}{q} \left(\frac{\|y - \mathcal{X}\tilde{\beta}\|^2}{\|y - \mathcal{X}\widehat{\beta}\|^2} - 1 \right)$$

$$= \frac{n - p}{q} \frac{(A\widehat{\beta} - a)^\top \{A(\mathcal{X}^\top \mathcal{X})^{-1} A^\top\}^{-1}(A\widehat{\beta} - a)}{(y - \mathcal{X}\widehat{\beta})^\top (y - \mathcal{X}\widehat{\beta})} \sim F_{q,n-p}.$$

Example 7.11 Let us continue with the "classic blue" pullovers. We can once more test if $\beta = 0$ in the regression of sales on prices. It holds that

$$\beta = 0 \quad \text{iff} \quad (0\ 1)\binom{\alpha}{\beta} = 0.$$

The LR statistic here is

$$-2 \log \lambda = 0.284$$

which is not significant for the χ_1^2 distribution. The F-test statistic

$$F = 0.231$$

is also not significant. Hence, we can assume independence of sales and prices (alone). Recall that this conclusion has to be revised if we consider the prices together with advertising costs and hours sales manager hours.

Recall the different conclusion that was made in Example 7.6 when we rejected $H_0 : \alpha = 211$ and $\beta = 0$. The rejection there came from the fact that the *pair of values* was rejected. Indeed, if $\beta = 0$ the estimator of α would be $\bar{y} = 172.70$ and this is too far from 211.

Example 7.12 Let us now consider the multivariate regression in the "classic blue" pullovers example. From Example 3.15, we know that the estimated parameters in the model

$$X_1 = \alpha + \beta_1 X_2 + \beta_2 X_3 + \beta_3 X_4 + \varepsilon$$

are

$$\widehat{\alpha} = 65.670, \quad \widehat{\beta}_1 = -0.216, \quad \widehat{\beta}_2 = 0.485, \quad \widehat{\beta}_3 = 0.844.$$

Hence, we could postulate the approximate relation:

$$\beta_1 \approx -\frac{1}{2}\beta_2,$$

which means in practice that augmenting the price by 20 EUR requires the advertising costs to increase by 10 EUR in order to keep the number of pullovers sold constant. Vice versa, reducing the price by 20 EUR yields the same result as before if we reduced the advertising costs by 10 EUR. Let us now test whether the hypothesis

$$H_0 : \beta_1 = -\frac{1}{2}\beta_2$$

is valid. This is equivalent to

$$\begin{pmatrix} 0 & 1 & \frac{1}{2} & 0 \end{pmatrix} \begin{pmatrix} \alpha \\ \beta_1 \\ \beta_2 \\ \beta_3 \end{pmatrix} = 0.$$

The LR statistic in this case is equal to (🔍MVAlrtest)

$$-2 \log \lambda = 0.012,$$

the F-statistic is

$$F = 0.007.$$

Hence, in both cases we will not reject the null hypothesis.

Comparison of Two Mean Vectors

In many situations, we want to compare two groups of individuals for whom a set of p characteristics has been observed. We have two random samples $\{x_{i1}\}_{i=1}^{n_1}$ and $\{x_{j2}\}_{j=1}^{n_2}$ from two distinct p-variate normal populations. Several testing issues can be addressed in this framework. In Test Problem 8, we will first test the hypothesis of equal mean vectors in the two groups under the assumption of equality of the two covariance matrices. This task can be solved by adapting Test Problem 2.

In Test Problem 9, a procedure for testing the equality of the two covariance matrices is presented. If the covariance matrices differ, the procedure of Test Problem 8 is no longer valid. If the equality of the covariance matrices is rejected, an easy rule for comparing two means with no restrictions on the covariance matrices is provided in Test Problem 10.

Test Problem 8 Assume that $X_{i1} \sim N_p(\mu_1, \Sigma)$, with $i = 1, \cdots , n_1$ and $X_{j2} \sim N_p(\mu_2, \Sigma)$, with $j = 1, \cdots , n_2$, where all the variables are independent.

$$H_0 : \mu_1 = \mu_2, \text{ versus } H_1 : \text{ no constraints.}$$

Both samples provide the statistics $\bar{x}_k$ and $\mathcal{S}_k$, $k = 1, 2$. Let $\delta = \mu_1 - \mu_2$. We have

$$(\bar{x}_1 - \bar{x}_2) \sim N_p \left(\delta, \frac{n_1 + n_2}{n_1 n_2} \Sigma \right) \tag{7.11}$$

$$n_1 \mathcal{S}_1 + n_2 \mathcal{S}_2 \sim W_p(\Sigma, n_1 + n_2 - 2). \tag{7.12}$$

Let $\mathcal{S} = (n_1 + n_2)^{-1}(n_1 \mathcal{S}_1 + n_2 \mathcal{S}_2)$ be the weighted mean of $\mathcal{S}_1$ and $\mathcal{S}_2$. Since the two samples are independent and since $\mathcal{S}_k$ is independent of $\bar{x}_k$ (for $k = 1, 2$) it follows that $\mathcal{S}$ is independent of $(\bar{x}_1 - \bar{x}_2)$. Hence, Theorem 5.8 applies and leads to a T^2-distribution:

$$\frac{n_1 n_2 (n_1 + n_2 - 2)}{(n_1 + n_2)^2} \{(\bar{x}_1 - \bar{x}_2) - \delta\}^\top \mathcal{S}^{-1} \{(\bar{x}_1 - \bar{x}_2) - \delta\}) \sim T^2_{p, n_1 + n_2 - 2} \tag{7.13}$$

or

$$\{(\bar{x}_1 - \bar{x}_2) - \delta\}^\top \mathcal{S}^{-1} \{(\bar{x}_1 - \bar{x}_2) - \delta\} \sim \frac{p(n_1 + n_2)^2}{(n_1 + n_2 - p - 1)(n_1 n_2)} F_{p, n_1 + n_2 - p - 1}.$$

This result, as in Test Problem 2, can be used to test $H_0 : \delta = 0$ or to construct a confidence region for $\delta \in \mathbb{R}^p$. The rejection region is given by

$$\frac{n_1 n_2 (n_1 + n_2 - p - 1)}{p(n_1 + n_2)^2} (\bar{x}_1 - \bar{x}_2)^\top \mathcal{S}^{-1} (\bar{x}_1 - \bar{x}_2) \geq F_{1-\alpha; p, n_1 + n_2 - p - 1}. \tag{7.14}$$

A $(1 - \alpha)$ confidence region for δ is given by the ellipsoid centered at $(\bar{x}_1 - \bar{x}_2)$

$$\{\delta - (\bar{x}_1 - \bar{x}_2)\}^\top \mathcal{S}^{-1} \{\delta - (\bar{x}_1 - \bar{x}_2)\} \leq \frac{p(n_1 + n_2)^2}{(n_1 + n_2 - p - 1)(n_1 n_2)} F_{1-\alpha; p, n_1 + n_2 - p - 1},$$

and the simultaneous confidence intervals for all linear combinations $a^\top \delta$ of the elements of δ are given by

$$a^\top \delta \in a^\top (\bar{x}_1 - \bar{x}_2) \pm \sqrt{\frac{p(n_1 + n_2)^2}{(n_1 + n_2 - p - 1)(n_1 n_2)} F_{1-\alpha; p, n_1 + n_2 - p - 1} a^\top \mathcal{S} a}.$$

In particular, we have at the $(1 - \alpha)$ level, for $j = 1, \ldots, p$,

$$\delta_j \in (\bar{x}_{1j} - \bar{x}_{2j}) \pm \sqrt{\frac{p(n_1 + n_2)^2}{(n_1 + n_2 - p - 1)(n_1 n_2)} F_{1-\alpha; p, n_1 + n_2 - p - 1} s_{jj}}. \tag{7.15}$$

Example 7.13 Let us come back to the questions raised in Example 7.5. We compare the means of assets (X_1) and of sales (X_2) for two sectors, energy (group 1) and manufacturing (group 2). With $n_1 = 15$, $n_2 = 10$, and $p = 2$, we obtain the statistics

$$\bar{x}_1 = \begin{pmatrix} 4084.0 \\ 2580.5 \end{pmatrix}, \ \bar{x}_2 = \begin{pmatrix} 4307.2 \\ 4925.2 \end{pmatrix}$$

and

$$S_1 = 10^7 \begin{pmatrix} 1.6635 & 1.2410 \\ 1.2410 & 1.3747 \end{pmatrix}, S_2 = 10^7 \begin{pmatrix} 1.2248 & 1.1425 \\ 1.1425 & 1.5112 \end{pmatrix},$$

so that

$$S = 10^7 \begin{pmatrix} 1.4880 & 1.2016 \\ 1.2016 & 1.4293 \end{pmatrix}.$$

The observed value of the test statistic (7.14) is $F = 2.7036$. Since $F_{0.95;2,22} = 3.4434$, the hypothesis of equal means of the two groups is not rejected although it would be rejected at a less severe level ($F > F_{0.90;2,22} = 2.5613$). By directly applying (7.15), the 95% simultaneous confidence intervals for the differences (QMVAsimcidif) are obtained as:

$$-4628.6 \leq \mu_{1a} - \mu_{2a} \leq 4182.2$$
$$-6662.4 \leq \mu_{1s} - \mu_{2s} \leq 1973.0.$$

Example 7.14 In order to illustrate the presented test procedures it is interesting to analyze some simulated data. This simulation will point out the importance of the covariances in testing means. We created 2 independent normal samples in $\mathbb{R}^4$ of sizes $n_1 = 30$ and $n_2 = 20$ with

$$\mu_1 = (8, 6, 10, 10)^{\top}$$

$$\mu_2 = (6, 6, 10, 13)^{\top}.$$

One may consider this as an example of $X = (X_1, \ldots, X_n)^{\top}$ being the students' scores from 4 tests, where the 2 groups of students were subjected to two different methods of teaching. First, we simulate the two samples with $\Sigma = \mathcal{I}_4$ and obtain the statistics

$$\bar{x}_1 = (7.607, 5.945, 10.213, 9.635)^{\top}$$
$$\bar{x}_2 = (6.222, 6.444, 9.560, 13.041)^{\top}$$

$$S_1 = \begin{pmatrix} 0.812 & -0.229 & -0.034 & 0.073 \\ -0.229 & 1.001 & 0.010 & -0.059 \\ -0.034 & 0.010 & 1.078 & -0.098 \\ 0.073 & -0.059 & -0.098 & 0.823 \end{pmatrix}$$

$$S_2 = \begin{pmatrix} 0.559 & -0.057 & -0.271 & 0.306 \\ -0.057 & 1.237 & 0.181 & 0.021 \\ -0.271 & 0.181 & 1.159 & -0.130 \\ 0.306 & 0.021 & -0.130 & 0.683 \end{pmatrix}.$$

The test statistic (7.14) takes the value $F = 60.65$, which is highly significant: the small variance allows the difference to be detected even with these relatively moderate sample sizes. We conclude (at the 95% level) that

$$
\begin{aligned}
0.6213 &\leq \delta_1 \leq 2.2691 \\
-1.5217 &\leq \delta_2 \leq 0.5241 \\
-0.3766 &\leq \delta_3 \leq 1.6830 \\
-4.2614 &\leq \delta_4 \leq -2.5494
\end{aligned}
$$

which confirms that the means for X_1 and X_4 are different.

Consider now a different simulation scenario, where the standard deviations are 4 times larger: $\Sigma = 16\mathcal{I}_4$. Here we obtain

$$
\bar{x}_1 = (7.312, 6.304, 10.840, 10.902)^\top
$$
$$
\bar{x}_2 = (6.353, 5.890, 8.604, 11.283)^\top
$$
$$
\mathcal{S}_1 = \begin{pmatrix}
21.907 & 1.415 & -2.050 & 2.379 \\
1.415 & 11.853 & 2.104 & -1.864 \\
-2.050 & 2.104 & 17.230 & 0.905 \\
2.379 & -1.864 & 0.905 & 9.037
\end{pmatrix}
$$
$$
\mathcal{S}_2 = \begin{pmatrix}
20.349 & -9.463 & 0.958 & -6.507 \\
-9.463 & 15.502 & -3.383 & -2.551 \\
0.958 & -3.383 & 14.470 & -0.323 \\
-6.507 & -2.551 & -0.323 & 10.311
\end{pmatrix}.
$$

Now the test statistic takes the value 1.54, which is no longer significant ($F_{0.95,4,45} = 2.58$). Now we cannot reject the null hypothesis (which we know to be false!) since the increase in variances prohibits the detection of differences of such magnitude.

The following situation illustrates once more the role of the covariances between covariates. Suppose that $\Sigma = 16\mathcal{I}_4$ as above but with $\sigma_{14} = \sigma_{41} = -3.999$ (this corresponds to a negative correlation $r_{41} = -0.9997$). We have

$$
\bar{x}_1 = (8.484, 5.908, 9.024, 10.459)^\top
$$
$$
\bar{x}_2 = (4.959, 7.307, 9.057, 13.803)^\top
$$
$$
\mathcal{S}_1 = \begin{pmatrix}
14.649 & -0.024 & 1.248 & -3.961 \\
-0.024 & 15.825 & 0.746 & 4.301 \\
1.248 & 0.746 & 9.446 & 1.241 \\
-3.961 & 4.301 & 1.241 & 20.002
\end{pmatrix}
$$
$$
\mathcal{S}_2 = \begin{pmatrix}
14.035 & -2.372 & 5.596 & -1.601 \\
-2.372 & 9.173 & -2.027 & -2.954 \\
5.596 & -2.027 & 9.021 & -1.301 \\
-1.601 & -2.954 & -1.301 & 9.593
\end{pmatrix}.
$$

The value of F is 3.853 which is significant at the 5% level (p-value = 0.0089). So the null hypothesis $\delta = \mu_1 - \mu_2 = 0$ is outside the 95% confidence ellipsoid. However, the simultaneous confidence intervals, which do not take the covariances into account are given by

$$-0.1837 \leq \delta_1 \leq 7.2343$$
$$-4.9452 \leq \delta_2 \leq 2.1466$$
$$-3.0091 \leq \delta_3 \leq 2.9438$$
$$-7.2336 \leq \delta_4 \leq 0.5450.$$

They contain the null value (see Remark 7.1) although they are very asymmetric for δ_1 and δ_4.

Example 7.15 Let us compare the vectors of means of the forged and the genuine banknotes. The matrices $\mathcal{S}_f$ and $\mathcal{S}_g$ were given in Example 3.1 and since here $n_f = n_g = 100$, S is the simple average of $\mathcal{S}_f$ and $\mathcal{S}_g$: $S = \frac{1}{2}(\mathcal{S}_f + \mathcal{S}_g)$.

$$\bar{x}_g = (214.97, 129.94, 129.72, 8.305, 10.168, 141.52)^\top$$
$$\bar{x}_f = (214.82, 130.3, 130.19, 10.53, 11.133, 139.45)^\top.$$

The test statistic is given by (7.14) and turns out to be $F = 391.92$ which is highly significant for $F_{6,193}$. The 95% simultaneous confidence intervals for the differences $\delta_j = \mu_{gj} - \mu_{fj}$, $j = 1, \ldots, p$ are

$$-0.0443 \leq \delta_1 \leq 0.3363$$
$$-0.5186 \leq \delta_2 \leq -0.1954$$
$$-0.6416 \leq \delta_3 \leq -0.3044$$
$$-2.6981 \leq \delta_4 \leq -1.7519$$
$$-1.2952 \leq \delta_5 \leq -0.6348$$
$$1.8072 \leq \delta_6 \leq 2.3268.$$

All of the components (except for the first one) show significant differences in the means. The main effects are taken by the lower border (X_4) and the diagonal (X_6).

The preceding test implicitly uses the fact that the two samples are extracted from two different populations with common variance Σ. In this case, the test statistic (7.14) measures the distance between the two centers of gravity of the two groups w.r.t. the common metric given by the pooled variance matrix S. If $\Sigma_1 \neq \Sigma_2$ no such matrix exists. There are no satisfactory test procedures for testing the equality of variance matrices which are robust with respect to normality assumptions of the populations. The following test extends Bartlett's test for equality of variances in the univariate case. But this test is known to be very sensitive to departures from normality.

Test Problem 9 (Comparison of Covariance Matrices)
Let $X_{ih} \sim N_p(\mu_h, \Sigma_h)$, $i = 1, \ldots, n_h$, $h = 1, \ldots, k$ be independent random variables,

$$H_0 : \Sigma_1 = \Sigma_2 = \cdots = \Sigma_k \text{ versus } H_1 : \text{no constraints.}$$

Each subsample provides S_h, an estimator of Σ_h, with

$$n_h S_h \sim W_p(\Sigma_h, n_h - 1).$$

Under H_0, $\sum_{h=1}^{k} n_h S_h \sim W_p(\Sigma, n - k)$ (Sect. 5.2), where Σ is the common covariance matrix X_{ih} and $n = \sum_{h=1}^{k} n_h$. Let $S = \frac{n_1 S_1 + \cdots + n_k S_k}{n}$ be the weighted average of the S_h (this is in fact the MLE of Σ when H_0 is true). The likelihood ratio test leads to the statistic

$$-2 \log \lambda = n \log |S| - \sum_{h=1}^{k} n_h \log |S_h| \qquad (7.16)$$

which under H_0 is approximately distributed as a $\mathcal{X}_m^2$ where $m = \frac{1}{2}(k-1)p(p+1)$.

Example 7.16 Let's come back to Example 7.13, where the mean of assets and sales have been compared for companies from the energy and manufacturing sector assuming that $\Sigma_1 = \Sigma_2$. The test of $\Sigma_1 = \Sigma_2$ leads to the value of the test statistic

$$-2 \log \lambda = 0.9076 \qquad (7.17)$$

which is not significant (p-value for a $\chi_3^2 = 0.82$). We cannot reject H_0 and the comparison of the means performed above is valid.

Example 7.17 Let us compare the covariance matrices of the forged and the genuine banknotes (the matrices S_f and S_g are shown in Example 3.1). A first look seems to suggest that $\Sigma_1 \neq \Sigma_2$. The pooled variance S is given by $S = \frac{1}{2}(S_f + S_g)$ since here $n_f = n_g$. The test statistic here is $-2 \log \lambda = 127.21$, which is highly significant χ^2 with 21 degrees of freedom. As expected, we reject the hypothesis of equal covariance matrices, and as a result the procedure for comparing the two means in Example 7.15 is not valid.

What can we do with unequal covariance matrices? When both n_1 and n_2 are large, we have a simple solution

---

Test Problem 10 (Comparison of two means, unequal covariance matrices, large samples)

Assume that $X_{i1} \sim N_p(\mu_1, \Sigma_1)$, with $i = 1, \cdots, n_1$ and $X_{j2} \sim N_p(\mu_2, \Sigma_2)$, with $j = 1, \cdots, n_2$ are independent random variables.

$$H_0 : \mu_1 = \mu_2 \text{ versus } H_1 : \text{ no constraints.}$$

Letting $\delta = \mu_1 - \mu_2$, we have

$$(\bar{x}_1 - \bar{x}_2) \sim N_p \left(\delta, \frac{\Sigma_1}{n_1} + \frac{\Sigma_2}{n_2} \right).$$

Therefore, by (5.4)

$$(\bar{x}_1 - \bar{x}_2)^\top \left(\frac{\Sigma_1}{n_1} + \frac{\Sigma_2}{n_2} \right)^{-1} (\bar{x}_1 - \bar{x}_2) \sim \chi_p^2.$$

Since S_i is a consistent estimator of Σ_i for $i = 1, 2$, we have

$$(\bar{x}_1 - \bar{x}_2)^\top \left(\frac{S_1}{n_1} + \frac{S_2}{n_2} \right)^{-1} (\bar{x}_1 - \bar{x}_2) \xrightarrow{\mathcal{L}} \chi_p^2. \tag{7.18}$$

This can be used in place of (7.13) for testing H_0, defining a confidence region for δ or constructing simultaneous confidence intervals for δ_j, $j = 1, \ldots, p$.

For instance, the rejection region at the level α will be

$$(\bar{x}_1 - \bar{x}_2)^\top \left(\frac{S_1}{n_1} + \frac{S_2}{n_2} \right)^{-1} (\bar{x}_1 - \bar{x}_2) > \chi_{1-\alpha;p}^2 \tag{7.19}$$

and the $(1 - \alpha)$ simultaneous confidence intervals for δ_j, $j = 1, \ldots, p$ are

$$\delta_j \in (\bar{x}_1 - \bar{x}_2) \pm \sqrt{\chi_{1-\alpha;p}^2 \left(\frac{s_{jj}^{(1)}}{n_1} + \frac{s_{jj}^{(2)}}{n_2} \right)} \tag{7.20}$$

where $s_{jj}^{(i)}$ is the (j, j) element of the matrix S_i. This may be compared to (7.15) where the pooled variance was used.

Remark 7.2 We see, by comparing the statistics (7.19) with (7.14), that we measure here the distance between $\bar{x}_1$ and $\bar{x}_2$ using the metric $\left(\frac{S_1}{n_1} + \frac{S_2}{n_2} \right)$. It should be noted that when $n_1 = n_2$, the two methods are essentially the same since then $S = \frac{1}{2}(S_1 + S_2)$. If the covariances are different but have the same eigenvectors (different eigenvalues), one can apply the common principal component (CPC) technique, see Chap. 11.

Example 7.18 Let us use the last test to compare the forged and the genuine banknotes again (n_1 and n_2 are both large). The test statistic (7.19) turns out to be 2436.8 which is again highly significant. The 95% simultaneous confidence intervals are

$$
\begin{aligned}
-0.0389 &\leq \delta_1 \leq 0.3309 \\
-0.5140 &\leq \delta_2 \leq -0.2000 \\
-0.6368 &\leq \delta_3 \leq -0.3092 \\
-2.6846 &\leq \delta_4 \leq -1.7654 \\
-1.2858 &\leq \delta_5 \leq -0.6442 \\
1.8146 &\leq \delta_6 \leq 2.3194
\end{aligned}
$$

showing that all the components except the first are different from zero, the largest difference coming from X_6 (length of the diagonal) and X_4 (lower border). The results are very similar to those obtained in Example 7.15. This is due to the fact that here $n_1 = n_2$ as we already mentioned in the remark above.

Profile Analysis

Another useful application of Test Problem 6 is the repeated measurements problem applied to two independent groups. This problem arises in practice when we observe repeated measurements of characteristics (or measures of the same type under different experimental conditions) on the different groups which have to be compared. It is important that the p measures (the "profiles") are comparable, and, in particular, are reported in the same units. For instance, they may be measures of blood pressure at p different points in time, one group being the control group and the other the group receiving a new treatment. The observations may be the scores obtained from p different tests of two different experimental groups. One is then interested in comparing the profiles of each group: the profile being just the vectors of the means of the p responses (the comparison may be visualized in a two-dimensional graph using the parallel coordinates plot introduced in Sect. 1.7).

We are thus in the same statistical situation as for the comparison of two means:

$$
X_{i1} \sim N_p\,(\mu_1,\,\Sigma) \quad i = 1, \ldots, n_1
$$

$$
X_{i2} \sim N_p\,(\mu_2,\,\Sigma) \quad i = 1, \ldots, n_2
$$

where all variables are independent. Suppose the two population profiles look like in Fig. 7.1.

The following questions are of interest:

1. Are the profiles similar in the sense of being parallel (which means no interaction between the treatments and the groups)?
2. If the profiles are parallel, are they at the same level?

Fig. 7.1 Example of population profiles 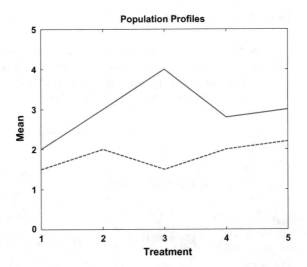 MVAprofil

3. If the profiles are parallel, is there any treatment effect, i.e., are the profiles horizontal (profiles remain the same no matter which treatment received)?

The above questions are easily translated into linear constraints on the means and a test statistic can be obtained accordingly.

Parallel Profiles

Let $\mathcal{C}$ be a $((p-1) \times p)$ matrix defined as $\mathcal{C} = \begin{pmatrix} 1 & -1 & 0 & \cdots & 0 \\ 0 & 1 & -1 & \cdots & 0 \\ \vdots & \vdots & \vdots & \vdots & \vdots \\ 0 & \cdots & 0 & 1 & -1 \end{pmatrix}$.

The hypothesis to be tested is

$$H_0^{(1)} : \mathcal{C}(\mu_1 - \mu_2) = 0.$$

From (7.11), (7.12), and Corollary 5.4 we know that under H_0:

$$\frac{n_1 n_2}{(n_1 + n_2)^2}(n_1 + n_2 - 2)\, \{\mathcal{C}(\bar{x}_1 - \bar{x}_2)\}^\top\, (\mathcal{C}\mathcal{S}\mathcal{C}^\top)^{-1}\mathcal{C}(\bar{x}_1 - \bar{x}_2) \sim T^2_{p-1,n_1+n_2-2}$$

(7.21)

where $\mathcal{S}$ is the pooled covariance matrix. The hypothesis is rejected if

$$\frac{n_1 n_2 (n_1 + n_1 - p)}{(n_1 + n_2)^2 (p-1)}\, (\mathcal{C}\bar{x})^\top\, (\mathcal{C}\mathcal{S}\mathcal{C}^\top)^{-1}\mathcal{C}\bar{x} > F_{1-\alpha;p-1,n_1+n_2-p}.$$

Equality of Two Levels

The question of equality of the two levels is meaningful only if the two profiles are parallel. In the case of interactions (rejection of $H_0^{(1)}$), the two populations react differently to the treatments and the question of the level has no meaning.
The equality of the two levels can be formalized as

$$H_0^{(2)} : 1_p^\top (\mu_1 - \mu_2) = 0$$

since

$$1_p^\top (\bar{x}_1 - \bar{x}_2) \sim N_1 \left(1_p^\top (\mu_1 - \mu_2), \frac{n_1 + n_2}{n_1 n_2} 1_p^\top \Sigma 1_p \right)$$

and

$$(n_1 + n_2) 1_p^\top \mathcal{S} 1_p \sim W_1 (1_p^\top \Sigma 1_p, n_1 + n_2 - 2).$$

Using Corollary 5.4 we have that

$$\frac{n_1 n_2}{(n_1 + n_2)^2} (n_1 + n_2 - 2) \frac{\left\{ 1_p^\top (\bar{x}_1 - \bar{x}_2) \right\}^2}{1_p^\top \mathcal{S} 1_p} \sim T_{1, n_1 + n_2 - 2}^2 \qquad (7.22)$$

$$= F_{1, n_1 + n_2 - 2}.$$

The rejection region is

$$\frac{n_1 n_2 (n_1 + n_2 - 2)}{(n_1 + n_2)^2} \frac{\left\{ 1_p^\top (\bar{x}_1 - \bar{x}_2) \right\}^2}{1_p^\top \mathcal{S} 1_p} > F_{1 - \alpha; 1, n_1 + n_2 - 2}.$$

Treatment Effect

If it is rejected that the profiles are parallel, then two independent analyses should be done on the two groups using the repeated measurement approach. But if it is accepted that they are parallel, then we can exploit the information contained in both groups (possibly at different levels) to test a treatment effect, i.e., if the two profiles are horizontal. This may be written as

$$H_0^{(3)} : \mathcal{C}(\mu_1 + \mu_2) = 0.$$

Consider the average profile $\bar{x}$

$$\bar{x} = \frac{n_1 \bar{x}_1 + n_2 \bar{x}_2}{n_1 + n_2}.$$

Clearly,

$$\bar{x} \sim N_p \left(\frac{n_1 \mu_1 + n_2 \mu_2}{n_1 + n_2}, \frac{1}{n_1 + n_2} \Sigma \right).$$

Now it is not hard to prove that $H_0^{(3)}$ with $H_0^{(1)}$ implies that

$$\mathcal{C} \left(\frac{n_1 \mu_1 + n_2 \mu_2}{n_1 + n_2} \right) = 0.$$

So under parallel, horizontal profiles we have

$$\sqrt{n_1 + n_2} \mathcal{C}\bar{x} \sim N_p(0, \mathcal{C}\Sigma\mathcal{C}^\top).$$

From Corollary 5.4, we again obtain

$$(n_1 + n_2 - 2)(\mathcal{C}\bar{x})^\top (\mathcal{C}\mathcal{S}\mathcal{C}^\top)^{-1} \mathcal{C}\bar{x} \sim T_{p-1, n_1 + n_2 - 2}^2. \tag{7.23}$$

This leads to the rejection region of $H_0^{(3)}$, namely,

$$\frac{n_1 + n_2 - p}{p - 1} (\mathcal{C}\bar{x})^\top (\mathcal{C}\mathcal{S}\mathcal{C}^\top)^{-1} \mathcal{C}\bar{x} > F_{1-\alpha; p-1, n_1 + n_2 - p}.$$

Example 7.19 Morrison (1990) proposed a test in which the results of 4 subtests of the Wechsler Adult Intelligence Scale (WAIS) are compared for 2 categories of people: group 1 contains $n_1 = 37$ people who do not have a senile factor and group 2 contains $n_2 = 12$ people who have a senile factor. The four WAIS subtests are X_1 (information), X_2 (similarities), X_3 (arithmetic), and X_4 (picture completion). The relevant statistics are

$$\bar{x}_1 = (12.57, 9.57, 11.49, 7.97)^\top$$
$$\bar{x}_2 = (8.75, 5.33, 8.50, 4.75)^\top$$
$$\mathcal{S}_1 = \begin{pmatrix} 11.164 & 8.840 & 6.210 & 2.020 \\ 8.840 & 11.759 & 5.778 & 0.529 \\ 6.210 & 5.778 & 10.790 & 1.743 \\ 2.020 & 0.529 & 1.743 & 3.594 \end{pmatrix}$$
$$\mathcal{S}_2 = \begin{pmatrix} 9.688 & 9.583 & 8.875 & 7.021 \\ 9.583 & 16.722 & 11.083 & 8.167 \\ 8.875 & 11.083 & 12.083 & 4.875 \\ 7.021 & 8.167 & 4.875 & 11.688 \end{pmatrix}.$$

The test statistic for testing if the two profiles are parallel is $F = 0.4634$, which is not significant (p-value $= 0.71$). Thus it is accepted that the two are parallel. The second test statistic (testing the equality of the levels of the 2 profiles) is $F = 17.21$, which is highly significant (p-value $\approx 10^{-4}$). The global level of the test for the non-senile people is superior to the senile group. The final test (testing the horizontality of the average profile) has the test statistic $F = 53.32$, which is also highly significant (p-value $\approx 10^{-14}$). This implies that there are substantial differences among the means of the different subtests.

Summary
$\hookrightarrow$ Hypotheses about μ can often be written as $\mathcal{A}\mu = a$, with matrix $\mathcal{A}$, and vector a.
$\hookrightarrow$ The hypothesis $H_0 : \mathcal{A}\mu = a$ for $X \sim N_p(\mu, \Sigma)$ with Σ known leads to $-2 \log \lambda = n(\mathcal{A}\bar{x} - a)^\top (\mathcal{A}\Sigma\mathcal{A}^\top)^{-1}(\mathcal{A}\bar{x} - a) \sim \chi_q^2$, where q is the number of elements in a.
$\hookrightarrow$ The hypothesis $H_0 : \mathcal{A}\mu = a$ for $X \sim N_p(\mu, \Sigma)$ with Σ unknown leads to $-2 \log \lambda = n \log\{1 + (\mathcal{A}\bar{x} - a)^\top (\mathcal{A}S\mathcal{A}^\top)^{-1}(\mathcal{A}\bar{x} - a)\} \xrightarrow{\mathcal{L}} \chi_q^2$, where q is the number of elements in a and we have an exact test $(n - 1)(\mathcal{A}\bar{x} - a)^\top (\mathcal{A}S\mathcal{A}^\top)^{-1}(\mathcal{A}\bar{x} - a) \sim T_{q,n-1}^2$.
$\hookrightarrow$ The hypothesis $H_0 : \mathcal{A}\beta = a$ for $Y_i \sim N_1(\beta^\top x_i, \sigma^2)$ with σ^2 unknown leads to $-2 \log \lambda = \frac{n}{2} \log \left(\frac{\|y - \mathcal{X}\tilde{\beta}\|^2}{\|y - \mathcal{X}\hat{\beta}\|^2} - 1 \right) \xrightarrow{\mathcal{L}} \chi_q^2$, with q being the length of a and with $$\frac{n - p}{q} \frac{(\mathcal{A}\hat{\beta} - a)^\top \left\{ \mathcal{A}\left(\mathcal{X}^\top\mathcal{X}\right)^{-1}\mathcal{A}^\top \right\}^{-1} (\mathcal{A}\hat{\beta} - a)}{\left(y - \mathcal{X}\hat{\beta}\right)^\top \left(y - \mathcal{X}\hat{\beta}\right)} \sim F_{q,n-p}.$$

7.3 Boston Housing

Returning to the Boston Housing data set, we are now in a position to test if the means of the variables vary according to their location, for example, when they are located in a district with high valued houses. In Chap. 1, we built 2 groups of observations according to the value of X_{14} being less than or equal to the median of X_{14} (a group of 256 districts) and greater than the median (a group of 250 districts). In what follows, we use the transformed variables motivated in Sect. 1.9.

Testing the equality of the means from the two groups was proposed in a multivariate setup, so we restrict the analysis to the variables X_1, X_5, X_8, X_{11}, and X_{13} to see if the differences between the two groups that were identified in Chap. 1 can be confirmed by a formal test. As in Test Problem 8, the hypothesis to be tested is

$$H_0 : \mu_1 = \mu_2, \text{ where } \mu_1 \in \mathbb{R}^5, n_1 = 256, \text{ and } n_2 = 250.$$

Σ is not known. The F-statistic given in (7.13) is equal to 126.30, which is much higher than the critical value $F_{0.95;5,500} = 2.23$. Therefore, we reject the hypothesis of equal means.

To see which component, X_1, X_5, X_8, X_{11}, or X_{13}, is responsible for this rejection, take a look at the simultaneous confidence intervals defined in (7.14):

$$\delta_1 \in (\quad 1.4020, \quad 2.5499)$$
$$\delta_5 \in (\quad 0.1315, \quad 0.2383)$$
$$\delta_8 \in (-0.5344, -0.2222)$$
$$\delta_{11} \in (\quad 1.0375, \quad 1.7384)$$
$$\delta_{13} \in (\quad 1.1577, \quad 1.5818).$$

These confidence intervals confirm that all of the δ_j are significantly different from zero (note there is a negative effect for X_8: weighted distances to employment centers), ⬛MVAsimcibh.

We could also check if the factor "being bounded by the river" (variable X_4) has some effect on the other variables. To do this, compare the means of $(X_5, X_8, X_9, X_{12}, X_{13}, X_{14})^\top$. There are two groups: $n_1 = 35$ districts bounded by the river and $n_2 = 471$ districts not bounded by the river. Test Problem 8 ($H_0 : \mu_1 = \mu_2$) is applied again with $p = 6$. The resulting test statistic, $F = 5.81$, is highly significant ($F_{0.95;6,499} = 2.12$). The simultaneous confidence intervals indicate that only X_{14} (the value of the houses) is responsible for the hypothesis being rejected at a significance level of 0.95:

$$\delta_5 \in (-0.0603, 0.1919)$$
$$\delta_8 \in (-0.5225, 0.1527)$$
$$\delta_9 \in (-0.5051, 0.5938)$$
$$\delta_{12} \in (-0.3974, 0.7481)$$
$$\delta_{13} \in (-0.8595, 0.3782)$$
$$\delta_{14} \in (\quad 0.0014, 0.5084).$$

Testing Linear Restrictions

In Chap. 3, a linear model was proposed that explained the variations of the price X_{14} by the variations of the other variables. Using the same procedure that was shown in Test Problem 7, we are in a position to test a set of linear restrictions on the vector of regression coefficients β.

The model we estimated in Sect. 3.7 provides the following: (⬛MVAlinregbh):

Recall that the estimated residuals $Y - \mathcal{X}\widehat{\beta}$ did not show a big departure from normality, which means that the testing procedure developed above can be used.

1. First a global test of significance for the regression coefficients is performed,

Variable	$\widehat{\beta}_j$	$SE(\widehat{\beta}_j)$	t	p-value
Constant	4.1769	0.3790	11.020	0.0000
X_1	−0.0146	0.0117	−1.254	0.2105
X_2	0.0014	0.0056	0.247	0.8051
X_3	−0.0127	0.0223	−0.570	0.5692
X_4	0.1100	0.0366	3.002	0.0028
X_5	−0.2831	0.1053	−2.688	0.0074
X_6	0.4211	0.1102	3.822	0.0001
X_7	0.0064	0.0049	1.317	0.1885
X_8	−0.1832	0.0368	−4.977	0.0000
X_9	0.0684	0.0225	3.042	0.0025
X_{10}	−0.2018	0.0484	−4.167	0.0000
X_{11}	−0.0400	0.0081	−4.946	0.0000
X_{12}	0.0445	0.0115	3.882	0.0001
X_{13}	−0.2626	0.0161	−16.320	0.0000

$$H_0 : (\beta_1, \ldots, \beta_{13}) = 0.$$

This is obtained by defining $\mathcal{A} = (0_{13}, \mathcal{I}_{13})$ and $a = 0_{13}$ so that H_0 is equivalent to $\mathcal{A}\beta = a$ where $\beta = (\beta_0, \beta_1, \ldots, \beta_{13})^\top$. Based on the observed values $F = 123.20$. This is highly significant ($F_{0.95;13,492} = 1.7401$), thus we reject H_0. Note that under H_0 $\widehat{\beta}_{H_0} = (3.0345, 0, \ldots, 0)$ where $3.0345 = \overline{y}$.

2. Since we are interested in the effect that being located close to the river has on the value of the houses, the second test is $H_0 : \beta_4 = 0$. This is done by fixing

$$\mathcal{A} = (0, 0, 0, 0, 1, 0, 0, 0, 0, 0, 0, 0, 0, 0)^\top$$

and $a = 0$ to obtain the equivalent hypothesis $H_0 : \mathcal{A}\beta = a$. The result is again significant: $F = 9.0125$ ($F_{0.95;1,492} = 3.8604$) with a p-value of 0.0028. Note that this is the same p-value as obtained in the individual test $\beta_4 = 0$ in Chap. 3, computed using a different setup.

3. A third test notices the fact that some of the regressors in the full model (3.57) appear to be insignificant (that is they have high individual p-values). It can be confirmed from a joint test if the corresponding reduced model, formulated by deleting the insignificant variables, is rejected by the data. We want to test $H_0 : \beta_1 = \beta_2 = \beta_3 = \beta_7 = 0$. Hence,

$$\mathcal{A} = \begin{array}{c} 0\,1\,0\,0\,0\,0\,0\,0\,0\,0\,0\,0\,0\,0 \\ 0\,0\,1\,0\,0\,0\,0\,0\,0\,0\,0\,0\,0\,0 \\ 0\,1\,0\,1\,0\,0\,0\,0\,0\,0\,0\,0\,0\,0 \\ 0\,1\,0\,0\,0\,0\,0\,1\,0\,0\,0\,0\,0\,0 \end{array}$$

and $a = 0_4$. The test statistic is 0.9344, which is not significant for $F_{4,492}$. Given that the p-value is equal to 0.44, we cannot reject the null hypothesis nor the corresponding reduced model. The value of $\widehat{\beta}$ under the null hypothesis is

Table 7.1 Linear regression for boston housing data set ◉MVAlinreg2bh

Variable	$\widehat{\beta}_j$	SE	t	p-value
Const	4.1582	0.3628	11.462	0.0000
X_4	0.1087	0.0362	2.999	0.0028
X_5	−0.3055	0.0973	−3.140	0.0018
X_6	0.4668	0.1059	4.407	0.0000
X_8	−0.1855	0.0327	−5.679	0.0000
X_9	0.0492	0.0183	2.690	0.0074
X_{10}	−0.2096	0.0446	−4.705	0.0000
X_{11}	−0.0410	0.0078	−5.280	0.0000
X_{12}	0.0481	0.0112	4.306	0.0000
X_{13}	−0.2588	0.0149	−17.396	0.0000

$$\widehat{\beta}_{H_0} = (4.16, 0, 0, 0, 0.11, -0.31, 0.47, 0, -0.19, 0.05, -0.20, -0.04, 0.05, -0.26)^\top.$$

A possible reduced model is

$$X_{14} = \beta_0 + \beta_4 X_4 + \beta_5 X_5 + \beta_6 X_6 + \beta_8 X_8 + \cdots + \beta_{13} X_{13} + \varepsilon.$$

Estimating this reduced model using OLS, as was done in Chap. 3, provides the results shown in Table 7.1.

Note that the reduced model has $r^2 = 0.763$ which is very close to $r^2 = 0.765$ obtained from the full model. Clearly, including variables X_1, X_2, X_3, and X_7 does not provide valuable information in explaining the variation of X_{14}, the price of the houses.

7.4 Exercises

Exercise 7.1 Use Theorem 7.1 to derive a test for testing the hypothesis that a dice is balanced, based on n tosses of that dice. (Hint: use the multinomial probability function.)

Exercise 7.2 Consider $N_3(\mu, \Sigma)$. Formulate the hypothesis $H_0 : \mu_1 = \mu_2 = \mu_3$ in terms of $\mathcal{A}\mu = a$.

Exercise 7.3 Simulate a normal sample with $\mu = \binom{1}{2}$ and $\Sigma = \begin{pmatrix} 1 & 0.5 \\ 0.5 & 2 \end{pmatrix}$ and test $H_0 : 2\mu_1 - \mu_2 = 0.2$ first with Σ known and then with Σ unknown. Compare the results.

Exercise 7.4 Derive expression (7.3) for the likelihood ratio test statistic in Test Problem 2.

Exercise 7.5 With the simulated data set of Example 7.14, test the hypothesis of equality of the covariance matrices.

Exercise 7.6 In the U.S. companies data set, test the equality of means between the energy and manufacturing sectors, taking the full vector of observations X_1 to X_6. Derive the simultaneous confidence intervals for the differences.

Exercise 7.7 Let $X \sim N_2(\mu, \Sigma)$ where Σ is known to be $\Sigma = \begin{pmatrix} 2 & -1 \\ -1 & 2 \end{pmatrix}$. We have an i.i.d. sample of size $n = 6$ providing $\bar{x}^\top = (1\ \frac{1}{2})$. Solve the following test problems ($\alpha = 0.05$):

a) $H_0 : \mu = \left(2, \frac{2}{3}\right)^\top\ H_1 : \mu \neq \left(2, \frac{2}{3}\right)^\top$
b) $H_0 : \mu_1 + \mu_2 = \frac{7}{2}\ H_1 : \mu_1 + \mu_2 \neq \frac{7}{2}$
c) $H_0 : \mu_1 - \mu_2 = \frac{1}{2}\ H_1 : \mu_1 - \mu_2 \neq \frac{1}{2}$
d) $H_0 : \mu_1 = 2\ H_1 : \mu_1 \neq 2$

For each case, represent the rejection region graphically (comment).

Exercise 7.8 Repeat the preceding exercise with Σ unknown and $S = \begin{pmatrix} 2 & -1 \\ -1 & 2 \end{pmatrix}$. Compare the results.

Exercise 7.9 Consider $X \sim N_3(\mu, \Sigma)$. An i.i.d. sample of size $n = 10$ provides

$$\bar{x} = (1, 0, 2)^\top$$

$$S = \begin{pmatrix} 3 & 2 & 1 \\ 2 & 3 & 1 \\ 1 & 1 & 4 \end{pmatrix}.$$

(a) Knowing that the eigenvalues of S are integers, describe a 95% confidence region for μ. (Hint: to compute eigenvalues use $|S| = \prod_{j=1}^{3} \lambda_j$ and $\mathrm{tr}(S) = \sum_{j=1}^{3} \lambda_j$).
(b) Calculate the simultaneous confidence intervals for μ_1, μ_2 and μ_3.
(c) Can we assert that μ_1 is an average of μ_2 and μ_3?

Exercise 7.10 Consider two independent i.i.d. samples, each of size 10, from two bivariate normal populations. The results are summarized below:

$$\bar{x}_1 = (3, 1)^\top; \ \bar{x}_2 = (1, 1)^\top$$

$$S_1 = \begin{pmatrix} 4 & -1 \\ -1 & 2 \end{pmatrix}; \ S_2 = \begin{pmatrix} 2 & -2 \\ -2 & 4 \end{pmatrix}.$$

Provide a solution to the following tests:
a) $H_0 : \mu_1 = \mu_2\ H_1 : \mu_1 \neq \mu_2$
b) $H_0 : \mu_{11} = \mu_{21}\ H_1 : \mu_{11} \neq \mu_{21}$

c) $H_0 : \mu_{12} = \mu_{22} \, H_1 : \mu_{12} \neq \mu_{22}$

Compare the solutions and comment.

Exercise 7.11 Prove expression (7.4) in the Test Problem 2 with log-likelihoods ℓ_0^* and ℓ_1^*. (Hint: use (2.29)).

Exercise 7.12 Assume that $X \sim N_p(\mu, \Sigma)$ where Σ is unknown.

(a) Derive the log-likelihood ratio test for testing the independence of the p components, that is $H_0 : \Sigma$ is a diagonal matrix. (Solution: $-2 \log \lambda = -n \log |\mathcal{R}|$ where $\mathcal{R}$ is the correlation matrix, which is asymptotically a $\chi^2_{\frac{1}{2}p(p-1)}$ under H_0).
(b) Assume that Σ is a diagonal matrix (all the variables are independent). Can an asymptotic test for $H_0 : \mu = \mu_0$ against $H_1 : \mu \neq \mu_0$ be derived? How would this compare to p independent univariate t-tests on each μ_j?
(c) Show an easy derivation of an asymptotic test for testing the equality of the p means (Hint: use $(C\bar{X})^\top (CSC^\top)^{-1} C\bar{X} \xrightarrow{\mathcal{L}} \chi^2_{p-1}$ where $S = \text{diag}(s_{11}, \dots, s_{pp})$ and C is defined as in (7.10)). Compare this to the simple ANOVA procedure used in Sect. 3.5.

Exercise 7.13 The yields of wheat have been measured in 30 parcels that have been randomly attributed to 3 lots prepared by one of 3 different fertilizers A, B, and C. The data are

Fertilizer yield	A	B	C
1	4	6	2
2	3	7	1
3	2	7	1
4	5	5	1
5	4	5	3
6	4	5	4
7	3	8	3
8	3	9	3
9	3	9	2
10	1	6	2

Using Exercise 7.12,

(a) test the independence between the 3 variables.
(b) test whether $\mu^\top = [2\ 6\ 4]$ and compare this to the 3 univariate t-tests.
(c) test whether $\mu_1 = \mu_2 = \mu_3$ using simple ANOVA and the χ^2 approximation.

Exercise 7.14 Consider an i.i.d. sample of size $n = 5$ from a bivariate normal distribution

$$X \sim N_2\left(\mu, \begin{pmatrix} 3 & \rho \\ \rho & 1 \end{pmatrix}\right)$$

where ρ is a known parameter. Suppose $\bar{x}^\top = (1\ 0)$. For what value of ρ would the hypothesis $H_0 : \mu^\top = (0\ 0)$ be rejected in favor of $H_1 : \mu^\top \neq (0\ 0)$ (at the 5% level)?

Exercise 7.15 Using Example 7.14, test the last two cases described there and test the sample number one ($n_1 = 30$), to see if they are from a normal population with $\Sigma = 4\mathcal{I}_4$ (the sample covariance matrix to be used is given by S_1).

Exercise 7.16 Consider the bank data set. For the counterfeit banknotes, we want to know if the length of the diagonal (X_6) can be predicted by a linear model in X_1 to X_5. Estimate the linear model and test if the coefficients are significantly different from zero.

Exercise 7.17 In Example 7.10, can you predict the vocabulary score of the children in the 11th grade, by knowing the results from grades 8, 9, and 10? Estimate a linear model and test its significance.

Exercise 7.18 Test the equality of the covariance matrices from the two groups in the WAIS subtest (Example 7.19).

Exercise 7.19 Prove expressions (7.21), (7.22) and (7.23).

Exercise 7.20 Using Theorem 6.3 and expression (7.16), construct an asymptotic rejection region of size α for testing, in a general model $f(x, \theta)$, with $\theta \in \mathbb{R}^k$, $H_0 : \theta = \theta_0$ against $H_1 : \theta \neq \theta_0$.

Exercise 7.21 Exercise 6.5 considered the pdf $f(x_1, x_2) = \frac{1}{\theta_1^2 \theta_2^2 x_2} e^{-\left(\frac{x_1}{\theta_1 x_2} + \frac{x_2}{\theta_1 \theta_2}\right)}$ $x_1, x_2 > 0$. Solve the problem of testing $H_0 : \theta^\top = (\theta_{01}, \theta_{02})$ from an i.i.d. sample of size n on $x = (x_1, x_2)^\top$, where n is large.

Exercise 7.22 In Olkin and Veath (1980), the evolution of citrate concentrations in plasma is observed at 3 different times of day, X_1 (8 am), X_2 (11 am) and X_3 (3 pm), for two groups of patients who follow different diets. (The patients were randomly attributed to each group under a balanced design $n_1 = n_2 = 5$).
The data are

Group	X_1(8 am)	X_2(11 am)	X_3(3 pm)
	125	137	121
	144	173	147
I	105	119	125
	151	149	128
	137	139	109
	93	121	107
	116	135	106
II	109	83	100
	89	95	83
	116	128	100

Test if the profiles of the groups are parallel, if they are at the same level and if they are horizontal.

References

R.D. Bock, *Multivariate Statistical Methods In Behavioral Research* (Mc Graw-Hill, New York, 1975)

D.F. Morrison, *Multivariate Statistical Methods* (McGraw-Hill, New York, 1990a)

I. Olkin, M. Veath, Maximum likelihood estimation in a two-way analysis with correlated errors in one classification. Biometrika **68**, 653–660 (1980)

Part III
Multivariate Techniques

Chapter 8
Regression Models

The main aim of regression models is to model the variation of a quantitative response variable y in terms of the variation of one or several explanatory variables $(x_1, \ldots, x_p)^\top$. We have already introduced such models in Chap. 3 and 7, where linear models were written in (3.50) as

$$y = \mathcal{X}\beta + \varepsilon,$$

where $y(n \times 1)$ is the vector of observation for the response variable, $\mathcal{X}(n \times p)$ is the data matrix of the p explanatory variables and ε are the errors. Linear models are not restricted to handle only linear relationships between y and x. Curvature is allowed by including appropriate higher order terms in the *design* matrix $\mathcal{X}$.

Example 8.1 If y represents response and x_1, x_2 are two factors that explain the variation of y via the quadratic response model:

$$y_i = \beta_0 + \beta_1 x_{i1} + \beta_2 x_{i2} + \beta_3 x_{i1}^2 + \beta_4 x_{i2}^2 + \beta_5 x_{i1} x_{i2} + \varepsilon_i, \quad i = 1, \ldots, n. \quad (8.1)$$

This model (8.1) belongs to the class of linear models because it is linear in β. The data matrix $\mathcal{X}$ is

$$\mathcal{X} = \begin{pmatrix} 1 & x_{11} & x_{12} & x_{11}^2 & x_{12}^2 & x_{11}x_{12} \\ 1 & x_{21} & x_{22} & x_{21}^2 & x_{22}^2 & x_{21}x_{22} \\ \vdots & \vdots & \vdots & \vdots & \vdots & \vdots \\ 1 & x_{n1} & x_{n2} & x_{n1}^2 & x_{n2}^2 & x_{n1}x_{n2} \end{pmatrix}$$

For a given value of β, the response surface can be represented in a three-dimensional plot as in Fig. 8.1, where we display $y = 20 + 1x_1 + 2x_2 - 8x_1^2 - 6x_2^2 + 6x_1x_2$, i.e., $\beta = (20, 1, 2, -8, -6, +6)^\top$.

© Springer Nature Switzerland AG 2019
W. K. Härdle and L. Simar, *Applied Multivariate Statistical Analysis*,
https://doi.org/10.1007/978-3-030-26006-4_8

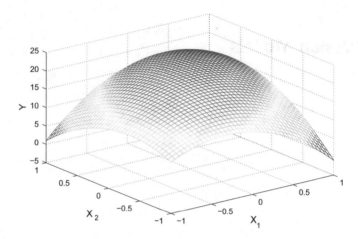

Fig. 8.1 A three-dimensional response surface 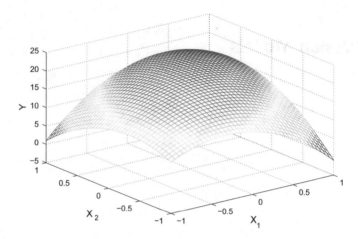 MVAresponsesurface

Note also that pure nonlinear models can sometimes be rewritten as a linear model by choosing an appropriate transformation of the coordinates of the variables. For instance the Cobb–Douglas production function

$$y_i = k \, x_{i1}^{\beta_1} \, x_{i2}^{\beta_2} \, x_{i3}^{\beta_3},$$

where y is the level of the production of a plant and $(x_1, x_2, x_3)^\top$ are 3 factors of production (e.g., labor, capital, and energy), can be transformed into a linear model in the log scale. We have indeed

$$\log y_i = \beta_0 + \beta_1 \log x_{i1} + \beta_2 \log x_{i2} + \beta_3 \log x_{i3},$$

where $\beta_0 = \log k$ and the β_j, $j = 1, \ldots, 3$ are the elasticities $\left(\beta_j = \frac{\partial \log y}{\partial \log x_j} \right)$.

Linear models are flexible and cover a wide class of models. If $\mathcal{X}$ has full rank, they can easily be estimated by least squares $\widehat{\beta} = (\mathcal{X}^\top \mathcal{X})^{-1} \mathcal{X}^\top y$ and linear restrictions on the β's can be tested using the tools developed in Chap. 7.

In Chap. 3, we saw that even qualitative explanatory variables can be used by defining appropriate coding of the nominal values of x. In this chapter, we will extend our toolbox by showing how to code these qualitative factors in a way which allows the introduction of several qualitative factors including the possibility of interactions. This covers more general ANOVA models than those introduced in Chap. 3. This includes the ANCOVA models where qualitative and quantitative variables are both present in the explanatory variables.

When the response variable is qualitative or categorical (for instance, an individual can be employed or unemployed, a company may be bankrupt or not, the opinion of one person relative to a particular issue can be "in favor", "against" or "indifferent to", etc...), linear models have to be adapted to this particular situation. The most useful models for these cases will be presented in the second part of the chapter; this covers the log-linear models for contingency tables (where we analyze the relations between several categorical variables) and the logit model for quantal or binomial responses where we analyze the probability of being in one state as a function of explanatory variables.

8.1 General ANOVA and ANCOVA Models

8.1.1 ANOVA Models

One-factor models

In Sect. 3.5, we introduced the example of analyzing the effect of one factor (3 possible marketing strategies) on the sales of a product (a pullover), see Table 3.2. The standard way to present one-factor ANOVA models with p levels, is as follows:

$$y_{k\ell} = \mu + \alpha_\ell + \varepsilon_{k\ell}, \ k = 1, \ldots, n_\ell, \ \text{and} \ \ell = 1, \ldots, p, \quad (8.2)$$

all the $\varepsilon_{k\ell}$ being independent. Here ℓ is the label which indicates the level of the factor and α_ℓ is the effect of the ℓth level: it measures the deviation from μ, the global mean of y, due to this level of the factor. In this notation, we need to impose the restriction $\sum_{\ell=1}^{p} \alpha_\ell = 0$ in order to identify μ as the mean of y. This presentation is equivalent, but slightly different, to the one presented in Chap. 3 (compare with Eq. (3.41)), but it allows for easier extension to the multiple factors case. Note also that here we allow different sample sizes for each level of the factor (an unbalanced design, more general than the balanced design presented in Chap. 3).

To simplify the presentation, assume as in the pullover example that $p = 3$. In this case, one could be tempted to write the model (8.2) under the general form of a linear model by using 3 indicator variables

$$y_i = \mu + \alpha_1 x_{i1} + \alpha_2 x_{i2} + \alpha_3 x_{i3} + \varepsilon_i,$$

where $x_{i\ell}$ is equal to 1 or 0 according to the i-th observation and belongs (or not) to the level ℓ of the factor. In matrix notation and letting, for simplicity, $n_1 = n_2 = n_3 = 2$ we have with $\beta = (\mu, \alpha_1, \alpha_2, \alpha_3)^\top$

$$y = \mathcal{X}\beta + \varepsilon, \quad (8.3)$$

where the design matrix $\mathcal{X}$ is given by:

$$X = \begin{pmatrix} 1 & 1 & 0 & 0 \\ 1 & 1 & 0 & 0 \\ 1 & 0 & 1 & 0 \\ 1 & 0 & 1 & 0 \\ 1 & 0 & 0 & 1 \\ 1 & 0 & 0 & 1 \end{pmatrix}.$$

Unfortunately, this type of coding is not useful because the matrix X is not of full rank (the sum of each row is equal to the same constant 2) and therefore the matrix $X^\top X$ is not invertible. One way to overcome this problem is to change the coding by introducing the additional constraint that the effects add up to zero. There are many ways to achieve this. Noting that $\alpha_3 = -\alpha_1 - \alpha_2$, we do not need to introduce α_3 explicitly in the model. The linear model could indeed be written as

$$y_i = \mu + \alpha_1 x_{i1} + \alpha_2 x_{i2} + \varepsilon_i,$$

with a design matrix defined as

$$X = \begin{pmatrix} 1 & 1 & 0 \\ 1 & 1 & 0 \\ 1 & 0 & 1 \\ 1 & 0 & 1 \\ 1 & -1 & -1 \\ 1 & -1 & -1 \end{pmatrix},$$

which automatically implies that $\alpha_3 = -(\alpha_1 + \alpha_2)$. The linear model (8.3) is now correct with $\beta = (\mu, \alpha_1, \alpha_2)^\top$. The least squares estimator $\widehat{\beta} = (X^\top X)^{-1} X^\top y$ can be computed providing the estimator of the ANOVA parameters μ and α_ℓ, $\ell = 1, \ldots, 3$. Any linear constraint on β can be tested by using the techniques described in Chap. 7. For instance, the null hypothesis of no factor effect $H_0 : \alpha_1 = \alpha_2 = \alpha_3 = 0$ can be written as $H_0 : A\beta = a$, where $A = \begin{pmatrix} 0 & 1 & 0 \\ 0 & 0 & 1 \end{pmatrix}$ and $a = (0 \ \ 0)^\top$.

Multiple factors models

The coding above can be extended to more general situations with many qualitative variables (factors) and with the possibility of interactions between the factors. Suppose that in a marketing example, the sales of a product can be explained by two factors: the marketing strategy with 3 levels (as in the pullover example) but also the location of the shop that may be either in a big shopping center or in a less commercial location (2 levels for this factor). We might also think that there is an interaction between the two factors: the marketing strategy might have a different effect in a shopping center than in a small quiet area. To fix the idea, the data are collected as in Table 8.1.

Table 8.1 A two-factor ANOVA data set, factor A, three levels of the marketing strategy and factor B, two levels for the location. The figures represent the resulting sales during the same period

	B_1	B_2
A_1	18	15
	15	20
		25
		30
A_2	5	10
	8	12
	8	
A_3	10	20
	14	25

The general two-factor model with interactions can be written as

$$y_{ijk} = \mu + \alpha_i + \gamma_j + (\alpha\gamma)_{ij} + \varepsilon_{ijk}; \ i = 1, \ldots, r, \ j = 1, \ldots, s, \ k = 1, \ldots, n_{ij} \tag{8.4}$$

where the identification constraints are

$$\sum_{i=1}^{r} \alpha_i = 0 \text{ and } \sum_{j=1}^{s} \gamma_j = 0$$

$$\sum_{i=1}^{r} (\alpha\gamma)_{ij} = 0, \ j = 1, \ldots, s \tag{8.5}$$

$$\sum_{j=1}^{s} (\alpha\gamma)_{ij} = 0, \ i = 1, \ldots, r.$$

In our example of Table 8.1, we have $r = 3$ and $s = 2$. The α's measure the effect of the marketing strategy (3 levels) and the γ's the effect of the location (2 levels). A positive (negative) value of one of these parameters would indicate a favorable (unfavorable) effect on the expected sales; the global average of sales being represented by the parameter μ. The interactions are measured by the parameters $(\alpha\gamma)_{ij}$, $i = 1, \ldots, r$, $j = 1, \ldots, s$, again identification constraints implies the $(r + s)$ constraints in (8.5) on the interactions terms.

For example, a positive value of $(\alpha\gamma)_{11}$ would indicate that the effect of the sale strategy A_1 (advertisement in local newspaper), if any, is more favorable on the sales in the location B_1 (in a big commercial center) than in the location B_2 (not a commercial center) with the relation $(\alpha\gamma)_{11} = -(\alpha\gamma)_{12}$. As another example, a negative value of $(\alpha\gamma)_{31}$ would indicate that the marketing strategy A_3 (luxury presentation in shop windows) has less effect, if any, in location type B_1 than in B_2: again $(\alpha\gamma)_{31} = -(\alpha\gamma)_{32}$, etc, ...

The nice thing is that it is easy to extend the coding rule for one-factor model to this general situation, in order to present the model a standard linear model with the appropriate design matrix $\mathcal{X}$. To build the columns of $\mathcal{X}$ for the effect of each factor, we will need, as above, $r - 1$ (and $s - 1$) variables for coding a qualitative variable with r (and s, respectively) levels with the convention defined above in the one-factor case. For the interactions between a r-level factor and a s-level factor, we will need $(r - 1) \times (s - 1)$ additional columns that will be obtained by performing the product, element by element, of the corresponding main effect columns. So, at the end, for a full model with all the interactions, we have $\{1 + r - 1 + s - 1 + (r - 1)(s - 1)\} = rs$ parameters where the first column of 1's is for the intercept (the constant μ). We illustrate this for our marketing example, where $r = 3$ and $s = 2$. We first describe a model without interactions.

1. *Model without interactions*

   Without the interactions (all the $(\alpha\gamma)_{ij} = 0$) the model could be written with $3 = (r - 1) + (s - 1)$ coded variables in a simple linear model form as in (8.3), with the matrices:

$$
y = \begin{pmatrix} 18 \\ 15 \\ 15 \\ 20 \\ 25 \\ 30 \\ 5 \\ 8 \\ 8 \\ 10 \\ 12 \\ 10 \\ 14 \\ 20 \\ 25 \end{pmatrix}, \quad
\mathcal{X} = \begin{pmatrix}
1 & 1 & 0 & 1 \\
1 & 1 & 0 & 1 \\
1 & 1 & 0 & -1 \\
1 & 1 & 0 & -1 \\
1 & 1 & 0 & -1 \\
1 & 1 & 0 & -1 \\
1 & 0 & 1 & 1 \\
1 & 0 & 1 & 1 \\
1 & 0 & 1 & 1 \\
1 & 0 & 1 & -1 \\
1 & 0 & 1 & -1 \\
1 & -1 & -1 & 1 \\
1 & -1 & -1 & 1 \\
1 & -1 & -1 & -1 \\
1 & -1 & -1 & -1
\end{pmatrix},
$$

   and $\beta = (\mu, \alpha_1, \alpha_2, \gamma_1)^\top$. Then, $\alpha_3 = -(\alpha_1 + \alpha_2)$ and $\gamma_2 = -\gamma_1$.

2. *Model with interactions*

   A model with interaction between A and B is obtained by adding new columns to the design matrix. We need $2 = (r - 1) \times (s - 1)$ new coding variables which are defined as the product, element by element, of the corresponding columns obtained for the main effects. For instance, for the interaction parameter $(\alpha\gamma)_{11}$, we multiply the column used for coding α_1 by the column defined for coding γ_1, where the product is element by element. The same is done for the parameter $(\alpha\gamma)_{21}$. No other columns are necessary, since the remaining interactions are derived from the identification constraints (8.5). We obtain

$$\mathcal{X} = \begin{pmatrix} 1 & 1 & 0 & 1 & 1 & 0 \\ 1 & 1 & 0 & 1 & 1 & 0 \\ 1 & 1 & 0 & -1 & -1 & 0 \\ 1 & 1 & 0 & -1 & -1 & 0 \\ 1 & 1 & 0 & -1 & -1 & 0 \\ 1 & 1 & 0 & -1 & -1 & 0 \\ 1 & 0 & 1 & 1 & 0 & 1 \\ 1 & 0 & 1 & 1 & 0 & 1 \\ 1 & 0 & 1 & 1 & 0 & 1 \\ 1 & 0 & 1 & -1 & 0 & -1 \\ 1 & 0 & 1 & -1 & 0 & -1 \\ 1 & -1 & -1 & 1 & -1 & -1 \\ 1 & -1 & -1 & 1 & -1 & -1 \\ 1 & -1 & -1 & -1 & 1 & 1 \\ 1 & -1 & -1 & -1 & 1 & 1 \end{pmatrix},$$

with $\beta = (\mu, \alpha_1, \alpha_2, \gamma_1, (\alpha\gamma)_{11}, (\alpha\gamma)_{21})^\top$. The other interactions can indeed be derived from (8.5)

$$(\alpha\gamma)_{12} = -(\alpha\gamma)_{11}$$
$$(\alpha\gamma)_{22} = -(\alpha\gamma)_{21}$$
$$(\alpha\gamma)_{31} = -((\alpha\gamma)_{11} + (\alpha\gamma)_{21})$$
$$(\alpha\gamma)_{32} = -(\alpha\gamma)_{31}.$$

The estimation of β is again simply given by the least squares solution $\hat{\beta} = (\mathcal{X}^\top \mathcal{X})^{-1} \mathcal{X}^\top y$.

Example 8.2 Let us come back to the marketing data provided by the two-way Table 8.1. The values of $\hat{\beta}$ in the full model, with interactions, are given in Table 8.2.

The p-values in the right column are for the individual tests: it appears that the interactions do not provide additional significant explanation of y, but the effect of the two factors seems significant.

Using the techniques of Chap. 7, we can test some reduced model corresponding to linear constraints on the β's. The full model is the model with all the parameters including all the interactions. The overall fit test H_0 : all the parameters, except μ, are equal to zero, gives the value $F_{observed} = 6.5772$ with a p-value of 0.0077 for a $F_{5,9}$, so that H_0 is rejected. In this case, the $RSS_{reduced} = 735.3333$. So there is some effect by the factors.

We then test a less reduced model. We can test if the interaction terms are significantly different to zero. This is a linear constraint on β with

$$A = \begin{pmatrix} 0 & 0 & 0 & 0 & 1 & 0 \\ 0 & 0 & 0 & 0 & 0 & 1 \end{pmatrix}; \quad a = \begin{pmatrix} 0 \\ 0 \end{pmatrix}.$$

Under the null we obtain:

Table 8.2 Estimation of the two-factor ANOVA model with data from Table 8.1

	$\hat{\beta}$	p-values
μ	15.25	
α_1	4.25	0.0218
α_2	−6.25	0.0033
γ_1	−3.42	0.0139
$(\alpha\gamma)_{11}$	0.42	0.7922
$(\alpha\gamma)_{21}$	1.42	0.8096
RSS_{full}	158.00	

$$\hat{\beta}_{H_0} = \begin{pmatrix} 15.3035 \\ 4.0975 \\ -6.0440 \\ -3.2972 \\ 0 \\ 0 \end{pmatrix},$$

and $RSS_{reduced} = 181.8019$. The observed value of $F = 0.6779$ which is not significant ($r = 11$, $f = 9$) the p-value $= P(F_{2,9} \geq 0.6779) = 0.5318$, confirming the absence of interactions.

Now taking the model without the interactions as the full model, we can test if one of the main effects α (marketing strategy) or γ (location) or both are significantly different from zero. We leave this as an exercise for the reader.

8.1.2 ANCOVA Models

ANCOVA (ANalysis of COVAriances) are mixed models, where some variables are qualitative and others are quantitative. The same coding of the ANOVA will be used for the qualitative variable. The design matrix $\mathcal{X}$ is completed by the columns for the quantitative explanatory variables x. Interactions between a qualitative variable (a factor with r levels) and a quantitative one x is also possible, this corresponds to situations where the effect of x on the response y is different according to the level of the factor. This is achieved by adding into the design matrix $\mathcal{X}$, a new column obtained by the product, element by element, of the quantitative variable with the coded variables for the factor ($r - 1$ interaction variables if the categorical variable has r levels).

Fig. 8.2 A model with interaction

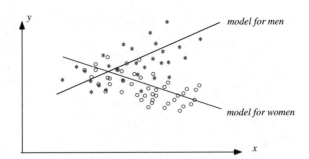

For instance, consider a simple model where a response y is explained by one explanatory variable x and one factor with 2 levels (for instance the gender level 1 for men and level 2 for women), we would have in the case $n_1 = n_2 = 3$

$$\mathcal{X} = \begin{pmatrix} 1 & x_1 & 1 & x_1 \\ 1 & x_2 & 1 & x_2 \\ 1 & x_3 & 1 & x_3 \\ 1 & x_4 & -1 & -x_4 \\ 1 & x_5 & -1 & -x_5 \\ 1 & x_6 & -1 & -x_6 \end{pmatrix},$$

with $\beta = (\beta_1, \beta_2, \beta_3, \beta_4)^\top$. The intercept and the slope are $(\beta_1 + \beta_3)$ and $(\beta_1 + \beta_4)$ for men and $(\beta_1 - \beta_3)$ and $(\beta_1 - \beta_4)$ for women. This situation is displayed in Fig. 8.2.

Example 8.3 Consider the Car data provided in Appendix B.3. We want to analyze the effect of the weight (W), the displacement (D) on the mileage (M). But we would like to test if the origin of the car (the factor C) has some effect on the response and if the effect of the continuous variables is different for the different levels of the factor. The linear regression model can be then written as

$$\log(\text{Mileage}_i) = \beta_0 + \beta_1 \log(\text{Weight}_i) + \beta_2 \log(\text{Displacement}_i)$$
$$+\beta_3 \, \text{I}(\text{Origin}_i = \text{Japan}) + \beta_4 \, \text{I}(\text{Origin}_i = \text{Europe}) + \varepsilon_i,$$

where the parameter β_0 is the absolute term for US cars, i.e., the column with "indicators of US cars" in the design matrix is omitted and thus the mileage of US cars used as a baseline. Such reparametrization is usually done automatically by software. The parameters β_3 and β_4 estimate the change for Japanese and European cars compared to the US baseline.

From the regression results in Table 8.3, we observe that the car weight has significant effect on the mileage. We reject the null hypothesis that there is no difference between the mileage of European and U.S. cars. However, in order to assess

Table 8.3 Linear regression output for Car example Q MVAcareffect

	$\hat{\beta}$	p-values
Intercept	10.210	0.000
log(weight)	−0.859	0.000
log(Displacement)	−0.056	0.481
Origin-Japan	−0.075	0.144
Origin-Europe	−0.156	0.003
RSS_{full}	158.00	

the significance of the factor variable "Origin", we need to test the null hypothesis $H_0 : \beta_3 = \beta_4 = 0$. This can be easily done using F-test. We obtain the origin, i.e., the location of company headquarters, has significant effect on the mileages.

8.1.3 Boston Housing

In Chaps. 3 and 7, linear models were used to analyze if the variations of the price (the variables were transformed in Sect. 1.9) could be explained by other variables. A reduced model was obtained in Sect. 7.3 with the results shown in Table 7.1, with $r^2 = 0.763$. The model was

$$X_{14} = \beta_0 + \beta_4 X_4 + \beta_5 X_5 + \beta_6 X_6 + \beta_8 X_8 + \beta_9 X_9 + \beta_{10} X_{10} + \beta_{11} X_{11}$$
$$+ \beta_{12} X_{12} + \beta_{13} X_{13}$$

One factor (X_4) was coded as a binary variable (1, if the house is close to the Charles River and 0 if it is not). Taking advantage of the ANCOVA models described above, we would like to add to a new factor built from the original quantitative variable $X_9=$ index of accessibility to radial highways. So we will transform X_4 as being 1 if close to the Charles River and −1 if not, and we will replace X_9 by a new factor coded $X_{15}= 1$ if $X_9 \geq$ median(X_9) and $X_{15} = -1$ if $X_9 <$ median(X_9). We also want to consider the interaction of X_4 with X_{12} (proportion of blacks) and the interaction of X_4 with the new factor X_{15}. The results are shown in Table 8.4.

Table 8.4 Estimation of the ANCOVA model using the Boston housing data **Q** MVAboshousing

	$\hat{\beta}$	p-values	$\tilde{\beta}$	p-values
β_0	32.27	0.00	27.65	0.00
β_4	1.54	0.00	−3.19	0.32
β_5	−17.59	0.00	−16.50	0.00
β_6	4.27	0.00	4.23	0.00
β_8	−1.13	0.00	−1.10	0.00
β_{10}	0.00	0.97	0.00	0.95
β_{11}	−0.97	0.00	−0.97	0.00
β_{12}	0.01	0.00	0.02	0.01
β_{13}	−0.54	0.00	−0.54	0.00
β_{15}	0.21	0.46	0.23	0.66
β_{4*14}			0.01	0.13
β_{4*15}			0.03	0.95

Summary
↪ ANOVA models can be dividend into one-factor models and multiple factor models.
↪ Multiple factor models analyze many qualitative variables and the interactions between them.
↪ ANCOVA models are mixed models with qualitative and quantitative variables, and can also incorporate the interaction between a qualitative and a quantitative variable.

8.2 Categorical Responses

8.2.1 Multinomial Sampling and Contingency Tables

In many applications, the response variable of interest is qualitative or categorical, in the sense that the response can take its nominal value in one of, say, K classes or categories. Often we observe counts y_k, the number of observations in category $k = 1, \ldots, K$. If the total number of observations $n = \sum_{k=1}^{K} y_k$ is fixed and we may assume independence of the observations, we obtain a multinomial sampling process.

If we denote by p_k the probability of observing the k-th category with $\sum_{k=1}^{K} p_k = 1$, we have $\mathsf{E}(Y_k) = m_k = np_k$. The likelihood of the sample can then be written as

$$L = \frac{n!}{\prod_{k=1}^{K} y_k!} \prod_{k=1}^{K} \left(\frac{m_k}{n}\right)^{y_k}. \tag{8.6}$$

In contingency tables, the categories are defined by several qualitative variables. For example in a $(J \times K)$ two-way table, the observations (counts) $y_{jk}, j = 1, \ldots, J$ and $k = 1, \ldots, K$ are reported for row j and column k. Here $n = \sum_{j=1}^{J} \sum_{k=1}^{K} y_{jk}$. Log-linear models introduce a linear structure on the logarithms of the expected frequencies $m_{jk} = \mathsf{E}(y_{jk}) = np_{jk}$, with $\sum_{j=1}^{J} \sum_{k=1}^{K} p_{jk} = 1$. Log-linear structures on m_{jk} will impose the same structure for the p_{jk}, the estimation of the model will then be obtained by constrained maximum likelihood. Three-way tables $(J \times K \times L)$ may be analyzed in the same way.

Sometimes additional information is available on explanatory variables x. In this case, the logit model will be appropriate when the categorical response is binary $(K = 2)$. We will introduce these models when the main response of interest is binary (for instance, tables $(2 \times K)$ or $(2 \times K \times L)$). Further, we will show how they can be adapted to the case of contingency tables. Contingency tables are also analyzed by multivariate descriptive tools in Chap. 15.

8.2.2 Log-Linear Models for Contingency Tables

8.2.2.1 Two-Way Tables

Consider a $(J \times K)$ two-way table, where y_{jk} is the number of observations having the nominal value j for the first qualitative character and nominal value k for the second character. Since the total number of observations is fixed $n = \sum_{j=1}^{J} \sum_{k=1}^{K} y_{jk}$, there are $JK - 1$ free cells in the table. The multinomial likelihood can be written as in (8.6)

$$L = \frac{n!}{\prod_{j=1}^{J} \prod_{k=1}^{K} y_{jk}!} \prod_{j=1}^{J} \prod_{k=1}^{K} \left(\frac{m_{jk}}{n} \right)^{y_{jk}}, \tag{8.7}$$

where we now introduce a log-linear structure to analyze the role of the rows and the columns to determine the parameters $m_{jk} = \mathsf{E}(y_{jk})$ (or p_{jk}).

1. *Model without interaction*
   Suppose that there is no interaction between the rows and the columns: this corresponds to the hypothesis of independence between the two qualitative characters. In other words, $p_{jk} = p_j p_k$ for all j, k. This implies the log-linear model:

$$\log m_{jk} = \mu + \alpha_j + \gamma_k \text{ for } j = 1, \ldots, J, \ k = 1, \ldots, K, \tag{8.8}$$

where, as in ANOVA models for identification purposes $\sum_{j=1}^{J} \alpha_j = \sum_{k=1}^{K} \gamma_k = 0$. Using the same coding devices as above, the model can be written as

$$\log m = \mathcal{X} \beta. \tag{8.9}$$

For a (2×3) table we have:

$$\log m = \begin{pmatrix} \log m_{11} \\ \log m_{12} \\ \log m_{13} \\ \log m_{21} \\ \log m_{22} \\ \log m_{23} \end{pmatrix}, \quad \mathcal{X} = \begin{pmatrix} 1 & 1 & 1 & 0 \\ 1 & 1 & 0 & 1 \\ 1 & 1 & -1 & -1 \\ 1 & -1 & 1 & 0 \\ 1 & -1 & 0 & 1 \\ 1 & -1 & -1 & -1 \end{pmatrix}, \quad \beta = \begin{pmatrix} \beta_0 \\ \beta_1 \\ \beta_2 \\ \beta_3 \end{pmatrix}$$

where the first column of $\mathcal{X}$ is for the constant term, the second column is the coded column for the 2-levels row effect, and the two last columns are the coded columns for the 3-levels column effect. The estimation is obtained by maximizing the log-likelihood which is equivalent to maximizing the function $L(\beta)$ in β:

$$L(\beta) = \sum_{j=1}^{J} \sum_{k=1}^{K} y_{jk} \log m_{jk}. \tag{8.10}$$

The maximization is under the constraint $\sum_{j,k} m_{jk} = n$. In summary, we have $1 + (J - 1) + (K - 1) - 1$ free parameters for $JK - 1$ free cells. The number of *degrees of freedom* in the model is the number of free cells minus the number of free parameters. It is given by

$$r = JK - 1 - (J - 1) - (K - 1) = (J - 1)(K - 1).$$

In the example above, we have, therefore, $(3 - 1) \times (2 - 1) = 2$ degrees of freedom.

The original parameters of the model can then be estimated as

$$\alpha_1 = \beta_1$$
$$\alpha_2 = -\beta_1$$
$$\gamma_1 = \beta_2$$
$$\gamma_2 = \beta_3$$
$$\gamma_3 = -(\beta_2 + \beta_3). \tag{8.11}$$

2. *Model with interactions*

In two-way tables, the interactions between the two variables are of interest. This corresponds to the general (full) model

$$\log m_{jk} = \mu + \alpha_j + \gamma_k + (\alpha\gamma)_{jk}, \quad j = 1, \ldots, J, \; k = 1 \ldots, K, \tag{8.12}$$

where in addition, we have the $J + K$ restrictions

$$\sum_{k=1}^{K} (\alpha\gamma)_{jk} = 0, \quad \text{for } j = 1, \dots, J$$

$$\sum_{j=1}^{J} (\alpha\gamma)_{jk} = 0, \quad \text{for } k = 1, \dots, K \tag{8.13}$$

As in the ANOVA model, the interactions may be coded by adding $(J - 1)(K - 1)$ columns to $\mathcal{X}$, obtained by the product of the corresponding coded variables. In our example for the (2×3) table, the design matrix $\mathcal{X}$ is completed with two more columns:

$$\mathcal{X} = \begin{pmatrix} 1 & 1 & 1 & 0 & 1 & 0 \\ 1 & 1 & 0 & 1 & 0 & 1 \\ 1 & 1 & -1 & -1 & -1 & -1 \\ 1 & -1 & 1 & 0 & -1 & 0 \\ 1 & -1 & 0 & 1 & 0 & -1 \\ 1 & -1 & -1 & -1 & 1 & 1 \end{pmatrix}, \quad \beta = \begin{pmatrix} \beta_0 \\ \beta_1 \\ \beta_2 \\ \beta_3 \\ \beta_4 \\ \beta_5 \end{pmatrix}.$$

Now the interactions are determined by using (8.13):

$$(\alpha\gamma)_{11} = \beta_4$$
$$(\alpha\gamma)_{12} = \beta_5$$
$$(\alpha\gamma)_{13} = -\{(\alpha\gamma)_{11} + (\alpha\gamma)_{12}\} = -(\beta_4 + \beta_5)$$
$$(\alpha\gamma)_{21} = -(\alpha\gamma)_{11} = -\beta_4$$
$$(\alpha\gamma)_{22} = -(\alpha\gamma)_{12} = -\beta_5$$
$$(\alpha\gamma)_{23} = -(\alpha\gamma)_{13} = \beta_4 + \beta_5$$

We have again a log-linear model as in (8.9) and the estimation of β goes through the maximization in β of $L(\beta)$ given by (8.10) under the same constraint.

The model with all the interaction terms is called the *saturated* model. In this model there are no degrees of freedom, the number of free parameters to be estimated equals the number of free cells. The parameters of interest are the interactions. In particular, we are interested in testing their significance. These issues will be addressed below.

8.2.2.2   Three-Way Tables

The models presented above for two-way tables can be extended to higher order tables but at a cost of notational complexity. We show how to adapt to three-way tables. This deserves special attention due to the presence of higher order interactions in the saturated model.

A $(J \times K \times L)$ three-way table may be constructed under multinomial sampling as follows: each of the n observations falls in one, and only one, category of each of three categorical variables having J, K, and L modalities, respectively. We end up with a three-dimensional table with JKL cells containing the counts $y_{jk\ell}$ where $n = \sum_{j,k,\ell} y_{jk\ell}$. The expected counts depend on the unknown probabilities $p_{jk\ell}$ in the usual way:

$$m_{jk\ell} = n\, p_{jk\ell}, \ j = 1\ldots, J, \ k = 1\ldots, K, \ \ell = 1, \ldots, L.$$

1. *The saturated model*
   A full saturated log-linear model reads as follows:

$$\log m_{jk\ell} = \mu + \alpha_j + \beta_k + \gamma_\ell + (\alpha\beta)_{jk} + (\alpha\gamma)_{j\ell} + (\beta\gamma)_{k\ell} + (\alpha\beta\gamma)_{jk\ell},$$
$$j = 1\ldots, J, \ k = 1\ldots, K, \ \ell = 1, \ldots, L. \tag{8.14}$$

The restrictions are the following (using the "dot" notation for summation on the corresponding indices):

$$\alpha_{(\bullet)} = \beta_{(\bullet)} = \gamma_{(\bullet)} = 0$$
$$(\alpha\beta)_{j\bullet} = (\alpha\gamma)_{j\bullet} = (\beta\gamma)_{k\bullet} = 0$$
$$(\alpha\beta)_{\bullet k} = (\alpha\gamma)_{\bullet\ell} = (\beta\gamma)_{\bullet\ell} = 0$$
$$(\alpha\beta\gamma)_{jk\bullet} = (\alpha\beta\gamma)_{j\bullet\ell} = (\alpha\beta\gamma)_{\bullet k\ell} = 0$$

The parameters $(\alpha\beta)_{jk}$, $(\alpha\gamma)_{j\ell}$, $(\beta\gamma)_{k\ell}$ are called *first-order interactions*. The *second-order interactions* are the parameters $(\alpha\beta\gamma)_{jk\ell}$, they allow to take into account heterogeneities in the interactions between two of the three variables. For instance, let ℓ stand for the two gender categories $(L = 2)$, if we suppose that $(\alpha\beta\gamma)_{jk1} = -(\alpha\beta\gamma)_{jk2} \neq 0$, we mean that the interactions between the variable J and K are not the same for both gender categories.

The estimation of the parameters of the saturated model are obtained through maximization of the log-likelihood. In the multinomial sampling scheme, it corresponds to maximizing the function:

$$L = \sum_{j,k,\ell} y_{jk\ell} \log m_{jk\ell},$$

under the constraint $\sum_{j,k,\ell} m_{jk\ell} = n$.
The number of degrees of freedom in the saturated model is again zero. Indeed, the number of free parameters in the model is

$$1 + (J-1) + (K-1) + (L-1) + (J-1)(K-1) + (J-1)(L-1)$$
$$+ (K-1)(L-1) + (J-1)(K-1)(L-1) - 1 = JKL - 1.$$

This is indeed equal to the number of free cells in the table and so, there is no degree of freedom.

2. *Hierarchical non-saturated models*

As illustrated above, a saturated model has no degrees of freedom. Non-saturated models correspond to reduced models where some parameters are fixed to be equal to zero. They are thus particular cases of the saturated model (8.14). The *hierarchical* non-saturated models that we will consider here, are models where once a set of parameters is set equal to zero, all the parameters of higher order containing the same indices are also set equal to zero.

For instance, if we suppose $\alpha_1 = 0$, we only consider non-saturated models where also $(\alpha\gamma)_{1\ell} = (\alpha\beta)_{1k} = (\alpha\beta\gamma)_{1k\ell} = 0$ for all values of k and ℓ. If we only suppose that $(\alpha\beta)_{12} = 0$, we also assume that $(\alpha\beta\gamma)_{12\ell} = 0$ for all ℓ.

Hierarchical models have the advantage of being more easily interpretable. Indeed without this hierarchy, the models would be difficult to interpret. What would be, for instance, the meaning of the parameter $(\alpha\beta\gamma)_{12\ell}$, if we know that $(\alpha\beta)_{12} = 0$? The estimation of the non-saturated models will be achieved by the usual way, i.e., by maximizing the log-likelihood function L as above but under the new constraints of the reduced model.

8.2.3   Testing Issues with Count Data

One of the main practical interests in regression models for contingency tables is to test restrictions on the parameters of a more complete model. These testing ideas are created in the same spirit as in Sect. 3.5 where we tested restrictions in ANOVA models.

In linear models, the test statistics is based on the comparison of the goodness of fit for the full model and for the reduced model. Goodness of fit is measured by the residual sum of squares (RSS). The idea here will be the same here but with a more appropriate measure for goodness of fit. Once a model has been estimated, we can compute the predicted value under that model for each cell of the table. We will denote, as above, the observed value in a cell by y_k and $\hat{m}_k$ will denote the expected value predicted by the model. The goodness of fit may be appreciated by measuring, in some way, the distance between the series of observed and of predicted values. Two statistics are proposed: the *Pearson chi-square* X^2 and the *Deviance* noted G^2. They are defined as follows:

$$X^2 = \sum_{k=1}^{K} \frac{(y_k - \hat{m}_k)^2}{\hat{m}_k} \tag{8.15}$$

$$G^2 = 2 \sum_{k=1}^{K} y_k \, \log \left(\frac{y_k}{\hat{m}_k} \right) \tag{8.16}$$

where K is the total number of cells of the table. The deviance is directly related to the log-likelihood ratio statistic and is usually preferred because it can be used to compare nested models as we usually do in this context.

Under the hypothesis that the model used to compute the predicted value is true, both statistics (for large samples) are approximately distributed as a χ^2 variable with degrees of freedom $d.f.$ depending on the model. The $d.f.$ can be computed as follows:

$$d.f. = \# \text{free cells} - \# \text{free parameters estimated.} \tag{8.17}$$

For saturated models, the fit is perfect: $X^2 = G^2 = 0$ with $d.f. = 0$.

Suppose now that we want to test a reduced model which is a restricted version of a full model. The deviance can then be used as the F statistics in linear regression. The test procedure is straightforward:

$$
\begin{aligned}
H_0 &: \quad \text{reduced model with } r \text{ degrees of freedom} \\
H_1 &: \quad \text{full model with } f \text{ degrees of freedom.}
\end{aligned}
\tag{8.18}
$$

Since, the full model contains more parameters, we expect the deviance to be smaller. We reject the H_0 if this reduction is significant, i.e., if $G^2_{H_0} - G^2_{H_1}$ is large enough. Under H_0 one has

$$G^2_{H_0} - G^2_{H_1} \sim \chi^2_{r-f}.$$

We reject H_0 if the p-value:

$$P\{\chi^2_{r-f} > (G^2_{H_0} - G^2_{H_1})\}.$$

is small. Suppose we want to test the independence in a $(J \times K)$ two-way table (no interaction). Here, the full model is the saturated one with no degrees of freedom ($f = 0$) and the restricted model has $r = (J - 1)(K - 1)$ degrees of freedom. We reject H_0 if the p-value of H_0 $P\{\chi^2_r > (G^2_{H_0})\}$ is too small.

This test is equivalent to the Pearson chi-square test for independence in two-way tables ($G^2_{H_0} \approx X^2_{H_0}$ when n is large).

Example 8.4 Everitt and Dunn (1998) provide a three-dimensional ($2 \times 2 \times 5$) count table of $n = 5833$ interviewed people. The count were on prescribed psychotropic drugs in the fortnight prior to the interview as a function of age and gender. The data are summarized in Table 8.5, where the categories for the 3 factors are M for

Table 8.5 A Three-way Contingency Table: top table for men and bottom table for women
MVAdrug

M	A1	A2	A3	A4	A5
DY	21	32	70	43	19
DN	683	596	705	295	99
F	A1	A2	A3	A4	A5
DY	46	89	169	98	51
DN	738	700	847	336	196

Table 8.6 Coefficient estimates based on the saturated model MVAdrug

	$\hat{\beta}$		$\hat{\beta}$
$\hat{\beta}_0$ intercept	5.0089	$\hat{\beta}_{10}$	0.0205
$\hat{\beta}_1$ gender: M	−0.2867	$\hat{\beta}_{11}$	0.0482
$\hat{\beta}_2$ drug: DY	−1.0660	$\hat{\beta}_{12}$ drug*age	−0.4983
$\hat{\beta}_3$ age	−0.0080	$\hat{\beta}_{13}$	−0.1807
$\hat{\beta}_4$	0.2151	$\hat{\beta}_{14}$	0.0857
$\hat{\beta}_5$	0.6607	$\hat{\beta}_{15}$	0.2766
$\hat{\beta}_6$	−0.0463	$\hat{\beta}_{16}$ gender*drug*age	−0.0134
$\hat{\beta}_7$ gender*drug	−0.1632	$\hat{\beta}_{17}$	−0.0523
$\hat{\beta}_8$ gender*age	0.0713	$\hat{\beta}_{18}$	−0.0112
$\hat{\beta}_9$	−0.0092	$\hat{\beta}_{19}$	−0.0102

male, F for female, DY for "yes" having taken drugs, DN for "no" not having taking drugs and the 5 age categories: A1 (16–29), A2 (30-44), A3 (45-64), A4 (65-74), A5 for over 74.

The table provides the observed frequencies $y_{jk\ell}$ in each of the cells of the three-way table: where j stands for gender, k for drug, and ℓ for age categories. The design matrix $\mathcal{X}$ for the full saturated model can be found in the quantlet MVAdrug.

The saturated model gives the estimates displayed in Table 8.6.

We see, for instance, that $\hat{\beta}_1 < 0$, so there are fewer men than women in the study, since $\hat{\beta}_7$ is also negative it seems that the tendency of men taking the drug is less important than for women. Also, note that $\hat{\beta}_{12}$ to $\hat{\beta}_{15}$ forms an increasing sequence, so that the age factor seems to increase the tendency to take the drug. Note that in this saturated model, there are no degrees of freedom and the fit is perfect, $\hat{m}_{jk\ell} = y_{jk\ell}$ for all the cells of the table.

The second-order interactions have a lower order of magnitude, so we want to test if they are significantly different to zero. We consider a restricted model where $(\alpha\beta\gamma)_{jk\ell}$ are all set to zero. This can be achieved by testing $H_0: \ \beta_{16} = \beta_{17} = \beta_{18} = \beta_{19} = 0$. The maximum likelihood estimators of the restricted model are obtained by deleting the last 4 columns in the design matrix $\mathcal{X}$. The results are given in Table 8.7.

We have $J = 2$, $K = 2$, and $L = 5$, this makes $JKL - 1 = 19$ free cells. The full model has $f = 0$ degrees of freedom and the reduced model has $r = 4$ degrees of freedom. The G^2 deviance is given by 2.3004; it has 4 degrees of freedom (the chi-square statistics is 2.3745). The p-value of the restricted model is 0.6807, so we do not reject the null hypothesis (the restricted model without second-order interaction). In others words, age does not interfere with the interactions between gender and drugs, or equivalently, gender does not interfere in the interactions between age and drugs. The reader can verify that the first-order interactions are significant, by taking, for instance, the model without interactions of the second order as the new full model and testing a reduced model where all the first-order interactions are all set to zero.

8.2.4 Logit Models

Logit models are useful to analyze how explanatory variables influence a binary response y. The response y may take the two values 1 and 0 to denote the presence or absence of a certain qualitative trait (a person can be employed or unemployed, a firm can be bankrupt or not, a patient can be affected by a certain disease or not, etc ...). Logit models are designed to estimate the probability of $y = 1$ as a logistic function of linear combinations of x. Logit models can be adapted to the analysis of contingency tables where one of the qualitative variables is binary. One obtains the probability of being in one of the two states of this binary variable as a function of the other variables. We concentrate here on $(2 \times K)$ and $(2 \times K \times L)$ tables.

8.2.4.1 Logit Models for Binary Response

Consider the vector y $(n \times 1)$ of observations on a binary response variable (a value of "1" indicating the presence of a particular qualitative trait and a value of "0", its

Table 8.7 Coefficients estimates based on the maximum likelihood method ⊙ MVAdrug3waysTab

	$\hat{\beta}$		$\hat{\beta}$
$\hat{\beta}_0$ intercept	5.0051	$\hat{\beta}_8$ gender*age	0.0795
$\hat{\beta}_1$ gender: M	−0.2919	$\hat{\beta}_9$	0.0321
$\hat{\beta}_2$ drug: DY	−1.0717	$\hat{\beta}_{10}$	0.0265
$\hat{\beta}_3$ age	−0.0030	$\hat{\beta}_{11}$	0.0534
$\hat{\beta}_4$	0.2358	$\hat{\beta}_{12}$ drug*age	−0.4915
$\hat{\beta}_5$	0.6649	$\hat{\beta}_{13}$	−0.1576
$\hat{\beta}_6$	−0.0425	$\hat{\beta}_{14}$	0.0917
$\hat{\beta}_7$ gender*drug	−0.1734	$\hat{\beta}_{15}$	0.2822

absence). The logit model makes the assumption that the probability for observing $y_i = 1$ given a particular value of $x_i = (x_{i1}, \ldots, x_{ip})^\top$ is given by the logistic function of a "score", a linear combination of x:

$$p\,(x_i) = \mathrm{P}(y_i = 1 \,|\, x_i) = \frac{\exp(\beta_0 + \sum_{j=1}^{p} \beta_j x_{ij})}{1 + \exp(\beta_0 + \sum_{j=1}^{p} \beta_j x_{ij})}. \tag{8.19}$$

This entails the probability of the absence of the trait:

$$1 - p\,(x_i) = \mathrm{P}(y_i = 0 \,|\, x_i) = \frac{1}{1 + \exp(\beta_0 + \sum_{j=1}^{p} \beta_j x_{ij})},$$

which implies

$$\log\left\{ \frac{p\,(x_i)}{1 - p\,(x_i)} \right\} = \beta_0 + \sum_{j=1}^{p} \beta_j x_{ij}. \tag{8.20}$$

This indicates that the logit model is equivalent to a log-linear model for the odds ratio $p\,(x_i)/\{1 - p\,(x_i)\}$. A positive value of β_j indicates an explanatory variable x_j that will favor the presence of the trait since it improves the odds. A zero value of β_j corresponds to the absence of an effect of this variable on the appearance of the qualitative trait. For i.i.d observations, the likelihood function is

$$L(\beta_0, \beta) = \prod_{i=1}^{n} p\,(x_i)^{y_i} \{1 - p\,(x_i)\}^{1-y_i}.$$

The maximum likelihood estimators of the β's are obtained as the solution of the nonlinear maximization problem $(\hat{\beta}_0, \hat{\beta}) = \arg\max_{\beta_0, \beta} \log L(\beta_0, \beta)$ where

$$\log L(\beta_0, \beta) = \sum_{i=1}^{n} \left[y_i \log p\,(x_i) + (1 - y_i) \log\{1 - p\,(x_i)\} \right].$$

The asymptotic theory of the MLE of Chap. 6 (see Theorem 6.3) applies and thus asymptotic inference on β is available (test of hypothesis or confidence intervals).

Example 8.5 In the bankruptcy data set, (see Appendix B.10), we have measures on 5 financial characteristics on 66 banks, 33 among them being bankrupt, and the other 33 still being solvent. The logit model can be used to evaluate the probability of bankruptcy as a function of these financial ratios. We obtain the results summarized in Table 8.8. We observe that only β_3 and β_4 are significant.

8.2.4.2 Logit Models for Contingency Tables

The logit model may contain quantitative and qualitative explanatory variables. In the latter case, the variable may be coded according to the rules described in the ANOVA/ANCOVA sections above. This enables a revisit to the contingency tables, where one of the variables is binary and is the variable of interest. How can the probability of taking one of the two nominal values be evaluated as a function of the other variables? We keep the notations of Sect. 8.1 and suppose, without loss of generality, that the first variable with $J = 2$ is the binary variable of interest. In the drug Example 8.4, we have a $(2 \times 2 \times 5)$ table and one is interested in the probability of taking a drug as a function of age and gender.

$(2 \times K)$ *tables with binomial sampling.*

In Table 8.9, we have displayed the situation.

Let p_k be the probability of falling into the first row for the k-th column, $k = 1, \ldots, K$. Since we are mainly interested in the probabilities p_k as a function of k, we suppose here that $y_{\bullet k}$ are fixed for $k = 1, \ldots, K$ (or we work conditionally on the observed value of these column totals), where $y_{\bullet k} = \sum_{j=1}^{J} y_{jk}$. Therefore, we have K independent binomial processes with parameters $(y_{\bullet k}, p_k)$. Since the column variable is nominal we can use an ANOVA model to analyze the effect of the column variable on p_k through the logs of the odds

$$\log \left(\frac{p_k}{1 - p_k} \right) = \eta_0 + \eta_k, \ k = 1, \ldots, K, \tag{8.21}$$

where $\sum_{k=1}^{K} \eta_k = 0$. As in the ANOVA models, one of the interests will be to test $H_0 : \eta_1 = \ldots = \eta_K = 0$. The log-linear model for the odds has its equivalent in a logit formulation for p_k

Table 8.8 Probabilities of the bankruptcies with the logit model 🔍 MVAbankrupt

	$\hat{\beta}$	p-values
β_0	3.6042	0.0660
β_3	−0.2031	0.0037
β_4	−0.0205	0.0183

Table 8.9 A $(2 \times K)$ contingency table

	1	$\cdots$	k	$\cdots$	K	Total
1	y_{11}	$\cdots$	y_{1k}	$\cdots$	y_{1K}	y_1
2	y_{21}	$\cdots$	y_{2k}	$\cdots$	y_{2K}	y_2
Total	$y_{\bullet 1}$	$\cdots$	$y_{\bullet k}$	$\cdots$	$y_{\bullet K}$	$y_{\bullet} = n$

$$p_k = \frac{\exp(\eta_0 + \eta_k)}{1 + \exp(\eta_0 + \eta_k)}, \quad k = 1, \ldots, K. \tag{8.22}$$

Note that we can code the RHS of (8.21) as a linear model $\mathcal{X}\theta$, where for instance, for a (2×4) table $(K = 4)$ we have

$$\mathcal{X} = \begin{pmatrix} 1 & 1 & 0 & 0 \\ 1 & 0 & 1 & 0 \\ 1 & 0 & 0 & 1 \\ 1 & -1 & -1 & -1 \end{pmatrix}, \quad \theta = \begin{pmatrix} \beta_0 \\ \beta_1 \\ \beta_2 \\ \beta_3 \end{pmatrix},$$

where $\eta_0 = \beta_0, \eta_1 = \beta_1, \eta_2 = \beta_2, \eta_3 = \beta_3$, and $\eta_4 = -(\beta_1 + \beta_2 + \beta_3)$. The logit model for $p_k, k = 1, \ldots, K$ can now be written, with some abuse of notation, as the K-vector

$$p = \frac{\exp(\mathcal{X}\theta)}{1 + \exp(\mathcal{X}\theta)},$$

where the division has to be understood as being element by element. The MLE of θ is obtained by maximizing the log-likelihood

$$L(\theta) = \sum_{k=1}^{K} \{y_{1k} \log p_k + y_{2k} \log(1 - p_k)\}, \tag{8.23}$$

where the p_k are elements of the K-vector p.

This logit model is a *saturated* model. Indeed the number of free parameters is K, the dimension of θ, and the number of free cells is also equal to K since we consider the column totals $y_{\bullet k}$ as being fixed. So, there are no degrees of freedom in this model. It can be proven that this logit model is equivalent to the saturated model for a table $(2 \times K)$ presented in Sect. 8.2.2 where all the interactions are present in the model. The hypothesis of all interactions $(\alpha\gamma)_{jk}$ being equal to zero (independence case) is equivalent to the hypothesis that the $\eta_k, k = 1, \ldots, K$ are all equal to zero (no column effect on the probabilities p_k).

The main interest of the logit presentation is its flexibility when the variable defining the column categories is a quantitative variable (age group, number of children, ...). Indeed, when this is the case, the logit model allows to quantify the effect of the column category by using less parameters and a more flexible relationship than a linear relation. Suppose that we could attach a representative value x_k to each column category for this class (for instance, it could be the median value, or the average value of the class category). We can then choose the following logit model for p_k:

$$p_k = \frac{\exp(\eta_0 + \eta_1 x_k)}{1 + \exp(\eta_0 + \eta_1 x_k)}, \quad k = 1, \ldots, K, \tag{8.24}$$

where we now have only two free parameters for K free cells, so we have $K - 2$ degrees of freedom. We could even introduce a quadratic term to allow some curvature effect of x on the odds

$$p_k = \frac{\exp(\eta_0 + \eta_1 x_k + \eta_2 x_k^2)}{1 + \exp(\eta_0 + \eta_1 x_k + \eta_2 x_k^2)}, \quad k = 1, \ldots, K.$$

In this latter case, we would still have $K - 3$ degrees of freedom.

We can follow the same idea for a three-way table when we want to model the behavior of the first binary variable as a function of the two other variables defining the table. In the drug example, one is interested in analyzing the tendency of taking a psychotropic drug as a function of the gender category and of the age. Fix the number of observations in each cell $k\ell$ (i.e. $y_{\bullet k\ell}$), so that we have a binomial sampling process with an unknown parameter $p_{k\ell}$ for each cell. As for the two-way case above, we can either use ANOVA-like models for the logarithm of the odds and ANCOVA-like models when one (or both) of the two qualitative variables defining the K and/or L categories is a quantitative variable.

One may study the following ANOVA model for the logarithms of the odds:

$$\log\left(\frac{p_{k\ell}}{1 - p_{k\ell}}\right) = \mu + \eta_k + \zeta_\ell, \quad k = 1, \ldots, K, \ \ell = 1, \ldots, L,$$

with $\eta = \zeta = 0$. As another example, if x_ℓ is a representative value (like the average age of the group) of the ℓth level of the third categorical variable, one might think of

$$\log\left(\frac{p_{k\ell}}{1 - p_{k\ell}}\right) = \mu + \eta_k + \zeta x_\ell, \quad k = 1, \ldots, K, \ \ell = 1, \ldots, L, \tag{8.25}$$

with the constraint $\eta = 0$. Here also, interactions and the curvature effect for x_ℓ can be introduced, as shown in the following example. Since the cell totals $y_{\bullet k\ell}$ are considered as fixed, the log-likelihood to be maximized is

$$\sum_{k=1}^{K}\sum_{\ell=1}^{L}\{y_{1k\ell} \log p_{k\ell} + y_{2k\ell} \log(1 - p_{k\ell})\}, \tag{8.26}$$

where $p_{k\ell}$ follows the appropriate logistic model.

Example 8.6 Consider again Example 8.4. One is interested in the influence of gender and age on drug prescription. Take the number of observations for each "gender–age group" combination, $y_{\bullet k\ell}$ as fixed. A logit model (8.25) can be used for the odds-ratios of the probability of taking drugs, where the value x_ℓ is the average age of the group. In the linear form, it may be written as one of the two following equivalent forms:

$$\log\left(\frac{p}{1-p}\right) = \mathcal{X}\theta,$$

$$p = \frac{\exp(\mathcal{X}\theta)}{1+\exp(\mathcal{X}\theta)},$$

where $\theta = (\beta_0, \beta_1, \beta_2)^\top$ and the design matrix $\mathcal{X}$ is given by

$$\mathcal{X} = \begin{pmatrix} 1.0 & 1.0 & 23.2 \\ 1.0 & 1.0 & 36.5 \\ 1.0 & 1.0 & 54.3 \\ 1.0 & 1.0 & 69.2 \\ 1.0 & 1.0 & 79.5 \\ 1.0 & -1.0 & 23.2 \\ 1.0 & -1.0 & 36.5 \\ 1.0 & -1.0 & 54.3 \\ 1.0 & -1.0 & 69.2 \\ 1.0 & -1.0 & 79.5 \end{pmatrix}$$

The first column of $\mathcal{X}$ is for the intercept, the second is the coded variable for the two gender categories, and the last column is the average of the ages for the corresponding age group. Then we estimate β by maximizing the log-likelihood function (8.26). We obtain

$$\hat{\beta}_0 = -3.5612$$
$$\hat{\beta}_1 = -0.3426$$
$$\hat{\beta}_2 = 0.0280,$$

the intercept for men is $\hat{\beta}_0 + \hat{\beta}_1 = -3.9038$ and for women is $\hat{\beta}_0 - \hat{\beta}_1 = -3.2186$, indicating a gender effect and the common slope for the positive age effect being $\hat{\beta}_2 = 0.0280$. The fit appears to be reasonably good. There are $K \times L = 2 \times 5 = 10$ free cells in the table. A saturated "full" model with 10 parameters and a zero degree of freedom would involve a constant (1 parameter) plus an effect for gender (1 parameter) plus an effect for age (4 parameters) and finally the interactions between gender and age (4 parameters). The model retained above is a "reduced model" with only 3 parameters, that can be tested against the most general saturated model. We obtain the value of the deviance $G^2_{H_0} = 11.5584$ with 7 degrees of freedom $(7 = 10 - 3)$, whereas, $G^2_{H_1} = 0$ with no degree of freedom. This gives a p-value $= 0.1160$, so we cannot reject the reduced model.

Figure 8.3 shows how well the model fits the data. It displays the fitted values of the log of the odds-ratios by the linear model for the men and the women along with the log of the odds-ratios computed from the observed corresponding frequencies. It seems that the age effect shows a curvature. So we fit a model introducing the square of the ages. This gives the following design matrix:

Fig. 8.3 Fit of the log of the odds-ratios for taking drugs: linear model for age effect with a "gender" effect (no interaction). Men are the stars and women are the circles 💠 MVAdruglogistic

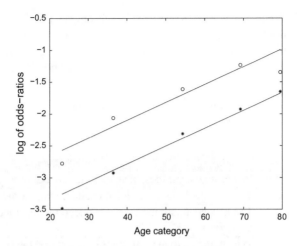

$$\mathcal{X} = \begin{pmatrix} 1.0 & 1.0 & 23.2 & 538.24 \\ 1.0 & 1.0 & 36.5 & 1332.25 \\ 1.0 & 1.0 & 54.3 & 2948.49 \\ 1.0 & 1.0 & 69.2 & 4788.64 \\ 1.0 & 1.0 & 79.5 & 6320.25 \\ 1.0 & -1.0 & 23.2 & 538.24 \\ 1.0 & -1.0 & 36.5 & 1332.25 \\ 1.0 & -1.0 & 54.3 & 2948.49 \\ 1.0 & -1.0 & 69.2 & 4788.64 \\ 1.0 & -1.0 & 79.5 & 6320.25 \end{pmatrix}$$

The maximum likelihood estimators are:

$$\hat{\beta}_0 = -4.4996$$
$$\hat{\beta}_1 = -0.3457$$
$$\hat{\beta}_2 = 0.0697$$
$$\hat{\beta}_3 = -0.0004.$$

💠 MVAdruglogistic.

The fit is better for this more flexible alternative, giving a deviance $G^2_{H_1} = 3.3251$ with 6 degrees of freedom ($6 = 10 - 4$). If we test H_0: no curvature for the age effect against H_1: curvature for the age effect, the reduction of the deviance is $G^2_{H_0} - G^2_{H_1} = 11.5584 - 3.3251 = 8.2333$ with one degree of freedom. The p-value $= 0.0041$, so we reject the reduced model (no curvature) in favor of the more general model with a curvature term.

We know already from Example 8.4 that second-order interactions are not significant for this data set (the influence of age on taking a drug is the same for both gender categories), so we can keep this model as a final reasonable model to analyze the probability of taking the drug as a function of the gender and of the age. To summarize this analysis, we end up saying that the probability of taking a psychotropic drug can be modeled as (with some abuse of notation)

$$\log\left(\frac{p}{1-p}\right) = \beta_0 + \beta_1 * Sex + \beta_2 * Age + \beta_3 * Age^2. \qquad (8.27)$$

Summary

↪ In contingency tables, the categories are defined by the qualitative variables.

↪ The saturated model has all of the interaction terms, and 0 degree of freedom.

↪ The non-saturated model is a reduced model since it fixes some parameters to be zero.

↪ Two statistics to test for the full model and the reduced model are:

$$X^2 = \sum_{k=1}^{K} (y_k - \hat{m}_k)^2 / \hat{m}_k$$

$$G^2 = 2 \sum_{k=1}^{K} y_k \, \log\left(y_k/\hat{m}_k\right)$$

↪ The logit models allow the column categories to be a quantitative variable, and quantify the effect of the column category by using fewer parameters and incorporating more flexible relationships than just a linear one.

↪ The logit model is equivalent to a log-linear model.

$$\log\left[p\left(x_i\right)/\{1 - p\left(x_i\right)\}\right] = \beta_0 + \sum_{j=1}^{p} \beta_j x_{ij}$$

8.3 Exercises

Exercise 8.1 For the one-factor ANOVA model, show that if the model is "balanced" ($n_1 = n_2 = n_3$), we have $\hat{\mu} = \bar{y}$. If the model is not balanced, show that $\bar{y} = \hat{\mu} + n_1\hat{\alpha}_1 + n_2\hat{\alpha}_2 + n_3\hat{\alpha}_3$.

Exercise 8.2 Redo the calculations of Example 8.2 and test if the main effects of the marketing strategy and of the location are significant.

Exercise 8.3 Redo the calculations of Example 8.3 with the Car data set.

Exercise 8.4 Calculate the prediction interval for "classic blue" pullover sales (Example 3.2) corresponding to price = 120.

Exercise 8.5 Redo the calculations of the Boston housing example in Sect. 8.1.3

Exercise 8.6 We want to analyze the variations in the consumption of packs of cigarettes per month as a function of the brand (A or B),of the price per pack and as a function of the gender of the smoker (M or F). The data are below.

y	Price	Gender	Brand
30	3.5	M	A
4	4	F	B
20	4.1	F	B
15	3.75	M	A
24	3.25	F	A
11	5	F	B
8	4.1	F	B
9	3.5	M	A
17	4.5	M	B
1	4	F	B
23	3.65	M	A
13	3.5	M	A

1. In addition to the effects of brand, price, and gender, test if there is an interaction between the brand and the price.
2. How would the design matrix of a full model with all the interactions between the variables appear? What would be the number of degrees of freedom of such a model?
3. We would like to introduce a curvature term for the price variable. How can we proceed? Test if this coefficient is significant.

Exercise 8.7 In the drug Example 8.4, test if the first-order interactions are significant.

Reference

B.S. Everitt, G. Dunn, *Applied Multivariate Data Analysis* (Edward Arnold, London, 1998)

Chapter 9
Variable Selection

Variable selection is very important in statistical modeling. We are frequently not only interested in using a model for prediction but also need to correctly identify the relevant variables, that is, to recover the correct model under given assumptions. It is known that under certain conditions, the ordinary least squares (OLS) method produces poor prediction results and does not yield a parsimonious model causing overfitting. Therefore the objective of the variable selection methods is to find the variables which are the most relevant for prediction. Such methods are particularly important when the true underlying model has a sparse representation (many parameters close to zero). The identification of relevant variables will reduce the noise and therefore improve the prediction performance of the fitted model.

Some popular regularization methods used are the ridge regression, subset selection, L_1-norm penalization, and their modifications and combinations. Ridge regression, for instance, which minimizes a penalized residual sum of squares using the squared L_2-norm penalty, is employed to improve the OLS estimate through a bias–variance trade-off. However, ridge regression has a drawback that it cannot yield a parsimonious model since it keeps all predictors in the model and therefore creates an interpretability problem. It also gives prediction errors close to those from the OLS model.

Another approach proposed for variable selection is the so-called "least absolute shrinkage and selection operator" (Lasso), aims at combining the features of ridge regression and subset selection either retaining (and shrinking) the coefficients or setting them to zero. This method received several extensions such as the Elastic net, a combination of Lasso and ridge regression or the Group Lasso used when predictors are divided into groups. This chapter describes the application of Lasso, Group Lasso as well as the Elastic net in linear regression models with continuous and binary response (logit model) variables.

© Springer Nature Switzerland AG 2019
W. K. Härdle and L. Simar, *Applied Multivariate Statistical Analysis*,
https://doi.org/10.1007/978-3-030-26006-4_9

9.1 Lasso

Tibshirani (1996) first introduced Lasso for generalized linear models, where the response variable y is continuous rather than categorical. Lasso has two important characteristics. First, it has an L_1-penalty term which performs shrinkage on coefficients in a way similar to ridge regression, where an L_2-penalty is used.

Second, unlike ridge regression, Lasso performs variable subset selection driving some coefficients to exactly zero due to the nature of the constraint, where the objective function may touch the quadratic constraint area at a corner. For this reason, the Lasso is able to produce sparse solutions and is therefore able to combine good features of both ridge regression and subset selection procedure. It yields interpretable models and has the stability of ridge regression.

9.1.1 Lasso in the Linear Regression Model

The linear regression model can be written as follows:

$$y = \mathcal{X}\beta + \varepsilon,$$

where y is an $(n \times 1)$ vector of observations for the response variable, $\mathcal{X} = (x_1^\top, \ldots, x_n^\top)^\top$, $x_i \in \mathbb{R}^p$, $i = 1, \ldots, n$ is a data matrix of p explanatory variables, and $\varepsilon = (\varepsilon_1, \ldots, \varepsilon_n)^\top$ is a vector of errors where $\mathsf{E}(\varepsilon_i) = 0$ and $\mathsf{Var}(\varepsilon_i) = \sigma^2$, $i = 1, \ldots, n$.

In this framework, $\mathsf{E}(y|\mathcal{X}) = \mathcal{X}\beta$ with $\beta = (\beta_1, \ldots, \beta_p)^\top$. Further assume that the columns of $\mathcal{X}$ are standardized such that $n^{-1} \sum_{i=1}^n x_{ij} = 0$ and $n^{-1} \sum_{i=1}^n x_{ij}^2 = 1$. The Lasso estimate $\widehat{\beta}$ can then be defined as follows:

$$\widehat{\beta} = \arg\min_{\beta} \left\{ \sum_{i=1}^n \left(y_i - x_i^\top \beta \right)^2 \right\}, \text{ subject to } \sum_{j=1}^p |\beta_j| \leq s, \qquad (9.1)$$

where $s \geq 0$ is the tuning parameter which controls the amount of shrinkage. For the OLS estimate $\widehat{\beta}^0 = (\mathcal{X}^\top \mathcal{X})^{-1} \mathcal{X}^\top y$ a choice of tuning parameter $s < s_0$, where $s_0 = \sum_{j=1}^p |\widehat{\beta}_j^0|$, will cause shrinkage of the solutions toward 0, and ultimately some coefficients may be exactly equal to 0. For values $s \geq s_0$ the Lasso coefficients are equal to the unpenalized OLS coefficients.

An alternative representation of (9.1) is

$$\widehat{\beta} = \arg\min_{\beta} \left\{ \sum_{i=1}^n \left(y_i - x_i^\top \beta \right)^2 + \lambda \sum_{j=1}^p |\beta_j| \right\}, \qquad (9.2)$$

with a tuning parameter $\lambda \geq 0$. As λ increases, the Lasso estimates are continuously shrunk toward zero. Then if λ is quite large, some coefficients are exactly zero. For $\lambda = 0$ the Lasso coefficients coincide with the OLS estimate. In fact, if the

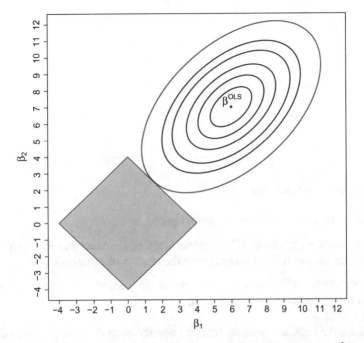

Fig. 9.1 Lasso in the general design case for $s = 4$ and OLS estimate $\widehat{\beta}^0 = (6, 7)^\top$
⊙MVAlassocontour

solution to (9.1) is denoted as $\widehat{\beta}_s$ and the solution to (9.2) as $\widehat{\beta}_\lambda$, then $\forall \lambda > 0$ and the
resulting solution $\widehat{\beta}_\lambda \, \exists s_\lambda$ such that $\widehat{\beta}_\lambda = \widehat{\beta}_{s_\lambda}$ and vice versa which implies a one-to-
one correspondence between these parameters. However, this does not hold if it is
required that $\lambda \geq 0$ only and not $\lambda > 0$, because if, for instance, $\lambda = 0$, then $\widehat{\beta}_\lambda$ is
the same for any $s \geq \|\widehat{\beta}\|_1$ and the correspondence is no longer one-to-one.

Geometrical Aspects in $\mathbb{R}^2$

The Lasso estimate under the least squares loss function solves a quadratic program-
ming problem with linear inequality constraints. The criterion $\sum_{i=1}^{n} \left(y_i - x_i^\top \beta \right)^2$
yields the quadratic form objective function

$$(\beta - \widehat{\beta}^0)^\top \mathcal{W}(\beta - \widehat{\beta}^0) \qquad (9.3)$$

with $\mathcal{W} = \mathcal{X}^\top \mathcal{X}$. For the special case when $p = 2$, $\beta = (\beta_1, \beta_2)^\top$, the resulting
elliptical contour lines are centered around the OLS estimate and the linear constraints
are represented by square (shaded area) shown in Fig. 9.1. The Lasso solution is the
first place that the contours touch the square, and this sometimes occurs at a corner,
corresponding to a zero coefficient. The nature of the Lasso shrinkage may not occur
completely obvious. In the work by Efron et al. (2004), the Least Angle Regression
algorithm with a Lasso modification was described which computes the whole path
of Lasso solutions and gives a better understanding of the shrinkage nature.

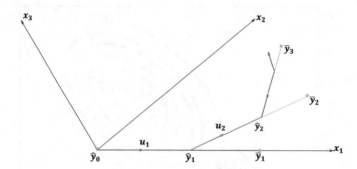

Fig. 9.2 Illustration of LAR algorithm

The LAR algorithm and Lasso solution paths

The least angle regression (LAR) algorithm may be introduced in the simple three-dimensional case as follows (assume that the number of covariates $p = 3$):

- first, standardize all the covariates to have mean 0 and unit length as well as make the response variable have mean zero;
- start with $\widehat{\beta} = 0$;
- initialize the algorithm with the first two covariates: let $\mathcal{X} = (x_1, x_2)$ and calculate the prediction vector $\widehat{y}_0 = \mathcal{X}\widehat{\beta} = 0$;
- calculate $\bar{y}_2$ the projection of y onto $\mathcal{L}(x_1, x_2)$, the linear space spanned by x_1 and x_2;
- compute the vector of current correlations between the covariates $\mathcal{X}$ and the two-dimensional current residual vector: $C^{\widehat{y}_0} = \mathcal{X}^{\top}(\bar{y}_2 - \widehat{y}_0) = (c_1^{\widehat{y}_0}, c_2^{\widehat{y}_0})^{\top}$. According to Fig. 9.2, the current residual $\bar{y}_2 - \widehat{y}_0$ makes a smaller angle with x_1, than with x_2, therefore $c_1^{\widehat{y}_0} > c_2^{\widehat{y}_0}$;
- augment $\widehat{y}_0$ in the direction of x_1 so that $\widehat{y}_1 = \widehat{y}_0 + \widehat{\gamma}_1 x_1$ with $\widehat{\gamma}_1$ chosen such that $c_1^{\widehat{y}_0} = c_2^{\widehat{y}_0}$ which means that the new current residual $\bar{y}_2 - \widehat{y}_1$ makes equal angles (is equiangular) with x_1 and x_2;
- suppose that another regressor x_3 enters the model: calculate a new projection $\bar{y}_3$ of y onto $\mathcal{L}(x_1, x_2, x_3)$;
- recompute the current correlations vector $C^{\widehat{y}_1} = (c_1^{\widehat{y}_1}, c_2^{\widehat{y}_1}, c_3^{\widehat{y}_1})^{\top}$ with $\mathcal{X} = (x_1, x_2, x_3)$, $\bar{y}_3$ and $\widehat{y}_1$;
- augment $\widehat{y}_1$ in the equiangular direction so that $\widehat{y}_2 = \widehat{y}_1 + \widehat{\gamma}_2 u_2$ with $\widehat{\gamma}_2$ chosen such that $c_1^{\widehat{y}_1} = c_2^{\widehat{y}_1} = c_3^{\widehat{y}_1}$, then the new current residual $\bar{y}_3 - \widehat{y}_2$ goes equiangularly between x_1, x_2 and x_3 (here u_2 is the unit vector lying along the equiangular direction $\widehat{y}_2$);
- the three-dimensional algorithm is terminated with the calculation of the final prediction vector $\widehat{y}_3 = \widehat{y}_2 + \widehat{\gamma}_3 u_3$ with $\widehat{\gamma}_3$ chosen such that $\widehat{y}_3 = \bar{y}_3$.

In the case of $p > 3$ covariates, $\widehat{y}_3$ would be smaller than $\bar{y}_3$ initiating another change of direction, as illustrated in Fig. 9.2. In this setup, it is important that the covariate vectors x_1, x_2, x_3 are linearly independent. The Least Angle Regression algorithm "moves" the variables' coefficients to their least squares values. So the

Lasso adjustment necessary for the sparse solution is that if a nonzero coefficient happens to return to zero, it should be dropped from the current ("active") set of variables and not be considered in further computations. The general LAR algorithm for p predictors can be summarized as follows:

Least angle regression (LAR) algorithm

1. The covariates are standardized to have mean 0 and unit length 1 and the response has mean 0:

$$\sum_{i=1}^{n} y_i = 0, \quad \sum_{i=1}^{n} x_{ij} = 0, \quad \sum_{i=1}^{n} x_{ij}^2 = 1; \quad j = 1, 2, \dots, p.$$

The task is to construct the fit $\widehat{\beta} = (\widehat{\beta}_1, \dots, \widehat{\beta}_p)^\top$ by iteratively changing the prediction vector $\widehat{y} = \sum_{j=1}^{p} x_j \widehat{\beta}_j = \mathcal{X}\widehat{\beta}$.

2. Denote $\mathcal{A}$ equal to a subset of the indices $\{1, 2, \dots, p\}$, begin with $\widehat{y}_\mathcal{A} = \widehat{y}_0 = 0$ and calculate the vector of current correlations

$$\widehat{c} = \mathcal{X}^\top (y - \widehat{y}_\mathcal{A}).$$

3. Then review the current set $\mathcal{A} = \{j : |\widehat{c}_j| = \widehat{C}\}$ as the set of indices corresponding to the covariates with the greatest absolute current correlations, where $\widehat{C} = \max_j\{|\widehat{c}_j|\}$; let $s_j = \text{sign}\{\widehat{c}_j\}$ for $j \in \mathcal{A}$ and compute the matrix $\mathcal{X}_\mathcal{A} = (s_j x_j)_{j \in \mathcal{A}}$, the scalar $A_\mathcal{A} = (1_\mathcal{A}^\top \mathcal{G}_\mathcal{A}^{-1} 1_\mathcal{A})^{-\frac{1}{2}}$ with $\mathcal{G}_\mathcal{A} = \mathcal{X}_\mathcal{A}^\top \mathcal{X}_\mathcal{A}$ and $1_\mathcal{A}^\top$ being a vector of ones of length $|\mathcal{A}|$, and the so-called equiangular vector $u_\mathcal{A} = \mathcal{X}_\mathcal{A} w_\mathcal{A}$ with $w_\mathcal{A} = A_\mathcal{A} \mathcal{G}_\mathcal{A}^{-1} 1_\mathcal{A}$ which makes equal angles, each less than 90°, with the columns of $\mathcal{X}_\mathcal{A}$.

4. Calculate the inner product vector $a \stackrel{\text{def}}{=} \mathcal{X}^\top u_\mathcal{A}$ and the direction

$$\widehat{\gamma} = \min_{j \in \mathcal{A}^c}^{+} \left\{ \frac{\widehat{C} - \widehat{c}_j}{A_\mathcal{A} - a_j}, \frac{\widehat{C} + \widehat{c}_j}{A_\mathcal{A} + a_j} \right\}$$

5. Define $\widehat{d}$ to be the m-vector equaling $s_j w_{\mathcal{A}_j}$ for $j \in \mathcal{A}$ and zero elsewhere and $\gamma_j = -\widehat{\beta}_j / \widehat{d}_j$ yielding $\widetilde{\gamma} = \min_{\gamma_j > 0} \{\gamma_j\}$

   a. If $\widetilde{\gamma} < \widehat{\gamma}$, calculate the next LAR step as

$$\widehat{y}_{\mathcal{A}_+} = \widehat{y}_\mathcal{A} + \widetilde{\gamma} u_\mathcal{A}$$

   with $\mathcal{A}_+ = \mathcal{A} - \{\widetilde{j}\}$.

   b. Else: calculate the next step as

$$\widehat{y}_{\mathcal{A}_+} = \widehat{y}_\mathcal{A} + \widehat{\gamma} u_\mathcal{A}$$

6. Iterate until all p predictors have entered, some of which are ultimately dropped from the active set $\mathcal{A}$.

This algorithm can be implemented on a grid from 0 to 1 of the standardized coefficients constraint s resulting in the complete paths of the Lasso coefficients and illustrating the nature of Lasso shrinkage.

Once the Lasso solution paths have been obtained, it is important to decide on a rule how to choose the "optimal" solution, or, equally, the regularization parameter λ. There are several existing methods to do this and the most popular examples are the K-fold cross-validation, generalized cross-validation, Schwarz's (Bayesian) Information Criterion (BIC). All these methods can be viewed as degrees of freedom adjustments to the residual sum of squares (RSS) which underestimates the true prediction error

$$\text{RSS} \overset{\text{def}}{=} \sum_{i=1}^{n} (y_i - \widehat{y}_i)^2.$$

Consider the generalized cross-validation statistic:

$$\text{GCV}(\lambda) = n^{-1}\text{RSS}_\lambda / \{1 - \text{df}(\lambda)/n\}^2, \tag{9.4}$$

where RSS_λ is the residual sum of squares for the constrained fit with a particular regularization parameter λ. An alternative is the Bayesian Information Criterion

$$\text{BIC} = n \log(\widehat{\sigma}^2) + \log(n) \cdot \text{df}(\lambda) \tag{9.5}$$

with the estimation of error variance $\widehat{\sigma}^2 = n^{-1} \sum_{i=1}^{n} (y_i - \widehat{y}_i)^2$.

The degrees of freedom of the predicted vector $\widehat{y}$ in the Lasso problem with the linear Gaussian model with normally distributed errors having zero expectation and variance σ^2, written $\varepsilon_i \sim N(0, \sigma^2)$, can be defined as follows:

$$\text{df}(\lambda) \overset{\text{def}}{=} \sigma^{-2} \sum_{i=1}^{n} \text{Cov}(\widehat{y}_i, y_i), \tag{9.6}$$

which can actually be used for both linear and nonlinear models. This expression for $\text{df}(\lambda)$ can be viewed as a quantitative measure of the prediction error bias dependence on how much each y_i affects its fitted value $\widehat{y}_i$. The estimate $\widehat{\beta}$ minimizing the GCV statistic can then be chosen. The following example shows how to compute $\text{df}(\lambda)$.

Example 9.1 Calculation of $\text{df}(\lambda)$.

As no closed-form solution exists for the Lasso problem, an approximation should be calculated. The constraint $\sum |\beta_j| \leq s$ can be rewritten as $\sum \beta_j^2/|\beta_j| \leq s$. Using the duality between the constrained and unconstrained problems and one-to-one correspondence between s and λ, the Lasso solution is computed as the ridge regression estimate

$$\widehat{\beta} = (\mathcal{X}^\top \mathcal{X} + \lambda B^{-1})^{-1} \mathcal{X}^\top y,$$

where $B = \text{diag}(|\widehat{\beta}_j|)$. Then it follows that

$$\widehat{y} = \mathcal{X}\widehat{\beta},$$
$$= \mathcal{X}(\mathcal{X}^\top \mathcal{X} + \lambda B^{-1})^{-1}\mathcal{X}^\top y.$$

Then, to calculate $\text{Cov}(\widehat{y}_i, y_i)$, one could use $\text{Cov}(\widehat{y}_i, y_i) = \text{Cov}(e_i^\top \widehat{y}, e_i^\top y) = e_i^\top \text{Cov}(\widehat{y}, y) e_i$, where e_i is a vector where the i-th entry is 1 and the rest are zero. Furthermore, each entry in the sum of (9.6) can be calculated to be

$$\text{Cov}(\widehat{y}_i, y_i) = e_i^\top \text{Cov}(\widehat{y}, y) e_i \tag{9.7}$$
$$= e_i^\top \mathcal{X}(\mathcal{X}^\top \mathcal{X} + \lambda B^{-1})^{-1}\mathcal{X}^\top \text{Cov}(y, y) e_i \tag{9.8}$$
$$= \sigma^2 (\mathcal{X}^\top e_i)^\top (\mathcal{X}^\top \mathcal{X} + \lambda B^{-1})^{-1}(\mathcal{X}^\top e_i) \tag{9.9}$$
$$= \sigma^2 x_i^\top (\mathcal{X}^\top \mathcal{X} + \lambda B^{-1})^{-1} x_i. \tag{9.10}$$

Using the fact that (9.10) are scalars for all i's as well as the properties of the trace of a matrix and matrix multiplication rules mentioned in Chap. 2, one obtains the final closed-form expression for the effective degrees of freedom in the Lasso problem:

$$\text{df}(\lambda) = \frac{1}{\sigma^2} \sum_{i=1}^{n} \text{tr}\left\{\sigma^2 x_i^\top (\mathcal{X}^\top \mathcal{X} + \lambda B^{-1})^{-1} x_i\right\}$$

$$= \sum_{i=1}^{n} \text{tr}\left\{x_i x_i^\top (\mathcal{X}^\top \mathcal{X} + \lambda B^{-1})^{-1}\right\}$$

$$= \text{tr}\left\{\left(\sum_{i=1}^{n} x_i x_i^\top\right)(\mathcal{X}^\top \mathcal{X} + \lambda B^{-1})^{-1}\right\}$$

$$= \text{tr}\left\{\mathcal{X}^\top \mathcal{X}(\mathcal{X}^\top \mathcal{X} + \lambda B^{-1})^{-1}\right\}$$

$$= \text{tr}\left\{\mathcal{X}(\mathcal{X}^\top \mathcal{X} + \lambda B^{-1})^{-1}\mathcal{X}^\top\right\}.$$

It should be noted that the formula for the effective degrees of freedom derived above is valid in the case of the underlying model with nonrandom regressors. When the random design is used and the set of nonzero predictors is not fixed, another estimator should be used.

Orthonormal Design Case

A computationally convenient special case is the so-called orthonormal design framework. In the orthonormal design case $\mathcal{X}^\top \mathcal{X}$ is a diagonal matrix, that $\mathcal{X}^\top \mathcal{X} = \mathcal{I}$. Here the explicit Lasso estimate is

$$\widehat{\beta}_j = \text{sign}\left(\widehat{\beta}_j^0\right)\left(|\widehat{\beta}_j^0| - \gamma\right)^+, \tag{9.11}$$

$$\gamma = \frac{\lambda}{2} \text{ subject to } \sum_{j=1}^{p} |\widehat{\beta}_j| = s. \tag{9.12}$$

The formula shows what was already mentioned at the beginning, namely that the Lasso estimate is a compromise between subset selection and ridge regression, the estimate is either shrunk by γ or is set to zero. As a consequence Lasso coefficients can take values between zero and $\widehat{\beta}_j^0$.

Example 9.2 Orthonormal design case for $p = 2$.
Let $\widehat{\beta} = \left(\widehat{\beta}_1, \widehat{\beta}_2\right)^{\mathsf{T}}$ w.l.o.g. be in the first quadrant, i.e., $\widehat{\beta}_1 \geq 0$ and $\widehat{\beta}_2 \geq 0$. This gives us the first condition. The orthonormal design ensures that the elliptical contour lines describe circles around the OLS estimate. Thus we get a linear function going through the point $\widehat{\beta}^0$ and being orthogonal (if possible) to the first condition. Equalizing both conditions

$$\widehat{\beta}_1 + \widehat{\beta}_2 = s \tag{9.13}$$

$$\widehat{\beta}_2 = \widehat{\beta}_1 + \left(\widehat{\beta}_2^0 - \widehat{\beta}_1^0\right) \tag{9.14}$$

the Lasso estimate can now be accurately determined:

$$\widehat{\beta}_1 = \left(\frac{s}{2} + \frac{\widehat{\beta}_1^0 - \widehat{\beta}_2^0}{2}\right)^+ \tag{9.15}$$

$$\widehat{\beta}_2 = \left(\frac{s}{2} - \frac{\widehat{\beta}_1^0 - \widehat{\beta}_2^0}{2}\right)^+. \tag{9.16}$$

For cases in which $\left(\frac{s}{2} + \frac{\widehat{\beta}_1^0 - \widehat{\beta}_2^0}{2}\right) \leq 0$ or $\left(\frac{s}{2} - \frac{\widehat{\beta}_1^0 - \widehat{\beta}_2^0}{2}\right) \leq 0$ the corresponding Lasso estimates will always be zero as the position of the $\widehat{\beta}_1^0$ and corresponding contour lines do not make it possible to get the orthogonality condition mentioned above. Let $\widehat{\beta}^0 = (6, 7)^{\mathsf{T}}$ and tuning parameter $s = 4$. In this case the Lasso estimator is given by, as shown in Fig. 9.3:

$$\widehat{\beta}_1 = \frac{4}{2} + \frac{6 - 7}{2} = 1.5, \tag{9.17}$$

$$\widehat{\beta}_2 = \frac{4}{2} - \frac{6 - 7}{2} = 2.5. \tag{9.18}$$

In terms of λ, the Lasso solution (9.11) in the orthonormal design case can be calculated in a usual unconstrained minimization problem. Note that in this case the least squares solution is given by

$$\widehat{\beta}^0 = (\mathcal{X}^{\mathsf{T}}\mathcal{X})^{-1}\mathcal{X}^{\mathsf{T}}y = \mathcal{X}^{\mathsf{T}}y.$$

Then the minimization problem is written as

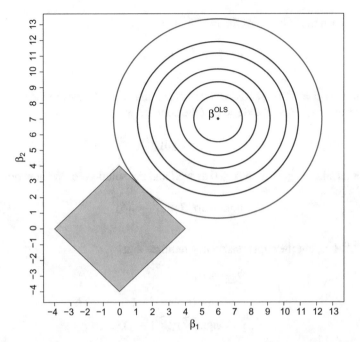

Fig. 9.3 Lasso in the orthonormal design case for $s = 4$ and OLS estimate $\widehat{\beta}^0 = (6, 7)^\top$

🔍MVAlassocontour

$$\widehat{\beta} = \arg\min_{\beta \in \mathbb{R}^p} \|y - \mathcal{X}\beta\|_2^2 + \lambda\|\beta\|_1$$

$$= \arg\min_{\beta \in \mathbb{R}^p} (y - \mathcal{X}\beta)^\top (y - \mathcal{X}\beta) + \lambda \sum_{j=1}^p |\beta_j|$$

$$= \arg\min_{\beta \in \mathbb{R}^p} -2y^\top \mathcal{X}\beta + \beta^\top \beta + \lambda \sum_{j=1}^p |\beta_j|$$

$$= \arg\min_{\beta \in \mathbb{R}^p} -2\widehat{\beta}^{0\top}\beta + \beta^\top \beta + \lambda \sum_{j=1}^p |\beta_j|$$

$$= \arg\min_{\beta \in \mathbb{R}^p} \sum_{j=1}^p \left(-2\widehat{\beta}_j^0 \beta_j + \beta_j^2 + \lambda|\beta_j| \right).$$

The objective function can now be minimized by separate minimization of its j-th element. To solve

$$\min_{\beta}(-2\widehat{\beta}^0\beta + \beta^2 - \lambda|\beta|), \tag{9.19}$$

where the index j was dropped for simplicity. Let's first assume that $\widehat{\beta}^0 > 0$, then $\beta \geq 0$, because a lower value for the objective function may be obtained by changing the sign. Then the solution of the modified problem

$$\min_{\beta}(-2\widehat{\beta}^0\beta + \beta^2 + \lambda\beta) \tag{9.20}$$

is, obviously, $\widehat{\beta} = \widehat{\beta}^0 - \gamma$, where $\gamma = \lambda/2$, as in (9.11). To ensure the sign consistency for this case, one could see that the solution is

$$\widehat{\beta} = (\widehat{\beta}^0 - \gamma)^+ = \text{sign}(\widehat{\beta}^0)(|\widehat{\beta}^0| - \gamma)^+. \tag{9.21}$$

Now let us take $\widehat{\beta}^0 \le 0$, then $\beta \le 0$ as well and the solution for the new problem

$$\min_{\beta}(-2\widehat{\beta}^0\beta + \beta^2 - \lambda\beta) \tag{9.22}$$

is $\widehat{\beta} = \widehat{\beta}^0 + \gamma$, but the sign consistency requires that

$$\begin{aligned}
\widehat{\beta} &= (\widehat{\beta}^0 + \gamma)^- \\
&= -(-\widehat{\beta}^0 - \gamma)^+ \\
&= \text{sign}(\widehat{\beta}^0)(|\widehat{\beta}^0| - \gamma)^+.
\end{aligned}$$

As the solutions are the same in both cases, the expression $\text{sign}(\widehat{\beta}^0)(|\widehat{\beta}^0| - \gamma)^+$ is indeed the solution to the original Lasso problem.

General Lasso solution

For a fixed $s \ge 0$, the Lasso estimation problem is a least squares problem subjected to 2^p linear inequality constraints as there are 2^p different possible signs for $\beta = (\beta_1, \ldots, \beta_p)^\top$. Lawson and Hansen (1974) suggested solving the least squares problem subject to a general linear inequality constraint $G\beta \le h$ where $G(m \times p)$ corresponds to the $m = 2^p$ constraints and $h = s1_m$. As m could be very large, this procedure is not very fast computationally. Therefore Lawson and Hansen (1974) introduced the inequality constraints sequentially in their algorithm, seeking a feasible solution.

Let $g(\beta) = \sum_{i=1}^n (y_i - x_i^\top\beta)^2$ and let $\delta_k, k = 1, \ldots, 2^p$, be column vectors of p-tuples of the form $(\pm1, \ldots, \pm1)$. It follows that the linear inequality condition can be equivalently described as $\delta_k^\top\beta \le s, \ k = 1, \ldots, 2^p$. Now let $E = \{k | \delta_k^\top\beta = s\}$ the equality set, m_E the number of elements of E and $G_E = (\delta_k^\top)_{k \in E}$ a matrix whose rows are all δ_k's for $k \in E$. Now the algorithm works as follows, see Tibshirani (1996):

1. Find OLS estimate $\widehat{\beta}^0$ and let $\delta_{k_0} = \text{sign}(\widehat{\beta}^0), E = \{k_0\}$.
2. Find $\widehat{\beta}$ to minimize $g(\beta)$ subject to $G_E\beta \le s1_{m_E}$.
3. If $\sum_{j=1}^p |\widehat{\beta}_j| \le s$ the computation is complete.
4. If $\sum_{j=1}^p |\widehat{\beta}_j| > s$ add k to the set E where $\delta_k = \text{sign}(\widehat{\beta})$ and go back to step 2.
5. The final iteration is a solution to the original problem.

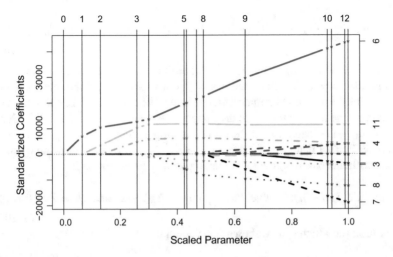

Fig. 9.4 Lasso estimates of standardized regression $\widehat{\beta}_j$ for car data with $n = 74$ and $p = 12$
MVAlassoregress

As the number of steps is limited by $m = 2^p$, the algorithm has to converge in finite time. The average number of iterations in practice is between $0.5p$ and $0.75p$.

Example 9.3 Let us consider the car data set (Sect. B.3) where $n = 74$. We want to study in what way the price (X_1) depends on the 12 other variables $(X_2), \ldots, (X_{13})$, which are represented by $j = 1, 2, \ldots, 12$, using Lasso regression. In Fig. 9.4 one can clearly see that coefficients become nonzero one at a time, that means the variables enter the regression equation sequentially as the scaled shrinkage parameter $\widehat{s} = s/\|\widehat{\beta}^0\|_1$ increases, in order $j = 6, 11, 9, 3, \ldots$ (representing $X_7, X_{12}, X_{10}, X_4, \ldots$), hence the L_1-penalty results in variable selection and the variables which are most relevant are shrunk less. In this example, an optimal $\widehat{s}$ can be found such that the fitted model gives the smallest residuals (see Exercise 9.3).

9.1.2 Lasso in High Dimensions

The problem with the algorithm by Tibshirani (1996) to calculate the Lasso solutions is that it is initialized from an OLS solution of the unconstrained problem, which does not correspond to the true model. Another problem is that for the case of $p > n$, this computation is infeasible. Therefore it may be optimal to start with a small initial guess for β and iterate through a different kind of an algorithm to obtain the Lasso solutions. Such an algorithm is based on the properties of the Lasso problem as a convex programming one. Osborne et al. (2000) showed that the original Lasso estimate problem (9.1) can be rewritten as

$$\widehat{\beta} = \arg \min_{\beta \in \mathbb{R}^p} \frac{1}{2}(y - \mathcal{X}\beta)^\top (y - \mathcal{X}\beta) \stackrel{\text{def}}{=} \frac{1}{2} r^\top r, \quad \text{subject to} \quad s - \|\beta\|_1 \geq 0,$$
$$(9.23)$$

where $r \stackrel{\text{def}}{=} (y - \mathcal{X}\beta)$. Let $\mathcal{J} = \{i_1, \ldots, i_p\}$ be the set of indices such that $|(\mathcal{X}^\top r)_{i_j}| = \|\mathcal{X}^\top r\|_\infty$, for $j = 1, \ldots, p$; so indices in $\mathcal{J}$ correspond to nonzero elements of β. Also let P be the permutation matrix that permutes the elements of the coefficient vector β so that the first elements are the nonzero elements: $\beta = P^\top (\beta_\mathcal{J}, 0)^\top$. Denote $\theta_\mathcal{J} = \text{sign}(\beta_\mathcal{J})$ be equal to 1 if the corresponding element of $\beta_\mathcal{J}$ is positive and -1 otherwise. Further denoting $f(\beta) = (y - \mathcal{X}\beta)^\top (y - \mathcal{X}\beta)$ the following optimization algorithm is based on the local linearization of (9.1) around β:

$$\widehat{\beta} = \arg \min_h f(\beta + h), \quad \text{subject to} \quad \theta_\mathcal{J}^\top (\beta_\mathcal{J} + h_\mathcal{J}) \leq s \quad \text{and} \quad h = P^\top (h_\mathcal{J}, 0)^\top,$$
$$(9.24)$$

the solution for which can be shown to be equal to

$$h_\mathcal{J} = (\mathcal{X}_\mathcal{J}^\top \mathcal{X}_\mathcal{J})^{-1} \{\mathcal{X}_\mathcal{J}^\top (y - \mathcal{X}_\mathcal{J}\beta_\mathcal{J}) - \mu \theta_\mathcal{J}\},$$

where

$$\mu = \max \left\{ 0, \frac{\theta_\mathcal{J}^\top (\mathcal{X}_\mathcal{J}^\top \mathcal{X}_\mathcal{J})^{-1} \mathcal{X}_\mathcal{J}^\top y - s}{\theta_\mathcal{J}^\top (\mathcal{X}_\mathcal{J}^\top \mathcal{X}_\mathcal{J})^{-1} \theta_\mathcal{J}} \right\}.$$

The procedure as a whole is implemented as shown in the "Lasso Solution-Path Optimization" algorithm. As shown in the algorithm, indices may enter and leave the set $\mathcal{J}$, which makes the Lasso problem similar to other subset selection techniques. Moreover, one can compute the whole path of Lasso solutions for $0 \leq s \leq s_0$, each time taking the solution for the previous s as a starting point for the next one.

9.1.3   Lasso in Logit Model

The Lasso model can be extended to generalized linear models, one of the most common of which is the logistic regression (logit) model. Coefficients in the logit model have probabilistic interpretation. In the logit model, the linear predictor $\mathcal{X}\beta$ is related to the conditional mean μ of the response variable y via the logit link $\log\{\mu/(1 - \mu)\}$. As the response variable is binary, it is binomial-distributed and $\mu = p(x_i)$. Therefore, as defined in (9.25), the logit model for $y \in \{0, 1\}$ of $(n \times 1)$ observations on a binary response variable and $x_i = (x_{i1}, \ldots, x_{ip})^\top$ is,

$$\log \left\{ \frac{p(x_i)}{1 - p(x_i)} \right\} = \sum_{j=1}^{p} \beta_j x_{ij},$$

where

Algorithm Lasso Solution-Path Optimization

1: **procedure** FIND(optimal β)
2:    Choose initial β and $\mathcal{J}$ (e.g., $\beta \leftarrow 0, \mathcal{J} \leftarrow \emptyset$)
3:    **repeat**
4:        Solve (9.23) to obtain h
5:        Set $\widehat{\beta} \leftarrow \beta + h$
6:        **if** $\text{sign}(\widehat{\beta}_{\mathcal{J}}) = \theta_{\mathcal{J}}$ **then**
7:            Obtain the solution $\beta = \widehat{\beta}$
8:        **else**
9:            **repeat**
10:                Find the smallest $\gamma, 0 < \gamma < 1, k \in \mathcal{J}$ such that $0 = \beta_k + \gamma h_k$
11:                Set $\beta = \beta + \gamma h$
12:                Set $\theta_k = -\theta_k$
13:                Solve (9.23) again to obtain a new h
14:                **if** $\theta_{\mathcal{J}}^{\top}(\beta_{\mathcal{J}} + h_{\mathcal{J}}) \leq s$ **then**
15:                    $\widehat{\beta} = \beta + h$
16:                **else**
17:                    Update $\mathcal{J} \leftarrow \mathcal{J}_{-k}$
18:                    Recompute $\beta_{\mathcal{J}}, \theta_{\mathcal{J}}, h$
19:                **end if**
20:            **until** $\text{sign}(\widehat{\beta}_{\mathcal{J}}) = \theta_{\mathcal{J}}$
21:        **end if**
22:        Compute $\widehat{v} \leftarrow \mathcal{X}^{\top}\widehat{r}/\|\mathcal{X}_{\mathcal{J}}^{\top}\widehat{r}\|_{\infty} = P^{\top}(\widehat{v}_1, \widehat{v}_2)^{\top}$ ▷ here $\widehat{r} = y - \mathcal{X}\widehat{\beta}$
23:        **if** $-1 \leq (\widehat{v}_2)_\iota \leq 1$ for $1 \leq \iota \leq p - |\mathcal{J}|$ **then**
24:            $\widehat{\beta}$ is a solution
25:        **else**
26:            Find $\jmath$ such that $|(\widehat{v}_2)_\jmath|$ is maximized
27:            Update $\mathcal{J} \leftarrow (\mathcal{J}, \jmath)$
28:            Update $\widehat{\beta}_{\mathcal{J}} \leftarrow (\widehat{\beta}_{\mathcal{J}}, 0)^{\top}$
29:            Update $\theta_{\mathcal{J}} \leftarrow (\theta_{\mathcal{J}}, \text{sign}(\widehat{v}_2)_\jmath)^{\top}$
30:        **end if**
31:        Set $\beta \leftarrow \widehat{\beta}$
32:    **until** $-1 \leq (\widehat{v}_2)_\iota \leq 1$ for $1 \leq \iota \leq p - |\mathcal{J}|$
33: **end procedure**

$$p(x_i) = P(y_i = 1 \mid x_i) = \frac{\exp(\sum_{j=1}^{p} \beta_j x_{ij})}{1 + \exp(\sum_{j=1}^{p} \beta_j x_{ij})}. \tag{9.25}$$

The Lasso estimate for the logit model is obtained by solving the following optimization problem:

$$\widehat{\beta} = \arg \min_{\beta} \left\{ \sum_{i=1}^{n} g\left(-y_i x_i^{\top} \beta\right) \right\}, \text{ subject to } \sum_{j=1}^{p} |\beta_j| \leq s, \tag{9.26}$$

with tuning parameter $s \geq 0$ and log-loss function $g(u) = \log\{1 + \exp(u)\}$. An alternative representation of the Lasso estimate $\widehat{\beta}$ in the logit model is

$$\arg\min_{\beta} \left\{ \sum_{i=1}^{n} g\left(-y_i x_i^\top \beta\right) + \lambda \sum_{j=1}^{p} |\beta_j| \right\}. \tag{9.27}$$

Shevade and Keerthi (2003) developed a simple and efficient algorithm to solve the optimization in (9.27) based on the Gauss–Seidel method using coordinate-wise descent approach. The algorithm is asymptotically convergent and easy to implement. First, define the following terms:

$$u_i = -y_i x_i^\top \beta,$$
$$F_j = \sum_{i=1}^{n} \frac{\exp(u_i)}{\exp(1 + u_i)} y_i x_{ij}. \tag{9.28}$$

The first-order optimality conditions for (9.27) are the following:

$$\begin{aligned}
F_j &= 0 && \text{if} \quad j = 0, \\
F_j &= \lambda && \text{if} \quad \beta_j > 0, \ j > 0, \\
F_j &= -\lambda && \text{if} \quad \beta_j < 0, \ j > 0, \\
-\lambda \leq F_j &\leq \lambda && \text{if} \quad \beta_j = 0, \ j > 0.
\end{aligned}$$

A new variable is defined

$$\begin{aligned}
v_j &= |F_j| && \text{if} \quad j = 0, \\
&= |\lambda - F_j| && \text{if} \quad \beta_j > 0, \ j > 0, \\
&= |\lambda + F_j| && \text{if} \quad \beta_j < 0, \ j > 0, \\
&= \psi_j && \text{if} \quad \beta_j = 0, \ j > 0.
\end{aligned}$$

where $\psi_j = \max\{(F_j - \lambda), (-\lambda - F_j), 0\}$. Thus, the first-order optimality conditions can be written as

$$v_j = 0 \quad \forall j. \tag{9.29}$$

It is difficult to obtain exact optimality condition, so the stopping criterion for (9.27) is defined as follows (for some small ε):

$$v_j \leq \varepsilon \quad \forall j. \tag{9.30}$$

To write the algorithm, let us define $I_z = \{j : \beta_j = 0, j > 0\}$ and $I_{nz} = \{j : \beta_j \neq 0, j > 0\}$ for sets of zero estimates and sets of nonzero estimates, respectively, and $I = I_z \cup I_{nz}$. The algorithm consists of two loops. The first loop runs over the variables in I_z to choose the maximum violator, v. In the second loop W is optimized with respect to β_v, therefore the set I_{nz} is modified and maximum violator in I_{nz} is obtained. The second loop is repeated until no violators are found in I_{nz}. The algorithm alternates between the first and second loop until no violators exist in both I_z and I_{nz}.

Algorithm Lasso in logit model

1: **procedure** FIND(optimal Lasso estimate $\widehat{\beta}$)
2:   Set $\beta_j = 0$ for all j
3:   **while** an optimality violator exists in I_z **do**
4:     Find the maximum violator (v) in I_z
5:     **repeat**
6:       Optimize W with respect to β_v
7:       Find the maximum violator (v) in I_{nz}
8:     **until** no violator exists in I_{nz}
9:   **end while**
10: **end procedure**

Another way to obtain the lasso estimate in the logit model is by maximizing the likelihood function of logit model with lasso constraint. The log-likelihood function of logit model is written as

$$\log L(\beta) = \sum_{i=1}^{n} \left[y_i \log p\,(x_i) + (1 - y_i) \log\{1 - p\,(x_i)\} \right]. \tag{9.31}$$

Suppose $\ell(\beta) = \log L(\beta)$, with $\beta = (\beta_1, \ldots, \beta_p)^\top$, the Lasso estimates are obtained by maximizing the penalized log-likelihood for logit model as follows:

$$\widehat{\beta} = \arg\max_{\beta} \left\{ n^{-1} \sum_{i=1}^{n} \ell(\beta) \right\}, \quad \text{subject to} \quad \sum_{j=1}^{p} |\beta_j| \leq s. \tag{9.32}$$

It can be solved by a general nonlinear programming procedure or by using iteratively reweighted least squares (IRLS). Friedman et al. (2010) developed an algorithm to solve the problem in (9.32). An alternative representation of the Lasso problem is defined as follows:

$$\widehat{\beta} = \arg\max_{\beta} \left\{ n^{-1} \sum_{i=1}^{n} \ell(\beta) - \lambda \sum_{j=1}^{p} |\beta_j| \right\}. \tag{9.33}$$

Example 9.4 Following the Example 9.3, the price (X_1) of car data set (Sect. B.3) has average 6192.28. We now define a new categorical variable which takes the value 0 if $X_1 \leq 6000$ and otherwise is equal to 1. We want to study in what way the price (X_1) depends on the 12 other variables ($X_2, \ldots, X_{13}$) using Lasso in logit model.

Fig. 9.5 Lasso estimates $\widehat{\beta}_j$
of logit model for car data
with $n = 74$ and $p = 12$

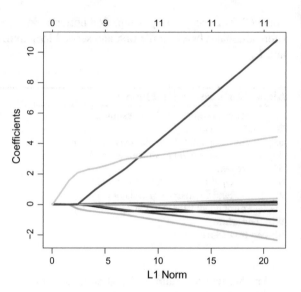

MVAlassologit

In Fig. 9.5, one can see that coefficients' dynamics depends on the shrinkage
parameter $s = \|\widehat{\beta}(\lambda)\|_1$, the L_1-norm of estimated coefficients. An optimal s can be
chosen such that the fitted model gives the smallest residuals (see Exercise 9.4).

9.2   Elastic Net

Although the Lasso is widely used in variable selection, it has several drawbacks.
Zou and Hastie (2005) stated that

1. if $p > n$, the Lasso selects at most n variables before it saturates;
2. if there is a group of variables which has very high correlation, then the Lasso
   tends to select only one variable from this group;
3. for usual $n > p$ condition, if there are high correlations between predictors,
   the prediction performance of the Lasso is dominated by ridge regression, see
   (Tibshirani 1996).

Zou and Hastie (2005) introduced the Elastic net which combines good features
of the L_1-norm and L_2-norm penalties. The Elastic net is a regularized regression
method which overcomes the limitations of the Lasso. This method is very useful
when $p \gg n$ or if there are many correlated variables. The advantages are: (i) a group
of correlated variables can be selected without arbitrary omissions and (ii) the number
of selected variables is no longer limited by the sample size.

9.2.1 Elastic Net in Linear Regression Model

We describe the Elastic net in linear regression model. For simplicity reason, we assume that the x_{ij} are standardized such that $\sum_{i=1}^{n} x_{ij} = 0$ and $n^{-1} \sum_{i=1}^{n} x_{ij}^2 = 1$. The Elastic net penalty $P_\alpha(\beta)$ leads to the following modification of the problem to obtain the estimator $\widehat{\beta}$

$$\arg\min_{\beta} \left\{ (2n)^{-1} \sum_{i=1}^{n} \left(y_i - x_i^\top \beta \right)^2 + \lambda P_\alpha(\beta) \right\}, \tag{9.34}$$

where

$$P_\alpha(\beta) = \frac{1}{2}(1 - \alpha) \, \|\beta\|_2^2 + \alpha \, \|\beta\|_1$$

$$= \sum_{j=1}^{p} \left\{ \frac{1}{2}(1 - \alpha)\beta_j^2 + \alpha|\beta_j| \right\}. \tag{9.35}$$

The penalty $P_\alpha(\beta)$ is a compromise between ridge regression and the Lasso. If $\alpha = 0$ then the criterion is the ridge regression and if $\alpha = 1$ the method will be the Lasso. Practically, for small $\varepsilon > 0$, the Elastic net with $\alpha = 1 - \varepsilon$ performs like the Lasso, but removes degeneracies and erratic variable selection behavior caused by extreme correlation. Given a specific λ, as α increases from 0 to 1, the sparsity of the Elastic net solutions increases monotonically from 0 to the sparsity of the Lasso solutions.

The Elastic net optimization problem can be represented as the usual Lasso problem, using modified $\mathcal{X}$ and y vectors, as shown in the following example.

Example 9.5 To turn the Elastic net optimization problem into the usual Lasso one, one should first augment y with p additional zeros to obtain $\tilde{y} = (y, 0)^\top$. Then, augment $\mathcal{X}$ with the multiple of the $p \times p$ identity matrix $\sqrt{\lambda\alpha}\mathcal{I}$ to get $\tilde{\mathcal{X}} = \left(\mathcal{X}^\top, \sqrt{\lambda\alpha}I \right)^\top$. Next, define $\tilde{\lambda} = \lambda(1 - \alpha)$ and solve the original Lasso minimization problem in terms of the new input $\tilde{y}$, $\tilde{\mathcal{X}}$ and $\tilde{\lambda}$. This new problem is equivalent to the original Elastic net problem:

$$\|\tilde{y} - \tilde{\mathcal{X}}\beta\|_2^2 + \tilde{\lambda}\|\beta\|_1 = \left\| \begin{bmatrix} y \\ 0 \end{bmatrix} - \begin{bmatrix} \mathcal{X}\beta \\ \sqrt{\lambda\alpha}\mathcal{I}\beta \end{bmatrix} \right\|_2^2 + \lambda(1 - \alpha)\|\beta\|_1,$$

$$= \|y - \mathcal{X}\beta\|_2^2 - \lambda\alpha\|\beta\|_2^2 + \lambda\|\beta\|_1 - \lambda\alpha\|\beta\|_1,$$

$$= \|y - \mathcal{X}\beta\|_2^2 + \lambda \left\{ \alpha\|\beta\|_2^2 + (1 - \alpha)\|\beta\|_1 \right\},$$

which is equivalent to the original Elastic net problem.

We follow the idea of Friedman et al. (2010) who used a coordinate descent algorithm to solve the optimization problem in (9.34). Let us suppose to have estimates

$\tilde{\beta}_k$ for $k \neq j$. Then we optimize (9.34) partially with respect to β_j by computing the gradient at $\beta_j = \tilde{\beta}_j$, which only exists if $\tilde{\beta}_j \neq 0$. Having the soft-thresholding operator $S(z, \gamma)$ as

$$\text{sign}(z)\,(|z| - \gamma)^+ = \begin{cases} z - \gamma & \text{if } z > 0 \ \text{and} \ \gamma < |z|, \\ z + \gamma & \text{if } z < 0 \ \text{and} \ \gamma < |z|, \\ 0 & \text{if } \ \ \gamma \geq |z|. \end{cases} \quad (9.36)$$

it can be shown that the coordinate-wise update has the following form:

$$\tilde{\beta}_j = \frac{S\left\{ n^{-1} \sum_{i=1}^n x_{ij} \left(y_i - \tilde{y}_i^{(j)} \right), \lambda\alpha \right\}}{1 + \lambda(1 - \alpha)}, \quad (9.37)$$

where $\tilde{y}_i^{(j)} = \sum_{k \neq j} x_{ik}\tilde{\beta}_k$ is a fitted value which excludes the contribution x_{ij}, therefore $y_i - \tilde{y}_i^{(j)}$ is partial residual for fitting β_j.

The algorithm computes the least squares estimate for the partial residual $y_i - \tilde{y}_i^{(j)}$, then applies the soft-thresholding rule to perform the Lasso contribution to the penalty $P_\alpha(\beta)$. Afterward, a proportional shrinkage is applied to ridge penalty. There are several methods used to update the current estimate $\tilde{\beta}$. We describe the simplest updating method, the so-called "naive" update.

The partial residual can be rewritten as follows:

$$\begin{aligned} y_i - \tilde{y}_i^{(j)} &= y_i - \widehat{y}_i + x_{ij}\tilde{\beta}_j \\ &= r_i + x_{ij}\tilde{\beta}_j, \end{aligned} \quad (9.38)$$

with $\widehat{y}_i$ being the current fit and r_i the current residual. As x_j is standardized, it holds

$$\frac{1}{n} \sum_{i=1}^n x_{ij} \left(y_i - \tilde{y}_i^{(j)} \right) = \frac{1}{n} \sum_{i=1}^n x_{ij} r_i + \tilde{\beta}_j. \quad (9.39)$$

Note that the first term on the right-hand side of the new partial residual is the gradient of the loss with respect to β_j.

9.2.2 Elastic Net in Logit Model

The Elastic net penalty can similarly be applied to the logit model. Recall the log-likelihood function of the logit model in (9.31),

$$\log L(\beta) = \sum_{i=1}^{n} \left[y_i \log p(x_i) + (1 - y_i) \log\{1 - p(x_i)\} \right].$$

Penalized log-likelihood for the logit model using Elastic net has the following form:

$$\max_{\beta} \left\{ n^{-1} \sum_{i=1}^{n} \ell(\beta) - \lambda P_{\alpha}(\beta) \right\}, \tag{9.40}$$

with $\ell(\beta) = \log L(\beta)$. The solution of (9.40) can be found by means of a Newton algorithm. For a fixed λ and a given current parameter $\tilde{\beta}$, the quadratic approximation (Taylor expansion) is updated about current estimates $\tilde{\beta}$ as follows:

$$\ell_Q(\beta) = -(2n)^{-1} \sum_{i=1}^{n} w_i (z_i - x_i^{\top} \beta)^2 + C(\tilde{\beta})^2, \tag{9.41}$$

where the working response and weight, respectively, are as follows:

$$z_i = x_i^{\top} \tilde{\beta} + \frac{y_i - \tilde{p}(x_i)}{\tilde{p}(x_i)\{1 - \tilde{p}(x_i)\}},$$
$$w_i = \tilde{p}(x_i) \{1 - \tilde{p}(x_i)\}.$$

A Newton update is obtained by minimizing $\ell_Q(\beta)$.

Friedman et al. (2010) proposed similar approach creating an outer loop for each value of λ, which computes a quadratic approximation in (9.41) about current estimates $\tilde{\beta}$. Afterward, a coordinate descent algorithm is used to solve the following penalized weigthed least squares problem (PWLS):

$$\min_{\beta} \left\{ -\ell_Q(\beta) + \lambda P_{\alpha}(\beta) \right\}. \tag{9.42}$$

This inner coordinate descent loop continues until the maximum change in (9.42) is less than a very small threshold.

9.3 Group Lasso

The Group Lasso was first introduced by Yuan and Lin (2006) and was motivated by the fact that the predictor variables can occur in several groups and one could want a parsimonious model which uses only a few of these groups. That is, assume that there are K groups and the vector of coefficients is structured as follows:

$$\beta^G = (\beta_1^{\top}, \dots, \beta_K^{\top})^{\top} \in \mathbb{R}^{\sum_k p_k},$$

where p_k is the coefficient vector dimension of the k-th group, $k = 1, \ldots, K$. A sparse set of groups is produced, although within each group either all entries of β_k, $k = 1, \ldots, K$ are zero or all of them are nonzero. The Group Lasso problem can be formulated in general as

$$\arg\min_{\beta \in \mathbb{R}^{\sum_k p_k}} n^{-1} \left\| y - \sum_{k=1}^{K} \mathcal{X}_k \beta_k \right\|_2^2 + \lambda \sum_{k=1}^{K} \sqrt{p_k} \|\beta_k\|_2, \qquad (9.43)$$

where $\mathcal{X}_k$ is the k-th component of the matrix $\mathcal{X}$ with columns corresponding to the predictors in the group k, β_k is the coefficient vector for that group and p_k is the cardinality of the group, i.e., the size of the coefficient vector which serves as a balancing weight in the case of widely differing group sizes. It is obvious that if groups consist of single elements, i.e., $p_k = 1$, $\forall k$, then the Group Lasso problem is reduced to the usual Lasso one.

The computation of the Group Lasso solution involves calculating the necessary and sufficient subgradient KKT conditions for $\widehat{\beta}^G = (\widehat{\beta}_1^\top, \ldots, \widehat{\beta}_K^\top)^\top$ to be a solution of (9.43):

$$- \mathcal{X}_k^\top \left(y - \sum_{k=1}^{K} \mathcal{X}_k \beta_k \right) + \frac{\lambda \beta_k \sqrt{p_k}}{\|\beta_k\|} = 0, \qquad (9.44)$$

if $\beta_k \neq 0$; otherwise, for $\beta_k = 0$, it holds that

$$\left\| \mathcal{X}_k^\top \left(y - \sum_{l \neq k} \mathcal{X}_l \widehat{\beta}_l \right) \right\| \leq \lambda \sqrt{p_k}. \qquad (9.45)$$

Expressions (9.44) and (9.45) allow to calculate the solution, the so-called "update step" which can be used to implement an iterative algorithm to solve the problem (9.43). The solution resulting from the KKT conditions is readily shown to be the following:

$$\widehat{\beta}_k = \left\{ \left(\lambda \sqrt{p_k} \|\widehat{\beta}_k\|^{-1} + \mathcal{X}_k^\top \mathcal{X}_k \right)^{-1} \right\}^+ \mathcal{X}_k^\top \widehat{r}_k, \qquad (9.46)$$

where the residual $\widehat{r}_k$ is defined as $\widehat{r}_k \stackrel{\text{def}}{=} y - \sum_{l \neq k} \mathcal{X}_l \widehat{\beta}_l$. As a special (orthonormal) case, when $\mathcal{X}_l^\top \mathcal{X}_l = \mathcal{I}$, the solution is simplified to the $\widehat{\beta}_k = (\lambda \sqrt{p_k} \|\widehat{\beta}_k\|^{-1} + 1) \mathcal{X}_k^\top \widehat{r}_k$. To obtain a full solution to this problem, Yuan and Lin (2006) suggest using a blockwise coordinate descent algorithm which iteratively applies the estimate (9.46) to $k = 1, \ldots, K$.

Meier et al. (2008) extended the Group Lasso to the case of logistic regression and demonstrated convergence of several algorithms for the computation of the solution as well as outlined consistency results for the Group Lasso logit estimator. The general setup for that model involves a binary response variable $y_i \in \{0, 1\}$ and K groups of predictor variables $x_i = (x_{i1}^\top, \ldots, x_{iK}^\top)^\top$, both x_i and y_i are i.i.d., $i = 1, \ldots, n$. Then the logistic linear regression model may be written as before:

$$\log\left\{\frac{p(x_i)}{1 - p(x_i)}\right\} = \eta(x_i) \stackrel{\text{def}}{=} \beta_0 + \sum_{k=1}^{K} x_{ik}^\top \beta_k, \tag{9.47}$$

where the conditional probability $p(x_i) = P(y_i = 1|x_i)$. The Group Lasso logit estimator $\widehat{\beta}$ then minimizes the objective function

$$\widehat{\beta} = \arg\min_{\beta \in \mathbb{R}^{p+1}}\left\{-\ell(\beta) + \lambda \sum_{k=1}^{K} \sqrt{p_k}\|\beta_k\|_2\right\}, \tag{9.48}$$

where $\ell(\cdot)$ is the log-likelihood function

$$\ell(\beta) = \sum_{i=1}^{n} y_i \eta(x_i) - \log[1 + \exp\{\eta(x_i)\}].$$

The problem is solved through a group-wise minimization of the penalized objective function by, for example, the block-coordinate descent method.

Example 9.6 The Group Lasso results can be illustrated by an application to the MEMset Donor data set of human donor splice sites with a sequence length of seven base pairs. The full data set (training and test parts) consists of 12.623 true ($y_i = 1$) and 269.155 false ($y_i = 0$) human donor sites. Each element of data represents a sequence of DNA within a window of the splice site which consists of the last three positions of the exon and first four positions of the intron; so the strings of length 7 are made up of four characters A, C, T, G, and therefore the predictor variables are seven factors, each having four levels. False splice sites are sequences on the DNA which match the consensus sequence at position four and five. Figure 9.6 shows how the Group Lasso does shrinkage on the level of groups built by DNA letters.

As can be seen from Example 9.6, the solution to the Group Lasso problem yields a sparse solution only regarding the "between" case, that is, it excludes some of the groups from the model but then all coefficients in the remaining groups are nonzero. To ensure both the sparsity of groups and within each group, Simon et al. (2013) proposed the so-called "sparse Group Lasso" which uses a more general penalty which yields sparsity on both inter- and intragroup level. The sparse Group Lasso estimate solves the problem

$$\widehat{\beta} = \arg\min_{\beta \in \mathbb{R}^p}\left\|y - \sum_{k=1}^{K} \mathcal{X}_k \beta_k\right\|_2^2 + \lambda_1 \sum_{k=1}^{K} \|\beta_k\|_2 + \lambda_2\|\beta\|_1, \tag{9.49}$$

where $\beta = (\beta_1, \beta_2, \ldots, \beta_K)^\top$ is the entire parameter vector.

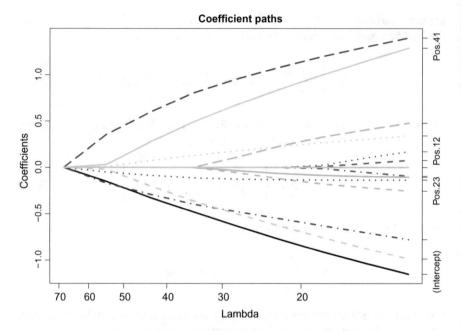

Fig. 9.6 Lasso estimates of standardized regression $\widehat{\beta}_j$ for car data with $n = 74$ and $p = 12$
🔵MVAgrouplasso

Summary
↪ Lasso gives a sparse solution. Lasso estimate combines best of both ridge regression and subset selection.
↪ If there is a group of variables which has very high correlation, then the Lasso tends to select only one variable from the group.
↪ The LAR algorithm computes the whole path of Lasso solutions and is feasible for the high-dimensional case $p \gg n$.
↪ Elastic net combines good features of L_1-norm and L_2-norm penalties.
↪ The Elastic net is very useful when $p \gg n$ or if there are many correlated variables.
↪ The sparse Group Lasso can perform shrinkage both on inter- and intragroup level.

9.4   Exercises

Exercise 9.1 Derive the explicit Lasso estimate in (9.11) for the orthonormal design case.

Exercise 9.2 Compare Lasso orthonormal design case for $p = 2$ graphically to ridge regression, i.e., to the problem $\widehat{\beta} = \mathrm{argmin} \left\{ \sum_{i=1}^{n} \left(y_i - x_i^{\top} \beta \right)^2 \right\}$ subject to

$\sum_{j=1}^{p} \beta_j^2 \leq s$. Why does Lasso perform variable selection and ridge regression does not?

Exercise 9.3 Optimize the value of s such that the fitted model in Example 9.3 produces the smallest residuals.

Exercise 9.4 Optimize the value of s such that the fitted model in Example 9.4 produces the smallest residuals.

References

B. Efron, T. Hastie, I. Johnstone, R. Tibshirani, Least angle regression (with discussion). Ann. Stat. **32**(2), 407–499 (2004)

J.H. Friedman, T. Hastie, R. Tibshirani, Regularization paths for generalized linear models via coordinate descent. J. Stat. Softw. **33** (2010)

C. Lawson, R. Hansen, *Solving Least Square Problems* (Prentice Hall, Englewood Cliffs, 1974)

L. Meier, S. van de Geer, P. Bühlmann, The group lasso for logistic regression. JRSSB **70**, 53–71 (2008)

M.R. Osborne, B. Presnell, B.A. Turlach, On the Lasso and Its Dual. J. Comput. Graph. Stat. **9**(2), 319–337 (2000)

S.K. Shevade, S.S. Keerthi, A simple and efficient algorithm for gene selection using sparse logistic regression. Bioinformatics **19**, 2246–2253 (2003)

N. Simon, J.H. Friedman, T. Hastie, R. Tibshirani, A sparse-group lasso. J. Comput. Graph. Stat. **22**(2), 231–245 (2013)

R. Tibshirani, Regression Shrinkage and Selection via the Lasso. J. Royal Stat. Soc. Ser. B **58**, 267–288 (1996)

M. Yuan, Y. Lin, Model selection and estimation in regression with grouped variables. J. Royal Stat. Soc. Ser. B **68**, 49–67 (2006)

H. Zou, T. Hastie, Regularization and variable selection via the elastic net. J. Royal Stat. Soc. Ser. B **67**, 301–320 (2005)

Chapter 10
Decomposition of Data Matrices by Factors

In Chap. 1, basic descriptive techniques were developed which provided tools for "looking" at multivariate data. They were based on adaptations of bivariate or univariate devices, which is used to reduce the dimensions of the observations. In the following three chapters, issues of reducing the dimension of a multivariate data set will be discussed. The perspectives will be different but the tools will be related.

In this chapter, we take a descriptive perspective and show how using a geometrical approach provides the "best" way of reducing the dimension of a data matrix. It is derived with respect to a least squares criterion. The result will be low dimensional graphical pictures of the data matrix. This involves the decomposition of the data matrix into "factors". These "factors" will be sorted in decreasing order of importance. The approach is very general and is the core idea of many multivariate techniques. We deliberately use the word "factor" here as a tool or transformation for structural interpretation in an exploratory analysis. In practice, the matrix to be decomposed will be some transformation of the original data matrix and as shown in the following chapters, these transformations provide easier interpretations of the obtained graphs in lower dimensional spaces.

Chapter 11 addresses the issue of reducing the dimensionality of a multivariate random variable by using linear combinations (the principal components). The identified principal components are ordered in decreasing order of importance. When applied in practice to a data matrix, the principal components will turn out to be the factors of a transformed data matrix (the data will be centered and eventually standardized).

Factor analysis is discussed in Chap. 12. The same problem of reducing the dimension of a multivariate random variable is addressed but in this case, the number of factors is fixed from the start. Each factor is interpreted as a latent characteristic of the individuals revealed by the original variables. The nonuniqueness of the solutions is dealt with by searching for the representation with the easiest interpretation for the analysis.

© Springer Nature Switzerland AG 2019
W. K. Härdle and L. Simar, *Applied Multivariate Statistical Analysis*,
https://doi.org/10.1007/978-3-030-26006-4_10

Summarizing, this chapter can be seen as a foundation since it develops a basic tool for reducing the dimension of a multivariate data matrix.

10.1   The Geometric Point of View

As a matter of introducing certain ideas, assume that the data matrix $\mathcal{X}(n \times p)$ is composed of n observations (or individuals) of p variables.

There are in fact two ways of looking at $\mathcal{X}$, row by row or column by column:

1. Each row (observation) is a vector $x_i^\top = (x_{i1}, \ldots, x_{ip}) \in \mathbb{R}^p$.
   From this point of view, our data matrix $\mathcal{X}$ is representable as a cloud of n points in $\mathbb{R}^p$ as shown in Fig. 10.1.
2. Each column (variable) is a vector $x_{[j]} = (x_{1j}, \ldots, x_{nj})^\top \in \mathbb{R}^n$.
   From this point of view, the data matrix $\mathcal{X}$ is a cloud of p points in $\mathbb{R}^n$ as shown in Fig. 10.2.

When n and/or p are large (larger than 2 or 3), we cannot produce interpretable graphs of these clouds of points. Therefore, the aim of the factorial methods to be developed here is twofold. We shall try to simultaneously approximate the column space $C(\mathcal{X})$ and the row space $C(\mathcal{X}^\top)$ with smaller subspaces. The hope is of course that this can be done without losing too much information about the variation and structure of the point clouds in both spaces. Ideally, this will provide insights into the structure of $\mathcal{X}$ through graphs in $\mathbb{R}$, $\mathbb{R}^2$, or $\mathbb{R}^3$. The main focus then is to find the dimension reducing factors.

Fig. 10.1 Cloud of n points in $\mathbb{R}^p$

Fig. 10.2 Cloud of p points in $\mathbb{R}^n$

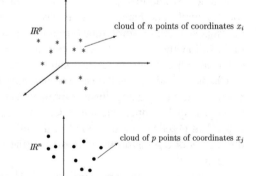

Summary
$\hookrightarrow$ Each row (individual) of $\mathcal{X}$ is a p-dimensional vector. From this point of view $\mathcal{X}$ can be considered as a cloud of n points in $\mathbb{R}^p$.
$\hookrightarrow$ Each column (variable) of $\mathcal{X}$ is a n-dimensional vector. From this point of view $\mathcal{X}$ can be considered as a cloud of p points in $\mathbb{R}^n$.

10.2 Fitting the p-Dimensional Point Cloud

Subspaces of Dimension 1

In this section, $\mathcal{X}$ is represented by a cloud of n points in $\mathbb{R}^p$ (considering each row). The question is how to project this point cloud onto a space of lower dimension. To begin consider the simplest problem, namely finding a subspace of dimension 1. The problem boils down to finding a straight line F_1 through the origin. The direction of this line can be defined by a unit vector $u_1 \in \mathbb{R}^p$. Hence, we are searching for the vector u_1 which gives the "best" fit of the initial cloud of n points. The situation is depicted in Fig. 10.3.

The representation of the i-th individual $x_i \in \mathbb{R}^p$ on this line is obtained by the projection of the corresponding point onto u_1, i.e., the projection point p_{x_i}. We know from (2.42) that the coordinate of x_i on F_1 is given by

$$p_{x_i} = x_i^\top \frac{u_1}{\|u_1\|} = x_i^\top u_1. \tag{10.1}$$

We define the *best line* F_1 in the following "least-squares" sense: Find $u_1 \in \mathbb{R}^p$ which minimizes

Fig. 10.3 Projection of point cloud onto u space of lower dimension

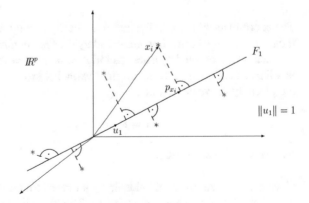

$$\sum_{i=1}^{n} \|x_i - p_{x_i}\|^2. \tag{10.2}$$

Since $\|x_i - p_{x_i}\|^2 = \|x_i\|^2 - \|p_{x_i}\|^2$ by Pythagoras's theorem, the problem of minimizing (10.2) is equivalent to maximizing $\sum_{i=1}^{n} \|p_{x_i}\|^2$. Thus the problem is to find $u_1 \in \mathbb{R}^p$ that maximizes $\sum_{i=1}^{n} \|p_{x_i}\|^2$ under the constraint $\|u_1\| = 1$. With (10.1) we can write

$$\begin{pmatrix} p_{x_1} \\ p_{x_2} \\ \vdots \\ p_{x_n} \end{pmatrix} = \begin{pmatrix} x_1^\top u_1 \\ x_2^\top u_1 \\ \vdots \\ x_n^\top u_1 \end{pmatrix} = \mathcal{X} u_1$$

and the problem can finally be reformulated as find $u_1 \in \mathbb{R}^p$ with $\|u_1\| = 1$ that maximizes the quadratic form $(\mathcal{X}u_1)^\top (\mathcal{X}u_1)$ or

$$\max_{u_1^\top u_1 = 1} u_1^\top (\mathcal{X}^\top \mathcal{X}) u_1. \tag{10.3}$$

The solution is given by Theorem 2.5 (using $\mathcal{A} = \mathcal{X}^\top \mathcal{X}$ and $\mathcal{B} = \mathcal{I}$ in the theorem).

Theorem 10.1 *The vector u_1 which minimizes (10.2) is the eigenvector of $\mathcal{X}^\top \mathcal{X}$ associated with the largest eigenvalue λ_1 of $\mathcal{X}^\top \mathcal{X}$.*

Note that if the data have been centered, i.e., $\bar{x} = 0$, then $\mathcal{X} = \mathcal{X}_c$, where $\mathcal{X}_c$ is the centered data matrix, and $\frac{1}{n}\mathcal{X}^\top \mathcal{X}$ is the covariance matrix. Thus Theorem 10.1 says that we are searching for a maximum of the quadratic form (10.3) with respect to the covariance matrix $\mathcal{S}_{\mathcal{X}} = n^{-1}\mathcal{X}^\top \mathcal{X}$.

Representation of the Cloud on F_1

The coordinates of the n individuals on F_1 are given by $\mathcal{X}u_1$. $\mathcal{X}u_1$ is called the *first factorial variable* or the *first factor* and u_1 the *first factorial axis*. The n individuals, x_i, are now represented by a new factorial variable $z_1 = \mathcal{X}u_1$. This factorial variable is a linear combination of the original variables $(x_{[1]}, \ldots, x_{[p]})$ whose coefficients are given by the vector u_1, i.e.,

$$z_1 = u_{11}x_{[1]} + \ldots + u_{p1}x_{[p]}. \tag{10.4}$$

Subspaces of Dimension 2

If we approximate the n individuals by a plane (dimension 2), it can be shown via Theorem 2.5 that this space contains u_1. The plane is determined by the best linear

Fig. 10.4 Representation of the individuals $x_1, \ldots, x_n$ as a two-dimensional point cloud

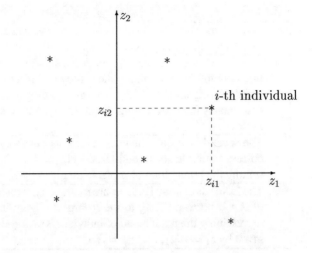

fit (u_1) and a unit vector u_2 orthogonal to u_1 which maximizes the quadratic form $u_2^\top (\mathcal{X}^\top \mathcal{X}) u_2$ under the constraints

$$\|u_2\| = 1, \text{ and } u_1^\top u_2 = 0.$$

Theorem 10.2 *The second factorial axis, u_2, is the eigenvector of $\mathcal{X}^\top \mathcal{X}$ corresponding to the second largest eigenvalue λ_2 of $\mathcal{X}^\top \mathcal{X}$.*

The unit vector u_2 characterizes a second line, F_2, on which the points are projected. The coordinates of the n individuals on F_2 are given by $z_2 = \mathcal{X} u_2$. The variable z_2 is called the *second factorial variable* or *the second factor*. The representation of the n individuals in two-dimensional space ($z_1 = \mathcal{X} u_1$ vs. $z_2 = \mathcal{X} u_2$) is shown in Fig. 10.4.

Subspaces of Dimension q $(q \leq p)$

In the case of q dimensions the task is again to minimize (10.2) but with projection points in a q-dimensional subspace. Following the same argument as above, it can be shown via Theorem 2.5 that this best subspace is generated by $u_1, u_2, \ldots, u_q$, the orthonormal eigenvectors of $\mathcal{X}^\top \mathcal{X}$ associated with the corresponding eigenvalues $\lambda_1 \geq \lambda_2 \geq \ldots \geq \lambda_q$. The coordinates of the n individuals on the k-th factorial axis, u_k, are given by the k-th factorial variable $z_k = \mathcal{X} u_k$ for $k = 1, \ldots, q$. Each factorial variable $z_k = (z_{1k}, z_{2k}, \ldots, z_{nk})^\top$ is a linear combination of the original variables $x_{[1]}, x_{[2]}, \ldots, x_{[p]}$ whose coefficients are given by the elements of the k-th vector u_k : $z_{ik} = \sum_{m=1}^{p} x_{im} u_{mk}$.

Summary
↪ The p-dimensional point cloud of individuals can be graphically represented by projecting each element onto spaces of smaller dimensions.
↪ The first factorial axis is u_1 and defines a line F_1 through the origin. This line is found by minimizing the orthogonal distances (10.2). The factor u_1 equals the eigenvector of $\mathcal{X}^\top \mathcal{X}$ corresponding to its largest eigenvalue. The coordinates for representing the point cloud on a straight line are given by $z_1 = \mathcal{X} u_1$.
↪ The second factorial axis is u_2, where u_2 denotes the eigenvector of $\mathcal{X}^\top \mathcal{X}$ corresponding to its second largest eigenvalue. The coordinates for representing the point cloud on a plane are given by $z_1 = \mathcal{X} u_1$ and $z_2 = \mathcal{X} u_2$.
↪ The factor directions $1, \ldots, q$ are $u_1, \ldots, u_q$, which denote the eigenvectors of $\mathcal{X}^\top \mathcal{X}$ corresponding to the q largest eigenvalues. The coordinates for representing the point cloud of individuals on a q-dimensional subspace are given by $z_1 = \mathcal{X} u_1, \ldots, z_q = \mathcal{X} u_q$.

10.3 Fitting the n-Dimensional Point Cloud

Subspaces of Dimension 1

Suppose that $\mathcal{X}$ is represented by a cloud of p points (variables) in $\mathbb{R}^n$ (considering each column). How can this cloud be projected into a lower dimensional space? We start as before with one dimension. In other words, we have to find a straight line G_1, which is defined by the unit vector $v_1 \in \mathbb{R}^n$, and which gives the best fit of the initial cloud of p points.

Algebraically, this is the same problem as above (replace $\mathcal{X}$ by $\mathcal{X}^\top$ and follow Sect. 10.2): the representation of the j-th variable $x_{[j]} \in \mathbb{R}^n$ is obtained by the projection of the corresponding point onto the straight line G_1 or the direction v_1. Hence we have to find v_1 such that $\sum_{j=1}^{p} \| p_{x_{[j]}} \|^2$ is maximized, or equivalently, we have to find the unit vector v_1 which maximizes $(\mathcal{X}^\top v_1)^\top (\mathcal{X}^\top v_1) = v_1^\top (\mathcal{X} \mathcal{X}^\top) v_1$. The solution is given by Theorem 2.5.

Theorem 10.3 v_1 *is the eigenvector of* $\mathcal{X} \mathcal{X}^\top$ *corresponding to the largest eigenvalue* μ_1 *of* $\mathcal{X} \mathcal{X}^\top$.

Representation of the Cloud on G_1

The coordinates of the p variables on G_1 are given by $w_1 = \mathcal{X}^\top v_1$, the first factorial axis. The p variables are now represented by a linear combination of the original individuals $x_1, \ldots, x_n$, whose coefficients are given by the vector v_1, i.e., for $j = 1, \ldots, p$

$$w_{1j} = v_{11} x_{1j} + \ldots + v_{1n} x_{nj}. \tag{10.5}$$

Fig. 10.5 Representation of
the variables $x_{[1]}, \ldots, x_{[p]}$ as
a two-dimensional point
cloud

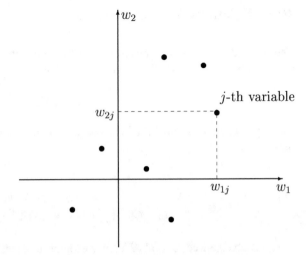

Subspaces of Dimension q $(q \leq n)$

The representation of the p variables in a subspace of dimension q is done in the
same manner as for the n individuals above. The best subspace is generated by
the orthonormal eigenvectors $v_1, v_2, \ldots, v_q$ of $\mathcal{X}\mathcal{X}^\top$ associated with the eigenvalues $\mu_1 \geq \mu_2 \geq \ldots \geq \mu_q$. The coordinates of the p variables on the k-th factorial
axis are given by the factorial variables $w_k = \mathcal{X}^\top v_k$, $k = 1, \ldots, q$. Each factorial
variable $w_k = (w_{k1}, w_{k2}, \ldots, w_{kp})^\top$ is a linear combination of the original individuals $x_1, x_2, \ldots, x_n$ whose coefficients are given by the elements of the k-th vector
$v_k : w_{kj} = \sum_{m=1}^{n} v_{km} x_{mj}$. The representation in a subspace of dimension $q = 2$ is
depicted in Fig. 10.5.

Summary
$\hookrightarrow$ The n-dimensional point cloud of variables can be graphically represented by projecting each element onto spaces of smaller dimensions.
$\hookrightarrow$ The first factor direction is v_1 and defines a line G_1 through the origin. The vector v_1 equals the eigenvector of $\mathcal{X}\mathcal{X}^\top$ corresponding to the largest eigenvalue of $\mathcal{X}\mathcal{X}^\top$. The coordinates for representing the point cloud on a straight line are $w_1 = \mathcal{X}^\top v_1$.
$\hookrightarrow$ The second factor direction is v_2, where v_2 denotes the eigenvector of $\mathcal{X}\mathcal{X}^\top$ corresponding to its second largest eigenvalue. The coordinates for representing the point cloud on a plane are given by $w_1 = \mathcal{X}^\top v_1$ and $w_2 = \mathcal{X}^\top v_2$.
$\hookrightarrow$ The factor directions $1, \ldots, q$ are $v_1, \ldots, v_q$, which denote the eigenvectors of $\mathcal{X}\mathcal{X}^\top$ corresponding to the q largest eigenvalues. The coordinates for representing the point cloud of variables on a q-dimensional subspace are given by $w_1 = \mathcal{X}^\top v_1, \ldots, w_q = \mathcal{X}^\top v_q$.

10.4   Relations Between Subspaces

The aim of this section is to present a duality relationship between the two approaches shown in Sects. 10.2 and 10.3. Consider the eigenvector equations in $\mathbb{R}^n$

$$(\mathcal{X}\mathcal{X}^\top)v_k = \mu_k v_k \tag{10.6}$$

for $k \leq r$, where $r = \text{rank}(\mathcal{X}\mathcal{X}^\top) = \text{rank}(\mathcal{X}) \leq \min(p, n)$. Multiplying by $\mathcal{X}^\top$, we have

$$\mathcal{X}^\top(\mathcal{X}\mathcal{X}^\top)v_k = \mu_k \mathcal{X}^\top v_k \tag{10.7}$$
$$\text{or} \quad (\mathcal{X}^\top\mathcal{X})(\mathcal{X}^\top v_k) = \mu_k(\mathcal{X}^\top v_k) \tag{10.8}$$

so that each eigenvector v_k of $\mathcal{X}\mathcal{X}^\top$ corresponds to an eigenvector $(\mathcal{X}^\top v_k)$ of $\mathcal{X}^\top\mathcal{X}$ associated with the same eigenvalue μ_k. This means that every nonzero eigenvalue of $\mathcal{X}\mathcal{X}^\top$ is an eigenvalue of $\mathcal{X}^\top\mathcal{X}$. The corresponding eigenvectors are related by

$$u_k = c_k \mathcal{X}^\top v_k,$$

where c_k is some constant.

Now consider the eigenvector equations in $\mathbb{R}^p$:

$$(\mathcal{X}^\top\mathcal{X})u_k = \lambda_k u_k \tag{10.9}$$

for $k \leq r$. Multiplying by $\mathcal{X}$, we have

$$(\mathcal{X}\mathcal{X}^\top)(\mathcal{X}u_k) = \lambda_k(\mathcal{X}u_k), \tag{10.10}$$

i.e., each eigenvector u_k of $\mathcal{X}^\top\mathcal{X}$ corresponds to an eigenvector $\mathcal{X}u_k$ of $\mathcal{X}\mathcal{X}^\top$ associated with the same eigenvalue λ_k. Therefore, every nonzero eigenvalue of $(\mathcal{X}^\top\mathcal{X})$ is an eigenvalue of $\mathcal{X}\mathcal{X}^\top$. The corresponding eigenvectors are related by

$$v_k = d_k \mathcal{X}u_k,$$

where d_k is some constant. Now, since $u_k^\top u_k = v_k^\top v_k = 1$ we have $c_k = d_k = \frac{1}{\sqrt{\lambda_k}}$. This leads to the following result:

Theorem 10.4 (Duality Relations) *Let r be the rank of $\mathcal{X}$. For $k \leq r$, the eigenvalues λ_k of $\mathcal{X}^\top\mathcal{X}$ and $\mathcal{X}\mathcal{X}^\top$ are the same and the eigenvectors (u_k and v_k, respectively) are related by*

$$u_k = \frac{1}{\sqrt{\lambda_k}} \mathcal{X}^\top v_k \qquad (10.11)$$

$$v_k = \frac{1}{\sqrt{\lambda_k}} \mathcal{X} u_k. \qquad (10.12)$$

Note that the projection of the p variables on the factorial axis v_k is given by

$$w_k = \mathcal{X}^\top v_k = \frac{1}{\sqrt{\lambda_k}} \mathcal{X}^\top \mathcal{X} u_k = \sqrt{\lambda_k}\, u_k. \qquad (10.13)$$

Therefore, the eigenvectors v_k do not have to be explicitly recomputed to get w_k.

Note that u_k and v_k provide the SVD of $\mathcal{X}$ (see Theorem 2.2). Letting $U = [u_1\ u_2\ \ldots\ u_r]$, $V = [v_1\ v_2\ \ldots\ v_r]$ and $\Lambda = \mathrm{diag}(\lambda_1, \ldots, \lambda_r)$ we have

$$\mathcal{X} = V\, \Lambda^{1/2}\, U^\top$$

so that

$$x_{ij} = \sum_{k=1}^{r} \lambda_k^{1/2}\, v_{ik}\, u_{jk}. \qquad (10.14)$$

In the following section, this method is applied in analyzing consumption behavior across different household types.

Summary

↪ The matrices $\mathcal{X}^\top \mathcal{X}$ and $\mathcal{X} \mathcal{X}^\top$ have the same nonzero eigenvalues $\lambda_1, \ldots, \lambda_r$, where $r = \mathrm{rank}(\mathcal{X})$.

↪ The eigenvectors of $\mathcal{X}^\top \mathcal{X}$ can be calculated from the eigenvectors of $\mathcal{X} \mathcal{X}^\top$ and vice versa:

$$u_k = \frac{1}{\sqrt{\lambda_k}} \mathcal{X}^\top v_k \quad \text{and} \quad v_k = \frac{1}{\sqrt{\lambda_k}} \mathcal{X} u_k.$$

↪ The coordinates representing the variables (columns) of $\mathcal{X}$ in a q-dimensional subspace can be easily calculated by $w_k = \sqrt{\lambda_k} u_k$.

10.5 Practical Computation

The practical implementation of the techniques introduced begins with the computation of the eigenvalues $\lambda_1 \geq \lambda_2 \geq \ldots \geq \lambda_p$ and the corresponding eigenvectors $u_1, \ldots, u_p$ of $\mathcal{X}^\top \mathcal{X}$. (Since p is usually less than n, this is numerically less involved than computing v_k directly for $k = 1, \ldots, p$). The representation of the n individuals on a plane is then obtained by plotting $z_1 = \mathcal{X} u_1$ versus $z_2 = \mathcal{X} u_2$ ($z_3 = \mathcal{X} u_3$

may eventually be added if a third dimension is helpful). Using the Duality Relation (10.13) representations for the p variables can easily be obtained. These representations can be visualized in a scatterplot of $w_1 = \sqrt{\lambda_1} u_1$ against $w_2 = \sqrt{\lambda_2} u_2$ (and eventually against $w_3 = \sqrt{\lambda_3} u_3$). Higher dimensional factorial resolutions can be obtained (by computing z_k and w_k for $k > 3$) but, of course, cannot be plotted.

A standard way of evaluating the quality of the factorial representations in a subspace of dimension q is given by the ratio

$$\tau_q = \frac{\lambda_1 + \lambda_2 + \ldots + \lambda_q}{\lambda_1 + \lambda_2 + \ldots + \lambda_p}, \tag{10.15}$$

where $0 \le \tau_q \le 1$. In general, the scalar product $y^\top y$ is called the inertia of $y \in \mathbb{R}^n$ with respect to the origin. Therefore, the ratio τ_q is usually interpreted as the percentage of the inertia explained by the first q factors. Note that $\lambda_j = (\mathcal{X} u_j)^\top (\mathcal{X} u_j) = z_j^\top z_j$. Thus, λ_j is the inertia of the j-th factorial variable with respect to the origin. The denominator in (10.15) is a measure of the total inertia of the p variables, $x_{[j]}$. Indeed, by (2.3)

$$\sum_{j=1}^{p} \lambda_j = \text{tr}(\mathcal{X}^\top \mathcal{X}) = \sum_{j=1}^{p} \sum_{i=1}^{n} x_{ij}^2 = \sum_{j=1}^{p} x_{[j]}^\top x_{[j]}.$$

Remark 10.1 It is clear that the sum $\sum_{j=1}^{q} \lambda_j$ is the sum of the inertia of the first q factorial variables $z_1, z_2, \ldots, z_q$.

Example 10.1 We consider the data set in Sect. B.6 which gives the food expenditures of various French families (manual workers = MA, employees = EM, managers = CA) with varying numbers of children (2, 3, 4, or 5 children). We are interested in investigating whether certain household types prefer certain food types. We can answer this question using the factorial approximations developed here.

The correlation matrix corresponding to the data is

$$\mathcal{R} = \begin{pmatrix} 1.00 & 0.59 & 0.20 & 0.32 & 0.25 & 0.86 & 0.30 \\ 0.59 & 1.00 & 0.86 & 0.88 & 0.83 & 0.66 & -0.36 \\ 0.20 & 0.86 & 1.00 & 0.96 & 0.93 & 0.33 & -0.49 \\ 0.32 & 0.88 & 0.96 & 1.00 & 0.98 & 0.37 & -0.44 \\ 0.25 & 0.83 & 0.93 & 0.98 & 1.00 & 0.23 & -0.40 \\ 0.86 & 0.66 & 0.33 & 0.37 & 0.23 & 1.00 & 0.01 \\ 0.30 & -0.36 & -0.49 & -0.44 & -0.40 & 0.01 & 1.00 \end{pmatrix}.$$

We observe a rather high correlation (0.98) between meat and poultry, whereas the correlation for expenditure for milk and wine (0.01) is rather small. Are there household types that prefer, say, meat over bread?

We shall now represent food expenditures and households simultaneously using two factors. First, note that in this particular problem the origin has no specific meaning (it represents a "zero" consumer). So it makes sense to compare the consumption of any family to that of an "average family" rather than to the origin. Therefore, the data is first centered (the origin is translated to the center of gravity, $\overline{x}$). Furthermore, since the dispersions of the seven variables are quite different each variable is standardized so that each has the same weight in the analysis (mean 0 and variance 1). Finally, for convenience, we divide each element in the matrix by $\sqrt{n} = \sqrt{12}$. (This will only change the scaling of the plots in the graphical representation.)

The data matrix to be analyzed is

$$\mathcal{X}_* = \frac{1}{\sqrt{n}}\mathcal{HXD}^{-1/2},$$

where $\mathcal{H}$ is the centering matrix and $\mathcal{D} = \text{diag}(s_{X_i X_i})$ (see Sect. 3.3). Note that by standardizing by $\sqrt{n}$, it follows that $\mathcal{X}_*^\top \mathcal{X}_* = \mathcal{R}$ where $\mathcal{R}$ is the correlation matrix of the original data. Calculating

$$\lambda = (4.33, 1.83, 0.63, 0.13, 0.06, 0.02, 0.00)^\top$$

shows that the directions of the first two eigenvectors play a dominant role ($\tau_2 = 88\%$), whereas the other directions contribute less than 15% of inertia. A two-dimensional plot should suffice for interpreting this data set.

The coordinates of the projected data points are given in the two lower windows of Fig. 10.6. Let us first examine the food expenditure window. In this window we see the representation of the $p = 7$ variables given by the first two factors. The plot shows the factorial variables w_1 and w_2 in the same fashion as Fig. 10.4. We see that the points for meat, poultry, vegetables, and fruits are close to each other in the lower left of the graph. The expenditures for bread and milk can be found in the upper left, whereas wine stands alone in the upper right. The first factor, w_1, may be interpreted as the meat/fruit factor of consumption, the second factor, w_2, as the bread/wine component.

In the lower window on the right-hand side, we show the factorial variables z_1 and z_2 from the fit of the $n = 12$ household types. Note that by the Duality Relations of Theorem 10.4, the factorial variables z_j are linear combinations of the factors w_k from the left window. The points displayed in the consumer window (graph on the right) are plotted relative to an average consumer represented by the origin. The manager families are located in the lower left corner of the graph whereas the manual workers and employees tend to be in the upper right. The factorial variables for CA5 (managers with five children) lie close to the meat/fruit factor. Relative to the average consumer this household type is a large consumer of meat/poultry and fruits/vegetables. In Chap. 11, we will return to these plots interpreting them in a much deeper way. At this stage, it suffices to notice that the plots provide a graphical representation in $\mathbb{R}^2$ of the information contained in the original, high-dimensional (12×7) data matrix.

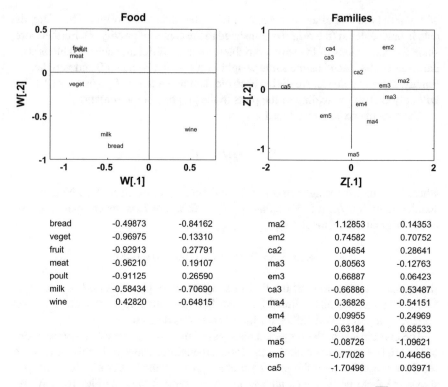

bread	-0.49873	-0.84162		ma2	1.12853	0.14353
veget	-0.96975	-0.13310		em2	0.74582	0.70752
fruit	-0.92913	0.27791		ca2	0.04654	0.28641
meat	-0.96210	0.19107		ma3	0.80563	-0.12763
poult	-0.91125	0.26590		em3	0.66887	0.06423
milk	-0.58434	-0.70690		ca3	-0.66886	0.53487
wine	0.42820	-0.64815		ma4	0.36826	-0.54151
				em4	0.09955	-0.24969
				ca4	-0.63184	0.68533
				ma5	-0.08726	-1.09621
				em5	-0.77026	-0.44656
				ca5	-1.70498	0.03971

Fig. 10.6 Representation of food expenditures and family types in two dimensions **Q** MVAdeco-food

Summary
$\hookrightarrow$ The practical implementation of factor decomposition of matrices consists of computing the eigenvalues $\lambda_1, \ldots, \lambda_p$ and the eigenvectors $u_1, \ldots, u_p$ of $\mathcal{X}^{\top}\mathcal{X}$. The representation of the n individuals is obtained by plotting $z_1 = \mathcal{X}u_1$ vs. $z_2 = \mathcal{X}u_2$ (and, if necessary, vs. $z_3 = \mathcal{X}u_3$). The representation of the p variables is obtained by plotting $w_1 = \sqrt{\lambda_1}u_1$ vs. $w_2 = \sqrt{\lambda_2}u_2$ (and, if necessary, vs. $w_3 = \sqrt{\lambda_3}u_3$).
$\hookrightarrow$ The quality of the factorial representation can be evaluated using τ_q which is the percentage of inertia explained by the first q factors.

10.6   Exercises

Exercise 10.1 Prove that $n^{-1}\mathcal{Z}^{\top}\mathcal{Z}$ is the covariance of the centered data matrix, where $\mathcal{Z}$ is the matrix formed by the columns $z_k = \mathcal{X}u_k$.

Exercise 10.2 Compute the SVD of the French food data (Sect. B.6).

Exercise 10.3 Compute τ_3, τ_4, ... for the French food data (Sect. B.6).

Exercise 10.4 Apply the factorial techniques to the Swiss bank notes (Sect. B.2).

Exercise 10.5 Apply the factorial techniques to the time budget data (Sect. B.12).

Exercise 10.6 Assume that you wish to analyze p independent identically distributed random variables. What is the percentage of the inertia explained by the first factor? What is the percentage of the inertia explained by the first q factors?

Exercise 10.7 Assume that you have p i.i.d. r.v.'s. What does the eigenvector corresponding to the first factor look like?

Exercise 10.8 Assume that you have two random variables, X_1 and $X_2 = 2X_1$. What do the eigenvalues and eigenvectors of their correlation matrix look like? How many eigenvalues are nonzero?

Exercise 10.9 What percentage of inertia is explained by the first factor in the previous exercise?

Exercise 10.10 How do the eigenvalues and eigenvectors in Example 10.1 change if we take the prices in USD instead of in EUR? Does it make a difference if some of the prices are in EUR and others in USD?

Chapter 11
Principal Components Analysis

Chapter 10 presented the basic geometric tools needed to produce a lower dimensional description of the rows and columns of a multivariate data matrix. Principal components analysis has the same objective with the exception that the rows of the data matrix $\mathcal{X}$ will now be considered as observations from a p-variate random variable X. The principle idea of reducing the dimension of X is achieved through linear combinations. Low-dimensional linear combinations are often easier to interpret and serve as an intermediate step in a more complex data analysis. More precisely, one looks for linear combinations which create the largest spread among the values of X. In other words, one is searching for linear combinations with the largest variances.

Section 11.1 introduces the basic ideas and technical elements behind principal components. No particular assumption will be made on X except that the mean vector and the covariance matrix exist. When reference is made to a data matrix $\mathcal{X}$ in Sect. 11.2, the empirical mean and covariance matrix will be used. Section 11.3 shows how to interpret the principal components by studying their correlations with the original components of X. Often analyses are performed in practice by looking at two-dimensional scatterplots. Section 11.4 develops inference techniques on principal components. This is particularly helpful in establishing the appropriate dimension reduction and thus in determining the quality of the resulting lower dimensional representations. Since principal component analysis is performed on covariance matrices, it is not scale invariant. Often, the measurement units of the components of X are quite different, so it is reasonable to standardize the measurement units. The normalized version of principal components is defined in Sect. 11.5. In Sect. 11.6, it is discovered that the empirical principal components are the factors of appropriate transformations of the data matrix. The classical way of defining principal components through linear combinations with respect to the largest variance is described here in geometric terms, i.e., in terms of the optimal fit within subspaces generated by the columns and/or the rows of $\mathcal{X}$ as was discussed in Chap. 10. Section 11.9 concludes with additional examples.

© Springer Nature Switzerland AG 2019
W. K. Härdle and L. Simar, *Applied Multivariate Statistical Analysis*,
https://doi.org/10.1007/978-3-030-26006-4_11

11.1   Standardized Linear Combination

The main objective of principal components analysis (PCA) is to reduce the dimension of the observations. The simplest way of dimension reduction is to take just one element of the observed vector and to discard all others. This is not a very reasonable approach, as we have seen in the earlier chapters, since strength may be lost in interpreting the data. In the bank notes example, we have seen that just one variable (e.g., $X_1 =$ length) had no discriminatory power in distinguishing counterfeit from genuine bank notes. An alternative method is to weight all variables equally, i.e., to consider the simple average $p^{-1} \sum_{j=1}^{p} X_j$ of all the elements in the vector $X = (X_1, \ldots, X_p)^\top$. This again is undesirable, since all of the elements of X are considered with equal importance (weight).

A more flexible approach is to study a weighted average, namely,

$$\delta^\top X = \sum_{j=1}^{p} \delta_j X_j, \quad \text{such that} \quad \sum_{j=1}^{p} \delta_j^2 = 1. \tag{11.1}$$

The weighting vector $\delta = (\delta_1, \ldots, \delta_p)^\top$ can then be optimized to investigate and to detect specific features. We call (11.1) a standardized linear combination (SLC). Which SLC should we choose? One aim is to maximize the variance of the projection $\delta^\top X$, i.e., to choose δ according to

$$\max_{\{\delta: \|\delta\|=1\}} \mathsf{Var}(\delta^\top X) = \max_{\{\delta: \|\delta\|=1\}} \delta^\top \mathsf{Var}(X)\delta. \tag{11.2}$$

The interesting "directions" of δ are found through the spectral decomposition of the covariance matrix. Indeed, from Theorem 2.5, the direction δ is given by the eigenvector γ_1 corresponding to the largest eigenvalue λ_1 of the covariance matrix $\Sigma = \mathsf{Var}(X)$.

Figures 11.1 and 11.2 show two such projections (SLCs) of the same data set with zero mean. In Fig. 11.1, an arbitrary projection is displayed. The upper window shows the data point cloud and the line onto which the data are projected. The middle window shows the projected values in the selected direction. The lower window shows the variance of the actual projection and the percentage of the total variance that is explained.

Figure 11.2 shows the projection that captures the majority of the variance in the data. This direction is of interest and is located along the main direction of the point cloud. The same line of thought can be applied to all data orthogonal to this direction leading to the second eigenvector. The SLC with the highest variance obtained from maximizing (11.2) is the first principal component (PC) $y_1 = \gamma_1^\top X$. Orthogonal to the direction γ_1, we find the SLC with the second highest variance: $y_2 = \gamma_2^\top X$, the second PC.

Proceeding in this way and writing in matrix notation, the result for a random variable X with $\mathsf{E}(X) = \mu$ and $\mathsf{Var}(X) = \Sigma = \Gamma \Lambda \Gamma^\top$ is the PC transformation which is defined as

Fig. 11.1 An arbitrary SLC
MVApcasimu

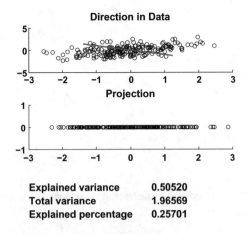

Fig. 11.2 The most
interesting SLC
MVApcasimu

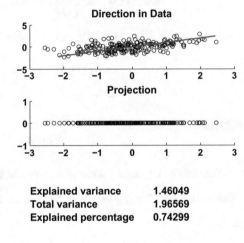

$$Y = \Gamma^\top (X - \mu). \tag{11.3}$$

Here we have centered the variable X in order to obtain a zero mean PC variable Y.

Example 11.1 Consider a bivariate normal distribution $N(0, \Sigma)$ with $\Sigma = \begin{pmatrix} 1 & \rho \\ \rho & 1 \end{pmatrix}$ and $\rho > 0$ (see Example 3.13). Recall that the eigenvalues of this matrix are $\lambda_1 = 1 + \rho$ and $\lambda_2 = 1 - \rho$ with corresponding eigenvectors

$$\gamma_1 = \frac{1}{\sqrt{2}} \begin{pmatrix} 1 \\ 1 \end{pmatrix}, \quad \gamma_2 = \frac{1}{\sqrt{2}} \begin{pmatrix} 1 \\ -1 \end{pmatrix}.$$

The PC transformation is thus

$$Y = \Gamma^\top (X - \mu) = \frac{1}{\sqrt{2}} \begin{pmatrix} 1 & 1 \\ 1 & -1 \end{pmatrix} X$$

or

$$\begin{pmatrix} Y_1 \\ Y_2 \end{pmatrix} = \frac{1}{\sqrt{2}} \begin{pmatrix} X_1 + X_2 \\ X_1 - X_2 \end{pmatrix}.$$

So the first principal component is

$$Y_1 = \frac{1}{\sqrt{2}}(X_1 + X_2)$$

and the second is

$$Y_2 = \frac{1}{\sqrt{2}}(X_1 - X_2).$$

Let us compute the variances of these PCs using formulas (4.22)–(4.26):

$$\begin{aligned}
\mathrm{Var}(Y_1) &= \mathrm{Var}\left\{\frac{1}{\sqrt{2}}(X_1 + X_2)\right\} = \frac{1}{2}\mathrm{Var}(X_1 + X_2) \\
&= \frac{1}{2}\{\mathrm{Var}(X_1) + \mathrm{Var}(X_2) + 2\mathrm{Cov}(X_1, X_2)\} \\
&= \frac{1}{2}(1 + 1 + 2\rho) = 1 + \rho \\
&= \lambda_1.
\end{aligned}$$

Similarly, we find that

$$\mathrm{Var}(Y_2) = \lambda_2.$$

This can be expressed more generally and is given in the next theorem.

Theorem 11.1 *For a given $X \sim (\mu, \Sigma)$, let $Y = \Gamma^\top(X - \mu)$ be the PC transformation. Then*

$$\mathrm{E}Y_j = 0, \quad j = 1, \ldots, p \tag{11.4}$$

$$\mathrm{Var}(Y_j) = \lambda_j, \qquad j = 1, \ldots, p \tag{11.5}$$

$$\mathrm{Cov}(Y_i, Y_j) = 0, \quad i \neq j \tag{11.6}$$

$$\mathrm{Var}(Y_1) \geq \mathrm{Var}(Y_2) \geq \cdots \geq \mathrm{Var}(Y_p) \geq 0 \tag{11.7}$$

$$\sum_{j=1}^{p} \mathrm{Var}(Y_j) = \mathrm{tr}(\Sigma) \tag{11.8}$$

$$\prod_{j=1}^{p} \mathrm{Var}(Y_j) = |\Sigma|. \tag{11.9}$$

Proof To prove (11.6), we use γ_i to denote the i-th column of Γ. Then

$$\mathrm{Cov}(Y_i, Y_j) = \gamma_i^\top \mathrm{Var}(X - \mu)\gamma_j = \gamma_i^\top \mathrm{Var}(X)\gamma_j.$$

As $\text{Var}(X) = \Sigma = \Gamma \Lambda \Gamma^\top$, $\Gamma^\top \Gamma = \mathcal{I}$, we obtain via the orthogonality of Γ:

$$\gamma_i^\top \Gamma \Lambda \Gamma^\top \gamma_j = \begin{cases} 0 & i \neq j, \\ \lambda_i & i = j. \end{cases}$$

In fact, as $Y_i = \gamma_i^\top (X - \mu)$ lies in the eigenvector space corresponding to γ_i, and eigenvector spaces corresponding to different eigenvalues are orthogonal to each other, we can directly see Y_i and Y_j are orthogonal to each other, so their covariance is 0.

The connection between the PC transformation and the search for the best SLC is made in the following theorem, which follows directly from (11.2) and Theorem 2.5.

Theorem 11.2 *There exists no SLC that has larger variance than $\lambda_1 = \text{Var}(Y_1)$.*

Theorem 11.3 *If $Y = a^\top X$ is an SLC that is not correlated with the first k PCs of X, then the variance of Y is maximized by choosing it to be the $(k+1)$-st PC.*

Summary
$\hookrightarrow$ A standardized linear combination (SLC) is a weighted average $\delta^\top X = \sum_{j=1}^{p} \delta_j X_j$ where δ is a vector of length 1.
$\hookrightarrow$ Maximizing the variance of $\delta^\top X$ leads to the choice $\delta = \gamma_1$, the eigenvector corresponding to the largest eigenvalue λ_1 of $\Sigma = \text{Var}(X)$. This is a projection of X into the one-dimensional space, where the components of X are weighted by the elements of γ_1. $Y_1 = \gamma_1^\top (X - \mu)$ is called the first principal component (PC).
$\hookrightarrow$ This projection can be generalized for higher dimensions. The PC transformation is the linear transformation $Y = \Gamma^\top (X - \mu)$, where $\Sigma = \text{Var}(X) = \Gamma \Lambda \Gamma^\top$ and $\mu = \mathbb{E}X$. $Y_1, Y_2, \ldots, Y_p$ are called the first, second,..., and p-th PCs.
$\hookrightarrow$ The PCs have zero means, variance $\text{Var}(Y_j) = \lambda_j$, and zero covariances. From $\lambda_1 \geq \ldots \geq \lambda_p$ it follows that $\text{Var}(Y_1) \geq \ldots \geq \text{Var}(Y_p)$. It holds that $\sum_{j=1}^{p} \text{Var}(Y_j) = \text{tr}(\Sigma)$ and $\prod_{j=1}^{p} \text{Var}(Y_j) = |\Sigma|$.
$\hookrightarrow$ If $Y = a^\top X$ is an SLC which is not correlated with the first k PCs of X then the variance of Y is maximized by choosing it to be the $(k+1)$-st PC.

11.2 Principal Components in Practice

In practice, the PC transformation has to be replaced by the respective estimators: μ becomes $\overline{x}$, Σ is replaced by $\mathcal{S}$, etc. If g_1 denotes the first eigenvector of $\mathcal{S}$, the first principal component is given by $y_1 = (\mathcal{X} - 1_n \overline{x}^\top) g_1$. More generally if $\mathcal{S} = \mathcal{G} \mathcal{L} \mathcal{G}^\top$

is the spectral decomposition of $\mathcal{S}$, then the PCs are obtained by

$$\mathcal{Y} = (\mathcal{X} - 1_n \bar{x}^\top)\mathcal{G}. \qquad (11.10)$$

Note that with the centering matrix $\mathcal{H} = \mathcal{I} - (n^{-1}1_n 1_n^\top)$ and $\mathcal{H}1_n \bar{x}^\top = 0$, we can write

$$\begin{aligned}
\mathcal{S}_y &= n^{-1}\mathcal{Y}^\top \mathcal{H}\mathcal{Y} = n^{-1}\mathcal{G}^\top (\mathcal{X} - 1_n \bar{x}^\top)^\top \mathcal{H}(\mathcal{X} - 1_n \bar{x}^\top)\mathcal{G} \\
&= n^{-1}\mathcal{G}^\top \mathcal{X}^\top \mathcal{H}\mathcal{X}\mathcal{G} = \mathcal{G}^\top \mathcal{S}\mathcal{G} = \mathcal{L}
\end{aligned} \qquad (11.11)$$

where $\mathcal{L} = \mathrm{diag}(\ell_1, \ldots, \ell_p)$ is the matrix of eigenvalues of $\mathcal{S}$. Hence, the variance of y_i equals the eigenvalue ℓ_i!

The PC technique is sensitive to scale changes. If we multiply one variable by a scalar we obtain different eigenvalues and eigenvectors. This is due to the fact that an eigenvalue decomposition is performed on the covariance matrix and not on the correlation matrix (see Sect. 11.5). The following warning is therefore important:

⚠ The PC transformation should be applied to data that have approximately the same scale in each variable.

Example 11.2 Let us apply this technique to the bank data set. In this example, we do not standardize the data. Figure 11.3 shows some PC plots of the bank data set. The genuine and counterfeit bank notes are marked by "o" and "+", respectively.

Recall that the mean vector of $\mathcal{X}$ is

$$\bar{x} = (214.9, 130.1, 129.9, 9.4, 10.6, 140.5)^\top.$$

The vector of eigenvalues of $\mathcal{S}$ is

$$\ell = (2.985, 0.931, 0.242, 0.194, 0.085, 0.035)^\top.$$

The eigenvectors g_j are given by the columns of the matrix

$$\mathcal{G} = \begin{pmatrix}
-0.044 & 0.011 & 0.326 & 0.562 & -0.753 & 0.098 \\
0.112 & 0.071 & 0.259 & 0.455 & 0.347 & -0.767 \\
0.139 & 0.066 & 0.345 & 0.415 & 0.535 & 0.632 \\
0.768 & -0.563 & 0.218 & -0.186 & -0.100 & -0.022 \\
0.202 & 0.659 & 0.557 & -0.451 & -0.102 & -0.035 \\
-0.579 & -0.489 & 0.592 & -0.258 & 0.085 & -0.046
\end{pmatrix}.$$

The first column of $\mathcal{G}$ is the first eigenvector and gives the weights used in the linear combination of the original data in the first PC.

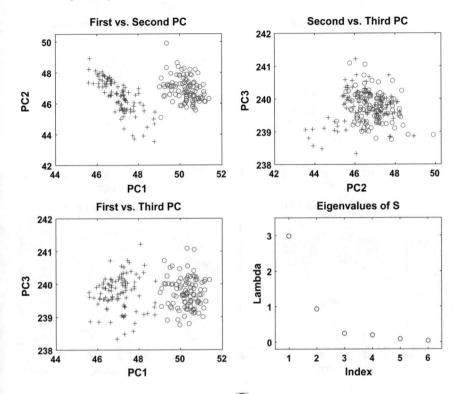

Fig. 11.3 Principal components of the bank data ⊙ MVApcabank

Example 11.3 To see how sensitive the PCs are to a change in the scale of the variables, assume that X_1, X_2, X_3 and X_6 are measured in *cm* and that X_4 and X_5 remain in *mm* in the bank data set. This leads to

$$\bar{x} = (21.49,\ 13.01,\ 12.99,\ 9.41,\ 10.65,\ 14.05)^{\top}.$$

The covariance matrix can be obtained from S in (3.4) by dividing rows 1, 2, 3, 6 and columns 1, 2, 3, 6 by 10. We obtain

$$\ell = (2.101,\ 0.623,\ 0.005,\ 0.002,\ 0.001,\ 0.0004)^{\top}$$

which clearly differs from Example 11.2. Only the first two eigenvectors are given:

$$g_1 = (-0.005,\ 0.011,\ 0.014,\ 0.992,\ 0.113,\ -0.052)^{\top}$$

$$g_2 = (-0.001,\ 0.013,\ 0.016,\ -0.117,\ 0.991,\ -0.069)^{\top}.$$

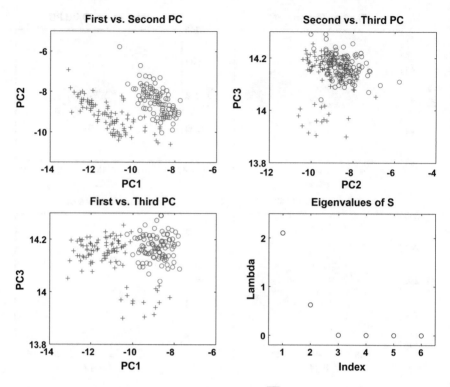

Fig. 11.4 Principal components of the rescaled bank data ⊙MVApcabankr

Comparing these results to the first two columns of $\mathcal{G}$ from Example 11.2, a completely different story is revealed. Here the first component is dominated by X_4 (lower margin) and the second by X_5 (upper margin), while all of the other variables have much less weight. The results are shown in Fig. 11.4. Section 11.5 will show how to select a reasonable standardization of the variables when the scales are too different.

Summary
↪   The scale of the variables should be roughly the same for PC transformations.
↪   For the practical implementation of principal components analysis (PCA) we replace μ by the mean $\bar{x}$ and Σ by the empirical covariance S. Then we compute the eigenvalues $\ell_1, \ldots, \ell_p$ and the eigenvectors $g_1, \ldots, g_p$ of S. The graphical representation of the PCs is obtained by plotting the first PC versus the second (and eventually vs. the third).
↪   The components of the eigenvectors g_i are the weights of the original variables in the PCs.

11.3 Interpretation of the PCs

Recall that the main idea of PC transformations is to find the most informative projections that maximize variances. The most informative SLC is given by the first eigenvector. In Sect. 11.2, the eigenvectors were calculated for the bank data. In particular, with centered x's, we had

$$y_1 = -0.044x_1 + 0.112x_2 + 0.139x_3 + 0.768x_4 + 0.202x_5 - 0.579x_6$$
$$y_2 = 0.011x_1 + 0.071x_2 + 0.066x_3 - 0.563x_4 + 0.659x_5 - 0.489x_6$$

and

$$x_1 = \text{length}$$
$$x_2 = \text{left height}$$
$$x_3 = \text{right height}$$
$$x_4 = \text{bottom frame}$$
$$x_5 = \text{top frame}$$
$$x_6 = \text{diagonal.}$$

Hence, the first PC is essentially the difference between the bottom frame variable and the diagonal. The second PC is best described by the difference between the top frame variable and the sum of bottom frame and diagonal variables.

The weighting of the PCs tells us in which directions, expressed in original coordinates, the best variance explanation is obtained. A measure of how well the first q PCs explain variation is given by the relative proportion:

$$\psi_q = \frac{\sum_{j=1}^{q} \lambda_j}{\sum_{j=1}^{p} \lambda_j} = \frac{\sum_{j=1}^{q} \text{Var}(Y_j)}{\sum_{j=1}^{p} \text{Var}(Y_j)}. \tag{11.12}$$

Referring to the bank data Example 11.2, the (cumulative) proportions of explained variance are given in Table 11.1. The first PC ($q = 1$) already explains 67% of the variation. The first three ($q = 3$) PCs explain 93% of the variation. Once again it should be noted that PCs are not scale invariant, e.g., the PCs derived from the correlation matrix give different results than the PCs derived from the covariance matrix (see Sect. 11.5).

A good graphical representation of the ability of the PCs to explain the variation in the data is given by the scree plot shown in the lower right-hand window of Fig. 11.3. The scree plot can be modified by using the relative proportions on the y-axis, as is shown in Fig. 11.5 for the bank data set.

Table 11.1 Proportion of variance of PC's

Eigenvalue	Proportion of variance	Cumulated proportion
2.985	0.67	0.67
0.931	0.21	0.88
0.242	0.05	0.93
0.194	0.04	0.97
0.085	0.02	0.99
0.035	0.01	1.00

Fig. 11.5 Relative proportion of variance explained by PCs

MVApcabanki

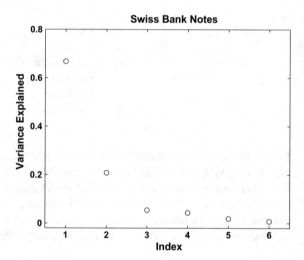

The covariance between the PC vector Y and the original vector X is calculated with the help of (11.4) as follows:

$$\begin{aligned}
\text{Cov}(X, Y) = \mathsf{E}(XY^\top) - \mathsf{E}X\mathsf{E}Y^\top &= \mathsf{E}(XY^\top) \\
&= \mathsf{E}(XX^\top \Gamma) - \mu\mu^\top \Gamma = \text{Var}(X)\Gamma \\
&= \Sigma\Gamma \\
&= \Gamma\Lambda\Gamma^\top\Gamma \\
&= \Gamma\Lambda.
\end{aligned} \tag{11.13}$$

Hence, the correlation, $\rho_{X_iY_j}$, between variable X_i and the PC Y_j is

$$\rho_{X_iY_j} = \frac{\gamma_{ij}\lambda_j}{(\sigma_{X_iX_i}\lambda_j)^{1/2}} = \gamma_{ij}\left(\frac{\lambda_j}{\sigma_{X_iX_i}}\right)^{1/2}. \tag{11.14}$$

Using actual data, this of course translates into

Fig. 11.6 The correlation of the original variable with the PCs 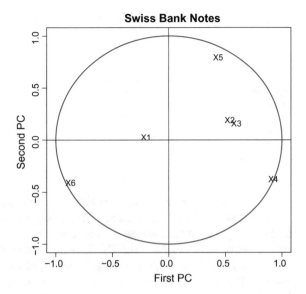MVApcabanki

$$r_{X_i Y_j} = g_{ij} \left(\frac{\ell_j}{s_{X_i X_i}} \right)^{1/2}. \tag{11.15}$$

The correlations can be used to evaluate the relations between the PCs Y_j where $j = 1, \ldots, q$, and the original variables X_i where $i = 1, \ldots, p$. Note that

$$\sum_{j=1}^{p} r_{X_i Y_j}^2 = \frac{\sum_{j=1}^{p} \ell_j g_{ij}^2}{s_{X_i X_i}} = \frac{s_{X_i X_i}}{s_{X_i X_i}} = 1. \tag{11.16}$$

Indeed, $\sum_{j=1}^{p} \ell_j g_{ij}^2 = g_i^\top \mathcal{L} g_i$ is the (i, i)-element of the matrix $\mathcal{G} \mathcal{L} \mathcal{G}^\top = \mathcal{S}$, so that $r_{X_i Y_j}^2$ may be seen as the proportion of variance of X_i explained by Y_j.

In the space of the first two PCs, we plot these proportions, i.e., $r_{X_i Y_1}$ versus $r_{X_i Y_2}$. Figure 11.6 shows this for the bank notes example. This plot shows which of the original variables are most strongly correlated with PC Y_1 and Y_2.

From (11.16), it obviously follows that $r_{X_i Y_1}^2 + r_{X_i Y_2}^2 \leq 1$ so that the points are always inside the circle of radius 1. In the bank notes example, the variables X_4, X_5 and X_6 correspond to correlations near the periphery of the circle and are thus well explained by the first two PCs. Recall that we have interpreted the first PC as being essentially the difference between X_4 and X_6. This is also reflected in Fig. 11.6 since the points corresponding to these variables lie on different sides of the vertical axis. An analogous remark applies to the second PC. We had seen that the second PC is well described by the difference between X_5 and the sum of X_4 and X_6. Now we are able to see this result again from Fig. 11.6 since the point corresponding to X_5 lies above the horizontal axis and the points corresponding to X_4 and X_6 lie below.

Table 11.2 Correlation between the original variables and the PCs

	$r_{X_i Y_1}$	$r_{X_i Y_2}$	$r^2_{X_i Y_1} + r^2_{X_i Y_2}$
X_1 length	-0.201	0.028	0.041
X_2 left h.	0.538	0.191	0.326
X_3 right h.	0.597	0.159	0.381
X_4 lower	0.921	-0.377	0.991
X_5 upper	0.435	0.794	0.820
X_6 diagonal	-0.870	-0.410	0.926

The correlations of the original variables X_i and the first two PCs are given in Table 11.2 along with the cumulated percentage of variance of each variable explained by Y_1 and Y_2. This table confirms the above results. In particular, it confirms that the percentage of variance of X_1 (and X_2, X_3) explained by the first two PCs is relatively small and so are their weights in the graphical representation of the individual bank notes in the space of the first two PCs (as can be seen in the upper left plot in Fig. 11.3). Looking simultaneously at Fig. 11.6 and the upper left plot of Fig. 11.3 shows that the genuine bank notes are roughly characterized by large values of X_6 and smaller values of X_4. The counterfeit bank notes show larger values of X_5 (see Example 7.15).

Summary

$\hookrightarrow$ The weighting of the PCs tells us in which directions, expressed in original coordinates, the best explanation of the variance is obtained. Note that the PCs are not scale invariant.

$\hookrightarrow$ A measure of how well the first q PCs explain variation is given by the relative proportion $\psi_q = \sum_{j=1}^{q} \lambda_j / \sum_{j=1}^{p} \lambda_j$. A good graphical representation of the ability of the PCs to explain the variation in the data is the scree plot of these proportions.

$\hookrightarrow$ The correlation between PC Y_j and an original variable X_i is $\rho_{X_i Y_j} = \gamma_{ij} \left(\frac{\lambda_j}{\sigma_{X_i X_i}} \right)^{1/2}$. For a data matrix this translates into $r^2_{X_i Y_j} = \frac{\ell_j g_{ij}^2}{s_{X_i X_i}}$. $r^2_{X_i Y_j}$ can be interpreted as the proportion of variance of X_i explained by Y_j. A plot of $r_{X_i Y_1}$ versus $r_{X_i Y_2}$ shows which of the original variables are most strongly correlated with the PCs, namely, those that are close to the periphery of the circle of radius 1.

11.4 Asymptotic Properties of the PCs

In practice, PCs are computed from sample data. The following theorem yields results on the asymptotic distribution of the sample PCs.

Theorem 11.4 *Let $\Sigma > 0$ with distinct eigenvalues, and let $S \sim n^{-1}W_p(\Sigma, n-1)$ with spectral decompositions $\Sigma = \Gamma \Lambda \Gamma^\top$, and $S = \mathcal{G}\mathcal{L}\mathcal{G}^\top$. Then*

(a) $\sqrt{n-1}(\ell - \lambda) \xrightarrow{\mathcal{L}} N_p(0, 2\Lambda^2)$,
  where $\ell = (\ell_1, \ldots, \ell_p)^\top$ and $\lambda = (\lambda_1, \ldots, \lambda_p)^\top$ are the diagonals of $\mathcal{L}$ and Λ,
(b) $\sqrt{n-1}(g_j - \gamma_j) \xrightarrow{\mathcal{L}} N_p(0, \mathcal{V}_j)$,
  with $\mathcal{V}_j = \lambda_j \sum_{k \neq j} \dfrac{\lambda_k}{(\lambda_k - \lambda_j)^2} \gamma_k \gamma_k^\top$,
(c) the elements in ℓ are asymptotically independent of the elements in $\mathcal{G}$.

Example 11.4 Since $nS \sim W_p(\Sigma, n-1)$ if $X_1, \ldots, X_n$ are drawn from $N(\mu, \Sigma)$, we have that

$$\sqrt{n-1}(\ell_j - \lambda_j) \xrightarrow{\mathcal{L}} N(0, 2\lambda_j^2), \quad j = 1, \ldots, p. \tag{11.17}$$

Since the variance of (11.17) depends on the true mean λ_j, a log transformation is useful. Consider $f(\ell_j) = \log(\ell_j)$. Then $\frac{d}{d\ell_j}f|_{\ell_j = \lambda_j} = \frac{1}{\lambda_j}$ and by the Transformation Theorem 4.11 we have from (11.17) that

$$\sqrt{n-1}(\log \ell_j - \log \lambda_j) \xrightarrow{\mathcal{L}} N(0, 2). \tag{11.18}$$

Hence,

$$\sqrt{\frac{n-1}{2}} (\log \ell_j - \log \lambda_j) \xrightarrow{\mathcal{L}} N(0, 1)$$

and a two-sided confidence interval at the $1 - \alpha = 0.95$ significance level is given by

$$\log(\ell_j) - 1.96\sqrt{\frac{2}{n-1}} \leq \log \lambda_j \leq \log(\ell_j) + 1.96\sqrt{\frac{2}{n-1}}.$$

In the bank data example, we have that

$$\ell_1 = 2.98.$$

Therefore,

$$\log(2.98) \pm 1.96\sqrt{\frac{2}{199}} = \log(2.98) \pm 0.1965.$$

It can be concluded for the true eigenvalue that

$$P\{\lambda_1 \in (2.448, 3.62)\} \approx 0.95.$$

Variance Explained by the First q PCs.

The variance explained by the first q PCs is given by

$$\psi = \frac{\lambda_1 + \cdots + \lambda_q}{\sum_{j=1}^{p} \lambda_j}.$$

In practice, this is estimated by

$$\widehat{\psi} = \frac{\ell_1 + \cdots + \ell_q}{\sum_{j=1}^{p} \ell_j}.$$

From Theorem 11.4, we know the distribution of $\sqrt{n-1}(\ell - \lambda)$. Since ψ is a non-linear function of λ, we can again apply the Transformation Theorem 4.11 to obtain that

$$\sqrt{n-1}(\widehat{\psi} - \psi) \xrightarrow{\mathcal{L}} N(0, \mathcal{D}^\top \mathcal{V} \mathcal{D})$$

where $\mathcal{V} = 2\Lambda^2$ (from Theorem 11.4) and $\mathcal{D} = (d_1, \ldots, d_p)^\top$ with

$$d_j = \frac{\partial \psi}{\partial \lambda_j} = \begin{cases} \dfrac{1 - \psi}{\text{tr}(\Sigma)} & \text{for } 1 \le j \le q, \\ \dfrac{-\psi}{\text{tr}(\Sigma)} & \text{for } q + 1 \le j \le p. \end{cases}$$

Given this result, the following theorem can be derived.

Theorem 11.5

$$\sqrt{n-1}(\widehat{\psi} - \psi) \xrightarrow{\mathcal{L}} N(0, \omega^2),$$

where

$$\omega^2 = \mathcal{D}^\top \mathcal{V} \mathcal{D} = \frac{2}{\{\text{tr}(\Sigma)\}^2} \left\{ (1 - \psi)^2 (\lambda_1^2 + \cdots + \lambda_q^2) + \psi^2 (\lambda_{q+1}^2 + \cdots + \lambda_p^2) \right\}$$

$$= \frac{2\,\text{tr}(\Sigma^2)}{\{\text{tr}(\Sigma)\}^2} (\psi^2 - 2\beta\psi + \beta)$$

and

$$\beta = \frac{\lambda_1^2 + \cdots + \lambda_q^2}{\lambda_1^2 + \cdots + \lambda_p^2}.$$

Example 11.5 From Sect. 11.3, it is known that the first PC for the Swiss bank notes resolves 67% of the variation. It can be tested whether the true proportion is actually 75%. Computing

$$\widehat{\beta} = \frac{\ell_1^2}{\ell_1^2 + \cdots + \ell_p^2} = \frac{(2.985)^2}{(2.985)^2 + (0.931)^2 + \cdots (0.035)^2} = 0.902$$

$$\text{tr}(\mathcal{S}) = 4.472$$

$$\text{tr}(\mathcal{S}^2) = \sum_{j=1}^{p} \ell_j^2 = 9.883$$

$$\widehat{\omega}^2 = \frac{2\,\text{tr}(\mathcal{S}^2)}{\{\text{tr}(\mathcal{S})\}^2}(\widehat{\psi}^2 - 2\widehat{\beta\psi} + \widehat{\beta})$$

$$= \frac{2 \cdot 9.883}{(4.472)^2}\{(0.668)^2 - 2(0.902)(0.668) + 0.902\} = 0.142.$$

Hence, a confidence interval at a significance of level $1 - \alpha = 0.95$ is given by

$$0.668 \pm 1.96\sqrt{\frac{0.142}{199}} = (0.615, 0.720).$$

Clearly, the hypothesis that $\psi = 75\%$ can be rejected!

Summary
↪ The eigenvalues ℓ_j and eigenvectors g_j are asymptotically, normally distributed, in particular $\sqrt{n-1}(\ell - \lambda) \overset{\mathcal{L}}{\longrightarrow} N_p(0, 2\Lambda^2)$.
↪ For the eigenvalues it holds that $\sqrt{\frac{n-1}{2}}\left(\log \ell_j - \log \lambda_j\right) \overset{\mathcal{L}}{\longrightarrow} N(0, 1)$.
↪ Given an asymptotic, normal distribution approximate confidence intervals and tests can be constructed for the proportion of variance which is explained by the first q PCs. The two-sided confidence interval at the $1 - \alpha = 0.95$ level is given by $\log(\ell_j) - 1.96\sqrt{\frac{2}{n-1}} \le \log \lambda_j \le \log(\ell_j) + 1.96\sqrt{\frac{2}{n-1}}$.
↪ It holds for $\widehat{\psi}$, the estimate of ψ (the proportion of the variance explained by the first q PCs) that $\sqrt{n-1}(\widehat{\psi} - \psi) \overset{\mathcal{L}}{\longrightarrow} N(0, \omega^2)$, where ω is given in Theorem 11.5.

11.5 Normalized Principal Components Analysis

In certain situations, the original variables can be heterogeneous w.r.t. their variances. This is particularly true when the variables are measured on heterogeneous scales (such as years, kilograms, dollars, etc.). In this case, a description of the information contained in the data needs to be provided which is robust w.r.t. the choice of scale. This can be achieved through a standardization of the variables, namely,

$$\mathcal{X}_S = \mathcal{H}\mathcal{X}\mathcal{D}^{-1/2} \tag{11.19}$$

where $\mathcal{D} = \text{diag}(s_{X_1 X_1}, \ldots, s_{X_p X_p})$. Note that $\overline{x}_S = 0$ and $\mathcal{S}_{\mathcal{X}_S} = \mathcal{R}$, the correlation matrix of $\mathcal{X}$. The PC transformations of the matrix $\mathcal{X}_S$ are referred to as the *Normalized Principal Components* (NPCs). The spectral decomposition of $\mathcal{R}$ is

$$\mathcal{R} = \mathcal{G}_{\mathcal{R}} \mathcal{L}_{\mathcal{R}} \mathcal{G}_{\mathcal{R}}^{\top}, \tag{11.20}$$

where $\mathcal{L}_{\mathcal{R}} = \text{diag}(\ell_1^{\mathcal{R}}, \ldots, \ell_p^{\mathcal{R}})$ and $\ell_1^{\mathcal{R}} \geq \ldots \geq \ell_p^{\mathcal{R}}$ are the eigenvalues of $\mathcal{R}$ with corresponding eigenvectors $g_1^{\mathcal{R}}, \ldots, g_p^{\mathcal{R}}$ (note that here $\sum_{j=1}^{p} \ell_j^{\mathcal{R}} = \text{tr}(\mathcal{R}) = p$).

The NPCs, Z_j, provide a representation of each individual and are given by

$$\mathcal{Z} = \mathcal{X}_S \mathcal{G}_{\mathcal{R}} = (z_1, \ldots, z_p). \tag{11.21}$$

After transforming the variables, once again, we have that

$$\overline{z} = 0, \tag{11.22}$$
$$\mathcal{S}_{\mathcal{Z}} = \mathcal{G}_{\mathcal{R}}^{\top} \mathcal{S}_{\mathcal{X}_S} \mathcal{G}_{\mathcal{R}} = \mathcal{G}_{\mathcal{R}}^{\top} \mathcal{R} \mathcal{G}_{\mathcal{R}} = \mathcal{L}_{\mathcal{R}}. \tag{11.23}$$

⚠️ The NPCs provide a perspective similar to that of the PCs, but in terms of the relative position of individuals, NPC gives each variable the same weight (with the PCs the variable with the largest variance received the largest weight).

Computing the covariance and correlation between X_i and Z_j is straightforward:

$$\mathcal{S}_{\mathcal{X}_S, \mathcal{Z}} = \frac{1}{n} \mathcal{X}_S^{\top} \mathcal{Z} = \mathcal{G}_{\mathcal{R}} \mathcal{L}_{\mathcal{R}}, \tag{11.24}$$

$$\mathcal{R}_{\mathcal{X}_S, \mathcal{Z}} = \mathcal{G}_{\mathcal{R}} \mathcal{L}_{\mathcal{R}} \mathcal{L}_{\mathcal{R}}^{-1/2} = \mathcal{G}_{\mathcal{R}} \mathcal{L}_{\mathcal{R}}^{1/2}. \tag{11.25}$$

The correlations between the original variables X_i and the NPCs Z_j are

$$r_{X_i Z_j} = \sqrt{\ell_j} g_{R,ij} \tag{11.26}$$

$$\sum_{j=1}^{p} r_{X_i Z_j}^2 = 1 \tag{11.27}$$

(compare this to (11.15) and (11.16)). The resulting NPCs, the Z_j, can be interpreted in terms of the original variables, and the role of each PC in explaining the variation in variable X_i can be evaluated.

11.6 Principal Components as a Factorial Method

The empirical PCs (normalized or not) turn out to be equivalent to the factors that one would obtain by decomposing the appropriate data matrix into its factors (see Chap. 10). It will be shown that the PCs are the factors representing the rows of the centered data matrix and that the NPCs correspond to the factors of the standardized data matrix. The representation of the columns of the standardized data matrix provides (at a scale factor) the correlations between the NPCs and the original variables. The derivation of the (N)PCs presented above will have a nice geometric justification here since they are the best fit in subspaces generated by the columns of the (transformed) data matrix $\mathcal{X}$. This analogy provides complementary interpretations of the graphical representations shown above.

Assume, as in Chap. 10, that we want to obtain representations of the individuals (the rows of $\mathcal{X}$) and of the variables (the columns of $\mathcal{X}$) in spaces of smaller dimension. To keep the representations simple, some prior transformations are performed. Since the origin has no particular statistical meaning in the space of individuals, we will first shift the origin to the center of gravity, $\bar{x}$, of the point cloud. This is the same as analyzing the centered data matrix $\mathcal{X}_C = \mathcal{H}\mathcal{X}$. Now all of the variables have zero means; thus, the technique used in Chap. 10 can be applied to the matrix $\mathcal{X}_C$. Note that the spectral decomposition of $\mathcal{X}_C^\top \mathcal{X}_C$ is related to that of $\mathcal{S}_X$, namely,

$$\mathcal{X}_C^\top \mathcal{X}_C = \mathcal{X}^\top \mathcal{H}^\top \mathcal{H} \mathcal{X} = n\mathcal{S}_X = n\mathcal{G}\mathcal{L}\mathcal{G}^\top. \tag{11.28}$$

The factorial variables are obtained by projecting $\mathcal{X}_C$ on $\mathcal{G}$,

$$\mathcal{Y} = \mathcal{X}_C \mathcal{G} = (y_1, \ldots, y_p). \tag{11.29}$$

These are the same principal components obtained above, see formula (11.10). (Note that the y's here correspond to the z's in Sect. 10.2.) Since $\mathcal{H}\mathcal{X}_C = \mathcal{X}_C$, it immediately follows that

$$\bar{y} = 0, \tag{11.30}$$
$$\mathcal{S}_Y = \mathcal{G}^\top \mathcal{S}_X \mathcal{G} = \mathcal{L} = \operatorname{diag}(\ell_1, \ldots, \ell_p). \tag{11.31}$$

The scatterplot of the individuals on the factorial axes are thus centered around the origin and are more spread out in the first direction (first PC has variance ℓ_1) than in the second direction (second PC has variance ℓ_2).

The representation of the variables can be obtained using the Duality Relations (10.11), and (10.12). The projections of the columns of $\mathcal{X}_C$ onto the eigenvectors v_k of $\mathcal{X}_C \mathcal{X}_C^\top$ are

$$\mathcal{X}_C^\top v_k = \frac{1}{\sqrt{n\ell_k}} \mathcal{X}_C^\top \mathcal{X}_C g_k = \sqrt{n\ell_k} g_k. \tag{11.32}$$

Thus, the projections of the variables on the first p axes are the columns of the matrix

$$\mathcal{X}_C^\top \mathcal{V} = \sqrt{n} \mathcal{G} \mathcal{L}^{1/2}. \tag{11.33}$$

Considering the geometric representation, there is a nice statistical interpretation of the angle between two columns of $\mathcal{X}_C$. Given that

$$x_{C[j]}^\top x_{C[k]} = n s_{X_j X_k}, \tag{11.34}$$

$$\|x_{C[j]}\|^2 = n s_{X_j X_j}, \tag{11.35}$$

where $x_{C[j]}$ and $x_{C[k]}$ denote the j-th and k-th column of $\mathcal{X}_C$, it holds that in the full space of the variables, if θ_{jk} is the angle between two variables, $x_{C[j]}$ and $x_{C[k]}$, then

$$\cos \theta_{jk} = \frac{x_{C[j]}^\top x_{C[k]}}{\|x_{C[j]}\| \, \|x_{C[k]}\|} = r_{X_j X_k}. \tag{11.36}$$

(Example 2.11 shows the general connection that exists between the angle and correlation of two variables). As a result, the relative positions of the variables in the scatterplot of the first columns of $\mathcal{X}_C^\top \mathcal{V}$ may be interpreted in terms of their correlations; the plot provides a picture of the correlation structure of the original data set. Clearly, one should take into account the percentage of variance explained by the chosen axes when evaluating the correlation.

The NPCs can also be viewed as a factorial method for reducing the dimension. The variables are again standardized so that each one has mean zero and unit variance and is independent of the scale of the variables. The factorial analysis of $\mathcal{X}_S$ provides the NPCs. The spectral decomposition of $\mathcal{X}_S^\top \mathcal{X}_S$ is related to that of $\mathcal{R}$, namely,

$$\mathcal{X}_S^\top \mathcal{X}_S = \mathcal{D}^{-1/2} \mathcal{X}^\top \mathcal{H} \mathcal{X} \mathcal{D}^{-1/2} = n\mathcal{R} = n\mathcal{G}_\mathcal{R} \mathcal{L}_\mathcal{R} \mathcal{G}_\mathcal{R}^\top.$$

The NPCs Z_j, given by (11.21), may be viewed as the projections of the rows of $\mathcal{X}_S$ onto $\mathcal{G}_R$.

The representation of the variables is again given by the columns of

$$\mathcal{X}_S^\top \mathcal{V}_\mathcal{R} = \sqrt{n} \mathcal{G}_\mathcal{R} \mathcal{L}_\mathcal{R}^{1/2}. \tag{11.37}$$

Comparing (11.37) and (11.25) we see that the projections of the variables in the factorial analysis provide the correlation between the NPCs $\mathcal{Z}_k$ and the original variables $x_{[j]}$ (up to the factor $\sqrt{n}$ which could be the scale of the axes).

This implies that a deeper interpretation of the representation of the individuals can be obtained by looking simultaneously at the graphs plotting the variables. Note that

$$x_{S[j]}^\top x_{S[k]} = n r_{X_j X_k}, \tag{11.38}$$

$$\|x_{S[j]}\|^2 = n, \tag{11.39}$$

where $x_{S[j]}$ and $x_{S[k]}$ denote the j-th and k-th column of $\mathcal{X}_S$. Hence, in the full space, all the standardized variables (columns of $\mathcal{X}_S$) are contained within the "sphere" in $\mathbb{R}^n$, which is centered at the origin and has radius $\sqrt{n}$ (the scale of the graph). As in (11.36), given the angle θ_{jk} between two columns $x_{S[j]}$ and $x_{S[k]}$, it holds that

$$\cos \theta_{jk} = r_{X_j X_k}. \qquad (11.40)$$

Therefore, when looking at the representation of the variables in the spaces of reduced dimension (for instance, the first two factors), we have a picture of the correlation structure between the original X_i's in terms of their angles. Of course, the quality of the representation in those subspaces has to be taken into account, which is presented in the next section.

Quality of the Representations

As said before, an overall measure of the quality of the representation is given by

$$\psi = \frac{\ell_1 + \ell_2 + \cdots + \ell_q}{\sum\limits_{j=1}^{p} \ell_j}.$$

In practice, q is chosen to be equal to 1, 2 or 3. Suppose, for instance, that $\psi = 0.93$ for $q = 2$. This means that the graphical representation in two dimensions captures 93% of the total variance. In other words, there is minimal dispersion in a third direction (no more than 7%).

It can be useful to check if each individual is well represented by the PCs. Clearly, the proximity of two individuals on the projected space may not necessarily coincide with the proximity in the full original space $\mathbb{R}^p$, which may lead to erroneous interpretations of the graphs. In this respect, it is worth computing the angle ϑ_{ik} between the representation of an individual i and the k-th PC or NPC axis. This can be done using (2.40), i.e.,

$$\cos \vartheta_{ik} = \frac{y_i^\top e_k}{\|y_i\|\|e_k\|} = \frac{y_{ik}}{\|x_{Ci}\|}$$

for the PCs or analogously

$$\cos \zeta_{ik} = \frac{z_i^\top e_k}{\|z_i\|\|e_k\|} = \frac{z_{ik}}{\|x_{Si}\|}$$

for the NPCs, where e_k denotes the k-th unit vector $e_k = (0, \ldots, 1, \ldots, 0)^\top$. An individual i will be represented on the k-th PC axis if its corresponding angle is small, i.e., if $\cos^2 \vartheta_{ik}$ for $k = 1, \ldots, p$ is close to one. Note that for each individual i,

$$\sum_{k=1}^{p} \cos^2 \vartheta_{ik} = \frac{y_i^\top y_i}{x_{Ci}^\top x_{Ci}} = \frac{x_{Ci}^\top \mathcal{G}\mathcal{G}^\top x_{Ci}}{x_{Ci}^\top x_{Ci}} = 1.$$

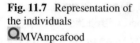

Fig. 11.7 Representation of
the individuals
MVApcafood

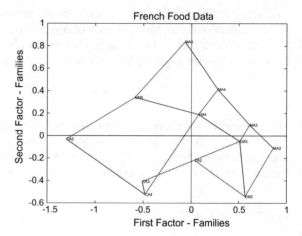

The values $\cos^2 \vartheta_{ik}$ are sometimes called the relative contributions of the k-th axis
to the representation of the i-th individual, e.g., if $\cos^2 \vartheta_{i1} + \cos^2 \vartheta_{i2}$ is large (near
one), we know that the individual i is well represented on the plane of the first two
principal axes since its corresponding angle with the plane is close to zero.

We already know that the quality of the representation of the variables can be
evaluated by the percentage of X_i's variance that is explained by a PC, which is
given by $r^2_{X_i Y_j}$ or $r^2_{X_i Z_j}$ according to (11.16) and (11.27), respectively.

Example 11.6 Let us return to the French food expenditure example, see
Appendix B.6. This yields a two-dimensional representation of the individuals as
shown in Fig. 11.7.

Calculating the matrix $\mathcal{G}_\mathcal{R}$, we have

$$\mathcal{G}_\mathcal{R} = \begin{pmatrix}
-0.240 & 0.622 & -0.011 & -0.544 & 0.036 & 0.508 \\
-0.466 & 0.098 & -0.062 & -0.023 & -0.809 & -0.301 \\
-0.446 & -0.205 & 0.145 & 0.548 & -0.067 & 0.625 \\
-0.462 & -0.141 & 0.207 & -0.053 & 0.411 & -0.093 \\
-0.438 & -0.197 & 0.356 & -0.324 & 0.224 & -0.350 \\
-0.281 & 0.523 & -0.444 & 0.450 & 0.341 & -0.332 \\
0.206 & 0.479 & 0.780 & 0.306 & -0.069 & -0.138
\end{pmatrix},$$

which gives the weights of the variables (milk, vegetables, etc.). The eigenvalues ℓ_j
and the proportions of explained variance are given in Table 11.5.

The interpretation of the principal components is best understood when looking
at the correlations between the original X_i's and the PCs. Since the first two PCs
explain 88.1% of the variance, we limit ourselves to the first two PCs. The results
are shown in Table 11.6.

The two-dimensional graphical representation of the variables in Fig. 11.8 is based
on the first two columns of Table 11.6.

Table 11.3 Eigenvalues and explained variance

Eigenvalues	Proportion of variance	Cumulated proportion
4.333	0.6190	61.9
1.830	0.2620	88.1
0.631	0.0900	97.1
0.128	0.0180	98.9
0.058	0.0080	99.7
0.019	0.0030	99.9
0.001	0.0001	100.0

Table 11.4 Correlations with PCs

	$r_{X_i Z_1}$	$r_{X_i Z_2}$	$r_{X_i Z_1}^2 + r_{X_i Z_2}^2$
X_1: bread	−0.499	0.842	0.957
X_2: vegetables	−0.970	0.133	0.958
X_3: fruits	−0.929	−0.278	0.941
X_4: meat	−0.962	−0.191	0.962
X_5: poultry	−0.911	−0.266	0.901
X_6: milk	−0.584	0.707	0.841
X_7: wine	0.428	0.648	0.604

Fig. 11.8 Representation of
the variables
MVAnpcafood

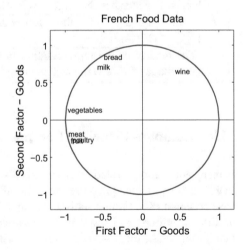

The plots are the projections of the variables into $\mathbb{R}^2$. Since the quality of the representation is good for all the variables (except maybe X_7), their relative angles give a picture of their original correlation: wine is negatively correlated with the vegetables, fruits, meat, and poultry groups ($\theta > 90°$), whereas taken individually this latter grouping of variables are highly positively correlated with each other

$(\theta \approx 0)$. Bread and milk are positively correlated but poorly correlated with meat, fruits and poultry $(\theta \approx 90°)$.

Now the representation of the individuals in Fig. 11.7 can be interpreted better. From Fig. 11.8 and Table 11.6, we can see that the first factor Z_1 is a vegetable–meat–poultry–fruit factor (with a negative sign), whereas the second factor is a milk–bread–wine factor (with a positive sign). Note that this corresponds to the most important weights in the first columns of $\mathcal{G}_\mathcal{R}$. In Fig. 11.7, lines were drawn to connect families of the same size and families of the same professional types. A grid can clearly be seen (with a slight deformation by the manager families) that shows the families with higher expenditures (higher number of children) on the left.

Considering both figures together explains what types of expenditures are responsible for similarities in food expenditures. Bread, milk, and wine expenditures are similar for manual workers and employees. Families of managers are characterized by higher expenditures on vegetables, fruits, meat, and poultry. Very often when analyzing NPCs (and PCs), it is illuminating to use such a device to introduce qualitative aspects of individuals in order to enrich the interpretations of the graphs.

Summary
↪ NPCs are PCs applied to the standardized (normalized) data matrix $\mathcal{X}_S$.
↪ The graphical representation of NPCs provides a similar type of picture as that of PCs, the difference being in the relative position of individuals, i.e., each variable in NPCs has the same weight (in PCs, the variable with the largest variance has the largest weight).
↪ The quality of the representation is evaluated by $\psi = (\sum_{j=1}^{p} \ell_j)^{-1}(\ell_1 + \ell_2 + \cdots + \ell_q)$.
↪ The quality of the representation of a variable can be evaluated by the percentage of X_i's variance that is explained by a PC, i.e., $r^2_{X_i Y_j}$.

11.7 Common Principal Components

In many applications, a statistical analysis is simultaneously done for groups of data. In this section, a technique is presented that allows us to analyze group elements that have common PCs. From a statistical point of view, estimating PCs simultaneously in different groups will result in a joint dimension reducing transformation. This multigroup PCA, the so-called common principle components analysis (CPCA), yields the joint eigenstructure across groups.

In addition to traditional PCA, the basic assumption of CPCA is that the space spanned by the eigenvectors is identical *across* several groups, whereas variances associated with the components are allowed to vary.

More formally, the hypothesis of common principle components can be stated in the following way, Flury (1988):

$$H_{CPC} : \Sigma_i = \Gamma \Lambda_i \Gamma^\top, \qquad i = 1, ..., k$$

where Σ_i is a positive definite $p \times p$ population covariance matrix for every i, $\Gamma = (\gamma_1, ..., \gamma_p)$ is an orthogonal $p \times p$ transformation matrix, and $\Lambda_i = \text{diag}(\lambda_{i1}, ..., \lambda_{ip})$ is the matrix of eigenvalues. Moreover, assume that all λ_i are distinct.

Let S be the (unbiased) sample covariance matrix of an underlying p-variate normal distribution $N_p(\mu, \Sigma)$ with sample size n. Then the distribution of nS has $n - 1$ degrees of freedom and is known as the Wishart distribution (Muirhead 1982, p. 86):

$$nS \sim \mathcal{W}_p(\Sigma, n - 1).$$

The density is given in (5.16). Hence, for a given Wishart matrix S_i with sample size n_i, the likelihood function can be written as

$$L(\Sigma_1, ..., \Sigma_k) = C \prod_{i=1}^{k} \exp\left[\text{tr}\left\{ -\frac{1}{2}(n_i - 1)\Sigma_i^{-1} S_i \right\} \right] |\Sigma_i|^{-\frac{1}{2}(n_i - 1)} \qquad (11.41)$$

where C is a constant independent of the parameters Σ_i. Maximizing the likelihood is equivalent to minimizing the function

$$g(\Sigma_1, ..., \Sigma_k) = \sum_{i=1}^{k}(n_i - 1)\left\{ \log|\Sigma_i| + \text{tr}(\Sigma_i^{-1} S_i) \right\}.$$

Assuming that H_{CPC} holds, i.e., in replacing Σ_i by $\Gamma \Lambda_i \Gamma^\top$, after some manipulations, one obtains

$$g(\Gamma, \Lambda_1, ..., \Lambda_k) = \sum_{i=1}^{k}(n_i - 1) \sum_{j=1}^{p}\left(\log\lambda_{ij} + \frac{\gamma_j^\top S_i \gamma_j}{\lambda_{ij}} \right).$$

As we know from Sect. 2.2, the vectors γ_j in Γ have to be orthogonal. Orthogonality of the vectors γ_j is achieved using the Lagrange method, i.e., we impose the p constraints $\gamma_j^\top \gamma_j = 1$ using the Lagrange multipliers μ_j, and the remaining $p(p - 1)/2$ constraints $\gamma_h^\top \gamma_j = 0$ for $h \neq j$ using the multiplier $2\mu_{hj}$ Flury (1988). This yields

$$g^*(\Gamma, \Lambda_1, ..., \Lambda_k) = g(\cdot) - \sum_{j=1}^{p} \mu_j(\gamma_j^\top \gamma_j - 1) - 2\sum_{h=1}^{p} \sum_{j=h+1}^{p} \mu_{hj} \gamma_h^\top \gamma_j.$$

Taking partial derivatives with respect to all λ_{im} and γ_m, it can be shown that the solution of the CPC model is given by the generalized system of characteristic equations

$$\gamma_m^\top \left\{ \sum_{i=1}^{k} (n_i - 1) \frac{\lambda_{im} - \lambda_{ij}}{\lambda_{im} \lambda_{ij}} \mathcal{S}_i \right\} \gamma_j = 0, \qquad m, j = 1, ..., p, \quad m \neq j. \quad (11.42)$$

This system can be solved using

$$\lambda_{im} = \gamma_m^\top \mathcal{S}_i \gamma_m, \qquad i = 1, ..., k, \quad m = 1, ..., p$$

under the constraints

$$\gamma_m^\top \gamma_j = \begin{cases} 0 & m \neq j \\ 1 & m = j \end{cases}.$$

Flury (1988) proves the existence and uniqueness of the maximum of the likelihood function, and Flury and Gautschi (1986) provide a numerical algorithm.

Example 11.7 As an example, we provide the data sets XFGvolsurf01, XFGvolsurf02, and XFGvolsurf03 that have been used in Fengler et al. (2003) to estimate common principle components for the implied volatility surfaces of the DAX 1999. The data has been generated by smoothing an implied volatility surface day by day. Next, the estimated grid points have been grouped into maturities of $\tau = 1$, $\tau = 2$ and $\tau = 3$ months and transformed into a vector of time series of the "smile", i.e., each element of the vector belongs to a distinct moneyness ranging from 0.85 to 1.10.

Figure 11.9 shows the first three eigenvectors in a parallel coordinates plot. The basic structure of the first three eigenvectors is not altered. We find a shift, a slope, and a twist structure. This structure is *common* to all maturity groups, i.e., when exploiting PCA as a dimension reducing tool, the same transformation applies to each group! However, by comparing the size of eigenvalues among groups we find that variability is decreasing across groups as we move from the short-term contracts to long-term contracts.

Before drawing conclusions we should convince ourselves that the CPC model is truly a good description of the data. This can be done by using a likelihood ratio test. The likelihood ratio statistic for comparing a restricted (the CPC) model against the unrestricted model (the model where all covariances are treated separately) is given by

$$T_{(n_1, n_2, ..., n_k)} = -2 \log \frac{L(\widehat{\Sigma}_1, ..., \widehat{\Sigma}_k)}{L(\mathcal{S}_1, ..., \mathcal{S}_k)}.$$

Inserting the likelihood function, we find that this is equivalent to

$$T_{(n_1, n_2, ..., n_k)} = \sum_{i=1}^{k} (n_i - 1) \frac{\det(\widehat{\Sigma}_i)}{\det(\mathcal{S}_i)},$$

Fig. 11.9 Factor loadings of the first (thick), the second (medium), and the third (thin) PC 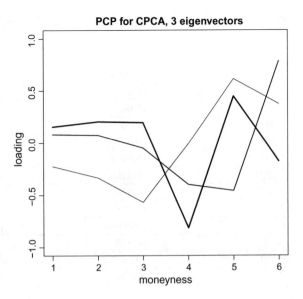MVAcpcaiv

which has a χ^2 distribution as $\min(n_i)$ tends to infinity with

$$k\left\{\frac{1}{2}p(p-1)+1\right\} - \left\{\frac{1}{2}p(p-1)+kp\right\} = \frac{1}{2}(k-1)p(p-1)$$

degrees of freedom.

The calculations yield $T_{(n_1,n_2,\ldots,n_k)} = 31.836$, which corresponds to the p-value $p = 0.37512$ for the $\chi^2(30)$ distribution. Hence, we cannot reject the CPC model against the unrestricted model, where PCA is applied to each maturity separately.

Using the methods in Sect. 11.3, we can estimate the amount of variability, ζ_l, explained by the first l principal components: (only a few factors, three at the most, are needed to capture a large amount of the total variability present in the data). Since the model now captures the variability in both the strike and maturity dimensions, this is a suitable starting point for a simplified VaR calculation for delta–gamma neutral option portfolios using Monte Carlo methods and is hence a valuable insight in risk management.

11.8 Boston Housing

A set of transformations were defined in Chap. 1 for the Boston housing data set that resulted in "regular" marginal distributions. The usefulness of principal component analysis with respect to such high-dimensional data sets will now be shown. The variable X_4 is dropped because it is a discrete 0–1 variable. It will be used later,

Table 11.5 Eigenvalues and percentage of explained variance for Boston housing data
MVAnpcahousi

Eigenvalue	Percentages	Cumulated percentages
7.2852	0.5604	0.5604
1.3517	0.1040	0.6644
1.1266	0.0867	0.7510
0.7802	0.0600	0.8111
0.6359	0.0489	0.8600
0.5290	0.0407	0.9007
0.3397	0.0261	0.9268
0.2628	0.0202	0.9470
0.1936	0.0149	0.9619
0.1547	0.0119	0.9738
0.1405	0.0108	0.9846
0.1100	0.0085	0.9931
0.0900	0.0069	1.0000

Table 11.6 Correlations of the first three PC's with the original variables MVAnpcahous

	PC_1	PC_2	PC_3
X_1	−0.9076	0.2247	0.1457
X_2	0.6399	−0.0292	0.5058
X_3	−0.8580	0.0409	−0.1845
X_5	−0.8737	0.2391	−0.1780
X_6	0.5104	0.7037	0.0869
X_7	−0.7999	0.1556	−0.2949
X_8	0.8259	−0.2904	0.2982
X_9	−0.7531	0.2857	0.3804
X_{10}	−0.8114	0.1645	0.3672
X_{11}	−0.5674	−0.2667	0.1498
X_{12}	0.4906	−0.1041	−0.5170
X_{13}	−0.7996	−0.4253	−0.0251
X_{14}	0.7366	0.5160	−0.1747

however, in the graphical representations. The scale difference of the remaining 13 variables motivates an NPCA based on the correlation matrix.

The eigenvalues and the percentage of explained variance are given in Table 11.5.

The first principal component explains 56% of the total variance and the first three components together explain more than 75%. These results imply that it is sufficient to look at 2, maximum 3, principal components.

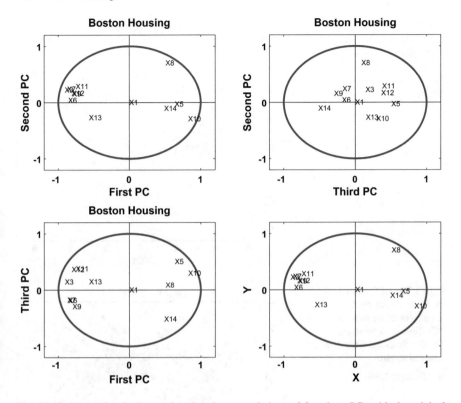

Fig. 11.10 NPCA for the Boston housing data, correlations of first three PCs with the original variables 🔍MVAnpcahousi

Table 11.6 provides the correlations between the first three PC's and the original variables. These can be seen in Fig. 11.10.

The correlations with the first PC show a very clear pattern. The variables X_2, X_6, X_8, X_{12}, and X_{14} are strongly positively correlated with the first PC, whereas the remaining variables are highly negatively correlated. The minimal correlation in the absolute value is 0.5. The first PC axis could be interpreted as a quality of life and house indicator. The second axis, given the polarities of X_{11} and X_{13} and of X_6 and X_{14}, can be interpreted as a social factor explaining only 10% of the total variance. The third axis is dominated by a polarity between X_2 and X_{12}.

The set of individuals from the first two PCs can be graphically interpreted if the plots are color coded with respect to some particular variable of interest. Figure 11.11 color codes X_{14} > median as red points. Clearly, the first and second PCs are related to house value. The situation is less clear in Fig. 11.12 where the color code corresponds to X_4, the Charles River indicator, i.e., houses near the river are colored red.

Fig. 11.11 NPC analysis for
the Boston housing data,
scatterplot of the first two
PCs. More expensive houses
are marked with red color
MVAnpcahous

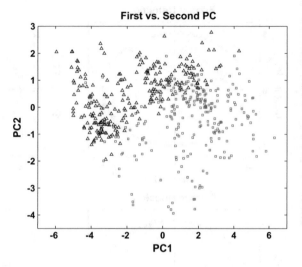

Fig. 11.12 NPC analysis for
the Boston housing data,
scatterplot of the first two
PCs. Houses close to the
Charles River are indicated
with red squares
MVAnpcahous

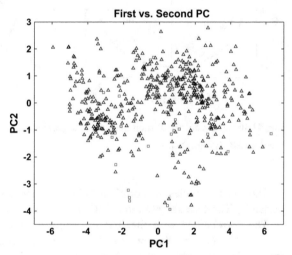

11.9 More Examples

Example 11.8 Let us now apply the PCA to the *standardized* bank data set (Sect. B.2).
Figure 11.13 shows some PC plots of the bank data set. The genuine and counterfeit
bank notes are marked by "o" and "+", respectively.

The vector of eigenvalues of $\mathcal{R}$ is

$$\ell = (2.946, 1.278, 0.869, 0.450, 0.269, 0.189)^{\top}.$$

The eigenvectors g_j are given by the columns of the matrix

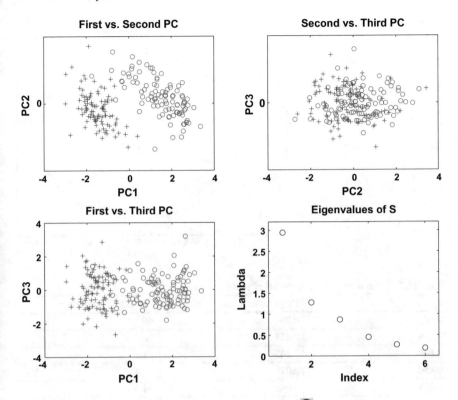

Fig. 11.13 Principal components of the *standardized* bank data ◯MVApcabank

$$
\mathcal{G} = \begin{pmatrix}
-0.007 & -0.815 & 0.018 & 0.575 & 0.059 & 0.031 \\
0.468 & -0.342 & -0.103 & -0.395 & -0.639 & -0.298 \\
0.487 & -0.252 & -0.123 & -0.430 & 0.614 & 0.349 \\
0.407 & 0.266 & -0.584 & 0.404 & 0.215 & -0.462 \\
0.368 & 0.091 & 0.788 & 0.110 & 0.220 & -0.419 \\
-0.493 & -0.274 & -0.114 & -0.392 & 0.340 & -0.632
\end{pmatrix}.
$$

Each original variable has the same weight in the analysis and the results are independent of the scale of each variable.

The proportions of explained variance are given in Table 11.5. It can be concluded that the representation in two dimensions should be sufficient. The correlations leading to Fig. 11.14 are given in Table 11.8. The picture is different from the one obtained in Sect. 11.3 (see Table 11.2). Here, the first factor is mainly a left–right versus diagonal factor and the second one is a length factor (with negative weight). Take another look at Fig. 11.13, where the individual bank notes are displayed. In the upper left graph it can be seen that the genuine bank notes are for the most part in the southeastern portion of the graph featuring a larger diagonal, smaller height ($Z_1 < 0$) and also

Fig. 11.14 The correlations of the original variable with the PCs 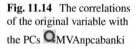 MVAnpcabanki

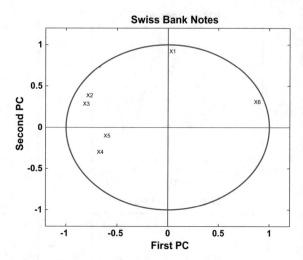

Table 11.7 Eigenvalues and proportions of explained variance

ℓ_j	Proportion of variances	Cumulated proportion
2.946	0.491	49.1
1.278	0.213	70.4
0.869	0.145	84.9
0.450	0.075	92.4
0.264	0.045	96.9
0.189	0.032	100.0

Table 11.8 Correlations with PCs

	$r_{X_i Z_1}$	$r_{X_i Z_2}$	$r^2_{X_i Z_1} + r^2_{X_i Z_2}$
X_1: length	−0.012	−0.922	0.85
X_2: left height	0.803	−0.387	0.79
X_3: right height	0.835	−0.285	0.78
X_4: lower	0.698	0.301	0.58
X_5: upper	0.631	0.104	0.41
X_6: diagonal	−0.847	−0.310	0.81

a larger length ($Z_2 < 0$). Note also that Fig. 11.14 gives an idea of the correlation structure of the original data matrix.

Example 11.9 Consider the data of 79 U.S. companies given in Sect. 22.5. The data is first standardized by subtracting the mean and dividing by the standard deviation. Note that the data set contains six variables: assets (X_1), sales (X_2), market value (X_3), profits (X_4), cash flow (X_5), and number of employees (X_6).

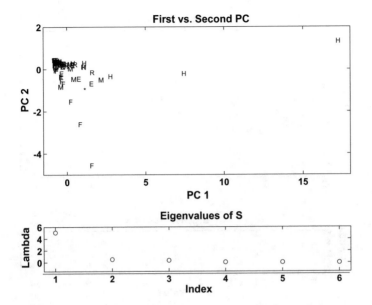

Fig. 11.15 Principal components of the U.S. company data ⭕MVAnpcausco

Calculating the corresponding vector of eigenvalues gives

$$\ell = (5.039, 0.517, 0.359, 0.050, 0.029, 0.007)^{\top}$$

and the matrix of eigenvectors is

$$\mathcal{G} = \begin{pmatrix} 0.340 & -0.849 & -0.339 & 0.205 & 0.077 & -0.006 \\ 0.423 & -0.170 & 0.379 & -0.783 & -0.006 & -0.186 \\ 0.434 & 0.190 & -0.192 & 0.071 & -0.844 & 0.149 \\ 0.420 & 0.364 & -0.324 & 0.156 & 0.261 & -0.703 \\ 0.428 & 0.285 & -0.267 & -0.121 & 0.452 & 0.667 \\ 0.397 & 0.010 & 0.726 & 0.548 & 0.098 & 0.065 \end{pmatrix}.$$

Using this information the graphical representations of the first two principal components are given in Fig. 11.15. The different sectors are marked by the following symbols:

H ... Hi Tech and Communication
E ... Energy
F ... Finance
M ... Manufacturing
R ... Retail
★ ... all other sectors.

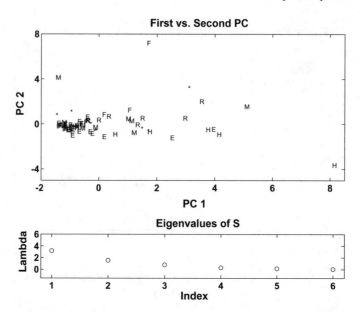

Fig. 11.16 Principal components of the U.S. company data (without IBM and General Electric)
MVAnpcausco2

The two outliers in the right-hand side of the graph are IBM and General Electric (GE), which differ from the other companies with their high market values. As can be seen in the first column of $\mathcal{G}$, market value has the largest weight in the first PC, adding to the isolation of these two companies. If IBM and GE were to be excluded from the data set, a completely different picture would emerge, as shown in Fig. 11.16. In this case, the vector of eigenvalues becomes

$$\ell = (3.191, 1.535, 0.791, 0.292, 0.149, 0.041)^{\top},$$

and the corresponding matrix of eigenvectors is

$$\mathcal{G} = \begin{pmatrix} 0.263 & -0.408 & -0.800 & -0.067 & 0.333 & 0.099 \\ 0.438 & -0.407 & 0.162 & -0.509 & -0.441 & -0.403 \\ 0.500 & -0.003 & -0.035 & 0.801 & -0.264 & -0.190 \\ 0.331 & 0.623 & -0.080 & -0.192 & 0.426 & -0.526 \\ 0.443 & 0.450 & -0.123 & -0.238 & -0.335 & 0.646 \\ 0.427 & -0.277 & 0.558 & 0.021 & 0.575 & 0.313 \end{pmatrix}.$$

The percentage of variation explained by each component is given in Table 11.9. The first two components explain almost 79% of the variance. The interpretation of the factors (the axes of Fig. 11.16) is given in the table of correlations (Table 11.10). The first two columns of this table are plotted in Fig. 11.17.

Table 11.9 Eigenvalues and proportions of explained variance

ℓ_j	Proportion of variance	Cumulated proportion
3.191	0.532	0.532
1.535	0.256	0.788
0.791	0.132	0.920
0.292	0.049	0.968
0.149	0.025	0.993
0.041	0.007	1.000

Table 11.10 Correlations with PCs

	$r_{X_i Z_1}$	$r_{X_i Z_2}$	$r_{X_i Z_1}^2 + r_{X_i Z_2}^2$
X_1: assets	0.47	−0.510	0.48
X_2: sales	0.78	−0.500	0.87
X_3: market value	0.89	−0.003	0.80
X_4: profits	0.59	0.770	0.95
X_5: cash flow	0.79	0.560	0.94
X_6: employees	0.76	−0.340	0.70

Fig. 11.17 The correlation of the original variables with the PCs MVAnpcausco2i

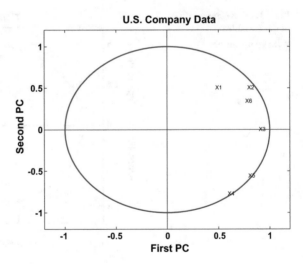

From Fig. 11.17 (and Table 11.10) it appears that the first factor is a "size effect", it is positively correlated with all the variables describing the size of the activity of the companies. It is also a measure of the economic strength of the firms. The second factor describes the "shape" of the companies ("profit–cash flow" vs. "assets-sales" factor), which is more difficult to interpret from an economic point of view.

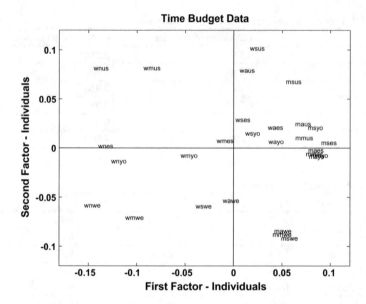

Fig. 11.18 Representation of the individuals 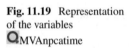MVAnpcatime

Fig. 11.19 Representation
of the variables
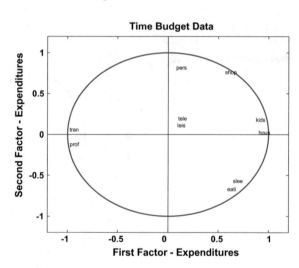MVAnpcatime

Example 11.10 Volle (1985) analyzes data on 28 individuals (Sect. B.12). For each
individual, the time spent (in hours) on 10 different activities has been recorded over
100 days, as well as informative statistics such as the individual's sex, country of
residence, professional activity, and matrimonial status. The results of an NPCA are
given below.

Table 11.11 Eigenvalues of correlation matrix for the time budget data

ℓ_j	Proportion of variance	Cumulated proportion
4.59	0.459	0.460
2.12	0.212	0.670
1.32	0.132	0.800
1.20	0.120	0.920
0.47	0.047	0.970
0.20	0.020	0.990
0.05	0.005	0.990
0.04	0.004	0.999
0.02	0.002	1.000
0.00	0.000	1.000

Table 11.12 Correlation of variables with PCs

		$r_{X_i W_1}$	$r_{X_i W_2}$	$r_{X_i W_3}$	$r_{X_i W_4}$
X_1:	prof	0.9772	−0.1210	−0.0846	0.0669
X_2:	tran	0.9798	0.0581	−0.0084	0.4555
X_3:	hous	−0.8999	0.0227	0.3624	0.2142
X_4:	kids .	−0.8721	0.1786	0.0837	0.2944
X_5:	shop	−0.5636	0.7606	−0.0046	−0.1210
X_6:	pers	−0.0795	0.8181	−0.3022	−0.0636
X_7:	eati	−0.5883	−0.6694	−0.4263	0.0141
X_8:	slee	−0.6442	−0.5693	−0.1908	−0.3125
X_9:	tele	−0.0994	0.1931	−0.9300	0.1512
X_{10}:	leis	−0.0922	0.1103	0.0302	−0.9574

The eigenvalues of the correlation matrix are given in Table 11.11. Note that the last eigenvalue is exactly zero since the correlation matrix is singular (the sum of all the variables is always equal to $2400 = 24 \times 100$). The results of the four first PCs are given in Tables 11.12 and 11.13.

From these tables (and Figs. 11.18 and 11.19), it appears that the professional and household activities are strongly contrasted in the first factor. Indeed on the horizontal axis of Fig. 11.18, it can be seen that all the active men are on the right and all the inactive women are on the left. Active women and/or single women are in between. The second factor contrasts meal/sleeping versus toilet/shopping (note the high correlation between meal and sleeping). Along the vertical axis of Fig. 11.18, we see near the bottom of the graph the people from Western-European countries, who spend more time on meals and sleeping than people from the U.S. (who can be found close to the top of the graph). The other categories are in between.

Table 11.13 PCs for time budget data

	Z_1	Z_2	Z_3	Z_4
maus	0.0633	0.0245	−0.0668	0.0205
waus	0.0061	0.0791	−0.0236	0.0156
wnus	−0.1448	0.0813	−0.0379	−0.0186
mmus	0.0635	0.0105	−0.0673	0.0262
wmus	−0.0934	0.0816	−0.0285	0.0038
msus	0.0537	0.0676	−0.0487	−0.0279
wsus	0.0166	0.1016	−0.0463	−0.0053
mawe	0.0420	−0.0846	−0.0399	−0.0016
wawe	−0.0111	−0.0534	−0.0097	0.0337
wnwe	−0.1544	−0.0583	−0.0318	−0.0051
mmwe	0.0402	−0.0880	−0.0459	0.0054
wmwe	−0.1118	−0.0710	−0.0210	0.0262
mswe	0.0489	−0.0919	−0.0188	−0.0365
wswe	−0.0393	−0.0591	−0.0194	−0.0534
mayo	0.0772	−0.0086	0.0253	−0.0085
wayo	0.0359	0.0064	0.0577	0.0762
wnyo	−0.1263	−0.0135	0.0584	−0.0189
mmyo	0.0793	−0.0076	0.0173	−0.0039
wmyo	−0.0550	−0.0077	0.0579	0.0416
msyo	0.0763	0.0207	0.0575	−0.0778
wsyo	0.0120	0.0149	0.0532	−0.0366
maes	0.0767	−0.0025	0.0047	0.0115
waes	0.0353	0.0209	0.0488	0.0729
wnes	−0.1399	0.0016	0.0240	−0.0348
mmes	0.0742	−0.0061	−0.0152	0.0283
wmes	−0.0175	0.0073	0.0429	0.0719
mses	0.0903	0.0052	0.0379	−0.0701
fses	0.0020	0.0287	0.0358	−0.0346

In Fig. 11.19, the variables television and other leisure activities hardly play any role (look at Table 11.12). The variable television appears in Z_3 (negatively correlated). Table 11.13 shows that this factor contrasts people from Eastern countries and Yugoslavia with men living in the U.S. The variable other leisure activities is the factor Z_4. It merely distinguishes between men and women in Eastern countries and in Yugoslavia. These last two factors are orthogonal to the preceding axes, and of course, their contribution to the total variation is less important.

11.10 Exercises

Exercise 11.1 Prove Theorem 11.1. (Hint: use (4.23).)

Exercise 11.2 Interpret the results of the PCA of the U.S. companies. Use the analysis of the bank notes in Sect. 11.3 as a guide. Compare your results with those in Example 11.9.

Exercise 11.3 Test the hypothesis that the proportion of variance explained by the first two PCs for the U.S. companies is $\psi = 0.75$.

Exercise 11.4 Apply the PCA to the car data (Sect. 22.7). Interpret the first two PCs. Would it be necessary to look at the third PC?

Exercise 11.5 Apply a PCA to $\Sigma = \begin{pmatrix} 1 & \rho \\ \rho & 1 \end{pmatrix}$, where $\rho > 0$. Now change the scale of X_1, i.e., consider the covariance of cX_1 and X_2. How do the PC directions change with the scree plot?

Exercise 11.6 Suppose that we have standardized some data using the Mahalanobis transformation. Would it be reasonable to apply a PCA?

Exercise 11.7 Apply an NPCA to the U.S. CRIME data set (Sect. B.8). Interpret the results. Would it be necessary to look at the third PC? Can you see any difference between the four regions? Redo the analysis excluding the variable "area of the state".

Exercise 11.8 Let U be a uniform r.v. on $[0, 1]$. Let $a \in \mathbb{R}^3$ be a vector of constants. Suppose that $X = Ua^\top = (X_1, X_2, X_3)$. What do you expect the NPCs of X to be?

Exercise 11.9 Let U_1 and U_2 be two independent uniform random variables on $[0, 1]$. Suppose that $X = (X_1, X_2, X_3, X_4)^\top$ where $X_1 = U_1$, $X_2 = U_2$, $X_3 = U_1 + U_2$ and $X_4 = U_1 - U_2$. Compute the correlation matrix P of X. How many PCs are of interest? Show that $\gamma_1 = \left(\frac{1}{\sqrt{2}}, \frac{1}{\sqrt{2}}, 1, 0 \right)^\top$ and $\gamma_2 = \left(\frac{1}{\sqrt{2}}, \frac{-1}{\sqrt{2}}, 0, 1 \right)^\top$ are eigenvectors of P corresponding to the nontrivial λ's. Interpret the first two NPCs obtained.

Exercise 11.10 Simulate a sample of size $n = 50$ for the r.v. X in Exercise 11.9 and analyze the results of an NPCA.

Exercise 11.11 Bouroche (1980) reported the data on the state expenses of France from the period 1872 to 1971 (24 selected years) by noting the percentage of 11 categories of expenses. Do an NPCA of this data set. Do the three main periods (before WWI, between WWI and WWII, and after WWII) indicate a change in behavior w.r.t. to state expenses?

References

J.-M. Bouroche, Saporta, G., *L'analyse des données*(Presses Universitaires de France, Paris, 1980)

M.R. Fengler, W. Härdle, C. Villa, The dynamics of implied volatilities: a common principal components approach. Rev. Deriv. Res. **6**, 179–202 (2003)

B. Flury, *Common Principle Components Analysis and Related Multivariate Models* (Wiley, New York, 1988)

B. Flury, W. Gautschi, An Algorithm for simultaneous orthogonal transformation of several positive definite symmetric matrices to nearly diagonal form. SIAM J. Sci. Stat. Comput. **7**, 169–184 (1986)

R.J. Muirhead, *Aspects of Multivariate Statistics* (Wiley, New York, 1982)

V. Volle, *Analyse des Données* (Economica, Paris, 1985)

Chapter 12
Factor Analysis

A frequently applied paradigm in analyzing data from multivariate observations is to model the relevant information (represented in a multivariate variable X) as coming from a limited number of latent factors. In a survey on household consumption, for example, the consumption levels, X, of p different goods during one month could be observed. The variations and covariations of the p components of X throughout the survey might in fact be explained by two or three main social behavior factors of the household. For instance, a basic desire of comfort or the willingness to achieve a certain social level or other social latent concepts might explain most of the consumption behavior. These unobserved factors are much more interesting to the social scientist than the observed quantitative measures (X) themselves, because they give a better understanding of the behavior of households. As shown in the examples below, the same kind of factor analysis is of interest in many fields such as psychology, marketing, economics, politic sciences, etc.

How can we provide a statistical model addressing these issues and how can we interpret the obtained model? This is the aim of factor analysis. As in Chaps. 10 and 11, the driving statistical theme of this chapter is to reduce the dimension of the observed data. The perspective used, however, is different: we assume that there is a model (it will be called the "Factor Model") stating that most of the covariances between the p elements of X can be explained by a limited number of latent factors. Section 12.1 defines the basic concepts and notations of the orthogonal factor model, stressing the non-uniqueness of the solutions. We show how to take advantage of this non-uniqueness to derive techniques which lead to easier interpretations. This will involve (geometric) rotations of the factors. Section 12.2 presents an empirical approach to factor analysis. Various estimation procedures are proposed and an optimal rotation procedure is defined. Many examples are used to illustrate the method.

© Springer Nature Switzerland AG 2019
W. K. Härdle and L. Simar, *Applied Multivariate Statistical Analysis*,
https://doi.org/10.1007/978-3-030-26006-4_12

12.1   The Orthogonal Factor Model

The aim of factor analysis is to explain the outcome of p variables in the data matrix $\mathcal{X}$ using fewer variables, the so-called *factors*. Ideally all the information in $\mathcal{X}$ can be reproduced by a smaller number of factors. These factors are interpreted as latent (unobserved) common characteristics of the observed $x \in \mathbb{R}^p$. The case just described occurs when every observed $x = (x_1, \ldots, x_p)^\top$ can be written as

$$x_j = \sum_{\ell=1}^{k} q_{j\ell} f_\ell + \mu_j, \; j = 1, ..., p. \tag{12.1}$$

Here f_ℓ, for $\ell = 1, \ldots, k$ denotes the factors. The number of factors, k, should always be much smaller than p. For instance, in psychology x may represent p results of a test measuring intelligence scores. One common latent factor explaining $x \in \mathbb{R}^p$ could be the overall level of "intelligence". In marketing studies, x may consist of p answers to a survey on the levels of satisfaction of the customers. These p measures could be explained by common latent factors like the attraction level of the product or the image of the brand, and so on. Indeed it is possible to create a representation of the observations that is similar to the one in (12.1) by means of principal components, but only if the last $p - k$ eigenvalues corresponding to the covariance matrix are equal to zero. Consider a p-dimensional random vector X with mean μ and covariance matrix $\mathsf{Var}(X) = \Sigma$. A model similar to (12.1) can be written for X in matrix notation, namely

$$X = QF + \mu, \tag{12.2}$$

where F is the k-dimensional vector of the k factors. When using the factor model (12.2) it is often assumed that the factors F are centered, uncorrelated and standardized: $\mathsf{E}(F) = 0$ and $\mathsf{Var}(F) = \mathcal{I}_k$. We will now show that if the last $p - k$ eigenvalues of Σ are equal to zero, we can easily express X by the factor model (12.2).

The spectral decomposition of Σ is given by $\Gamma \Lambda \Gamma^\top$. Suppose that only the first k eigenvalues are positive, i.e., $\lambda_{k+1} = \ldots = \lambda_p = 0$. Then the (singular) covariance matrix can be written as

$$\Sigma = \sum_{\ell=1}^{k} \lambda_\ell \gamma_\ell \gamma_\ell^\top = (\Gamma_1 \Gamma_2) \begin{pmatrix} \Lambda_1 & 0 \\ 0 & 0 \end{pmatrix} \begin{pmatrix} \Gamma_1^\top \\ \Gamma_2^\top \end{pmatrix}.$$

In order to show the connection to the factor model (12.2), recall that the PCs are given by $Y = \Gamma^\top (X - \mu)$. Rearranging we have $X - \mu = \Gamma Y = \Gamma_1 Y_1 + \Gamma_2 Y_2$, where the components of Y are partitioned according to the partition of Γ above, namely

$$Y = \begin{pmatrix} Y_1 \\ Y_2 \end{pmatrix} = \begin{pmatrix} \Gamma_1^\top \\ \Gamma_2^\top \end{pmatrix} (X - \mu), \text{ where } \begin{pmatrix} \Gamma_1^\top \\ \Gamma_2^\top \end{pmatrix} (X - \mu) \sim \left(0, \begin{pmatrix} \Lambda_1 & 0 \\ 0 & 0 \end{pmatrix} \right).$$

In other words, Y_2 has a singular distribution with mean and covariance matrix equal to zero. Therefore, $X - \mu = \Gamma_1 Y_1 + \Gamma_2 Y_2$ implies that $X - \mu$ is equivalent to $\Gamma_1 Y_1$, which can be written as

$$X = \Gamma_1 \Lambda_1^{1/2} \Lambda_1^{-1/2} Y_1 + \mu.$$

Defining $Q = \Gamma_1 \Lambda_1^{1/2}$ and $F = \Lambda_1^{-1/2} Y_1$, we obtain the factor model (12.2).

Note that the covariance matrix of model (12.2) can be written as

$$\Sigma = \mathsf{E}(X - \mu)(X - \mu)^\top = Q\mathsf{E}(FF^\top)Q^\top = QQ^\top = \sum_{j=1}^{k} \lambda_j \gamma_j \gamma_j^\top. \qquad (12.3)$$

We have just shown how the variable X can be completely determined by a weighted sum of k (where $k < p$) uncorrelated factors. The situation used in the derivation, however, is too idealistic. In practice the covariance matrix is rarely singular.

It is common praxis in factor analysis to split the influences of the factors into common and specific ones. There are, for example, highly informative factors that are common to all of the components of X and factors that are specific to certain components. The factor analysis model used in praxis is a generalization of (12.2):

$$X = QF + U + \mu, \qquad (12.4)$$

where Q is a $(p \times k)$ matrix of the (non-random) *loadings* of the *common factors* $F(k \times 1)$ and U is a $(p \times 1)$ matrix of the (random) *specific factors*. It is assumed that the factor variables F are uncorrelated random vectors and that the specific factors are uncorrelated and have zero covariance with the common factors. More precisely, it is assumed that:

$$
\begin{aligned}
\mathsf{E}F &= 0, \\
\mathsf{Var}(F) &= \mathcal{I}_k, \\
\mathsf{E}U &= 0, \\
\mathsf{Cov}(U_i, U_j) &= 0, \quad i \neq j \\
\mathsf{Cov}(F, U) &= 0.
\end{aligned}
\qquad (12.5)
$$

Define

$$\mathsf{Var}(U) = \Psi = \mathrm{diag}(\psi_{11}, \ldots, \psi_{pp}).$$

The generalized factor model (12.4) together with the assumptions given in (12.5) constitute the *orthogonal factor model*.

Orthogonal Factor Model

$$X \quad = \quad Q \qquad F \quad + \quad U \quad + \quad \mu$$
$$(p \times 1) \quad (p \times k) \, (k \times 1) \quad (p \times 1) \quad (p \times 1)$$

$$\mu_j \quad = \quad \text{mean of variable } j$$
$$U_j \quad = \quad j\text{-th specific factor}$$
$$F_\ell \quad = \quad \ell\text{-th common factor}$$
$$q_{j\ell} \quad = \quad \text{loading of the } j\text{-th variable on the } \ell\text{-th factor}$$

The random vectors F and U are unobservable and uncorrelated.

Note that (12.4) implies for the components of $X = (X_1, \ldots, X_p)^\top$ that

$$X_j = \sum_{\ell=1}^{k} q_{j\ell} F_\ell + U_j + \mu_j, \quad j = 1, \ldots, p. \tag{12.6}$$

Using (12.5) we obtain $\sigma_{X_j X_j} = \mathsf{Var}(X_j) = \sum_{\ell=1}^{k} q_{j\ell}^2 + \psi_{jj}$. The quantity $h_j^2 = \sum_{\ell=1}^{k} q_{j\ell}^2$ is called the *communality* and ψ_{jj} the *specific variance*. Thus the covariance of X can be rewritten as

$$\begin{aligned} \Sigma = \mathsf{E}(X - \mu)(X - \mu)^\top &= \mathsf{E}(QF + U)(QF + U)^\top \\ &= Q\mathsf{E}(FF^\top)Q^\top + \mathsf{E}(UU^\top) = Q\mathsf{Var}(F)Q^\top + \mathsf{Var}(U) \\ &= QQ^\top + \Psi. \end{aligned} \tag{12.7}$$

In a sense, the factor model explains the variations of X for the most part by a small number of latent factors F common to its p components and entirely explains all the correlation structure between its components, plus some "noise" U which allows specific variations of each component to enter. The specific factors adjust to capture the individual variance of each component. Factor analysis relies on the assumptions presented above. If the assumptions are not met, the analysis could be spurious. Although principal components analysis and factor analysis might be related (this was hinted at in the derivation of the factor model), they are quite different in nature. PCs are linear transformations of X arranged in decreasing order of variance and used to reduce the dimension of the data set, whereas in factor analysis, we try to model the variations of X using a linear transformation of a fixed, limited number of latent factors. The objective of factor analysis is to find the loadings Q and the specific variance Ψ. Estimates of Q and Ψ are deduced from the covariance structure (12.7).

Interpretation of the Factors

Assume that a factor model with k factors was found to be reasonable, i.e., most of the (co) variations of the p measures in X were explained by the k fixed latent factors. The next natural step is to try to understand what these factors represent. To interpret F_ℓ, it makes sense to compute its correlations with the original variables X_j first. This is done for $\ell = 1, \ldots, k$ and for $j = 1, \ldots, p$ to obtain the matrix P_{XF}. The sequence of calculations used here are in fact the same that were used to interprete the PCs in the principal components analysis.

The following covariance between X and F is obtained via (12.5),

$$\Sigma_{XF} = \mathsf{E}\{(QF + U)F^\top\} = Q.$$

The correlation is

$$P_{XF} = D^{-1/2}Q, \tag{12.8}$$

where $D = \operatorname{diag}(\sigma_{X_1 X_1}, \ldots, \sigma_{X_p X_p})$. Using (12.8) it is possible to construct a figure analogous to Fig. 11.6 and thus to consider which of the original variables $X_1, \ldots, X_p$ play a role in the unobserved common factors $F_1, \ldots, F_k$.

Returning to the psychology example where X are the observed scores to p different intelligence tests (the WAIS data set in Exrecise 7.19 provides an example), we would expect a model with one factor to produce a factor that is positively correlated with all of the components in X. For this example the factor represents the overall level of intelligence of an individual. A model with two factors could produce a refinement in explaining the variations of the p scores. For example, the first factor could be the same as before (overall level of intelligence), whereas the second factor could be positively correlated with some of the tests, X_j, that are related to the individual's ability to think abstractly and negatively correlated with other tests, X_i, that are related to the individual's practical ability. The second factor would then concern a particular dimension of the intelligence stressing the distinctions between the "theoretical" and "practical" abilities of the individual. If the model is true, most of the information coming from the p scores can be summarized by these two latent factors. Other practical examples are given below.

Invariance of Scale

What happens if we change the scale of X to $Y = CX$ with $C = \operatorname{diag}(c_1, \ldots, c_p)$? If the k-factor model (12.6) is true for X with $Q = Q_X$, $\Psi = \Psi_X$, then, since

$$\mathsf{Var}(Y) = C\Sigma C^\top = CQ_X Q_X^\top C^\top + C\Psi_X C^\top,$$

the same k-factor model is also true for Y with $Q_Y = CQ_X$ and $\Psi_Y = C\Psi_X C^\top$. In many applications, the search for the loadings Q and for the specific variance Ψ will be done by the decomposition of the correlation matrix of X rather than the covariance matrix Σ. This corresponds to a factor analysis of a linear transformation of X (i.e., $Y = D^{-1/2}(X - \mu)$). The goal is to try to find the loadings Q_Y and the specific variance Ψ_Y such that

$$P = Q_Y \, Q_Y^\top + \Psi_Y. \tag{12.9}$$

In this case the interpretation of the factors F immediately follows from (12.8) given the following correlation matrix:

$$P_{XF} = P_{YF} = Q_Y. \tag{12.10}$$

Because of the scale invariance of the factors, the loadings and the specific variance of the model, where X is expressed in its original units of measure, are given by

$$\begin{aligned} Q_X &= D^{1/2} Q_Y \\ \Psi_X &= D^{1/2} \Psi_Y D^{1/2}. \end{aligned}$$

It should be noted that although the factor analysis model (12.4) enjoys the scale invariance property, the actual estimated factors could be scale dependent. We will come back to this point later when we discuss the method of principal factors.

Non-uniqueness of Factor Loadings

The factor loadings are not unique! Suppose that G is an orthogonal matrix. Then X in (12.4) can also be written as

$$X = (QG)(G^\top F) + U + \mu.$$

This implies that, if a k-factor of X with factors F and loadings Q is true, then the k-factor model with factors $G^\top F$ and loadings QG is also true. In practice, we will take advantage of this non-uniqueness. Indeed, referring back to Sect. 2.6 we can conclude that premultiplying a vector F by an orthogonal matrix corresponds to a rotation of the system of axis, the direction of the first new axis being given by the first row of the orthogonal matrix. It will be shown that choosing an appropriate rotation will result in a matrix of loadings QG that will be easier to interpret. We have seen that the loadings provide the correlations between the factors and the original variables, therefore, it makes sense to search for rotations that give factors that are maximally correlated with various groups of variables.

From a numerical point of view, the non-uniqueness is a drawback. We have to find loadings Q and specific variances Ψ satisfying the decomposition $\Sigma = QQ^\top + \Psi$, but no straightforward numerical algorithm can solve this problem due to the multiplicity of the solutions. An acceptable technique is to impose some chosen constraints in order to get—in the best case—an unique solution to the decomposition. Then, as suggested above, once we have a solution we will take advantage of the rotations in order to obtain a solution that is easier to interprete.

An obvious question is: what kind of constraints should we impose in order to eliminate the non-uniqueness problem? Usually, we impose additional constraints where

$$Q^\top \Psi^{-1} Q \quad \text{is diagonal} \tag{12.11}$$

or

$$Q^\top D^{-1} Q \quad \text{is diagonal.} \tag{12.12}$$

How many parameters does the model (12.7) have without constraints?

$$Q(p \times k) \quad \text{has} \quad p \cdot k \quad \text{parameters, and}$$
$$\Psi(p \times p) \quad \text{has} \quad p \quad \text{parameters.}$$

Hence we have to determine $pk + p$ parameters! Conditions (12.11) respectively (12.12) introduce $\frac{1}{2}\{k(k-1)\}$ constraints, since we require the matrices to be diagonal. Therefore, the degrees of freedom of a model with k factors is:

$$d = (\# \text{ parameters for } \Sigma \text{ unconstrained}) - (\# \text{ parameters for } \Sigma \text{ constrained})$$
$$= \tfrac{1}{2}p(p+1) - (pk + p - \tfrac{1}{2}k(k-1))$$
$$= \tfrac{1}{2}(p-k)^2 - \tfrac{1}{2}(p+k).$$

If $d < 0$, then the model is undetermined: there are infinitely many solutions to (12.7). This means that the number of parameters of the factorial model is larger than the number of parameters of the original model, or that the number of factors k is "too large" relative to p. In some cases $d = 0$: there is a unique solution to the problem (except for rotation). In practice we usually have that $d > 0$: there are more equations than parameters, thus an exact solution does not exist. In this case approximate solutions are used. An approximation of Σ, for example, is $QQ^\top + \Psi$. The last case is the most interesting since the factorial model has less parameters than the original one. Estimation methods are introduced in the next section.

Evaluating the degrees of freedom, d, is particularly important, because it already gives an idea of the upper bound on the number of factors we can hope to identify in a factor model. For instance, if $p = 4$, we could not identify a factor model with 2 factors (this results in $d = -1$ which has infinitly many solutions). With $p = 4$, only a one factor model gives an approximate solution ($d = 2$). When $p = 6$, models with 1 and 2 factors provide approximate solutions and a model with 3 factors results in an unique solution (up to the rotations) since $d = 0$. A model with 4 or more factors would not be allowed, but of course, the aim of factor analysis is to find suitable models with a small number of factors, i.e., smaller than p. The next two examples give more insights into the notion of degrees of freedom.

Example 12.1 Let $p = 3$ and $k = 1$, then $d = 0$ and

$$\Sigma = \begin{pmatrix} \sigma_{11} & \sigma_{12} & \sigma_{13} \\ \sigma_{21} & \sigma_{22} & \sigma_{23} \\ \sigma_{31} & \sigma_{32} & \sigma_{33} \end{pmatrix} = \begin{pmatrix} q_1^2 + \psi_{11} & q_1 q_2 & q_1 q_3 \\ q_1 q_2 & q_2^2 + \psi_{22} & q_2 q_3 \\ q_1 q_3 & q_2 q_3 & q_3^2 + \psi_{33} \end{pmatrix}$$

with $\mathcal{Q} = \begin{pmatrix} q_1 \\ q_2 \\ q_3 \end{pmatrix}$ and $\Psi = \begin{pmatrix} \psi_{11} & 0 & 0 \\ 0 & \psi_{22} & 0 \\ 0 & 0 & \psi_{33} \end{pmatrix}$. Note that here the constraint (12.11) is automatically verified since $k = 1$. We have

$$q_1^2 = \frac{\sigma_{12}\sigma_{13}}{\sigma_{23}}; \quad q_2^2 = \frac{\sigma_{12}\sigma_{23}}{\sigma_{13}}; \quad q_3^2 = \frac{\sigma_{13}\sigma_{23}}{\sigma_{12}}$$

and

$$\psi_{11} = \sigma_{11} - q_1^2; \quad \psi_{22} = \sigma_{22} - q_2^2; \quad \psi_{33} = \sigma_{33} - q_3^2.$$

In this particular case ($k = 1$), the only rotation is defined by $\mathcal{G} = -1$, so the other solution for the loadings is provided by $-\mathcal{Q}$.

Example 12.2 Suppose now $p = 2$ and $k = 1$, then $d < 0$ and

$$\Sigma = \begin{pmatrix} 1 & \rho \\ \rho & 1 \end{pmatrix} = \begin{pmatrix} q_1^2 + \psi_{11} & q_1 q_2 \\ q_1 q_2 & q_2^2 + \psi_{22} \end{pmatrix}.$$

We have infinitely many solutions: for any α ($\rho < \alpha < 1$), a solution is provided by

$$q_1 = \alpha; \quad q_2 = \rho/\alpha; \quad \psi_{11} = 1 - \alpha^2; \quad \psi_{22} = 1 - (\rho/\alpha)^2.$$

The solution in Example 12.1 may be unique (up to a rotation), but it is not proper in the sense that it cannot be interpreted statistically. Exercise 12.5 gives an example where the specific variance ψ_{11} is negative.

⚠ Even in the case of a unique solution ($d = 0$), the solution may be inconsistent with statistical interpretations.

Summary

↪ The factor analysis model aims to describe how the original p variables in a data set depend on a small number of latent factors $k < p$, i.e., it assumes that $X = \mathcal{Q}F + U + \mu$. The ($k$-dimensional) random vector F contains the common factors, the (p-dimensional) U contains the specific factors and $\mathcal{Q}(p \times k)$ contains the factor loadings

↪ It is assumed that F and U are uncorrelated and have zero means, i.e., $F \sim (0, \mathcal{I})$, $U \sim (0, \Psi)$ where Ψ is diagonal matrix and $\text{Cov}(F, U) = 0$. This leads to the covariance structure $\Sigma = \mathcal{Q}\mathcal{Q}^\top + \Psi$

↪ The interpretation of the factor F is obtained through the correlation $P_{XF} = D^{-1/2}\mathcal{Q}$

Continued on next page

Summary (continue)
$\hookrightarrow$ A normalized analysis is obtained by the model $P = QQ^\top + \Psi$. The interpretation of the factors is given directly by the loadings Q : $P_{XF} = Q$
$\hookrightarrow$ The factor analysis model is scale invariant. The loadings are not unique (only up to multiplication by an orthogonal matrix)
$\hookrightarrow$ Whether a model has an unique solution or not is determined by the degrees of freedom $d = 1/2(p-k)^2 - 1/2(p+k)$

12.2 Estimation of the Factor Model

In practice, we have to find estimates $\widehat{Q}$ of the loadings Q and estimates $\widehat{\Psi}$ of the specific variances Ψ such that analogously to (12.7)

$$S = \widehat{Q}\widehat{Q}^\top + \widehat{\Psi},$$

where S denotes the empirical covariance of $\mathcal{X}$. Given an estimate $\widehat{Q}$ of Q, it is natural to set

$$\widehat{\psi}_{jj} = s_{X_j X_j} - \sum_{\ell=1}^{k} \widehat{q}_{j\ell}^2.$$

We have that $\widehat{h}_j^2 = \sum_{\ell=1}^{k} \widehat{q}_{j\ell}^2$ is an estimate for the communality h_j^2.

In the ideal case $d = 0$, there is an exact solution. However, d is usually greater than zero, therefore we have to find $\widehat{Q}$ and $\widehat{\Psi}$ such that S is approximated by $\widehat{Q}\widehat{Q}^\top + \widehat{\Psi}$. As mentioned above, it is often easier to compute the loadings and the specific variances of the standardized model.

Define $\mathcal{Y} = \mathcal{H}\mathcal{X}\mathcal{D}^{-1/2}$, the standardization of the data matrix $\mathcal{X}$, where $\mathcal{D} = \mathrm{diag}(s_{X_1 X_1}, \ldots, s_{X_p X_p})$ and the centering matrix $\mathcal{H} = \mathcal{I} - n^{-1} 1_n 1_n^\top$ (recall from Chap. 2 that $S = \frac{1}{n} \mathcal{X}^\top \mathcal{H} \mathcal{X}$). The estimated factor loading matrix $\widehat{Q}_Y$ and the estimated specific variance $\widehat{\Psi}_Y$ of $\mathcal{Y}$ are

$$\widehat{Q}_Y = \mathcal{D}^{-1/2} \widehat{Q}_X \quad \text{and} \quad \widehat{\Psi}_Y = \mathcal{D}^{-1} \widehat{\Psi}_X.$$

For the correlation matrix $\mathcal{R}$ of $\mathcal{X}$, we have that

$$\mathcal{R} = \widehat{Q}_Y \widehat{Q}_Y^\top + \widehat{\Psi}_Y.$$

The interpretations of the factors are formulated from the analysis of the loadings $\widehat{Q}_Y$.

Example 12.3 Let us calculate the matrices just defined for the car data given in Sect. 22.7. This data set consists of the averaged marks (from $1 = $ low to $6 = $ high) for 24 car types. Considering the three variables price, security and easy handling, we get the following correlation matrix:

$$\mathcal{R} = \begin{pmatrix} 1 & 0.975 & 0.613 \\ 0.975 & 1 & 0.620 \\ 0.613 & 0.620 & 1 \end{pmatrix}.$$

We will first look for one factor, i.e., $k = 1$. Note that (# number of parameters of Σ unconstrained – # parameters of Σ constrained) is equal to $\frac{1}{2}(p - k)^2 - \frac{1}{2}(p + k) = \frac{1}{2}(3 - 1)^2 - \frac{1}{2}(3 + 1) = 0$. This implies that there is an exact solution! The equation

$$\begin{pmatrix} 1 & r_{X_1 X_2} & r_{X_1 X_3} \\ r_{X_1 X_2} & 1 & r_{X_2 X_3} \\ r_{X_1 X_3} & r_{X_2 X_3} & 1 \end{pmatrix} = \mathcal{R} = \begin{pmatrix} \widehat{q}_1^2 + \widehat{\psi}_{11} & \widehat{q}_1 \widehat{q}_2 & \widehat{q}_1 \widehat{q}_3 \\ \widehat{q}_1 \widehat{q}_2 & \widehat{q}_2^2 + \widehat{\psi}_{22} & \widehat{q}_2 \widehat{q}_3 \\ \widehat{q}_1 \widehat{q}_3 & \widehat{q}_2 \widehat{q}_3 & \widehat{q}_3^2 + \widehat{\psi}_{33} \end{pmatrix}$$

yields the communalities $\widehat{h}_i^2 = \widehat{q}_i^2$, where

$$\widehat{q}_1^2 = \frac{r_{X_1 X_2} r_{X_1 X_3}}{r_{X_2 X_3}}, \qquad \widehat{q}_2^2 = \frac{r_{X_1 X_2} r_{X_2 X_3}}{r_{X_1 X_3}} \quad \text{and} \quad \widehat{q}_3^2 = \frac{r_{X_1 X_3} r_{X_2 X_3}}{r_{X_1 X_2}}.$$

Combining this with the specific variances $\widehat{\psi}_{11} = 1 - \widehat{q}_1^2$, $\widehat{\psi}_{22} = 1 - \widehat{q}_2^2$ and $\widehat{\psi}_{33} = 1 - \widehat{q}_3^2$, we obtain the following solution

$$\begin{array}{lll} \widehat{q}_1 = 0.982 & \widehat{q}_2 = 0.993 & \widehat{q}_3 = 0.624 \\ \widehat{\psi}_{11} = 0.035 & \widehat{\psi}_{22} = 0.014 & \widehat{\psi}_{33} = 0.610. \end{array}$$

Since the first two communalities $(\widehat{h}_i^2 = \widehat{q}_i^2)$ are close to one, we can conclude that the first two variables, namely price and security, are explained by the single factor quite well. This factor can be interpreted as a "price+security" factor.

The Maximum Likelihood Method

Recall from Chap. 6 the log-likelihood function ℓ for a data matrix $\mathcal{X}$ of observations of $X \sim N_p(\mu, \Sigma)$:

$$\ell(\mathcal{X}; \mu, \Sigma) = -\frac{n}{2} \log | 2\pi \Sigma | - \frac{1}{2} \sum_{i=1}^{n} (x_i - \mu) \Sigma^{-1} (x_i - \mu)^{\top}$$

$$= -\frac{n}{2} \log | 2\pi \Sigma | - \frac{n}{2} \text{tr}(\Sigma^{-1} S) - \frac{n}{2} (\overline{x} - \mu) \Sigma^{-1} (\overline{x} - \mu)^{\top}.$$

This can be rewritten as

$$\ell(\mathcal{X}; \widehat{\mu}, \Sigma) = -\frac{n}{2} \left\{ \log | 2\pi \Sigma | + \text{tr}(\Sigma^{-1} S) \right\}.$$

Replacing μ by $\widehat{\mu} = \bar{x}$ and substituting $\Sigma = \mathcal{Q}\mathcal{Q}^\top + \Psi$ this becomes

$$\ell(\mathcal{X}; \widehat{\mu}, \mathcal{Q}, \Psi) = -\frac{n}{2} \left[\log\{|\, 2\pi(\mathcal{Q}\mathcal{Q}^\top + \Psi)\,|\} + \operatorname{tr}\{(\mathcal{Q}\mathcal{Q}^\top + \Psi)^{-1}\mathcal{S}\} \right]. \quad (12.13)$$

Even in the case of a single factor ($k = 1$), these equations are rather complicated and iterative numerical algorithms have to be used (for more details see Mardia et al. 1979, p. 263). A practical computation scheme is also given in Supplement 9A of Johnson and Wichern (1998).

Likelihood Ratio Test for the Number of Common Factors

Using the methodology of Chap. 7, it is easy to test the adequacy of the factor analysis model by comparing the likelihood under the null (factor analysis) and alternative (no constraints on covariance matrix) hypotheses.

Assuming that $\widehat{\mathcal{Q}}$ and $\widehat{\Psi}$ are the maximum likelihood estimates corresponding to (12.13), we obtain the following LR test statistic:

$$-2\log\left(\frac{\text{maximized likelihood under } H_0}{\text{maximized likelihood}}\right) = n\log\left(\frac{|\widehat{\mathcal{Q}}\widehat{\mathcal{Q}}^\top + \widehat{\Psi}|}{|\mathcal{S}|}\right), \quad (12.14)$$

which asymptotically has the $\chi^2_{\frac{1}{2}\{(p-k)^2-p-k\}}$ distribution.

The χ^2 approximation can be improved if we replace n by $n - 1 - (2p + 4k + 5)/6$ in (12.14) (Bartlett 1954). Using Bartlett's correction, we reject the factor analysis model at the α level if

$$\{n - 1 - (2p + 4k + 5)/6\} \log\left(\frac{|\widehat{\mathcal{Q}}\widehat{\mathcal{Q}}^\top + \widehat{\Psi}|}{|\mathcal{S}|}\right) > \chi^2_{1-\alpha;\{(p-k)^2-p-k\}/2}, \quad (12.15)$$

and if the number of observations n is large and the number of common factors k is such that the χ^2 statistic has a positive number of degrees of freedom.

The Method of Principal Factors

The *method of principal factors* concentrates on the decomposition of the correlation matrix $\mathcal{R}$ or the covariance matrix $\mathcal{S}$. For simplicity, only the method for the correlation matrix $\mathcal{R}$ will be discussed. As pointed out in Chap. 11, the spectral decompositions of $\mathcal{R}$ and $\mathcal{S}$ yield different results and therefore, the method of principal factors may result in different estimators. The method can be motivated as follows: Suppose we know the exact Ψ, then the constraint (12.12) implies that the columns of $\mathcal{Q}$ are orthogonal since $\mathcal{D} = \mathcal{I}$ and it implies that they are eigenvectors of $\mathcal{Q}\mathcal{Q}^\top = \mathcal{R} - \Psi$. Furthermore, assume that the first k eigenvalues are positive. In this case we could calculate $\mathcal{Q}$ by means of a spectral decomposition of $\mathcal{Q}\mathcal{Q}^\top$ and k would be the number of factors.

The principal factors algorithm is based on good preliminary estimators $\widetilde{h}_j^2$ of the communalities h_j^2, for $j = 1, \ldots, p$. There are two traditional proposals:

- $\widetilde{h}_j^2$, defined as the square of the multiple correlation coefficient of X_j with (X_l), for $l \neq j$, i.e., $\rho^2(V, W\widehat{\beta})$ with $V = X_j$, $W = (X_\ell)_{\ell \neq j}$ and where $\widehat{\beta}$ is the least squares regression parameter of a regression of V on W.
- $\widetilde{h}_j^2 = \max\limits_{\ell \neq j} |r_{X_j X_\ell}|$, where $\mathcal{R} = (r_{X_j X_\ell})$ is the correlation matrix of $\mathcal{X}$.

Given $\widetilde{\psi}_{jj} = 1 - \widetilde{h}_j^2$ we can construct the *reduced correlation matrix*, $\mathcal{R} - \widetilde{\Psi}$. The Spectral Decomposition Theorem says that

$$\mathcal{R} - \widetilde{\Psi} = \sum_{\ell=1}^{p} \lambda_\ell \gamma_\ell \gamma_\ell^\top ,$$

with eigenvalues $\lambda_1 \geq \cdots \geq \lambda_p$. Assume that the first k eigenvalues $\lambda_1, \ldots, \lambda_k$ are positive and large compared to the others. Then we can set

$$\widehat{q}_\ell = \sqrt{\lambda_\ell}\, \gamma_\ell, \quad \ell = 1, \ldots, k$$

or

$$\widehat{\mathcal{Q}} = \Gamma_1 \Lambda_1^{1/2}$$

with

$$\Gamma_1 = (\gamma_1, \ldots, \gamma_k) \quad \text{and} \quad \Lambda_1 = \mathrm{diag}(\lambda_1, \ldots, \lambda_k).$$

In the next step set

$$\widehat{\psi}_{jj} = 1 - \sum_{\ell=1}^{k} \widehat{q}_{j\ell}^2, \quad j = 1, \ldots, p.$$

Note that the procedure can be iterated: from $\widehat{\psi}_{jj}$ we can compute a new reduced correlation matrix $\mathcal{R} - \widehat{\Psi}$ following the same procedure. The iteration usually stops when the $\widehat{\psi}_{jj}$ have converged to a stable value.

Example 12.4 Consider once again the car data given in Sect. 22.7. From Exercise 11.4 we know that the first PC is mainly influenced by X_2–X_7. Moreover, we know that most of the variance is already captured by the first PC. Thus we can conclude that the data are mainly determined by one factor ($k = 1$).
   The eigenvalues of $\mathcal{R} - \widehat{\Psi}$ for $\widehat{\Psi} = (\max\limits_{j \neq i} |r_{X_i X_j}|)$ are

$$(4.628, 1.340, 1.201, 1.045, 1.007, 0.993, 0.980, -4.028)^\top .$$

It would suffice to choose only one factor. Nevertheless, we have computed two factors. The result (the factor loadings for two factors) is shown in Fig. 12.1.
   We can clearly see a cluster of points to the right, which contain the factor loadings for the variables X_2–X_7. This shows, as did the PCA, that these variables are highly dependent and are thus more or less equivalent. The factor loadings for X_1 (economy)

and X_8 (easy handling) are separate, but note the different scales on the horizontal and vertical axes! Although there are two or three sets of variables in the plot, the variance is already explained by the first factor, the "price+security" factor.

The Principal Component Method

The *principal factor method* involves finding an approximation $\widetilde{\Psi}$ of Ψ, the matrix of specific variances, and then correcting $\mathcal{R}$, the correlation matrix of X, by $\widetilde{\Psi}$. The *principal component method* starts with an approximation $\widehat{Q}$ of Q, the factor loadings matrix. The sample covariance matrix is diagonalized, $\mathcal{S} = \Gamma \Lambda \Gamma^{\top}$. Then the first k eigenvectors are retained to build

$$\widehat{Q} = (\sqrt{\lambda_1}\gamma_1, \ldots, \sqrt{\lambda_k}\gamma_k). \tag{12.16}$$

The estimated specific variances are provided by the diagonal elements of the matrix $\mathcal{S} - \widehat{Q}\widehat{Q}^{\top}$,

$$\widehat{\Psi} = \begin{pmatrix} \widehat{\psi}_{11} & & & 0 \\ & \widehat{\psi}_{22} & & \\ & & \ddots & \\ 0 & & & \widehat{\psi}_{pp} \end{pmatrix} \quad \text{with } \widehat{\psi}_{jj} = s_{X_j X_j} - \sum_{\ell=1}^{k} \widehat{q}_{j\ell}^2. \tag{12.17}$$

By definition, the diagonal elements of $\mathcal{S}$ are equal to the diagonal elements of $\widehat{Q}\widehat{Q}^{\top} + \widehat{\Psi}$. The off-diagonal elements are not necessarily estimated. How good then is this approximation? Consider the residual matrix

$$\mathcal{S} - (\widehat{Q}\widehat{Q}^{\top} + \widehat{\Psi})$$

Fig. 12.1 Loadings of the evaluated car qualities, factor analysis with $k = 2$

🔍 MVAfactcarm

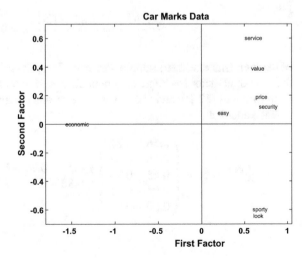

resulting from the principal component solution. Analytically we have that

$$\sum_{i,j}(\mathcal{S} - \widehat{\mathcal{Q}}\widehat{\mathcal{Q}}^{\top} - \widehat{\Psi})_{ij}^{2} \le \lambda_{k+1}^{2} + \ldots + \lambda_{p}^{2}.$$

This implies that a small value of the neglected eigenvalues can result in a small approximation error. A heuristic device for selecting the number of factors is to consider the proportion of the total sample variance due to the jth factor. This quantity is in general equal to

(A) $\lambda_j / \sum_{j=1}^{p} s_{jj}$ for a factor analysis of $\mathcal{S}$,
(B) λ_j / p for a factor analysis of $\mathcal{R}$.

Example 12.5 This example uses a consumer-preference study from Johnson and Wichern (1998). Customers were asked to rate several attributes of a new product. The responses were tabulated and the following correlation matrix $\mathcal{R}$ was constructed:

	Attribute (Variable)		
Taste	1	1.00 0.02 **0.96** 0.42 0.01
Good buy for money	2	0.02 1.00 0.13 0.71 **0.85**
Flavor	3	0.96 0.13 1.00 0.50 0.11
Suitable for snack	4	0.42 0.71 0.50 1.00 **0.79**
Provides lots of energy	5	0.01 **0.85** 0.11 0.79 1.00

The bold entries of $\mathcal{R}$ show that variables 1 and 3 and variables 2 and 5 are highly correlated. Variable 4 is more correlated with variables 2 and 5 than with variables 1 and 3. Hence, a model with 2 (or 3) factors seems to be reasonable.

The first two eigenvalues $\lambda_1 = 2.85$ and $\lambda_2 = 1.81$ of $\mathcal{R}$ are the only eigenvalues greater than one. Moreover, $k = 2$ common factors account for a cumulative proportion

$$\frac{\lambda_1 + \lambda_2}{p} = \frac{2.85 + 1.81}{5} = 0.93$$

of the total (standardized) sample variance. Using the principal component method, the estimated factor loadings, communalities, and specific variances, are calculated from formulas (12.16) and (12.17), and the results are given in Table 12.1.

Take a look at:

$$\widehat{\mathcal{Q}}\widehat{\mathcal{Q}}^{\top} + \widehat{\Psi} = \begin{pmatrix} 0.56 & 0.82 \\ 0.78 & -0.53 \\ 0.65 & 0.75 \\ 0.94 & -0.11 \\ 0.80 & -0.54 \end{pmatrix} \begin{pmatrix} 0.56 & 0.78 & 0.65 & 0.94 & 0.80 \\ 0.82 & -0.53 & 0.75 & -0.11 & -0.54 \end{pmatrix} +$$

Table 12.1 Estimated factor loadings, communalities, and specific variances

Variables	Estimated factor loadings		Communalities	Specific variances
	$\widehat{q}_1$	$\widehat{q}_2$	$\widehat{h}_j^2$	$\widehat{\psi}_{jj} = 1 - \widehat{h}_j^2$
1. Taste	0.56	0.82	0.98	0.02
2. Good buy for money	0.78	−0.53	0.88	0.12
3. Flavor	0.65	0.75	0.98	0.02
4. Suitable for snack	0.94	−0.11	0.89	0.11
5. Provides lots of energy	0.80	−0.54	0.93	0.07
Eigenvalues	2.85	1.81		
Cumulative proportion of total (standardized) sample variance	0.571	0.932		

$$+ \begin{pmatrix} 0.02 & 0 & 0 & 0 & 0 \\ 0 & 0.12 & 0 & 0 & 0 \\ 0 & 0 & 0.02 & 0 & 0 \\ 0 & 0 & 0 & 0.11 & 0 \\ 0 & 0 & 0 & 0 & 0.07 \end{pmatrix} = \begin{pmatrix} 1.00 & 0.01 & 0.97 & 0.44 & 0.00 \\ 0.01 & 1.00 & 0.11 & 0.79 & 0.91 \\ 0.97 & 0.11 & 1.00 & 0.53 & 0.11 \\ 0.44 & 0.79 & 0.53 & 1.00 & 0.81 \\ 0.00 & 0.91 & 0.11 & 0.81 & 1.00 \end{pmatrix}.$$

This nearly reproduces the correlation matrix $\mathcal{R}$. We conclude that the two-factor model provides a good fit of the data. The communalities $(0.98, 0.88, 0.98, 0.89, 0.93)$ indicate that the two factors account for a large percentage of the sample variance of each variable. Due to the nonuniqueness of factor loadings, the interpretation might be enhanced by rotation. This is the topic of the next subsection.

Rotation

The constraints (12.11) and (12.12) are given as a matter of mathematical convenience (to create unique solutions) and can therefore complicate the problem of interpretation. The interpretation of the loadings would be very simple if the variables could be split into disjoint sets, each being associated with one factor. A well known analytical algorithm to rotate the loadings is given by the *varimax rotation method* proposed by Kaiser (1985). In the simplest case of $k = 2$ factors, a rotation matrix $\mathcal{G}$ is given by

$$\mathcal{G}(\theta) = \begin{pmatrix} \cos \theta & \sin \theta \\ -\sin \theta & \cos \theta \end{pmatrix},$$

representing a clockwise rotation of the coordinate axes by the angle θ. The corresponding rotation of loadings is calculated via $\widehat{Q}^* = \widehat{Q}\mathcal{G}(\theta)$. The idea of the *varimax method* is to find the angle θ that maximizes the sum of the variances of the squared loadings $\widehat{q}_{ij}^*$ within each column of $\widehat{Q}^*$. More precisely, defining $\widetilde{q}_{jl}^* = \widehat{q}_{jl}^*/\widehat{h}_j^*$, the *varimax criterion* chooses θ so that

$$V = \frac{1}{p} \sum_{\ell=1}^{k} \left[\sum_{j=1}^{p} (\widetilde{q}_{jl}^{*})^4 - \left\{ \frac{1}{p} \sum_{j=1}^{p} (\widetilde{q}_{jl}^{*})^2 \right\}^2 \right]$$

is maximized.

Example 12.6 Let us return to the marketing example of Johnson and Wichern (1998) (Example 12.5). The basic factor loadings given in Table 12.1 of the first factor and a second factor are almost identical making it difficult to interpret the factors. Applying the varimax rotation we obtain the loadings $\widetilde{q}_1 = (0.02, \mathbf{0.94}, 0.13, \mathbf{0.84}, \mathbf{0.97})^\top$ and $\widetilde{q}_2 = (\mathbf{0.99}, -0.01, \mathbf{0.98}, 0.43, -0.02)^\top$. The high loadings, indicated as bold entries, show that variables 2, 4, 5 define factor 1, a nutricional factor. Variables 1 and 3 define factor 2 which might be referred to as a taste factor.

Summary
$\hookrightarrow$ In practice, Q and Ψ have to be estimated from $S = \widehat{Q}\widehat{Q}^\top + \widehat{\Psi}$. The number of parameters is $d = \frac{1}{2}(p-k)^2 - \frac{1}{2}(p+k)$
$\hookrightarrow$ If $d = 0$, then there exists an exact solution. In practice, d is usually greater than 0, thus approximations must be considered
$\hookrightarrow$ The maximum-likelihood method assumes a normal distribution for the data. A solution can be found using numerical algorithms
$\hookrightarrow$ The method of principal factors is a two-stage method which calculates $\widehat{Q}$ from the reduced correlation matrix $\mathcal{R} - \widetilde{\Psi}$, where $\widetilde{\Psi}$ is a pre-estimate of Ψ. The final estimate of Ψ is found by $\widehat{\psi}_{ii} = 1 - \sum_{j=1}^{k} \widehat{q}_{ij}^2$
$\hookrightarrow$ The principal component method is based on an approximation, $\widehat{Q}$, of Q
$\hookrightarrow$ Often a more informative interpretation of the factors can be found by rotating the factors
$\hookrightarrow$ The varimax rotation chooses a rotation θ that maximizes $V = \frac{1}{p} \sum_{\ell=1}^{k} \left[\sum_{j=1}^{p} \left(\widetilde{q}_{jl}^{*}\right)^4 - \left\{ \frac{1}{p} \sum_{j=1}^{p} \left(\widetilde{q}_{jl}^{*}\right)^2 \right\}^2 \right]$

12.3   Factor Scores and Strategies

Up to now strategies have been presented for factor analysis that have concentrated on the estimation of loadings and communalities and on their interpretations. This was a logical step since the factors F were considered to be normalized random sources of information and were explicitely addressed as nonspecific (common factors). The estimated values of the factors, called the *factor scores*, may also be useful in the interpretation as well as in the diagnostic analysis. To be more precise, the factor scores are estimates of the unobserved random vectors F_l, $l = 1, \ldots, k$, for each individual x_i, $i = 1, \ldots, n$. (Johnson and Wichern, 1998) describe three methods which in practice yield very similar results. Here, we present the regression method which has the advantage of being the simplest technique and is easy to implement.

The idea is to consider the joint distribution of $(X - \mu)$ and F, and then to proceed with the regression analysis presented in Chap. 5. Under the factor model (12.4), the joint covariance matrix of $(X - \mu)$ and F is:

$$\mathrm{Var}\begin{pmatrix} X - \mu \\ F \end{pmatrix} = \begin{pmatrix} \mathcal{Q}\mathcal{Q}^\top + \Psi & \mathcal{Q} \\ \mathcal{Q}^\top & \mathcal{I}_k \end{pmatrix}. \tag{12.18}$$

Note that the upper left entry of this matrix equals Σ and that the matrix has size $(p + k) \times (p + k)$.

Assuming joint normality, the conditional distribution of $F|X$ is multinormal, see Theorem 5.1, with

$$\mathsf{E}(F|X = x) = \mathcal{Q}^\top \Sigma^{-1}(X - \mu) \tag{12.19}$$

and using (5.7) the covariance matrix can be calculated:

$$\mathrm{Var}(F|X = x) = \mathcal{I}_k - \mathcal{Q}^\top \Sigma^{-1} \mathcal{Q}. \tag{12.20}$$

In practice, we replace the unknown $\mathcal{Q}$, Σ and μ by corresponding estimators, leading to the estimated individual factor scores:

$$\widehat{f}_i = \widehat{\mathcal{Q}}^\top S^{-1}(x_i - \overline{x}). \tag{12.21}$$

We prefer to use the original sample covariance matrix S as an estimator of Σ, instead of the factor analysis approximation $\widehat{\mathcal{Q}}\widehat{\mathcal{Q}}^\top + \widehat{\Psi}$, in order to be more robust against incorrect determination of the number of factors.

The same rule can be followed when using $\mathcal{R}$ instead of S. Then (12.18) remains valid when standardized variables, i.e., $Z = \mathcal{D}_\Sigma^{-1/2}(X - \mu)$, are considered if $\mathcal{D}_\Sigma = \mathrm{diag}(\sigma_{11}, \ldots, \sigma_{pp})$. In this case the factors are given by

$$\widehat{f}_i = \widehat{\mathcal{Q}}^\top \mathcal{R}^{-1}(z_i), \tag{12.22}$$

where $z_i = \mathcal{D}_S^{-1/2}(x_i - \overline{x})$, $\widehat{\mathcal{Q}}$ is the loading obtained with the matrix $\mathcal{R}$, and $\mathcal{D}_S = \mathrm{diag}(s_{11}, \ldots, s_{pp})$.

If the factors are rotated by the orthogonal matrix $\mathcal{G}$, the factor scores have to be rotated accordingly, that is

$$\widehat{f}_i^* = \mathcal{G}^\top \widehat{f}_i. \tag{12.23}$$

A practical example is presented in Sect. 12.4 using the Boston Housing data.

Practical Suggestions

No one method outperforms another in the practical implementation of factor analysis. However, by applying a *tâtonnement* process, the factor analysis view of the data can be stabilized. This motivates the following procedure.

1. Fix a reasonable number of factors, say $k = 2$ or 3, based on the correlation structure of the data and/or screeplot of eigenvalues.
2. Perform several of the presented methods, including rotation. Compare the loadings, communalities, and factor scores from the respective results.
3. If the results show significant deviations, check for outliers (based on factor scores), and consider changing the number of factors k.

For larger data sets, cross-validation methods are recommended. Such methods involve splitting the sample into a training set and a validation data set. On the training sample one estimates the factor model with the desired methodology and uses the obtained parameters to predict the factor scores for the validation data set. The predicted factor scores should be comparable to the factor scores obtained using only the validation data set. This stability criterion may also involve the loadings and communalities.

Factor Analysis Versus PCA

Factor analysis and principal component analysis use the same set of mathematical tools (spectral decomposition, projections, ...). One could conclude, on first sight, that they share the same view and strategy and therefore yield very similar results. This is not true. There are substantial differences between these two data analysis techniques that we would like to describe here.

The biggest difference between PCA and factor analysis comes from the model philosophy. Factor analysis imposes a strict structure of a fixed number of common (latent) factors whereas the PCA determines p factors in decreasing order of importance. The most important factor in PCA is the one that maximizes the projected variance. The most important factor in factor analysis is the one that (after rotation) gives the maximal interpretation. Often this is different from the direction of the first principal component.

From an implementation point of view, the PCA is based on a well-defined, unique algorithm (spectral decomposition), whereas fitting a factor analysis model involves a variety of numerical procedures. The non-uniqueness of the factor analysis procedure opens the door for subjective interpretation and yields therefore a spectrum of results. This data analysis philosophy makes factor analysis difficult especially if the model specification involves cross-validation and a data-driven selection of the number of factors.

Table 12.2 Estimated factor loadings, communalities, and specific variances, MLM MVAfacthous

	Estimated factor loadings			Communalities	Specific variances
	$\widehat{q}_1$	$\widehat{q}_2$	$\widehat{q}_3$	$\widehat{h}_j^2$	$\widehat{\psi}_{jj} = 1 - \widehat{h}_j^2$
1. Crime	0.9295	0.1653	0.1107	0.9036	0.0964
2. Large lots	−0.5823	0.0379	0.2902	0.4248	0.5752
3. Nonretail acres	0.8192	−0.0296	−0.1378	0.6909	0.3091
4. Nitric oxides	0.8789	0.0987	−0.2719	0.8561	0.1439
5. Rooms	−0.4447	0.5311	−0.0380	0.4812	0.5188
6. Prior 1940	0.7837	−0.0149	−0.3554	0.7406	0.2594
7. Empl. centers	−0.8294	−0.1570	0.4110	0.8816	0.1184
8. Accessibility	0.7955	0.3062	0.4053	0.8908	0.1092
9. Tax-rate	0.8262	0.1401	0.2906	0.7867	0.2133
10. Pupil/Teacher	0.5051	−0.1850	0.1553	0.3135	0.6865
11. African American	0.4701	−0.0227	−0.1627	0.2480	0.7520
12. Lower status	0.7601	−0.5059	−0.0070	0.8337	0.1663
13. Value	−0.6942	0.5904	−0.1798	0.8628	0.1371

Table 12.3 Estimated factor loadings, communalities, and specific variances, MLM, varimax rotation MVAfacthous

	Estimated factor loadings			Communalities	Specific variances
	$\widehat{q}_1$	$\widehat{q}_2$	$\widehat{q}_3$	$\widehat{h}_j^2$	$\widehat{\psi}_{jj} = 1 - \widehat{h}_j^2$
1. Crime	0.7247	−0.2705	−0.5525	0.9036	0.0964
2. Large lots	−0.1570	0.2377	0.5858	0.4248	0.5752
3. Nonretail acres	0.4195	−0.3566	−0.6287	0.6909	0.3091
5. Nitric oxides	0.4141	−0.2468	−0.7896	0.8561	0.1439
6. Rooms	−0.0799	0.6691	0.1644	0.4812	0.5188
7. Prior 1940	0.2518	−0.2934	−0.7688	0.7406	0.2594
8. Empl. centers	−0.3164	0.1515	0.8709	0.8816	0.1184
9. Accessibility	0.8932	−0.1347	−0.2736	0.8908	0.1092
10. Tax-rate	0.7673	−0.2772	−0.3480	0.7867	0.2133
11. Pupil/Teacher	0.3405	−0.4065	−0.1800	0.3135	0.6865
12. African American	−0.3917	0.2483	0.1813	0.2480	0.7520
13. Lower status	0.2586	−0.7752	−0.4072	0.8337	0.1663
14. Value	−0.3043	0.8520	0.2111	0.8630	0.1370

12.4 Boston Housing

To illustrate how to implement factor analysis we will use the Boston Housing data set and the by now well known set of transformations. Once again, the variable X_4 (Charles River indicator) will be excluded. As before, standardized variables are used and the analysis is based on the correlation matrix.

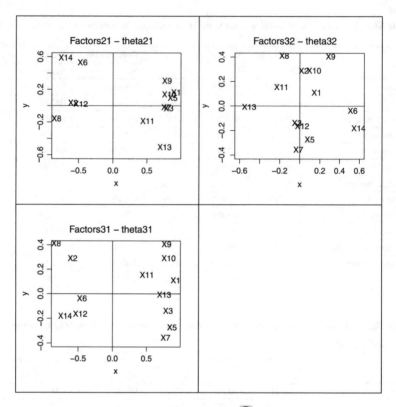

Fig. 12.2 Factor analysis for Boston housing data, MLM 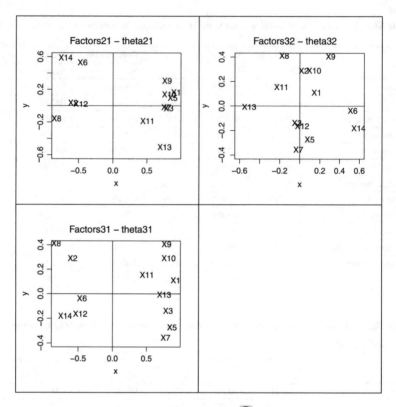MVAfacthous

In Sect. 12.3, we described a practical implementation of factor analysis. Based on principal components, three factors were chosen and factor analysis was applied using the maximum likelihood method (MLM), the principal factor method (PFM), and the principal component method (PCM). For illustration, the MLM will be presented with and without varimax rotation.

Table 12.2 gives the MLM factor loadings without rotation and Table 12.3 gives the varimax version of this analysis. The corresponding graphical representations of the loadings are displayed in Figs. 12.2 and 12.3. We can see that the varimax does not significantly change the interpretation of the factors obtained by the MLM. Factor 1 can be roughly interpreted as a "quality of life factor" because it is positively correlated with variables like X_{11} and negatively correlated with X_8, both having low specific variances. The second factor may be interpreted as a "residential factor", since it is highly correlated with variables X_6, and X_{13}. The most striking difference between the results with and without varimax rotation can be seen by comparing the lower left corners of Figs. 12.2 and 12.3. There is a clear separation of the variables in the varimax version of the MLM. Given this arrangement of the variables in Fig. 12.3,

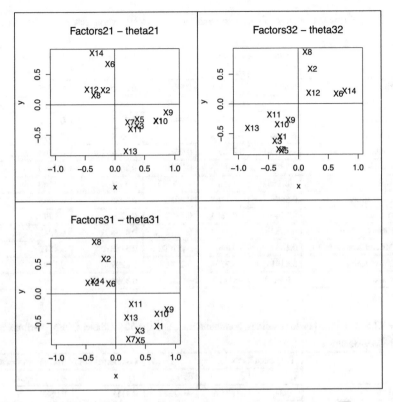

Fig. 12.3 Factor analysis for Boston housing data, MLM after varimax rotation 🔴MVAfacthous

we can interpret factor 3 as an employment factor, since we observe high correlations with X_8 and X_5.

We now turn to the PCM and PFM analyses. The results are presented in Tables 12.4 and 12.5 and in Figs. 12.4 and 12.5. We would like to focus on the PCM, because this 3-factor model yields only one specific variance (unexplained variation) above 0.5. Looking at Fig. 12.4, it turns out that factor 1 remains a "quality of life factor" which is clearly visible from the clustering of X_5, X_3, X_{10} and X_1 on the right-hand side of the graph, while the variables X_8, X_2, X_{14}, X_{12} and X_6 are on the left-hand side. Again, the second factor is a "residential factor", clearly demonstrated by the location of variables X_6, X_{14}, X_{11}, and X_{13}. The interpretation of the third factor is more difficult because all of the loadings (except for X_{12}) are very small.

Table 12.4 Estimated factor loadings, communalities, and specific variances, PCM, varimax rotation **Q**MVAfacthous

	Estimated factor loadings			Communalities	Specific variances
	$\widehat{q}_1$	$\widehat{q}_2$	$\widehat{q}_3$	$\widehat{h}_j^2$	$\widehat{\psi}_{jj} = 1 - \widehat{h}_j^2$
1. Crime	0.6034	−0.2456	0.6864	0.8955	0.1045
2. Large lots	−0.7722	0.2631	0.0270	0.6661	0.3339
3. Nonretail acres	0.7183	−0.3701	0.3449	0.7719	0.2281
5. Nitric oxides	0.7936	−0.2043	0.4250	0.8521	0.1479
6. Rooms	−0.1601	0.8585	0.0218	0.7632	0.2368
7. Prior 1940	0.7895	−0.2375	0.2670	0.7510	0.2490
8. Empl. centers	−0.8562	0.1318	−0.3240	0.8554	0.1446
9. Accessibility	0.3681	−0.1268	0.8012	0.7935	0.2065
10. Tax-rate	0.3744	−0.2604	0.7825	0.8203	0.1797
11. Pupil/Teacher	0.1982	−0.5124	0.3372	0.4155	0.5845
12. African American	0.1647	0.0368	−0.7002	0.5188	0.4812
13. Lower status	0.4141	−0.7564	0.2781	0.8209	0.1791
14. Value	−0.2111	0.8131	−0.3671	0.8394	0.1606

Table 12.5 Estimated factor loadings, communalities, and specific variances, PFM, varimax rotation **Q**MVAfacthous

	Estimated factor loadings			Communalities	Specific variances
	$\widehat{q}_1$	$\widehat{q}_2$	$\widehat{q}_3$	$\widehat{h}_j^2$	$\widehat{\psi}_{jj} = 1 - \widehat{h}_j^2$
1. Crime	0.5477	−0.2558	−0.7387	0.9111	0.0889
2. Large lots	−0.6148	0.2668	0.1281	0.4655	0.5345
3. Nonretail acres	0.6523	−0.3761	−0.3996	0.7266	0.2734
4. Nitric oxides	0.7723	−0.2291	−0.4412	0.8439	0.1561
5. Rooms	−0.1732	0.6783	0.1296	0.0699	0.5046
6. Prior 1940	0.7390	−0.2723	−0.2909	0.7049	0.2951
7. Empl. centers	−0.8565	0.1485	0.3395	0.8708	0.1292
8. Accessibility	0.2855	−0.1359	−0.8460	0.8156	0.1844
9. Tax-rate	0.3062	−0.2656	−0.8174	0.8325	0.1675
10. Pupil/Teacher	0.2116	−0.3943	−0.3297	0.3090	0.6910
11. African American	0.1994	0.0666	0.4217	0.2433	0.7567
12. Lower status	0.4005	−0.7743	−0.2706	0.8333	0.1667
13. Value	−0.1885	0.8400	0.3473	0.8611	0.1389

12.5   Exercises

Exercise 12.1 In Example 12.4 we have computed $\widehat{Q}$ and $\widehat{\Psi}$ using the method of principal factors. We used a two-step iteration for $\widehat{\Psi}$. Perform the third iteration step and compare the results (i.e., use the given $\widehat{Q}$ as a pre-estimate to find the final Ψ).

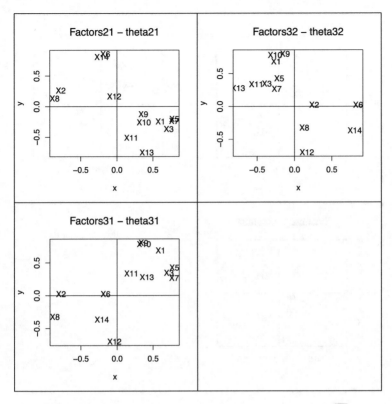

Fig. 12.4 Factor analysis for Boston housing data, PCM after varimax rotation MVAfacthous

Exercise 12.2 Using the bank data set, how many factors can you find with the Method of Principal Factors?

Exercise 12.3 Repeat Exercise 12.2 with the U.S. company data set!

Exercise 12.4 Generalize the two-dimensional rotation matrix in Sect. 12.2 to n-dimensional space.

Exercise 12.5 Compute the orthogonal factor model for

$$\Sigma = \begin{pmatrix} 1 & 0.9 & 0.7 \\ 0.9 & 1 & 0.4 \\ 0.7 & 0.4 & 1 \end{pmatrix}.$$

[Solution: $\psi_{11} = -0.575, q_{11} = 1.255$]

Exercise 12.6 Perform a factor analysis on the type of families in the French food data set. Rotate the resulting factors in a way which provides the most reasonable interpretation. Compare your result with the varimax method.

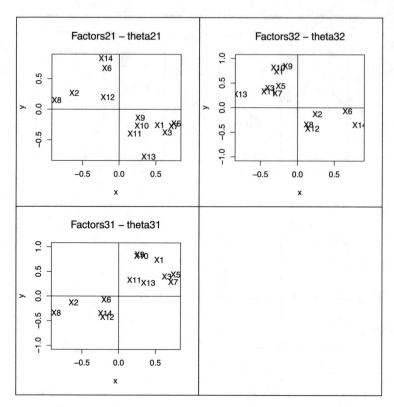

Fig. 12.5 Factor analysis for Boston housing data, PFM after varimax rotation 🔵MVAfacthous

Exercise 12.7 Perform a factor analysis on the variables X_3–X_9 in the U.S. crime data set (Sect. B.8). Would it make sense to use all of the variables for the analysis?

Exercise 12.8 Perform a factor analysis and estimate the factor scores using the U.S. crime data in Sect. B.8. Compare the estimated factor scores of the two data sets.

Exercise 12.9 Analyze the vocabulary data given in Sect. B.13.

References

M.S. Bartlett, A note on multiplying factors for various chi-squared approximations. JRSSB **16**, 296–298 (1954)

R.A. Johnson, D.W. Wichern, *Applied Multivariate Analysis*, 4th edn. (Prentice Hall, Englewood Cliffs, NJ, 1998)

H.F. Kaiser, The varimax criterion for analytic rotation in factor analysis. Psychometrika **23**, 187–200 (1985)

K.V. Mardia, J.T. Kent, J.M. Bibby, *Multivariate Analysis* (Academic Press, Duluth, London, 1979)

Chapter 13
Cluster Analysis

The next two chapters address classification issues from two varying perspectives. When considering groups of objects in a multivariate data set, two situations can arise. Given a data set containing measurements on individuals, in some cases we want to see if some natural groups or classes of individuals exist, and in other cases, we want to classify the individuals according to a set of existing groups. Cluster analysis develops tools and methods concerning the former case. That is, given data containing multivariate measurements on a large number of individuals (or objects), the objective is to build some natural subgroups or clusters of individuals. This is done by grouping individuals that are "similar" according to some appropriate criterion.

Once the clusters are obtained, it is generally useful to describe each group using some descriptive tool from Chaps. 1, 10, and 11 to create a better understanding of the differences that exist among the formulated groups. Cluster analysis is in today's language terms often called unsupervised learning. Here, "learning" refers to the fact that the statistician calibrates models to data or vice versa. The term "unsupervised" points to the act of creating a model (here the separation of individuals into groups) out of the observed data. Once groups or clusters are defined one applies *supervised learning*, where the term "supervised" refers to the fact that the group membership is known a priori.

Cluster analysis is applied in many fields such as the natural sciences, the medical sciences, economics, and marketing. In marketing, for instance, it is useful to build and describe the different segments of a market from a survey on potential consumers. An insurance company, on the other hand, might be interested in the distinction among classes of potential customers so that it can derive optimal prices for its services. Other examples are provided below.

Discriminant analysis presented in Chap. 14 addresses the other issue of classification. It focuses on situations where the different groups are known *a priori*. Decision rules are provided in classifying a multivariate observation into one of the known groups. In today's lingo one speaks of *unsupervised learning* for cluster analysis

© Springer Nature Switzerland AG 2019
W. K. Härdle and L. Simar, *Applied Multivariate Statistical Analysis*,
https://doi.org/10.1007/978-3-030-26006-4_13

since the data has to tell the group membership. By contrast *supervised learning* is the task used for discriminant analysis since the group membership is given.

Section 13.1 states the problem of cluster analysis where the criterion chosen to measure the similarity among objects clearly plays an important role. Section 13.2 shows how to precisely measure the proximity between objects. Section 13.3 explains some traditional clustering hierarchical and partitioning clustering techniques, while Sects. 13.4 and 13.5 are focused on recently introduced unsupervised learning techniques: adaptive weights clustering (AWC) and spectral clustering. Finally, Sect. 13.6 provides some real data applications of described clustering methods.

13.1   The Problem

Cluster analysis is a set of tools for building groups (clusters) from multivariate data objects. The aim is to construct groups with homogeneous properties out of heterogeneous large samples. The groups or clusters should be as homogeneous as possible and the differences among the various groups as large as possible. Cluster analysis can be divided into two fundamental steps.

1. *Choice of a proximity measure*:
   One checks each pair of observations (objects) for the similarity of their values. A similarity (proximity) measure is defined to measure the "closeness" of the objects. The "closer" they are, the more homogeneous they are.
2. *Choice of group-building algorithm*:
   On the basis of the proximity measures, the objects assigned to groups so that differences between groups become large and observations in a group become as close as possible.

In marketing, for example, cluster analysis is used to select test markets. Other applications include the classification of companies according to their organizational structures, technologies, and types. In psychology, cluster analysis is used to find types of personalities on the basis of questionnaires. In archeology, it is applied to classify art objects in different time periods. Other scientific branches that use cluster analysis are medicine, sociology, linguistics, and biology. In each case, a heterogeneous sample of objects is analyzed with the aim to identify homogeneous subgroups.

Summary
↪   Cluster analysis is a set of tools for building groups (clusters) from multivariate data objects.
↪   The methods used are usually divided into two fundamental steps: The choice of a proximity measure and the choice of a group-building algorithm.

13.2   The Proximity Between Objects

The starting point of a cluster analysis is a data matrix $\mathcal{X}(n \times p)$ with n measurements (objects) of p variables. The proximity (similarity) among objects is described by a matrix $\mathcal{D}(n \times n)$

$$
\mathcal{D} = \begin{pmatrix} d_{11} & d_{12} & \cdots & \cdots & \cdots & d_{1n} \\ \vdots & d_{22} & & & & \vdots \\ \vdots & \vdots & \ddots & & & \vdots \\ \vdots & \vdots & & \ddots & & \vdots \\ \vdots & \vdots & & & \ddots & \vdots \\ d_{n1} & d_{n2} & \cdots & \cdots & \cdots & d_{nn} \end{pmatrix} . \tag{13.1}
$$

The matrix $\mathcal{D}$ contains measures of similarity or dissimilarity among the n objects. If the values d_{ij} are distances, then they measure dissimilarity. The greater the distance, the less similar the objects are. If the values d_{ij} are proximity measures, then the opposite is true, i.e., the greater the proximity value, the more similar the objects are. A distance matrix, for example, could be defined by the L_2-norm: $d_{ij} = \|x_i - x_j\|_2$, where x_i and x_j denote the rows of the data matrix $\mathcal{X}$. Distance and similarity are of course dual. If d_{ij} is a distance, then $d'_{ij} = \max_{i,j}\{d_{ij}\} - d_{ij}$ is a proximity measure.

The nature of the observations plays an important role in the choice of proximity measure. Nominal values (like binary variables) lead in general to proximity values, whereas metric values lead (in general) to distance matrices. We first present possibilities for $\mathcal{D}$ in the binary case and then consider the continuous case.

Similarity of Objects with Binary Structure

In order to measure the similarity between objects we always compare pairs of observations (x_i, x_j) where $x_i^\top = (x_{i1}, \ldots, x_{ip})$, $x_j^\top = (x_{j1}, \ldots, x_{jp})$, and $x_{ik}, x_{jk} \in \{0, 1\}$. Obviously there are four cases:

$$
\begin{aligned}
x_{ik} &= x_{jk} = 1, \\
x_{ik} &= 0, x_{jk} = 1, \\
x_{ik} &= 1, x_{jk} = 0, \\
x_{ik} &= x_{jk} = 0.
\end{aligned}
$$

Define

$$
a_1 = \sum_{k=1}^{p} \mathrm{I}(x_{ik} = x_{jk} = 1),
$$

$$
a_2 = \sum_{k=1}^{p} \mathrm{I}(x_{ik} = 0, x_{jk} = 1),
$$

$$a_3 = \sum_{k=1}^{p} I(x_{ik} = 1, x_{jk} = 0),$$

$$a_4 = \sum_{k=1}^{p} I(x_{ik} = x_{jk} = 0).$$

Note that each a_l, $l = 1, \ldots, 4$, depends on the pair (x_i, x_j).

The following proximity measures are used in practice:

$$d_{ij} = \frac{a_1 + \delta a_4}{a_1 + \delta a_4 + \lambda(a_2 + a_3)} \qquad (13.2)$$

where δ and λ are weighting factors. Table 13.1 shows some similarity measures for given weighting factors.

These measures provide alternative ways of weighting mismatchings and positive (presence of a common character) or negative (absence of a common character) matchings. In principle, we could also consider the Euclidean distance. However, the disadvantage of this distance is that it treats the observations 0 and 1 in the same way. If $x_{ik} = 1$ denotes, say, knowledge of a certain language, then the contrary, $x_{ik} = 0$ (not knowing the language) should eventually be treated differently.

Example 13.1 Let us consider binary variables computed from the Car Marks data set (Sect. 22.7). We define the new binary data by

$$y_{ik} = \begin{cases} 1 & \text{if } x_{ik} > \bar{x}_k, \\ 0 & \text{otherwise,} \end{cases}$$

for $i = 1, \ldots, n$ and $k = 1, \ldots, p$. This means that we transform the observations of the k-th variable to 1 if it is larger than the mean value of all observations of the k-th variable. Let us only consider the data points 17–19 (Renault 19, Rover and Toyota Corolla) which lead to (3×3) distance matrices. The Jaccard measure gives the similarity matrix

Table 13.1 The common similarity coefficients

Name	δ	λ	Definition
Jaccard	0	1	$\frac{a_1}{a_1 + a_2 + a_3}$
Tanimoto	1	2	$\frac{a_1 + a_4}{a_1 + 2(a_2 + a_3) + a_4}$
Simple Matching (M)	1	1	$\frac{a_1 + a_4}{p}$
Russel and Rao (RR)	–	–	$\frac{a_1}{p}$
Dice	0	0.5	$\frac{2a_1}{2a_1 + (a_2 + a_3)}$
Kulczynski	–	–	$\frac{a_1}{a_2 + a_3}$

$$D = \begin{pmatrix} 1.000 & 0.000 & 0.400 \\ & 1.000 & 0.167 \\ & & 1.000 \end{pmatrix},$$

the Tanimoto measure yields

$$D = \begin{pmatrix} 1.000 & 0.000 & 0.455 \\ & 1.000 & 0.231 \\ & & 1.000 \end{pmatrix},$$

whereas the Simple Matching measure gives

$$D = \begin{pmatrix} 1.000 & 0.000 & 0.625 \\ & 1.000 & 0.375 \\ & & 1.000 \end{pmatrix}.$$

Distance Measures for Continuous Variables

A wide variety of distance measures can be generated by the L_r-norms, $r \geq 1$,

$$d_{ij} = ||x_i - x_j||_r = \left\{ \sum_{k=1}^{p} |x_{ik} - x_{jk}|^r \right\}^{1/r}. \qquad (13.3)$$

Here x_{ik} denotes the value of the k-th variable on object i. It is clear that $d_{ii} = 0$ for $i = 1, \ldots, n$. The class of distances (13.3) for varying r measures the dissimilarity of different weights. The L_1-metric, for example, gives less weight to outliers than the L_2-norm (Euclidean norm). It is common to consider the squared L_2-norm.

Example 13.2 Suppose we have $x_1 = (0, 0)$, $x_2 = (1, 0)$ and $x_3 = (5, 5)$. Then the distance matrix for the L_1-norm is

$$D_1 = \begin{pmatrix} 0 & 1 & 10 \\ 1 & 0 & 9 \\ 10 & 9 & 0 \end{pmatrix},$$

and for the squared L_2- or Euclidean norm

$$D_2 = \begin{pmatrix} 0 & 1 & 50 \\ 1 & 0 & 41 \\ 50 & 41 & 0 \end{pmatrix}.$$

One can see that the third observation x_3 receives much more weight in the squared L_2-norm than in the L_1-norm.

An underlying assumption in applying distances based on L_r-norms is that the variables are measured on the same scale. If this is not the case, a standardization should first be applied. This corresponds to using a more general L_2- or Euclidean norm with a metric $\mathcal{A}$, where $\mathcal{A} > 0$ (see Sect. 2.6):

$$d_{ij}^2 = \|x_i - x_j\|_{\mathcal{A}} = (x_i - x_j)^\top \mathcal{A}(x_i - x_j). \tag{13.4}$$

L_2-norms are given by $\mathcal{A} = \mathcal{I}_p$, but if a standardization is desired, then the weight matrix $\mathcal{A} = \mathrm{diag}(s_{X_1 X_1}^{-1}, \ldots, s_{X_p X_p}^{-1})$ may be suitable. Recall that $s_{X_k X_k}$ is the variance of the k-th component. Hence we have

$$d_{ij}^2 = \sum_{k=1}^{p} \frac{(x_{ik} - x_{jk})^2}{s_{X_k X_k}}. \tag{13.5}$$

Here each component has the same weight in the computation of the distances and the distances do not depend on a particular choice of the units of measure.

Example 13.3 Consider the French Food expenditures (Sect. B.6). The Euclidean distance matrix (squared L_2-norm) is

$$\mathcal{D} = 10^4 \cdot \begin{pmatrix} 0.00 & 5.82 & 58.19 & 3.54 & 5.15 & 151.44 & 16.91 & 36.15 & 147.99 & 51.84 & 102.56 & 271.83 \\ & 0.00 & 41.73 & 4.53 & 2.93 & 120.59 & 13.52 & 25.39 & 116.31 & 43.68 & 76.81 & 226.87 \\ & & 0.00 & 44.14 & 40.10 & 24.12 & 29.95 & 8.17 & 25.57 & 20.81 & 20.30 & 88.62 \\ & & & 0.00 & 0.76 & 127.85 & 5.62 & 21.70 & 124.98 & 31.21 & 72.97 & 231.57 \\ & & & & 0.00 & 121.05 & 5.70 & 19.85 & 118.77 & 30.82 & 67.39 & 220.72 \\ & & & & & 0.00 & 96.57 & 48.16 & 1.80 & 60.52 & 28.90 & 29.56 \\ & & & & & & 0.00 & 9.20 & 94.87 & 11.07 & 42.12 & 179.84 \\ & & & & & & & 0.00 & 46.95 & 6.17 & 18.76 & 113.03 \\ & & & & & & & & 0.00 & 61.08 & 29.62 & 31.86 \\ & & & & & & & & & 0.00 & 15.83 & 116.11 \\ & & & & & & & & & & 0.00 & 53.77 \\ & & & & & & & & & & & 0.00 \end{pmatrix} \cdot$$

Taking the weight matrix $\mathcal{A} = \mathrm{diag}(s_{X_1 X_1}^{-1}, \ldots, s_{X_7 X_7}^{-1})$, we obtain the distance matrix (squared L_2-norm)

$$\mathcal{D} = \begin{pmatrix} 0.00 & 6.85 & 10.04 & 1.68 & 2.66 & 24.90 & 8.28 & 8.56 & 24.61 & 21.55 & 30.68 & 57.48 \\ & 0.00 & 13.11 & 6.59 & 3.75 & 20.12 & 13.13 & 12.38 & 15.88 & 31.52 & 25.65 & 46.64 \\ & & 0.00 & 8.03 & 7.27 & 4.99 & 9.27 & 3.88 & 7.46 & 14.92 & 15.08 & 26.89 \\ & & & 0.00 & 0.64 & 20.06 & 2.76 & 3.82 & 19.63 & 12.81 & 19.28 & 45.01 \\ & & & & 0.00 & 17.00 & 3.54 & 3.81 & 15.76 & 14.98 & 16.89 & 39.87 \\ & & & & & 0.00 & 17.51 & 9.79 & 1.58 & 21.32 & 11.36 & 13.40 \\ & & & & & & 0.00 & 1.80 & 17.92 & 4.39 & 9.93 & 33.61 \\ & & & & & & & 0.00 & 10.50 & 5.70 & 7.97 & 24.41 \\ & & & & & & & & 0.00 & 24.75 & 11.02 & 13.07 \\ & & & & & & & & & 0.00 & 9.13 & 29.78 \\ & & & & & & & & & & 0.00 & 9.39 \\ & & & & & & & & & & & 0.00 \end{pmatrix} \cdot \tag{13.6}$$

When applied to contingency tables, a χ^2-metric is suitable to compare (and cluster) rows and columns of a contingency table.

If $\mathcal{X}$ is a contingency table, row i is characterized by the conditional frequency distribution $\frac{x_{ij}}{x_{i\bullet}}$, where $x_{i\bullet} = \sum_{j=1}^{p} x_{ij}$ indicates the marginal distributions over the rows: $\frac{x_{i\bullet}}{x_{\bullet\bullet}}$, $x_{\bullet\bullet} = \sum_{i=1}^{n} x_{i\bullet}$. Similarly, column j of $\mathcal{X}$ is characterized by the conditional frequencies $\frac{x_{ij}}{x_{\bullet j}}$, where $x_{\bullet j} = \sum_{i=1}^{n} x_{ij}$. The marginal frequencies of the columns are $\frac{x_{\bullet j}}{x_{\bullet\bullet}}$.

The distance between two rows, i_1 and i_2, corresponds to the distance between their respective frequency distributions. It is common to define this distance using the χ^2-metric:

$$d^2(i_1, i_2) = \sum_{j=1}^{p} \frac{1}{\left(\frac{x_{\bullet j}}{x_{\bullet\bullet}}\right)} \left(\frac{x_{i_1 j}}{x_{i_1 \bullet}} - \frac{x_{i_2 j}}{x_{i_2 \bullet}}\right)^2. \tag{13.7}$$

Note that this can be expressed as a distance between the vectors $x_1 = \left(\frac{x_{i_1 j}}{x_{\bullet\bullet}}\right)$ and $x_2 = \left(\frac{x_{i_2 j}}{x_{\bullet\bullet}}\right)$ as in (13.4) with weighting matrix $\mathcal{A} = \left\{\text{diag}\left(\frac{x_{\bullet j}}{x_{\bullet\bullet}}\right)\right\}^{-1}$. Similarly, if we are interested in clusters among the columns, we can define:

$$d^2(j_1, j_2) = \sum_{i=1}^{n} \frac{1}{\left(\frac{x_{i\bullet}}{x_{\bullet\bullet}}\right)} \left(\frac{x_{ij_1}}{x_{\bullet j_1}} - \frac{x_{ij_2}}{x_{\bullet j_2}}\right)^2.$$

Example 13.4 Cluster Analysis for the US Crime data. This is a data set consisting of 50 measurements of 7 variables. It states for 1 year (1985) the reported number of crimes in the 50 states of the United States classified according to 7 categories $(X_3 - X_9)$:

Variable	Description
X_1	Land area (land)
X_2	Population 1985 (popu 1985)
X_3	Murder (murd)
X_4	Rape
X_5	Robbery (robb)
X_6	Assault (assa)
X_7	Burglary (burg)
X_8	Larceny (larc)
X_9	Auto theft (auto)
X_{10}	U.S. states region number (reg)
X_{11}	U.S. states division number (div)

For our example we used just the seven crime variables. The matrix is interpreted as contingency table. The distance matrix $\mathcal{D}$ is given in (13.8).

$$\mathcal{D} = \begin{pmatrix} 0.000\ 0.004\ 0.002\ 0.230\ 0.142\ 0.034\ \ldots\ 0.009 \\ 0.000\ 0.011\ 0.172\ 0.098\ 0.014\ \ldots\ 0.001 \\ 0.000\ 0.272\ 0.176\ 0.051\ \ldots\ 0.019 \\ 0.000\ 0.010\ 0.087\ \ldots\ 0.146 \\ 0.000\ 0.037\ \ldots\ 0.079 \\ 0.000\ \ldots\ 0.008 \\ \ddots\quad \vdots \\ 0.000 \end{pmatrix} \quad (13.8)$$

Q MVAclususcrime

Apart from the Euclidean and the L_r-norm measures one can use a proximity measure such as the correlation coefficient

$$d_{ij} = \frac{\sum_{k=1}^{p}(x_{ik} - \overline{x}_i)(x_{jk} - \overline{x}_j)}{\left\{\sum_{k=1}^{p}(x_{ik} - \overline{x}_i)^2 \sum_{k=1}^{p}(x_{jk} - \overline{x}_j)^2\right\}^{1/2}}. \quad (13.9)$$

Here $\overline{x}_i$ denotes the mean over the variables $(x_{i1}, \ldots, x_{ip})$. Please note that (13.9) is the same as cosine similarity as defined in Sect. 2.6.

Summary
↪ The proximity between data points is measured by a distance or similarity matrix $\mathcal{D}$ whose components d_{ij} give the similarity coefficient or the distance between two points x_i and x_j.
↪ A variety of similarity (distance) measures exist for binary data (e.g., Jaccard, Tanimoto, and Simple Matching coefficients) and for continuous data (e.g., L_r-norms).
↪ The nature of the data could impose the choice of a particular metric $\mathcal{A}$ in defining the distances (standardization, χ^2-metric etc.).

13.3   Cluster Algorithms

There are essentially three traditional clustering methods: hierarchical and partitioning algorithms. The hierarchical algorithms can be divided into agglomerative and splitting procedures. The first type of hierarchical clustering starts from the finest partition possible (each observation forms a cluster) and groups them. The second type starts with the coarsest partition possible: one cluster contains all of the observations. It proceeds by splitting the single cluster up into smaller sized clusters.

The partitioning algorithms start from a given group definition and proceed by exchanging elements between groups until a certain score is optimized. The main difference between the two clustering techniques is that in hierarchical clustering once groups are found and elements are assigned to the groups, this assignment cannot be changed. In partitioning techniques, on the other hand, the assignment of objects into groups may change during the algorithm application.

Partitioning Algorithms

Partitional clustering indicates a popular class of methods to find clusters in a set of data points. Here, the number of clusters k is fixed a priori. The points are embedded in a metric space, so that each vertex is a point and a distance measure is defined between pairs of points in the space. The distance is a measure of dissimilarity between vertices. The goal is to separate the points in k clusters such to maximize or minimize a given objective function based on distances between points and/or from points to centroids.

The most popular partitional technique in the literature is *k-means* clustering . The *k-means* standard algorithm is iterative and is starting from random partitions/points. The objective function for this algorithm is the total intra-cluster distance or squared error function. The equation to be minimized is given in (13.10).

$$\hat{S} = \underset{S}{\text{argmin}} \sum_{j=1}^{k} \sum_{i \in S_j} \|x_i - \mu_j\|^2 \qquad (13.10)$$

with respect to $S = \{S_1, \ldots, S_k\}$, $\bigcup_{j=1}^{k} S_j = \{1, 2, \ldots, n\}$, where S_j indicates the subset of points in cluster j and μ_j its centroid, k is the number of clusters, n is the number of data points. The *k-means* problem can be simply solved with the standard Lloyd's Algorithm 13.1. The solution found can be not optimal, and it

---

Algorithm 13.1 *k-means* Clustering Standard Algorithm

Fix an initial set $\{\mu_j^{(t)}\}_{j=1}^{k}$, $t = 1$

Assign $\hat{j}(i) = \underset{j}{\text{argmin}} \|x_i - \mu_j^{(t)}\|^2$

x_i belongs then to cluster $\hat{j}(i)$ resulting in (new) partition

$$\bigcup_{j=1}^{k} S_j^{(t)} = \{1, \ldots, n\}$$

Update $\mu_j^{(t+1)} = \left(\#S_j^{(t)}\right)^{-1} \sum_{i \in S_j^{(t)}} x_i$

repeat
    assign, update
until convergence in terms of (13.10).

---

strongly depends on the initial choice of the centroids. The result can be improved by performing more runs starting from different initial conditions, and picking the solution which yields the minimum value of the total intra-cluster distance. Other methods claiming to choose better initial centroids are *k-means ++*, *intelligent k-means* and *genetic k-means* (Arthur and Vassilvitskii 2007).

Fig. 13.1 8 points—*k-means*—clustering

MVAclus8km

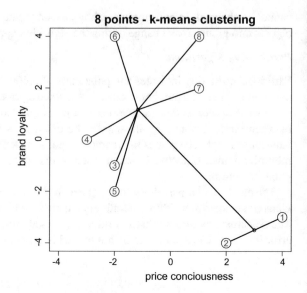

Fig. 13.2 8 points—*k-median*—clustering

MVAclus8km

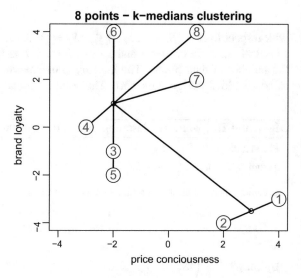

Example 13.5 Figure 13.1 demonstrates partition to two clusters of 8 points performed by *k-means* method with the standard (Lloyd) algorithm (Lloyd 1982).

Some of other popular partitioning methods are listed below:

- *k-medoids*: for each cluster j one defines a reference point with position x_j, the most centrally located object in the cluster—medoid. Therefore, the difference between *k-means* and *k-medoids* is *k-means* can select the k virtual centroids, for *k-medoids* centroids should be k representatives of real objects (Fig. 13.3).

Fig. 13.3 8 points—*k-medoids*—clustering

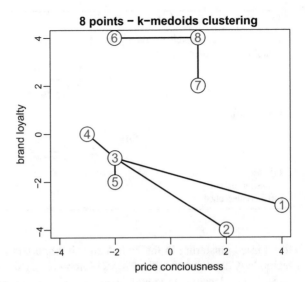

⬤MVAclus8km

- *k-mode*: is an extension to handle with categorical data sets, by replacing means of clusters with modes.
- *k-median*: algorithm was developed to overcome the issue of *k-means'* high sensitivity to outliers, because empirical mean is easily influenced by extremes. Instead of calculating the mean for each cluster to determine its centroid, one calculates the median with respect to coordinates (Fig. 13.2). The objective function uses Manhattan distance instead of squared Euclidean distance and therefore (13.10) can be reformulated for this algorithm as follows:

$$\hat{S} = \underset{S}{\operatorname{argmin}} \sum_{j=1}^{k} \sum_{i \in S_j} |x_i - med_j| \tag{13.11}$$

Another popular technique, similar to *k-means* clustering, is *fuzzy k-means* . This method that allows each data point to belong to multiple clusters with varying degrees of membership (Bezdek 1981). The associated objective function to be minimized is given in (13.12).

$$\hat{S} = \underset{S}{\operatorname{argmin}} \sum_{j=1}^{k} \sum_{i \in S_j} u_{i,j} \|x_i - \mu_j\|^2 \tag{13.12}$$

where u_{ij} is the membership matrix, which measures the degree of membership of point i (with position x_i) in cluster j.

The *fuzzy k-means* problem can be solved with the Algorithm 13.2.

It should be noted though that there are some drawbacks of partition clustering methods. The number of clusters k must be preset at the beginning, which might lead

---

Algorithm 13.2 *fuzzy k-means algorithm*

Fix an initial membership matrix $u_{ij}^{(t)}$, $t = 1$
Compute the cluster centers

$$\mu_j = \frac{\sum_{i=1}^{n} u_{ij}^{(t)} x_i}{\sum_{i=1}^{n} u_{ij}^{(t)}}$$

Update $u_{ij}^{(t+1)}$

$$u_{ij}^{(t+1)} = \frac{1}{\sum_{l=1}^{k} \left(\frac{\|x_i - \mu_j\|}{\|x_i - \mu_l\|} \right)^{\frac{2}{t-1}}}$$

repeat
    assign, update
until convergence

---

to a biased clustering in the end. Also it is important to realize that the presented technology cannot deal with clusters of non-convex shape.

Recently developed method *Adaptive Weights Clustering (AWC)*, which is close in spirit *fuzzy* clustering, allows to overcome mentioned limitations of partitioning methods.

Hierarchical Algorithms, Agglomerative Techniques

Agglomerative algorithms are used quite frequently in practice. The algorithm consists of the following steps:

Algorithm 13.3 Hierarchical Algorithms—Agglomerative Technique

---
1: Construct the finest partition
2: Compute the distance matrix $\mathcal{D}$.
3: **repeat**
4:     Find the two clusters with the closest distance
5:     Put those two clusters into one cluster
6:     Compute the distance between the new groups and obtain a reduced distance matrix $\mathcal{D}$
7: **until** all clusters are agglomerated into $\mathcal{X}$

If two objects or groups say, P and Q, are united, one computes the distance between this new group (object) $P + Q$ and group R using the following distance function:

$$d(R, P + Q) = \delta_1 d(R, P) + \delta_2 d(R, Q) + \delta_3 d(P, Q) + \delta_4 |d(R, P) - d(R, Q)|. \tag{13.13}$$

The δ_j's are weighting factors that lead to different agglomerative algorithms as described in Table 13.2. Here $n_P = \sum_{i=1}^{n} I(x_i \in P)$ is the number of objects in group P. The values of n_Q and n_R are defined analogously.

For the most commonly used Single and Complete linkages, below are the modified agglomerative algorithm steps:

Table 13.2 Computations of group distances

Name	δ_1	δ_2	δ_3	δ_4
Single linkage	1/2	1/2	0	$-1/2$
Complete linkage	1/2	1/2	0	1/2
Average linkage (unweighted)	1/2	1/2	0	0
Average linkage (weighted)	$\dfrac{n_P}{n_P + n_Q}$	$\dfrac{n_Q}{n_P + n_Q}$	0	0
Centroid	$\dfrac{n_P}{n_P + n_Q}$	$\dfrac{n_Q}{n_P + n_Q}$	$-\dfrac{n_P n_Q}{(n_P + n_Q)^2}$	0
Median	1/2	1/2	$-1/4$	0
Ward	$\dfrac{n_R + n_P}{n_R + n_P + n_Q}$	$\dfrac{n_R + n_Q}{n_R + n_P + n_Q}$	$-\dfrac{n_R}{n_R + n_P + n_Q}$	0

Algorithm 13.4 Modified Hierarchical Algorithms—Agglomerative Technique

---

1: Construct the finest partition
2: Compute the distance matrix $\mathcal{D}$.
3: **repeat**
4:   Find the smallest (Single linkage)/ largest (Complete linkage) value d (between objects m and n) in $\mathcal{D}$
5:   If m and n are not in the same cluster, combine the clusters m and n belonging to together, and delete the smallest value
6: **until** all clusters are agglomerated into $\mathcal{X}$ or the value d exceeds the preset level

---

As instead of computing new distance matrices every step, a linear search in the original distance matrix is enough for clustering in the modified algorithm, it is more efficient in practice.

Example 13.6 Let us examine the agglomerative algorithm for the three points in Example 13.2, $x_1 = (0, 0)$, $x_2 = (1, 0)$ and $x_3 = (5, 5)$, and the squared Euclidean distance matrix with single linkage weighting. The algorithm starts with $N = 3$ clusters: $P = \{x_1\}$, $Q = \{x_2\}$ and $R = \{x_3\}$. The distance matrix $\mathcal{D}_2$ is given in Example 13.2. The smallest distance in $\mathcal{D}_2$ is the one between the clusters P and Q. Therefore, applying step 4 in the above algorithm we combine these clusters to form $P + Q = \{x_1, x_2\}$. The single linkage distance between the remaining two clusters is from Table 13.2 and (13.13) equal to

$$d(R, P + Q) = \frac{1}{2}d(R, P) + \frac{1}{2}d(R, Q) - \frac{1}{2}|d(R, P) - d(R, Q)| \quad (13.14)$$

$$= \frac{1}{2}d_{13} + \frac{1}{2}d_{23} - \frac{1}{2} \cdot |d_{13} - d_{23}|$$

$$= \frac{50}{2} + \frac{41}{2} - \frac{1}{2} \cdot |50 - 41|$$

$$= 41.$$

Fig. 13.4 The 8-point
example MVAclus8p

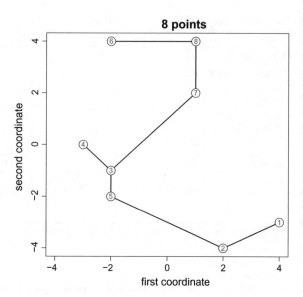

The reduced distance matrix is then $\begin{pmatrix} 0 & 41 \\ 41 & 0 \end{pmatrix}$. The next and last step is to unite the clusters R and $P + Q$ into a single cluster $\mathcal{X}$, the original data matrix.

When there are more data points than in the example above, a visualization of the implication of clusters is desirable. A graphical representation of the sequence of clustering is called a *dendrogram*. It displays the observations, the sequence of clusters and the distances between the clusters. The vertical axis displays the indices of the points, whereas the horizontal axis gives the distance between the clusters. Large distances indicate the clustering of heterogeneous groups. Thus, if we choose to "cut the tree" at a desired level, the branches describe the corresponding clusters.

Example 13.7 Here we describe the single linkage algorithm for the eight data points displayed in Fig. 13.4. The distance matrix (squared L_2-norms) is

$$\mathcal{D} = \begin{pmatrix} 0 & 5 & 40 & 58 & 37 & 85 & 34 & 58 \\ & 0 & 25 & 41 & 20 & 80 & 37 & 65 \\ & & 0 & 2 & 1 & 25 & 18 & 34 \\ & & & 0 & 5 & 17 & 20 & 32 \\ & & & & 0 & 36 & 25 & 45 \\ & & & & & 0 & 13 & 9 \\ & & & & & & 0 & 4 \\ & & & & & & & 0 \end{pmatrix}$$

and the dendrogram is shown in Fig. 13.5.

If we decide to cut the tree at the level 10, three clusters are defined: $\{1, 2\}$, $\{3, 4, 5\}$, and $\{6, 7, 8\}$.

Fig. 13.5 The dendrogram for the 8-point example, Single linkage algorithm

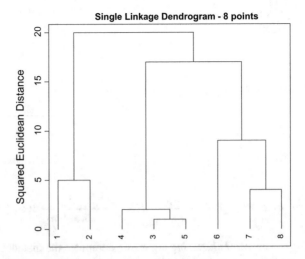

○MVAclus8p

The *single linkage* algorithm defines the distance between two groups as the smallest value of the individual distances. Table 13.2 shows that in this case

$$d(R, P + Q) = \min\{d(R, P), d(R, Q)\}. \tag{13.15}$$

This algorithm is also called the *Nearest Neighbor* algorithm. As a consequence of its construction, single linkage tends to build large groups. Groups that differ but are not well separated may thus be classified into one group as long as they have two approximate points. The *complete linkage* algorithm tries to correct this kind of grouping by considering the largest (individual) distances. Indeed, the complete linkage distance can be written as

$$d(R, P + Q) = \max\{d(R, P), d(R, Q)\}. \tag{13.16}$$

It is also called the *Farthest Neighbor* algorithm. This algorithm will cluster groups where all the points are proximate, since it compares the largest distances. The *average linkage* algorithm (weighted or unweighted) proposes a compromise between the two preceding algorithms, in that it computes an average distance:

$$d(R, P + Q) = \frac{n_P}{n_P + n_Q} d(R, P) + \frac{n_Q}{n_P + n_Q} d(R, Q). \tag{13.17}$$

The *centroid* algorithm is quite similar to the average linkage algorithm and uses the natural geometrical distance between R and the weighted center of gravity of P and Q (see Fig. 13.6):

Fig. 13.6 The centroid
algorithm

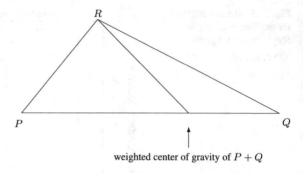

weighted center of gravity of $P + Q$

$$d(R, P + Q) = \frac{n_P}{n_P + n_Q} d(R, P) + \frac{n_Q}{n_P + n_Q} d(R, Q) - \frac{n_P n_Q}{(n_P + n_Q)^2} d(P, Q).$$

$$(13.18)$$

The *Ward clustering* algorithm computes the distance between groups according to the formula in Table 13.2. The main difference between this algorithm and the linkage procedures is in the unification procedure. The Ward algorithm does not put together groups with smallest distance. Instead, it joins groups that do not increase a given measure of heterogeneity "too much". The aim of the Ward procedure is to unify groups such that the variation inside these groups does not increase too drastically: the resulting groups are as homogeneous as possible.

The heterogeneity of group R is measured by the inertia inside the group. This inertia is defined as follows:

$$I_R = \frac{1}{n_R} \sum_{i=1}^{n_R} d^2(x_i, \bar{x}_R) \qquad (13.19)$$

where $\bar{x}_R$ is the center of gravity (mean) over the groups. I_R clearly provides a scalar measure of the dispersion of the group around its center of gravity. If the usual Euclidean distance is used, then I_R represents the sum of the variances of the p components of x_i inside group R.

When two objects or groups P and Q are joined, the new group $P + Q$ has a larger inertia I_{P+Q}. It can be shown that the corresponding increase of inertia is given by

$$\Delta(P, Q) = \frac{n_P n_Q}{n_P + n_Q} \, d^2(P, Q). \qquad (13.20)$$

In this case, the Ward algorithm is defined as an algorithm that "joins the groups that give the smallest increase in $\Delta(P, Q)$". It is easy to prove that when P and Q are joined, the new criterion values are given by (13.13) along with the values of δ_i given in Table 13.2, when the centroid formula is used to modify $d^2(R, P + Q)$. So, the Ward algorithm is related to the centroid algorithm, but with an "inertial" distance Δ rather than the "geometric" distance d^2.

Fig. 13.7 PCA for 20 randomly chosen bank notes

MVAclusbank

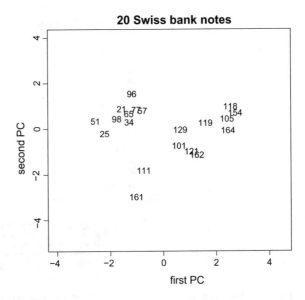

As pointed out in Sect. 13.2, all the algorithms above can be adjusted by the choice of the metric $\mathcal{A}$ defining the geometric distance d^2. If the results of a clustering algorithm are illustrated as graphical representations of individuals in spaces of low dimension (using principal components (normalized or not) or using a correspondence analysis for contingency tables), it is important to be coherent in the choice of the metric used.

Example 13.8 As an example we randomly select 20 observations from the bank notes data and apply the Ward technique using Euclidean distances. Figure 13.7 shows the first two PCs of these data, Fig. 13.8 displays the dendrogram.

Example 13.9 Consider the French food expenditures data set. As in Chap. 11 we use the standardized data which is equivalent to $\mathcal{A} = \text{diag}(s_{X_1 X_1}^{-1}, \ldots, s_{X_7 X_7}^{-1})$ as the weight matrix in the L_2-norm. The NPCA plot of the individuals was given in Fig. 11.7. The Euclidean distance matrix is of course given by (13.6). The dendrogram obtained by using the Ward algorithm is shown in Fig. 13.9.

If the aim was to have only two groups, as can be seen in Fig. 13.9, they would be {CA2, CA3, CA4, CA5, EM5} and {MA2, MA3, MA4, MA5, EM2, EM3, EM4}. Clustering three groups is somewhat arbitrary (the levels of the distances are too similar). If we were interested in four groups, we would obtain {CA2, CA3, CA4}, {EM2, MA2, EM3, MA3}, {EM4, MA4, MA5}, and {EM5, CA5}. This grouping shows a balance between socio-professional levels and size of the families in determining the clusters. The four groups are clearly well represented in the NPCA plot in Fig. 11.7.

Fig. 13.8 The dendrogram
for the 20 bank notes, Ward
algorithm 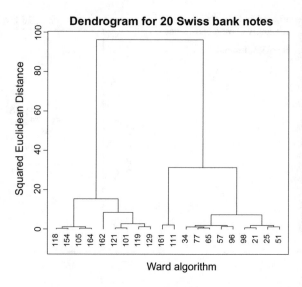MVAclusbank

Fig. 13.9 The dendrogram
for the French food
expenditures, Ward
algorithm 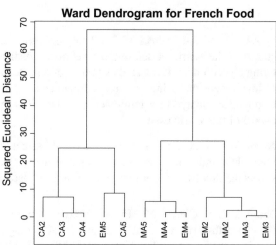MVAclusfood

Summary

↪   The class of clustering algorithms can be divided into two types: hierarchical
    and partitioning algorithms. Hierarchical algorithms start with the finest
    (coarsest) possible partition and put groups together (split groups apart)
    step by step. Partitioning algorithms start from a preliminary clustering and
    exchange group elements until a certain score is reached.

Continued on next page

Summary (continue)
$\hookrightarrow$ Hierarchical agglomerative techniques are frequently used in practice. They start from the finest possible structure (each data point forms a cluster), compute the distance matrix for the clusters and join the clusters that have the smallest distance. This step is repeated until all points are united in one cluster.
$\hookrightarrow$ The agglomerative procedure depends on the definition of the distance between two clusters. Single linkage, complete linkage, and Ward distance are frequently used distances.
$\hookrightarrow$ The process of the unification of clusters can be graphically represented by a dendrogram.

13.4   Adaptive Weights Clustering

An alternative clustering technique may be based on nonparametric ideas of finding cluster structure through a separation approach via a homogeneity detection test. We follow here the exposition of Efimov et al. (2017). The method is fully adaptive and does not require to specify the number of clusters or their structure. The clustering results are not sensitive to noise and outliers, the procedure is able to recover different clusters with sharp edges or manifold structure.

Let $\{X_i\}_{i=1}^{n} \subset \mathbb{R}^p$ be the observations. Here the dimension p can be very large or even infinite. Calculate for any pair (X_i, X_j) the distance $d(X_i, X_j)$ between X_i and X_j e.g., the Euclidean norm $\|X_i - X_j\|$. The proposed procedure operates with the distance (or similarity) matrix $\mathcal{D} = \big(d(X_i, X_j)\big)_{i,j=1}^{n}$ only. For describing the clustering structure of the data, we introduce a $n \times n$ matrix of weights $W = (w_{ij})$, $i, j = 1 \ldots n$. Usually the weights w_{ij} are binary and $w_{ij} = 1$ means that X_i and X_j are in the same cluster, while $w_{ij} = 0$ indicates that these points are in different clusters. The matrix W should be symmetric and each block of ones describes one cluster. Below we do not require a block structure which allows to incorporate even overlapping clusters. For every fixed i, the associated cluster C_i is given by the collection of positive weights (w_{ij}) overall j.

The proposed procedure attempts to recover the weights w_{ij} from the data, which explains the name "adaptive weights clustering". The procedure is sequential. It starts with very local clustering structure $C_i^{(0)}$, that is, the starting positive weights $w_{ij}^{(0)}$ are limited to the closest neighbors X_j of the point X_i in terms of the distance $d(X_i, X_j)$. At each step $k \geq 1$, the weights $w_{ij}^{(k)}$ are recomputed by means of statistical tests of "no gap" between $C_i^{(k-1)}$ and $C_j^{(k-1)}$, the local clusters on step $k - 1$ for points x_i and x_j correspondingly. Only the neighbor pairs X_i, X_j with $d(X_i, X_j) \leq h_k$ are checked, however the locality parameter h_k and the number of neighbors X_j for each fixed point X_i grows in each step. The resulting matrix of weights W is used for the final clustering. The core element of the method is the way how the weights $w_{ij}^{(k)}$ are recomputed.

Sequence of radii:

One looks at a growing sequence of radii $h_1 \leq h_2 \leq \ldots \leq h_K$ which determines how fast the algorithm will evolve from considering local structures to large-scale objects. Each value h_k can be viewed as a resolution (scale) of the method at step k. For theoretical reasons, the rule has to ensure that the average number of screened neighbors for each X_i at step k grows at most exponentially with $k \geq 1$.

Initialization of weights:

On initialization step we connect each point with its n_0 closest neighbors:

$$w_{ij}^{(0)} = \mathbb{I}\left[d(X_i, X_j) \leq \max\{h_0(X_i), h_0(X_j)\}\right], \tag{13.21}$$

where $h_0(X_i)$ is the distance between X_i and its n_0 closest neighbor, our default choice $n_0 = 2p + 2$.

Updates at step k:

Suppose that the first $k - 1$ steps of the iterative procedure have been carried out. This results in collection of weights $\{w_{ij}^{(k-1)}, \ j = 1, \ldots, n\}$ for each point X_i. These weights describe a local "cluster" associated with X_i. By construction, only those weights $w_{ij}^{(k-1)}$ can be positive for which X_j belongs to the ball $B(X_i, h_{k-1}) = \{x: d(X_i, x) \leq h_{k-1}\}$, or, equivalently, $d(X_i, X_j) \leq h_{k-1}$. At the next step k we pick up a larger radius h_k and recompute the weights $w_{ij}^{(k)}$ using the previous results. Again, only points with $d(X_i, X_j) \leq h_k$ have to be screened at step k. The basic idea behind the definition of $w_{ij}^{(k)}$ is to check for each pair i, j with $d(X_i, X_j) \leq h_k$ whether the related clusters are well separated or they can be aggregated into one homogeneous region. The test compares the data density in the union and overlap of two clusters for points X_i and X_j. We treat the points X_i and X_j as fixed and compute the test statistic $T_{ij}^{(k)}$ using the weights $w_{ij}^{(k-1)}$ from the preceding step. The formal definition involves the weighted empirical mass of the overlap and the weighted empirical mass of the union of two balls $B(X_i, h_{k-1})$ and $B(X_j, h_{k-1})$ shown on Fig. 13.10. *The empirical mass of the overlap $N_{i \wedge j}^{(k)}$ can be naturally defined as*

$$N_{i \wedge j}^{(k)} = \sum_{l \neq i, j} w_{il}^{(k-1)} w_{jl}^{(k-1)}. \tag{13.22}$$

In the case of indicator weights $w_{ij}^{(k-1)}$, this value is indeed equal to the number of points in the overlap of $B(X_i, h_{k-1})$ and $B(X_j, h_{k-1})$ except X_i, X_j. Similarly, the *mass of the complement* is defined as

$$N_{i \triangle j}^{(k)} = \sum_{l \neq i, j} \left\{ w_{il}^{(k-1)} \mathbb{I}(X_l \notin B(X_j, h_{k-1})) + w_{jl}^{(k-1)} \mathbb{I}(X_l \notin B(X_i, h_{k-1})) \right\}. \tag{13.23}$$

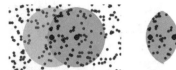

Fig. 13.10 Test of "no gap between local clusters". From left: Homogeneous case; $N_{i\wedge j}^{(k)}$; $N_{i\triangle j}^{(k)}$; $N_{i\vee j}^{(k)}$

Note that $N_{i\triangle j}^{(k)}$ is nearly the number of points in $C_i^{(k-1)}$ and $C_j^{(k-1)}$ which do not belong to the overlap $B(X_i, h_{k-1}) \cap B(X_j, h_{k-1})$. Finally, *mass of the union* $N_{i\vee j}^{(k)}$ can be defined as the sum of the mass of overlap and the mass of the complement:

$$N_{i\vee j}^{(k)} = N_{i\wedge j}^{(k)} + N_{i\triangle j}^{(k)}. \tag{13.24}$$

To measure the gap, consider the ratio of these two masses:

$$\tilde{\theta}_{ij}^{(k)} = N_{i\wedge j}^{(k)} / N_{i\vee j}^{(k)}. \tag{13.25}$$

This value can be viewed as an estimate of the value θ_{ij} which measures the ratio of the averaged density in the overlap of two local regions C_i and C_j relative to the average density. Under local homogeneity one can suppose that the density in the union of two balls is nearly constant. In this case, the estimate $\tilde{\theta}_{ij}^{(k)}$ should be close to the ratio of the volume of overlap and the volume of union of these balls:

$$\tilde{\theta}_{ij}^{(k)} \approx q_{ij}^{(k)} = \frac{V_\cap(d_{ij}, h_{k-1})}{2V(h_{k-1}) - V_\cap(d_{ij}, h_{k-1})}, \tag{13.26}$$

where $V(h)$ is the volume of a ball with radius h and $V_\cap(d, h)$ is the volume of the intersection of two balls with radius h and the distance d between centers, $d = d_{ij} = d(X_i, X_j)$. The new value $w_{ij}^{(k)}$ can be viewed as a randomized test of the null hypothesis H_{ij} of no gap between X_i and X_j against the alternative of a significant gap. The gap is significant if $\tilde{\theta}_{ij}^{(k)}$ is significantly smaller than $q_{ij}^{(k)}$. The construction is illustrated by Fig. 13.11 for the homogeneous situation and for a situation with a gap.

To quantify the notion of significance, we consider the statistical likelihood ratio test of "no gap" between two local clusters, that is $\tilde{\theta}_{ij}^{(k)} > q_{ij}^{(k)}$ vs $\tilde{\theta}_{ij}^{(k)} \le q_{ij}^{(k)}$:

$$T_{ij}^{(k)} = N_{i\vee j}^{(k)} KL(\tilde{\theta}_{ij}^{(k)}, q_{ij}^{(k)}) \{I(\tilde{\theta}_{ij}^{(k)} \le q_{ij}^{(k)}) - I(\tilde{\theta}_{ij}^{(k)} > q_{ij}^{(k)})\}. \tag{13.27}$$

$KL(\theta, \eta)$ is the *Kullback–Leibler (KL) divergence* between two Bernoulli laws with parameters θ and η:

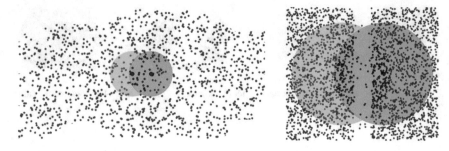

Fig. 13.11 Left: Homogeneous case. Right: "Gap" case

$$KL(\theta, \eta) = \theta \log \frac{\theta}{\eta} + (1 - \theta) \log \frac{1 - \theta}{1 - \eta}. \tag{13.28}$$

In the end, we update the weights $w_{ij}^{(k)}$ for all pairs of points X_i and X_j with distance $d(X_i, X_j) \leq h_k$:

$$w_{ij}^{(k)} = I\left(T_{ij}^{(k)} \leq \lambda\right). \tag{13.29}$$

The first indicator allows us to recompute only $n \times n_k$ weights, where n_k is the average number of neighbors in the h_k neighborhood.

Tests $T_{ij}^{(k)}$ are scaled by a global constant λ which is the only tuning parameter of the method. The parameter λ has an important influence on the performance of the method. Large λ-values can lead to aggregation of the in-homogeneous regions. On the contrary, small λ increases the sensitivity of the methods to in-homogeneity but may lead to artificial segmentation.

Parameter tuning

A heuristic choice of λ is proposed based on the effective cluster size given by the total sum of final weights w_{ij}^K.

Let $w_{ij}^K(\lambda)$ be the collection of final weights obtained by the procedure with the parameter λ. Define

$$S(\lambda) = \sum_{i,j=1}^{n} w_{ij}^K(\lambda). \tag{13.30}$$

An increase of λ yields larger homogeneous blocks and thus, a larger value $S(\lambda)$. A proposal is to pick up the λ-value corresponding to a point right before observing a huge jump in graph of $S(\lambda)$. In the case of a complex cluster structure, several jump points can be observed with the corresponding λ-value for each jump. In this case all those λ-values should be checked and compared the obtained clustering results afterward. All steps of *AWC* procedure are summarized in Algorithm 13.5.

Algorithm 13.5 AWC

1: **Fix** a sequence of radii $h_1 \leq h_2 \leq \ldots \leq h_K$
2: **Initialization of weights**: $w_{ij}^{(0)} = \mathrm{I}\left(d(X_i, X_j) \leq \max(h_0(X_i), h_0(X_j))\right)$
3: **Updates at step** k :
4:    Compute $T_{ij}^{(k)}$ using 13.27
5:    $w_{ij}^{(k)} = \mathrm{I}\left(d(X_i, X_j) \leq h_k\right) \mathrm{I}\left(T_{ij}^{(k)} \leq \lambda\right)$
6: **Repeat** until $k = K$.

13.5 Spectral Clustering

Spectral clustering is based on a graph theoretic approach to divide the data points into homogeneous groups (clusters). All techniques are based on the observations $\{x_i\}_{i=1}^n \in \mathbb{R}^p$ that are collected in the data matrix $\mathcal{X}(n \times p)$ and their similarity (13.2), between all data points. For the presentation of this clustering technique we follow (Luxburg 2007).

The graph theoretic point of view on $\mathcal{D}$ is based on the undirected graph $\mathcal{G} = (\mathcal{V}, \mathcal{E})$, where the vertices $v_i \in \mathcal{V}$ correspond to the data points x_i. Two v's are connected if $d_i > 0$ and the edge between v_i and v_j is weighted by d_i. Clustering now boils down to find a partition of $\mathcal{G}$ such that the edges e_{ij} between different groups have low weights, in term of values of the $\mathcal{D}$ matrix. The similarity matrix

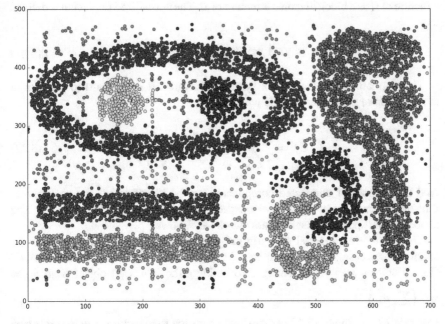

Fig. 13.12 AWC performance for $\lambda = 15$

can be also called an adjacency matrix $\mathcal{W}$ of $\mathcal{G}$, since with element w_{ij} of $\mathcal{W}$ the fact that $w_{ij} = 0$ corresponds the nodes v_i and v_j being not connected (or not similar) (Fig. 13.12).

Since clusters are different through local closeness of data points, it makes sense to study the connectedness through the degree $d_i = \sum_{j=1}^{n} w_{ij}$. Note that $d_{ii} = w_{ii}$ makes x_i a singleton, since it is not connected to any other vector. Define $\mathcal{D} = diag(d_i)$ and indicator vector $1_A = (f_1, \ldots f_n)^\top$, $f_i = \mathrm{I}(x_i \in A)$, $A \subset \mathcal{V}$. The size of subset $A \subset \mathcal{V}$ is measured via $|A| \overset{\text{def}}{=} \#$ of vertices in A $vol(A) \overset{\text{def}}{=} \sum_{i \in A} d_i$. If $\overline{A} = \mathcal{V} \backslash A$ then we can define A to be connected if there no points w_{ij} for $i \in A$, $j \in \overline{A}$.

The basic idea of spectral clustering is a skillful eigenanalysis of $\mathcal{L} = \mathcal{D} - -\mathcal{W}$, the so-called Laplacian matrix. Some properties are given in Luxburg (2007). Laplacian matrix properties:

1. $\forall f \in \mathbb{R}^n$ $f^\top \mathcal{L} = \frac{1}{2} \sum_{i,j=1}^{n} w_{ij} \left(f_i - f_j \right)^2$
2. 0 is the smallest eigenvalue of L with eigenvector 1_n
3. $\mathcal{L}$ is symmetric and positive semi-definite
4. $\mathcal{L}$ has n eigenvalues $0 = \lambda_1 \leq \ldots \leq \lambda_n$

The unnormalized Laplacian gives us a tool to detect connected components.

Proposition 2 (Luxburg 2007) (Number of connected components and the spectrum of $\mathcal{L}$). Let $\mathcal{G}$ be an undirected graph with nonnegative weights. Then the multiplivitz k of the eigenvalue 0 of $\mathcal{L}$ equals the number of connected components $A_1, \ldots, A_k$ in the graph. The eigenspace of eigenvalue 0 is spanned by the indicator vectors $1_{A_1}, \ldots, 1_{A_k}$ of those components. In fact one can describe k-component structure of $\mathcal{G}$ as a block diagonal structure of $\mathcal{L}$. The normalized Laplacian is defined as

$$\mathcal{L}_1 \overset{\text{def}}{=} \mathcal{D}^{-\frac{1}{2}} \mathcal{L} \mathcal{D}^{-\frac{1}{2}} = \mathcal{I}_n - \mathcal{D}^{-\frac{1}{2}} \mathcal{W} \mathcal{D}^{-\frac{1}{2}} \tag{13.31}$$

The normalized Laplacian is a symmetric matrix. It is not hard to see that in modification of properties 1 and 2 we have:

$$f^\top \mathcal{L} f = \frac{1}{2} \sum_{i,j=1}^{n} w_{ij} \left(\frac{f_i}{\sqrt{d_i}} - \frac{f_j}{\sqrt{d_j}} \right)^2. \tag{13.32}$$

And the number of connected components can be identified via the multiplying k of the eigenvalues of $\mathcal{L}_1$ with eigenspace spanned by $\mathcal{D}^{\frac{1}{2}} 1_{A_i}, i = 1, \ldots, k$. The unnormalized spectral clustering now works as follows:

The normalized spectral clustering algorithm is built the same way as for an unnormalized case, except that at *step 3* one solves the eigenvalue problem for $\mathcal{L}_1$ (13.31) and on *step 4* gets changed to building the $\mathcal{U}(n \times p)$ object from $u_{ij} = v_{ij} / \left(\sum_k u_{ij}^2 \right)^{\frac{1}{2}}$.

Example 13.10 Figure 13.13 demonstrates partition to two clusters of 8 points performed by *spectral clustering algorithm*.

Algorithm 13.6 Spectral clustering algorithm

1: Construct from the proximity matrix the adjacency matrix $\mathcal{W}$
2: Compute unnormalized Laplacian $\mathcal{L}$.
3: Compute the first k eigenvectors $\nu_1, \ldots, \nu_k$ of $\mathcal{L}$
4: Define the $(n \times k)$ matrix $\mathcal{V} = (\nu_1, \nu_2 \ldots, \nu_k)$ with ν_i as columns
5: Take $y_i \in \mathbb{R}^k$ a the i-th row of $\mathcal{V}$, yielding the data matrix $\mathcal{Y}(n \times p)$
6: Cluster $\mathcal{Y}$ according to the *k-means* into $C_1, \ldots C_k$.
7: Output: Clusters $A_1, \ldots, A_k$ with $A_i = \{j : y_j \in C_j\}$

Fig. 13.13 The 8 points
example; second smallest
eigenvalue: -0.98 and
corresponding eigenvector
$(-0.0000, 0.0000, -0.0001,$
$0.0006, 0.0000, -0.1605,$
$-0.6857, 0.7099)^\top$

QMVAclus8psc

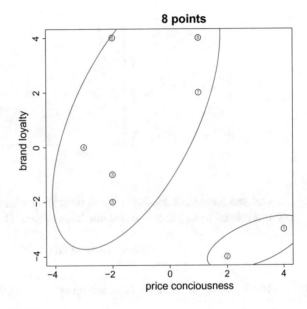

The graph cut point of view on clustering can be most easily explained for $k = 2$.
Define

$$cut(A, B) = \sum_{i \in A, j \in B} w_{i,j}$$

for two sets and

$$cut(A_1, \ldots, A_k) = \sum_{i=1}^{k} cut(A_i, \overline{A}_i)$$

In order to avoid singletons as group, one likes to create reasonably large group.
This is achieved by measuring the weights of the edges via

$$Ncut = \sum_{i=1}^{k} \frac{cut(A_i, \overline{A}_i)}{vol(A_i)} \tag{13.33}$$

Fig. 13.14 Parcellation
results for the simulated data
into four clusters by NCut
algorithm based on the
Euclidean distance
⬛MVAspecclust

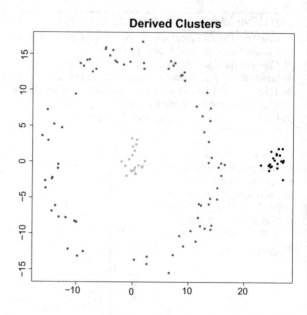

A related measure RatioCut is based on scaling with $|A_i|$. How are these penal-
izations related to $\mathcal{L}_1$? Suppose that one likes to build ($k = 2$)

$$\underset{A\subset\mathcal{V}}{\text{argmin}}\, RatioCut(A, \overline{A}).$$

The Objective function can be rewritten as

$$f^\top \mathcal{L} f = |\mathcal{V}| \cdot RatioCut(A, \overline{A}) \tag{13.34}$$

It turns out that this is a too hard problem to tackle one therefore moves to more
relaxed problem yielding f as eigenvector corresponding to the second smallest
eigenvalue of $\mathcal{L}$. Very similar arguments lead to optimizing the Ncut criterion leading
to the second smallest eigenvector of $\mathcal{L}_1$. The Fig. 13.14 illustrates clusters, defined
by Ncut algorithm for simulated data set.

13.6   Boston Housing

Presented multivariate techniques are now applied to the Boston housing data. We
focus our attention to 14 transformed and standardized variables, see, e.g., Fig. 13.15
that provides descriptive statistics via boxplots for two clusters, as discussed in the
sequel. A dendrogram for 13 variables (excluding the dummy variable $\tilde{X}_4$—Charles
River indicator) using the Ward method is displayed in Fig. 13.16. One observes two

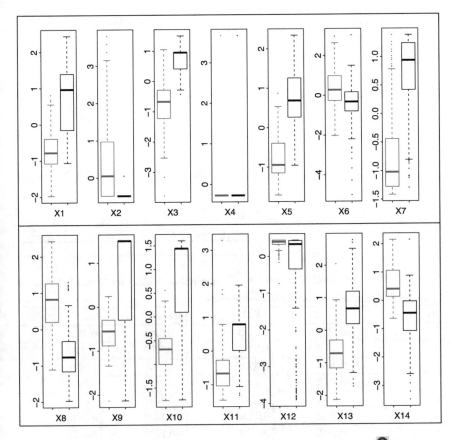

Fig. 13.15 Boxplots of the 14 standardized variables of the Boston housing data. ℚMVAclusbh

dominant clusters. A further refinement of say, four clusters, could be considered at a lower level of distance.

To interpret the two clusters, we present the mean values and their respective standard errors of the 13 $\widetilde{X}$ variables by groups in Table 13.3. Comparison of the mean values for both groups shows that all the differences in the means are individually significant. Moreover, cluster one corresponds to housing districts with better living quality and higher house prices, whereas cluster two corresponds to less favored districts in Boston. This can be confirmed, for instance, by a lower crime rate, a higher proportion of residential land, lower proportion of African American, etc. for cluster one. Cluster two is identified by a higher proportion of older houses, a higher pupil/teacher ratio and a higher percentage of the lower status population.

This interpretation is underlined by visual inspection of all the variables via scatterplot matrices, see, e.g., Figs. 13.17 and 13.18. For example, the lower right boxplot of Fig. 13.15 and the correspondingly colored clusters in the last row of Fig. 13.18 confirm the role of each variable in determining the clusters. This interpretation perfectly

Fig. 13.16 Dendrogram of
the Boston housing data
using the Ward algorithm
MVAclusbh

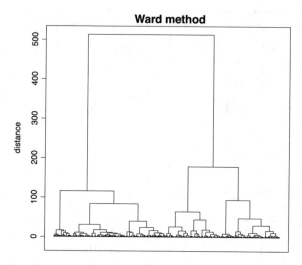

Table 13.3 Means and standard errors of the 13 standardized variables for Cluster 1 (251 observations) and Cluster 2 (255 observations). MVAclusbh

Variable	Mean C1	SE C1	Mean C2	SE C2
1	−0.7105	0.0332	0.6994	0.0535
2	0.4848	0.0786	−0.4772	0.0047
3	−0.7665	0.0510	0.7545	0.0279
5	−0.7672	0.0365	0.7552	0.0447
6	0.4162	0.0571	−0.4097	0.0576
7	−0.7730	0.0429	0.7609	0.0378
8	0.7140	0.0472	−0.7028	0.0417
9	−0.5429	0.0358	0.5344	0.0656
10	−0.6932	0.0301	0.6823	0.0569
11	−0.5464	0.0469	0.5378	0.0582
12	0.3547	0.0080	−0.3491	0.0824
13	−0.6899	0.0401	0.6791	0.0509
14	0.5996	0.0431	−0.5902	0.0570

coincides with the previous PC analysis (Fig. 11.11). The quality of life factor is
clearly visible in Fig. 13.19, where cluster membership is distinguished by the shape
and color of the points graphed according to the first two principal components.
Clearly, the first PC completely separates the two clusters and corresponds, as we
have discussed in Chap. 11, to a quality of life and house indicator.

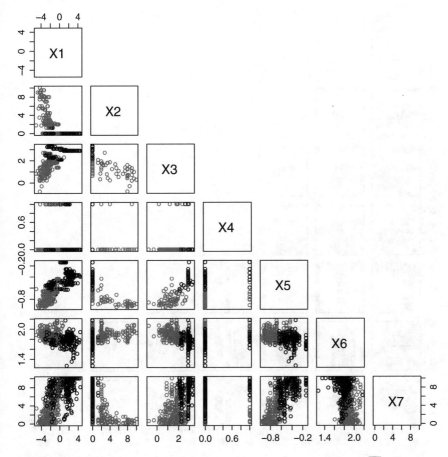

Fig. 13.17 Scatterplot matrix for variables $\widetilde{X}_1$ to $\widetilde{X}_7$ of the Boston housing data 🔴 MVAclusbh

13.7 Exercises

Exercise 13.1 Prove formula (13.20).

Exercise 13.2 Prove that $I_R = \mathrm{tr}(\mathcal{S}_R)$, where $\mathcal{S}_R$ denotes the empirical covariance matrix of the observations contained in R.

Exercise 13.3 Prove that

$$\Delta(R, P + Q) = \frac{n_R + n_P}{n_R + n_P + n_Q} \, \Delta(R, P)+$$

$$+\frac{n_R + n_Q}{n_R + n_P + n_Q} \, \Delta(R, Q) - \frac{n_R}{n_R + n_P + n_Q} \, \Delta(P, Q),$$

when the centroid formula is used to define $d^2(R, P + Q)$.

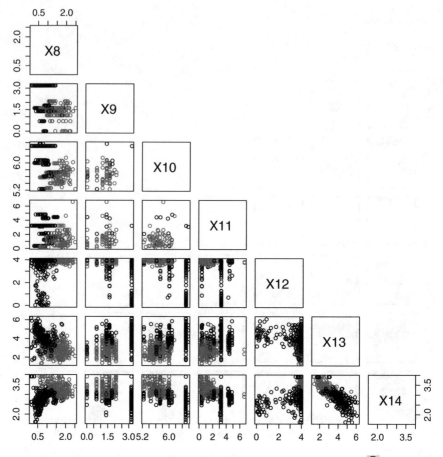

Fig. 13.18 Scatterplot matrix for variables $\widetilde{X}_8$ to $\widetilde{X}_{14}$ of the Boston housing data 🔲MVAclusbh

Exercise 13.4 Repeat the 8-point example (Example 13.7) using the complete linkage and the Ward algorithm. Explain the difference to single linkage.

Exercise 13.5 Explain the differences between various proximity measures by means of an example.

Exercise 13.6 Repeat the bank notes example (Example 13.8) with another random sample of 20 notes.

Exercise 13.7 Repeat the bank notes example (Example 13.8) with another clustering algorithm.

Exercise 13.8 Repeat the bank notes example (Example 13.8) or the 8-point example (Example 13.7) with the L_1-norm.

Fig. 13.19 Scatterplot of the first two PCs displaying the two clusters 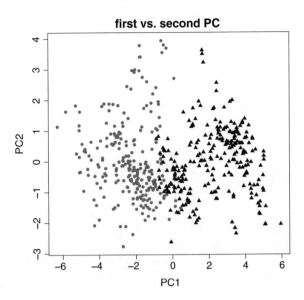MVAclusbh

Exercise 13.9 Analyze the U.S. companies example (Sect. 22.5) using the Ward algorithm and the L_2-norm.

Exercise 13.10 Analyze the U.S. crime data set (Sect. B.8) with the Ward algorithm and the L_2-norm on standardized variables (use only the crime variables).

Exercise 13.11 Repeat Exercise 13.10 with the U.S. health data set (use only the number of deaths variables).

Exercise 13.12 Redo Exercise 13.10 with the χ^2-metric. Compare the results.

Exercise 13.13 Redo Exercise 13.11 with the χ^2-metric and compare the results.

References

D. Arthur, S. Vassilvitskii, K-means++: the advantages of careful seeding, in *Proceedings of the Eighteenth Annual ACM-SIAM Symposium on Discrete Algorithms, SODA '07*, Philadelphia, PA, USA (2007), pp. 1027–1035. Society for Industrial and Applied Mathematics. ISBN 978-0-898716-24-5. http://dl.acm.org/citation.cfm?id=1283383.1283494

K. Efimov, L. Adamyan, V. Spokoiny, Adaptive weights clustering. J. R. Stat. Soc. (2017)

C. James, *Bezdek* (Pattern Recognition with Fuzzy Objective Function Algorithms. Kluwer Academic Publishers, Norwell, MA, USA, 1981). ISBN 0306406713

S.P. Lloyd, Least squares quantization in PCM. IEEE Trans. Inf. Theory **28**, 129–137 (1982)

U. Luxburg, A tutorial on spectral clustering. *Stat. Comput.* **17**(4), 395–416 (2007). ISSN 0960-3174. http://dx.doi.org/10.1007/s11222-007-9033-z

Chapter 14
Discriminant Analysis

Discriminant analysis is used in situations where the clusters are known *a priori*. The aim of discriminant analysis is to classify an observation, or several observations, into these known groups. For instance, in credit scoring, a bank knows from past experience that there are good customers (who repay their loan without any problems) and bad customers (who showed difficulties in repaying their loan). When a new customer asks for a loan, the bank has to decide whether or not to give the loan. The past records of the bank provides two data sets: multivariate observations x_i on the two categories of customers (including, for example, age, salary, marital status, the amount of the loan, etc.). The new customer is a new observation x with the same variables. The discrimination rule has to classify the customer into one of the two existing groups and the discriminant analysis should evaluate the risk of a possible "bad decision".

Many other examples are described below, and in most applications, the groups correspond to natural classifications or to groups known from history (like in the credit scoring example). These groups could have been formed by a cluster analysis performed on past data.

Section 14.1 presents the allocation rules when the populations are known, i.e., when we know the distribution of each population. As described in Sect. 14.2, in practice, the population characteristics have to be estimated from history. The methods are illustrated in several examples.

14.1 Allocation Rules for Known Distributions

Discriminant analysis is a set of methods and tools used to distinguish between groups of populations Π_j and to determine how to allocate new observations into groups. In one of our running examples, we are interested in discriminating between

© Springer Nature Switzerland AG 2019
W. K. Härdle and L. Simar, *Applied Multivariate Statistical Analysis*,
https://doi.org/10.1007/978-3-030-26006-4_14

counterfeit and true bank notes on the basis of measurements of these bank notes, see Sect. B.2. In this case, we have two groups (counterfeit and genuine bank notes) and we would like to establish an algorithm (rule) that can allocate a new observation (a new bank note) into one of the groups.

Another example is the detection of "fast" and "slow" consumers of a newly introduced product. Using a consumer's characteristics like education, income, family size, and amount of previous brand switching, we want to classify each consumer into the two groups just identified.

In poetry and literary studies, the frequencies of spoken or written words and lengths of sentences indicate profiles of different artists and writers. It can be of interest to attribute unknown literary or artistic works to certain writers with a specific profile. Anthropological measures on ancient sculls help in discriminating between male and female bodies. Good and poor credit risk ratings constitute a discrimination problem that might be tackled using observations on income, age, number of credit cards, family size, etc.

In general, we have populations $\Pi_j, j = 1, 2, ..., J$ and we have to allocate an observation x to one of these groups. A *discriminant rule* is a separation of the sample space (in general $\mathbb{R}^p$) into sets R_j such that if $x \in R_j$, it is identified as a member of population Π_j.

The main task of discriminant analysis is to find "good" regions R_j such that the error of misclassification is small. In the following we describe such rules when the population distributions are known.

Maximum Likelihood Discriminant Rule

Denote the densities of each population Π_j by $f_j(x)$. The *maximum likelihood discriminant rule* (ML rule) is given by allocating x to Π_j maximizing the likelihood $L_j(x) = f_j(x) = \arg\max_i f_i(x)$.

If several f_i give the same maximum then any of them may be selected. Mathematically, the sets R_j given by the ML discriminant rule are defined as

$$R_j = \{x : L_j(x) > L_i(x) \text{ for } i = 1, \ldots, J, i \neq j\}. \tag{14.1}$$

By classifying the observation into a certain group we may encounter a misclassification error. For $J = 2$ groups the probability of putting x into group 2 although it is from population 1 can be calculated as

$$p_{21} = P(X \in R_2|\Pi_1) = \int_{R_2} f_1(x)dx. \tag{14.2}$$

Similarly the conditional probability of classifying an object as belonging to the first population Π_1 although it actually comes from Π_2 is

$$p_{12} = P(X \in R_1|\Pi_2) = \int_{R_1} f_2(x)dx. \tag{14.3}$$

The misclassified observations create a cost $C(i|j)$ when a Π_j observation is assigned to R_i. In the credit risk example, this might be the cost of a "sour" credit. The cost structure can be pinned down in a cost matrix:

	Classified population	
	Π_1	Π_2
True population Π_1	0	$C(2|1)$
Π_2	$C(1|2)$	0

Let π_j be the prior probability of population Π_j, where "prior" means the *a priori* probability that an individual selected at random belongs to Π_j (i.e., before looking to the value x). Prior probabilities should be considered if it is clear ahead of time that an observation is more likely to stem from a certain population Π_j. An example is the classification of musical tunes. If it is known that during a certain period of time a majority of tunes were written by a certain composer, then there is a higher probability that a certain tune was composed by this composer. Therefore, he should receive a higher prior probability when tunes are assigned to a specific group.

The *expected cost of misclassification* (*ECM*) is given by

$$ECM = C(2|1)p_{21}\pi_1 + C(1|2)p_{12}\pi_2. \tag{14.4}$$

We will be interested in classification rules that keep the *ECM* small or minimize it over a class of rules. The discriminant rule minimizing the *ECM* (14.4) for two populations is given below.

Theorem 14.1 *For two given populations, the rule minimizing the ECM is given by*

$$R_1 = \left\{x : \frac{f_1(x)}{f_2(x)} \geq \left(\frac{C(1|2)}{C(2|1)}\right)\left(\frac{\pi_2}{\pi_1}\right)\right\}$$

$$R_2 = \left\{x : \frac{f_1(x)}{f_2(x)} < \left(\frac{C(1|2)}{C(2|1)}\right)\left(\frac{\pi_2}{\pi_1}\right)\right\}$$

The ML discriminant rule is thus a special case of the ECM rule for equal misclassification costs and equal prior probabilities. For simplicity the unity cost case, $C(1|2) = C(2|1) = 1$, and equal prior probabilities, $\pi_2 = \pi_1$, are assumed in the following.

Theorem 14.1 will be proven by an example from credit scoring.

Example 14.1 Suppose that Π_1 represents the population of bad clients who create the cost $C(2|1)$ if they are classified as good clients. Analogously, define $C(1|2)$ as the cost of loosing a good client classified as a bad one. Let γ denote the gain of the bank for the correct classification of a good client. The total gain of the bank is then

$$G(R_2) = -C(2|1)\pi_1 \int I(x \in R_2) f_1(x) dx$$

$$-C(1|2)\pi_2 \int \{1 - I(x \in R_2)\} f_2(x) dx + \gamma \pi_2 \int I(x \in R_2) f_2(x) dx$$

$$= -C(1|2)\pi_2 + \int I(x \in R_2)\{-C(2|1)\pi_1 f_1(x)$$

$$+(C(1|2) + \gamma)\pi_2 f_2(x)\} dx$$

Since the first term in this equation is constant, the maximum is obviously obtained for

$$R_2 = \{x : -C(2|1)\pi_1 f_1(x) + \{C(1|2) + \gamma\}\pi_2 f_2(x) \geq 0\}.$$

This is equivalent to

$$R_2 = \left\{ x : \frac{f_2(x)}{f_1(x)} \geq \frac{C(2|1)\pi_1}{\{C(1|2) + \gamma\}\pi_2} \right\},$$

which corresponds to the set R_2 in Theorem 14.1 for a gain of $\gamma = 0$.

Example 14.2 Suppose $x \in \{0, 1\}$ and

$$\Pi_1 : P(X = 0) = P(X = 1) = \frac{1}{2}$$

$$\Pi_2 : P(X = 0) = \frac{1}{4} = 1 - P(X = 1).$$

The sample space is the set $\{0, 1\}$. The ML discriminant rule is to allocate $x = 0$ to Π_1 and $x = 1$ to Π_2, defining the sets $R_1 = \{0\}$, $R_2 = \{1\}$ and $R_1 \cup R_2 = \{0, 1\}$.

Example 14.3 Consider two normal populations

$$\Pi_1 : N(\mu_1, \sigma_1^2),$$

$$\Pi_2 : N(\mu_2, \sigma_2^2).$$

Then

$$L_i(x) = (2\pi\sigma_i^2)^{-1/2} \exp\left\{ -\frac{1}{2} \left(\frac{x - \mu_i}{\sigma_i} \right)^2 \right\}.$$

Hence x is allocated to Π_1 ($x \in R_1$) if $L_1(x) \geq L_2(x)$. Note that $L_1(x) \geq L_2(x)$ is equivalent to

$$\frac{\sigma_2}{\sigma_1} \exp\left[-\frac{1}{2} \left\{ \left(\frac{x - \mu_1}{\sigma_1} \right)^2 - \left(\frac{x - \mu_2}{\sigma_2} \right)^2 \right\} \right] \geq 1$$

or

$$x^2 \left(\frac{1}{\sigma_1^2} - \frac{1}{\sigma_2^2} \right) - 2x \left(\frac{\mu_1}{\sigma_1^2} - \frac{\mu_2}{\sigma_2^2} \right) + \left(\frac{\mu_1^2}{\sigma_1^2} - \frac{\mu_2^2}{\sigma_2^2} \right) \leq 2 \log \frac{\sigma_2}{\sigma_1}. \tag{14.5}$$

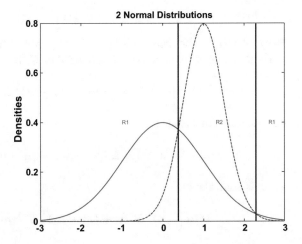

Fig. 14.1 Maximum likelihood rule for normal distributions 🔍 MVAdisnorm

Suppose that $\mu_1 = 0$, $\sigma_1 = 1$ and $\mu_2 = 1$, $\sigma_2 = \frac{1}{2}$. Formula (14.5) leads to

$$R_1 = \left\{ x : x \le \frac{1}{3} \left(4 - \sqrt{4 + 6\log(2)} \right) \text{ or } x \ge \frac{1}{3} \left(4 + \sqrt{4 + 6\log(2)} \right) \right\},$$
$$R_2 = \mathbb{R} \setminus R_1.$$

This situation is shown in Fig. 14.1.

The situation simplifies in the case of equal variances $\sigma_1 = \sigma_2$. The discriminant rule (14.5) is then (for $\mu_1 < \mu_2$)

$$\begin{aligned} x &\to \Pi_1, \text{ if } x \in R_1 = \{x : x \le \tfrac{1}{2}(\mu_1 + \mu_2)\}, \\ x &\to \Pi_2, \text{ if } x \in R_2 = \{x : x > \tfrac{1}{2}(\mu_1 + \mu_2)\}. \end{aligned} \quad (14.6)$$

Theorem 14.2 shows that the ML discriminant rule for multinormal observations is intimately connected with the Mahalanobis distance. The discriminant rule is based on linear combinations and belongs to the family of linear discriminant analysis (LDA) methods.

Theorem 14.2 *Suppose* $\Pi_i = N_p(\mu_i, \Sigma)$.

(a) The ML rule allocates x to Π_j, where $j \in \{1, \dots, J\}$ is the value minimizing the square Mahalanobis distance between x and μ_i:

$$\delta^2(x, \mu_i) = (x - \mu_i)^\top \Sigma^{-1} (x - \mu_i) , \quad i = 1, \dots, J .$$

(b) In the case of $J = 2$,

$$x \in R_1 \iff \alpha^\top (x - \mu) \ge 0 ,$$

where $\alpha = \Sigma^{-1}(\mu_1 - \mu_2)$ and $\mu = \frac{1}{2}(\mu_1 + \mu_2)$.

Proof. Part (a) of the theorem follows directly from comparison of the likelihoods. For $J = 2$, part (a) says that x is allocated to Π_1 if

$$(x - \mu_1)^\top \Sigma^{-1}(x - \mu_1) \leq (x - \mu_2)^\top \Sigma^{-1}(x - \mu_2)$$

Rearranging terms leads to

$$-2\mu_1^\top \Sigma^{-1}x + 2\mu_2^\top \Sigma^{-1}x + \mu_1^\top \Sigma^{-1}\mu_1 - \mu_2^\top \Sigma^{-1}\mu_2 \leq 0,$$

which is equivalent to

$$2(\mu_2 - \mu_1)^\top \Sigma^{-1}x + (\mu_1 - \mu_2)^\top \Sigma^{-1}(\mu_1 + \mu_2) \leq 0,$$

$$(\mu_1 - \mu_2)^\top \Sigma^{-1}\{x - \frac{1}{2}(\mu_1 + \mu_2)\} \geq 0,$$

$$\alpha^\top(x - \mu) \geq 0.$$

Bayes Discriminant Rule

We have seen an example where prior knowledge on the probability of classification into Π_j was assumed. Denote the prior probabilities by π_j and note that $\sum_{j=1}^{J} \pi_j = 1$. The Bayes rule of discrimination allocates x to the Π_j that gives the largest value of $\pi_i f_i(x)$, $\pi_j f_j(x) = \max_i \pi_i f_i(x)$. Hence, the discriminant rule is defined by $R_j = \{x : \pi_j f_j(x) \geq \pi_i f_i(x)$ for $i = 1, \ldots, J\}$. Obviously the Bayes rule is identical to the ML discriminant rule for $\pi_j = 1/J$.

A further modification is to allocate x to Π_j with a certain probability $\phi_j(x)$, such that $\sum_{j=1}^{J} \phi_j(x) = 1$ for all x. This is called a *randomized discriminant rule*. A randomized discriminant rule is a generalization of deterministic discriminant rules since

$$\phi_j(x) = \begin{cases} 1 & \text{if } \pi_j f_j(x) = \max_i \pi_i f_i(x), \\ 0 & \text{otherwise} \end{cases}$$

reflects the deterministic rules.

Which discriminant rules are good? We need a measure of comparison. Denote

$$p_{ij} = \int \phi_i(x) f_j(x) dx \tag{14.7}$$

as the probability of allocating x to Π_i if it in fact belongs to Π_j. A discriminant rule with probabilities p_{ij} is as good as any other discriminant rule with probabilities p'_{ij} if

$$p_{ii} \geq p'_{ii} \quad \text{for all } i = 1, \ldots, J. \tag{14.8}$$

We call the first rule better if the strict inequality in (14.8) holds for at least one i. A discriminant rule is called *admissible* if there is no better discriminant rule.

Theorem 14.3 *All Bayes discriminant rules (including the ML rule) are admissible.*

Probability of Misclassification for the ML rule ($J = 2$)

Suppose that $\Pi_i = N_p(\mu_i, \Sigma)$. In the case of two groups, it is not difficult to derive the probabilities of misclassification for the ML discriminant rule. Consider for instance $p_{12} = P(x \in R_1 \mid \Pi_2)$. By part (b) in Theorem 14.2 we have

$$p_{12} = P\{\alpha^\top (x - \mu) > 0 \mid \Pi_2\}.$$

If $X \in R_2$, $\alpha^\top (X - \mu) \sim N\left(-\frac{1}{2}\delta^2, \delta^2\right)$ where $\delta^2 = (\mu_1 - \mu_2)^\top \Sigma^{-1}(\mu_1 - \mu_2)$ is the squared Mahalanobis distance between the two populations, we obtain

$$p_{12} = \Phi\left(-\frac{1}{2}\delta\right).$$

Similarly, the probability of being classified into population 2 although x stems from Π_1 is equal to $p_{21} = \Phi\left(-\frac{1}{2}\delta\right)$.

Classification with Different Covariance Matrices

The minimum *ECM* depends on the ratio of the densities $\frac{f_1(x)}{f_2(x)}$ or equivalently on the difference $\log\{f_1(x)\} - \log\{f_2(x)\}$. When the covariance for both density functions differ, the allocation rule becomes more complicated:

$$R_1 = \left\{x : -\frac{1}{2}x^\top (\Sigma_1^{-1} - \Sigma_2^{-1})x + (\mu_1^\top \Sigma_1^{-1} - \mu_2^\top \Sigma_2^{-1})x - k \geq \log\left[\left(\frac{C(1|2)}{C(2|1)}\right)\left(\frac{\pi_2}{\pi_1}\right)\right]\right\},$$

$$R_2 = \left\{x : -\frac{1}{2}x^\top (\Sigma_1^{-1} - \Sigma_2^{-1})x + (\mu_1^\top \Sigma_1^{-1} - \mu_2^\top \Sigma_2^{-1})x - k < \log\left[\left(\frac{C(1|2)}{C(2|1)}\right)\left(\frac{\pi_2}{\pi_1}\right)\right]\right\}$$

where $k = \frac{1}{2}\log\left(\frac{|\Sigma_1|}{|\Sigma_2|}\right) + \frac{1}{2}(\mu_1^\top \Sigma_1^{-1}\mu_1 - \mu_2^\top \Sigma_2^{-1}\mu_2)$. The classification regions are defined by *quadratic* functions. Therefore they belong to the family of Quadratic Discriminant Analysis (QDA) methods. This *quadratic* classification rule coincides with the rules used when $\Sigma_1 = \Sigma_2$, since the term $\frac{1}{2}x^\top (\Sigma_1^{-1} - \Sigma_2^{-1})x$ disappears.

Summary
$\hookrightarrow$ Discriminant analysis is a set of methods used to distinguish among groups in data and to allocate new observations into the existing groups.
$\hookrightarrow$ Given that data are from populations Π_j with densities $f_j, j = 1, \ldots, J$, the maximum likelihood discriminant rule (ML rule) allocates an observation x to that population Π_j which has the maximum likelihood $L_j(x) = f_j(x) = \max_i f_i(x)$.

Continued on next page

Summary (continue)
$\hookrightarrow$ Given prior probabilities π_j for populations Π_j, Bayes discriminant rule allocates an observation x to the population Π_j that maximizes $\pi_i f_i(x)$ with respect to i. All Bayes discriminant rules (incl. the ML rule) are admissible.
$\hookrightarrow$ For the ML rule and $J = 2$ normal populations, the probabilities of misclassification are given by $p_{12} = p_{21} = \Phi\left(-\frac{1}{2}\delta\right)$ where δ is the Mahalanobis distance between the two populations.
$\hookrightarrow$ Classification of two normal populations with different covariance matrices (ML rule) leads to regions defined by a quadratic function.
$\hookrightarrow$ Desirable discriminant rules have a low expected cost of misclassification (*ECM*).

14.2 Discrimination Rules in Practice

The ML rule is used if the distribution of the data is known up to parameters. Suppose for example that the data come from multivariate normal distributions $N_p(\mu_j, \Sigma)$. If we have J groups with n_j observations in each group, we use $\bar{x}_j$ to estimate μ_j, and S_j to estimate Σ. The common covariance may be estimated by

$$ S_u = \sum_{j=1}^{J} n_j \left(\frac{S_j}{n - J} \right), \tag{14.9} $$

with $n = \sum_{j=1}^{J} n_j$. Thus the empirical version of the ML rule of Theorem 14.2 is to allocate a new observation x to Π_j such that j minimizes

$$ (x - \bar{x}_i)^\top S_u^{-1} (x - \bar{x}_i) \quad \text{for} \quad i \in \{1, \ldots, J\}. $$

Example 14.4 Let us apply this rule to the Swiss bank notes. The 20 randomly chosen bank notes which we had clustered into 2 groups in Example 13.8 are used. First, the covariance Σ is estimated by the average of the covariances of Π_1 (cluster 1) and Π_2 (cluster 2). The hyperplane $\hat{\alpha}^\top (x - \bar{x}) = 0$ which separates the two populations is given by

$$ \hat{\alpha} = S_u^{-1}(\bar{x}_1 - \bar{x}_2) = (-12.18, 20.54, -19.22, -15.55, -13.06, 21.43)^\top, $$
$$ \bar{x} = \frac{1}{2}(\bar{x}_1 + \bar{x}_2) = (214.79, 130.05, 129.92, 9.23, 10.48, 140.46)^\top. $$

Now let us apply the discriminant rule to the entire bank notes data set. Counting the number of misclassifications by

$$\sum_{i=1}^{100} I\{\widehat{\alpha}^\top (x_i - \bar{x}) < 0\}, \sum_{i=101}^{200} I\{\widehat{\alpha}^\top (x_i - \bar{x}) > 0\},$$

we obtain 1 misclassified observation for the conterfeit bank notes and 0 misclassification for the genuine bank notes. When $J = 3$ groups, the allocation regions can be calculated using

$$h_{12}(x) = (\bar{x}_1 - \bar{x}_2)^\top S_u^{-1} \left\{ x - \frac{1}{2}(\bar{x}_1 + \bar{x}_2) \right\}$$

$$h_{13}(x) = (\bar{x}_1 - \bar{x}_3)^\top S_u^{-1} \left\{ x - \frac{1}{2}(\bar{x}_1 + \bar{x}_3) \right\}$$

$$h_{23}(x) = (\bar{x}_2 - \bar{x}_3)^\top S_u^{-1} \left\{ x - \frac{1}{2}(\bar{x}_2 + \bar{x}_3) \right\}.$$

The rule is to allocate x to

$$\begin{cases} \Pi_1 & \text{if } h_{12}(x) \geq 0 \text{ and } h_{13}(x) \geq 0 \\ \Pi_2 & \text{if } h_{12}(x) < 0 \text{ and } h_{23}(x) \geq 0 \\ \Pi_3 & \text{if } h_{13}(x) < 0 \text{ and } h_{23}(x) < 0. \end{cases}$$

Estimation of the Probabilities of Misclassifications

Misclassification probabilities are given by (14.7) and can be estimated by replacing the unknown parameters by their corresponding estimators.

For the ML rule for two normal populations we obtain

$$\hat{p}_{12} = \hat{p}_{21} = \Phi\left(-\frac{1}{2}\hat{\delta}\right)$$

where $\hat{\delta}^2 = (\bar{x}_1 - \bar{x}_2)^\top S_u^{-1}(\bar{x}_1 - \bar{x}_2)$ is the estimator for δ^2.

The probabilities of misclassification may also be estimated by the *re-substitution method*. We reclassify each original observation x_i, $i = 1, \cdots, n$ into $\Pi_1, \cdots, \Pi_J$ according to the chosen rule. Then denoting the number of individuals coming from Π_j which have been classified into Π_i by n_{ij}, we have $\hat{p}_{ij} = \frac{n_{ij}}{n_j}$, an estimator of p_{ij}. Clearly, this method leads to too optimistic estimators of p_{ij}, but it provides a rough measure of the quality of the discriminant rule. The matrix $(\hat{p}_{ij})$ is called the *confusion matrix* in Johnson and Wichern (1998).

Example 14.5 In the above classification problem for the Swiss bank notes (Sect. B.2), we have the following confusion matrix: **Q**MVAaper

	true membership	
	genuine (Π_1)	counterfeit (Π_2)
predicted Π_1	100	1
Π_2	0	99

The *apparent error rate* (APER) is defined as the fraction of observations that are misclassified. The APER, expressed as a percentage, is

$$\text{APER} = \left(\frac{1}{200}\right)100\% = 0.5\%.$$

For the calculation of the APER we use the observations twice: the first time to construct the classification rule and the second time to evaluate this rule. An APER of 0.5% might therefore be too optimistic. An approach that corrects for this bias is based on the holdout procedure of Lachenbruch and Mickey (1968). For two populations this procedure is as follows:

1. Start with the first population Π_1. Omit one observation and develop the classification rule based on the remaining $n_1 - 1$, n_2 observations.
2. Classify the "holdout" observation using the discrimination rule in Step 1.
3. Repeat Steps 1 and 2 until all of the Π_1 observations are classified. Count the number n'_{21} of misclassified observations.
4. Repeat Steps 1 through 3 for population Π_2. Count the number n'_{12} of misclassified observations.

Estimates of the misclassification probabilities are given by

$$\hat{p}'_{12} = \frac{n'_{12}}{n_2}$$

and

$$\hat{p}'_{21} = \frac{n'_{21}}{n_1}.$$

A more realistic estimator of the actual error rate (AER) is given by

$$\frac{n'_{12} + n'_{21}}{n_2 + n_1}. \tag{14.10}$$

Statisticians favor the AER (for its unbiasedness) over the APER. In large samples, however, the computational costs might counterbalance the statistical advantage. This is not a real problem since the two misclassification measures are asymptotically equivalent.

Fisher's Linear Discrimination Function

Another approach stems from R. A. Fisher. His idea was to base the discriminant rule on a projection $a^\top x$ such that a good separation was achieved. This LDA projection method is called *Fisher's linear discrimination function*. If

$$y = \mathcal{X}a$$

denotes a linear combination of observations, then the total sum of squares of y, $\sum_{i=1}^{n}(y_i - \bar{y})^2$, is equal to

$$y^\top \mathcal{H}y = a^\top \mathcal{X}^\top \mathcal{H}\mathcal{X}a = a^\top \mathcal{T}a \qquad (14.11)$$

with the centering matrix $\mathcal{H} = \mathcal{I} - n^{-1}1_n 1_n^\top$ and $\mathcal{T} = \mathcal{X}^\top \mathcal{H}\mathcal{X}$.

Suppose we have samples $\mathcal{X}_j$, $j = 1, \ldots, J$, from J populations. Fisher's suggestion was to find the linear combination $a^\top x$ which maximizes the ratio of the *between-group-sum of squares* to the *within-group-sum of squares*.

The within-group-sum of squares is given by

$$\sum_{j=1}^{J} y_j^\top \mathcal{H}_j y_j = \sum_{j=1}^{J} a^\top \mathcal{X}_j^\top \mathcal{H}_j \mathcal{X}_j a = a^\top \mathcal{W}a, \qquad (14.12)$$

where $\mathcal{Y}_j$ denotes the j-th sub-matrix of $\mathcal{Y}$ corresponding to observations of group j and $\mathcal{H}_j$ denotes the $(n_j \times n_j)$ centering matrix. The within-group-sum of squares measures the sum of variations within each group.

The between-group-sum of squares is

$$\sum_{j=1}^{J} n_j (\bar{y}_j - \bar{y})^2 = \sum_{j=1}^{J} n_j \{a^\top (\bar{x}_j - \bar{x})\}^2 = a^\top \mathcal{B}a, \qquad (14.13)$$

where $\bar{y}_j$ and $\bar{x}_j$ denote the means of $\mathcal{Y}_j$ and $\mathcal{X}_j$ and $\bar{y}$ and $\bar{x}$ denote the sample means of $\mathcal{Y}$ and $\mathcal{X}$. The between-group-sum of squares measures the variation of the means across groups.

The total sum of squares (14.11) is the sum of the within-group-sum of squares and the between-group-sum of squares, i.e.,

$$a^\top \mathcal{T}a = a^\top \mathcal{W}a + a^\top \mathcal{B}a.$$

Fisher's idea was to select a projection vector a that maximizes the ratio

$$\frac{a^\top \mathcal{B}a}{a^\top \mathcal{W}a}. \qquad (14.14)$$

The solution is found by applying Theorem 2.5.

Theorem 14.4 *The vector a that maximizes (14.14) is the eigenvector of $W^{-1}B$ that corresponds to the largest eigenvalue.*

Now a discrimination rule is easy to obtain:
classify x into group j where $a^\top \bar{x}_j$ is closest to $a^\top x$, i.e.,

$$x \rightarrow \Pi_j \text{ where } j = \arg \min_i |a^\top (x - \bar{x}_i)|.$$

When $J = 2$ groups, the discriminant rule is easy to compute. Suppose that group 1 has n_1 elements and group 2 has n_2 elements. In this case

$$B = \left(\frac{n_1 n_2}{n}\right) dd^\top,$$

where $d = (\bar{x}_1 - \bar{x}_2)$. $W^{-1}B$ has only one eigenvalue which equals

$$\text{tr}(W^{-1}B) = \left(\frac{n_1 n_2}{n}\right) d^\top W^{-1} d,$$

and the corresponding eigenvector is $a = W^{-1}d$. The corresponding discriminant rule is

$$\begin{array}{lll} x \rightarrow \Pi_1 & \text{if} & a^\top \{x - \frac{1}{2}(\bar{x}_1 + \bar{x}_2)\} > 0, \\ x \rightarrow \Pi_2 & \text{if} & a^\top \{x - \frac{1}{2}(\bar{x}_1 + \bar{x}_2)\} \leq 0. \end{array} \qquad (14.15)$$

The Fisher LDA is closely related to projection pursuit (Chap. 20) since the statistical technique is based on a *one-dimensional* index $a^\top x$.

Example 14.6 Consider the bank notes data again. Let us use the subscript "g" for the genuine and "f" for the conterfeit bank notes, e.g., $\mathcal{X}_g$ denotes the first 100 observations of $\mathcal{X}$ and $\mathcal{X}_f$ the second 100. In the context of the bank data set the "between-group-sum of squares" is defined as

$$100 \left\{ (\bar{y}_g - \bar{y})^2 + (\bar{y}_f - \bar{y})^2 \right\} = a^\top Ba \qquad (14.16)$$

for some matrix B. Here, $\bar{y}_g$ and $\bar{y}_f$ denote the means for the genuine and counterfeit bank notes and $\bar{y} = \frac{1}{2}(\bar{y}_g + \bar{y}_f)$. The "within-group-sum of squares" is

$$\sum_{i=1}^{100} \{(y_g)_i - \bar{y}_g\}^2 + \sum_{i=1}^{100} \{(y_f)_i - \bar{y}_f\}^2 = a^\top Wa, \qquad (14.17)$$

with $(y_g)_i = a^\top x_i$ and $(y_f)_i = a^\top x_{i+100}$ for $i = 1, \ldots, 100$.

The resulting discriminant rule consists of allocating an observation x_0 to the genuine sample space if

Fig. 14.2 Densities of projections of genuine and counterfeit bank notes by Fisher's discrimination rule

🔍MVAdisfbank

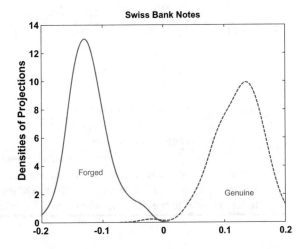

$$a^{\top}(x_0 - \bar{x}) > 0,$$

with $a = W^{-1}(\bar{x}_g - \bar{x}_f)$ (see Exercise 14.8) and of allocating x_0 to the counterfeit sample space when the opposite is true. In our case

$$a = (0.000, 0.029, -0.029, -0.039, -0.041, 0.054)^{\top}.$$

One genuine and no counterfeit bank notes are misclassified. Figure 14.2 shows the estimated densities for $y_g = a^{\top}\mathcal{X}_g$ and $y_f = a^{\top}\mathcal{X}_f$. They are separated better than those of the diagonals in Fig. 1.9.

Note that the allocation rule (14.15) is exactly the same as the ML rule for $J = 2$ groups and for normal distributions with the same covariance. For $J = 3$ groups this rule will be different, except for the special case of collinear sample means.

Summary
↪ A discriminant rule is a separation of the sample space into sets R_j. An observation x is classified as coming from population Π_j if it lies in R_j.
↪ The expected cost of misclassification (ECM) for two populations is given by ECM $= C(2\|1)p_{21}\pi_1 + C(1\|2)p_{12}\pi_2$.
↪ The ML rule is applied if the distributions in the populations are known up to parameters, e.g., for normal distributions $N_p(\mu_j, \Sigma)$.
↪ The ML rule allocates x to the population that exhibits the smallest Mahalanobis distance $$\delta^2(x; \mu_i) = (x - \mu_i)^{\top}\Sigma^{-1}(x - \mu_i).$$

Continued on next page

Summary (continue)
$\hookrightarrow$ The probability of misclassification is given by $$p_{12} = p_{21} = \Phi\left(-\frac{1}{2}\delta\right),$$ where δ is the Mahalanobis distance between μ_1 and μ_2.
$\hookrightarrow$ Classification for different covariance structures in the two populations leads to quadratic discrimination rules.
$\hookrightarrow$ A different approach is Fisher's linear discrimination rule which finds a linear combination $a^\top x$ that maximizes the ratio of the "between-group-sum of squares" and the "within-group-sum of squares". This rule turns out to be identical to the ML rule when $J = 2$ for normal populations.

14.3   Boston Housing

One interesting application of discriminant analysis with respect to the Boston housing data is the classification of the districts according to the house values. The rationale behind this is that certain observables must determine the value of a district, as in Sect. 3.7 where the house value was regressed on the other variables. Two groups are defined according to the median value of houses $\widetilde{X}_{14}$: in group Π_1 the value of $\widetilde{X}_{14}$ is greater than or equal to the median of $\widetilde{X}_{14}$ and in group Π_2 the value of $\widetilde{X}_{14}$ is less than the median of $\widetilde{X}_{14}$.

The linear discriminant rule, defined on the remaining 12 variables (excluding $\widetilde{X}_4$ and $\widetilde{X}_{14}$) is applied. After reclassifying the 506 observations, we obtain an apparent error rate of 0.146. The details are given in Table 14.1. The more appropriate error rate, given by the AER, is 0.160 (see Table 14.2).

Let us now turn to a group definition suggested by the Cluster Analysis in Sect. 13.6. Group Π_1 was defined by higher quality of life and house. We define the linear discriminant rule using the 13 variables from $\widetilde{\mathcal{X}}$ excluding $\widetilde{X}_4$. Then we reclassify the 506 observations and we obtain an APER of 0.0395. Details are summarized in Table 14.3. The AER turns out to be 0.0415 (see Table 14.4).

Table 14.1 APER for price of Boston houses MVAdiscbh

		True	
		Π_1	Π_2
Predicted	Π_1	216	40
	Π_2	34	216

Table 14.2 AER for price of Boston houses 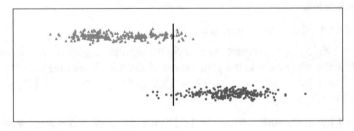 MVAaerbh

		True	
		Π_1	Π_2
	Π_1	211	42
Predicted			
	Π_2	39	214

Table 14.3 APER for clusters of Boston houses MVAdiscbh

		True	
		Π_1	Π_2
	Π_1	244	13
Predicted			
	Π_2	7	242

Table 14.4 AER for clusters of Boston houses MVAaerbh

		True	
		Π_1	Π_2
	Π_1	244	14
Predicted			
	Π_2	7	241

Fig. 14.3 Discrimination scores for the two clusters created from the Boston housing data
MVAdiscbh

Figure 14.3 displays the values of the linear discriminant scores (see Theorem 14.2) for all of the 506 observations, colored by groups. One can clearly see the APER is derived from the 7 observations from group Π_1 with a negative score and the 13 observations from group Π_2 with positive score.

14.4 Exercises

Exercise 14.1 Prove Theorem 14.2 (a) and 14.2 (b).

Exercise 14.2 Apply the rule from Theorem 14.2 (b) for $p = 1$ and compare the result with that of Example 14.3.

Exercise 14.3 Calculate the ML discrimination rule based on observations of a one-dimensional variable with an exponential distribution.

Exercise 14.4 Calculate the ML discrimination rule based on observations of a two-dimensional random variable, where the first component has an exponential distribution and the other has an alternative distribution. What is the difference between the discrimination rule obtained in this exercise and the Bayes discrimination rule?

Exercise 14.5 Apply the Bayes rule to the car data (Sect. B.3) in order to discriminate between Japanese, European, and U.S. cars, i.e., $J = 3$. Consider only the "miles per gallon" variable and take the relative frequencies as prior probabilities.

Exercise 14.6 Compute Fisher's linear discrimination function for the 20 bank notes from Example 13.8. Apply it to the entire bank data set. How many observations are misclassified?

Exercise 14.7 Use the Fisher's linear discrimination function on the WAIS data set (Exrecise 7.19) and evaluate the results by re-substitution the probabilities of misclassification.

Exercise 14.8 Show that in Example 14.6

(a) $\mathcal{W} = 100\,(\mathcal{S}_g + \mathcal{S}_f)$, where $\mathcal{S}_g$ and $\mathcal{S}_f$ denote the empirical covariances (3.6) and (3.5) with respect to the genuine and counterfeit bank notes,
(b) $\mathcal{B} = 100\left\{(\overline{x}_g - \overline{x})(\overline{x}_g - \overline{x})^\top + (\overline{x}_f - \overline{x})(\overline{x}_f - \overline{x})^\top\right\}$, where $\overline{x} = \frac{1}{2}(\overline{x}_g + \overline{x}_f)$,
(c) $a = \mathcal{W}^{-1}(\overline{x}_g - \overline{x}_f)$.

Exercise 14.9 Recalculate Example 14.3 with the prior probability $\pi_1 = \frac{1}{3}$ and $C(2|1) = 2C(1|2)$.

Exercise 14.10 Explain the effect of changing π_1 or $C(1|2)$ on the relative location of the region $R_j, j = 1, 2$.

Exercise 14.11 Prove that Fisher's linear discrimination function is identical to the ML rule when the covariance matrices are identical ($J = 2$).

Exercise 14.12 Suppose that $x \in \{0, 1, 2, 3, 4, 5, 6, 7, 8, 9, 10\}$ and

$$\begin{aligned} \Pi_1 &: X \sim \text{Bi}(10, 0.2) && \text{with the prior probability } \pi_1 = 0.5; \\ \Pi_2 &: X \sim \text{Bi}(10, 0.3) && \text{with the prior probability } \pi_2 = 0.3; \\ \Pi_3 &: X \sim \text{Bi}(10, 0.5) && \text{with the prior probability } \pi_3 = 0.2. \end{aligned}$$

Determine the sets R_1, R_2 and R_3. (Use the Bayes discriminant rule.)

References

R.A. Johnson, D.W. Wichern, *Applied Multivariate Analysis*, 4th edn. (Prentice Hall, Englewood Cliffs, New Jersey, 1998)

P.A. Lachenbruch, M.R. Mickey, Estimation of error rates in discriminant analysis. Technometrics **10**, 1–11 (1968)

Chapter 15
Correspondence Analysis

Correspondence analysis provides tools for analyzing the associations between rows
and columns of contingency tables. A contingency table is a two-entry frequency table
where the joint frequencies of two qualitative variables are reported. For instance, a
(2×2) table could be formed by observing from a sample of n individuals' two qualitative variables: the individual's sex and whether the individual smokes. The table
reports the observed joint frequencies. In general $(n \times p)$ tables may be considered.

The main idea of correspondence analysis is to develop simple indices that will
show the relations between the row and the columns categories. These indices will
tell us simultaneously which column categories have more weight in a row category
and vice versa. Correspondence analysis is also related to the issue of reducing the
dimension of the table, similar to principal component analysis in Chap. 11, and to the
issue of decomposing the table into its factors as discussed in Chap. 10. The idea is to
extract the indices in decreasing order of importance so that the main information of
the table can be summarized in spaces with smaller dimensions. For instance, if only
two factors (indices) are used, the results can be shown in two-dimensional graphs,
showing the relationship between the rows and the columns of the table.

Section 15.1 defines the basic notation and motivates the approach and Sect. 15.2
gives the basic theory. The indices will be used to describe the χ^2 statistic measuring
the associations in the table. Several examples in Sect. 15.3 show how to provide
and interpret, in practice, the two-dimensional graphs displaying the relationship
between the rows and the columns of a contingency table.

15.1 Motivation

The aim of correspondence analysis is to develop simple indices that show relations between the rows and columns of a contingency table. Contingency tables are
a very useful tool for the description of the association between two variables in

© Springer Nature Switzerland AG 2019
W. K. Härdle and L. Simar, *Applied Multivariate Statistical Analysis*,
https://doi.org/10.1007/978-3-030-26006-4_15

very general situations. The two variables can be qualitative (nominal), in which case they are also referred to as categorical variables. Each row and each column in the table represents one category of the corresponding variable. The entry x_{ij} in the table $\mathcal{X}$ (with dimension $(n \times p)$) is the number of observations in a sample which simultaneously fall in the i-th row category and the j-th column category, for $i = 1, \ldots, n$ and $j = 1, \ldots, p$. Sometimes a "category" of a nominal variable is also called a "modality" of the variable.

The variables of interest can also be discrete quantitative variables, such as the number of family members or the number of accidents an insurance company had to cover during 1 year. Here, each possible value that the variable can have defines a row or a column category. Continuous variables may be taken into account by defining the categories in terms of intervals or classes of values which the variable can take on. Thus, contingency tables can be used in many situations, implying that correspondence analysis is a very useful tool in many applications.

The graphical relationships between the rows and the columns of the table $\mathcal{X}$ that result from correspondence analysis are based on the idea of representing all the row and column categories and interpreting the relative positions of the points in terms of the weights corresponding to the column and the row. This is achieved by deriving a system of simple indices providing the coordinates of each row and each column. These row and column coordinates are simultaneously represented in the same graph. It is then clear to see which column categories are more important in the row categories of the table (and the other way around).

As was already eluded to, the construction of the indices is based on an idea similar to that of PCA. Using PCA, the total variance is partitioned into independent contributions stemming from the principal components. Correspondence analysis, on the other hand, decomposes a measure of association, typically the total χ^2 value used in testing independence, rather than decomposing the total variance.

Example 15.1 The French "baccalauréat" frequencies have been classified into regions and different baccalauréat categories, see Appendix, Sect. B.14. Altogether $n = 202100$ baccalauréats were observed. The joint frequency of the region *Ile-de-France* and the modality *Philosophy*, for example, is 9724. That is, 9724 baccalauréats were in Ile-de-France and the category Philosophy.

The question is whether certain regions prefer certain baccalauréat types. If we consider, for instance, the region *Lorraine*, we have the following percentages:

A	B	C	D	E	F	G	H
20.5	7.6	15.3	19.6	3.4	14.5	18.9	0.2

The total percentages of the different modalities of the variable baccalauréat are as follows:

A	B	C	D	E	F	G	H
22.6	10.7	16.2	22.8	2.6	9.7	15.2	0.2

One might argue that the region *Lorraine* seems to prefer the modalities E, F, G and dislike the specializations A, B, C, D relative to the overall frequency of baccalauréat type.

In correspondence analysis, we try to develop an index for the regions so that this over- or underrepresentation can be measured in just one single number. Simultaneously we try to weight the regions so that we can see in which region certain baccalauréat types are preferred.

Example 15.2 Consider n types of companies and p locations of these companies. Is there a certain type of company that prefers a certain location? Or is there a location index that corresponds to a certain type of company?

Assume that $n = 3$, $p = 3$, and that the frequencies are as follows:

$$\mathcal{X} = \begin{pmatrix} 4 & 0 & 2 \\ 0 & 1 & 1 \\ 1 & 1 & 4 \end{pmatrix} \begin{matrix} \leftarrow \text{Finance} \\ \leftarrow \text{Energy} \\ \leftarrow \text{HiTech} \end{matrix}$$

$$\begin{matrix} \uparrow \text{Frankfurt} \\ \uparrow \text{Berlin} \\ \uparrow \text{Munich} \end{matrix}$$

The frequencies imply that four type three companies (HiTech) are in location 3 (Munich), and so on. Suppose there is a (company) weight vector $r = (r_1, \ldots, r_n)^\top$, such that a location index s_j could be defined as

$$s_j = c \sum_{i=1}^n r_i \frac{x_{ij}}{x_{\bullet j}}, \tag{15.1}$$

where $x_{\bullet j} = \sum_{i=1}^n x_{ij}$ is the number of companies in location j and c is a constant. s_1, for example, would give the average weighted frequency (by r) of companies in location 1 (Frankfurt).

Given a location weight vector $s^* = (s_1^*, \ldots, s_p^*)^\top$, we can define a company index in the same way as

$$r_i^* = c^* \sum_{j=1}^p s_j^* \frac{x_{ij}}{x_{i\bullet}}, \tag{15.2}$$

where c^* is a constant and $x_{i\bullet} = \sum_{j=1}^p x_{ij}$ is the sum of the i-th row of $\mathcal{X}$, i.e., the number of type i companies. Thus r_2^*, for example, would give the average weighted frequency (by s^*) of energy companies.

If (15.1) and (15.2) can be solved simultaneously for a "row weight" vector $r = (r_1, \ldots, r_n)^\top$ and a "column weight" vector $s = (s_1, \ldots, s_p)^\top$, we may represent each row category by r_i, $i = 1, \ldots, n$, and each column category by s_j, $j = 1, \ldots, p$, in a one-dimensional graph. If in this graph r_i and s_j are in close proximity (far from the origin), this would indicate that the i-th row category has an important conditional frequency $x_{ij}/x_{\bullet j}$ in (15.1) and that the j-th column category has an important conditional frequency $x_{ij}/x_{i\bullet}$ in (15.2). This would indicate a positive association between the i-th row and the j-th column. A similar line of argument could be used if r_i was very far away from s_j (and far from the origin). This would indicate a small conditional frequency contribution, or a negative association between the i-th row and the j-th column.

Summary
$\hookrightarrow$ The aim of correspondence analysis is to develop simple indices that show relations among qualitative variables in a contingency table.
$\hookrightarrow$ The joint representation of the indices reveals relations among the variables.

15.2 Chi-Square Decomposition

An alternative way of measuring the association between the row and column categories is a decomposition of the value of the χ^2-test statistic. The well known χ^2-test for independence in a two-dimensional contingency table consists of two steps. First, the expected value of each cell of the table is estimated under the hypothesis of independence. Second, the corresponding observed values are compared to the expected values using the statistic

$$t = \sum_{i=1}^{n} \sum_{j=1}^{p} (x_{ij} - E_{ij})^2 / E_{ij}, \tag{15.3}$$

where x_{ij} is the observed frequency in cell (i, j) and E_{ij} is the corresponding estimated expected value under the assumption of independence, i.e.,

$$E_{ij} = \frac{x_{i\bullet} x_{\bullet j}}{x_{\bullet\bullet}}. \tag{15.4}$$

Here $x_{\bullet\bullet} = \sum_{i=1}^{n} x_{i\bullet}$. Under the hypothesis of independence, t has a $\chi^2_{(n-1)(p-1)}$ distribution. In the industrial location example introduced above, the value of $t = 6.26$ is almost significant at the 5% level. It is therefore worth investigating the special reasons for departure from independence.

The method of χ^2 decomposition consists of finding the SVD of the $(n \times p)$ matrix $\mathcal{C}$ with elements

$$c_{ij} = (x_{ij} - E_{ij})/E_{ij}^{1/2}. \tag{15.5}$$

The elements c_{ij} may be viewed as measuring the (weighted) departure between the observed x_{ij} and the theoretical values E_{ij} under independence. This leads to the factorial tools of Chap. 10 which describe the rows and the columns of $\mathcal{C}$.

For simplification, define the matrices $\mathcal{A}$ $(n \times n)$ and $\mathcal{B}$ $(p \times p)$ as

$$\mathcal{A} = \text{diag}(x_{i\bullet}) \quad \text{and} \quad \mathcal{B} = \text{diag}(x_{\bullet j}). \tag{15.6}$$

These matrices provide the marginal row frequencies, a $(n \times 1)$, and the marginal column frequencies, b $(p \times 1)$:

$$a = \mathcal{A}1_n \quad \text{and} \quad b = \mathcal{B}1_p. \tag{15.7}$$

It is easy to verify that

$$\mathcal{C}\sqrt{b} = 0 \quad \text{and} \quad \mathcal{C}^\top \sqrt{a} = 0, \tag{15.8}$$

where the square root of the vector is taken element by element and $R = \text{rank}(\mathcal{C}) \leq \min\{(n-1), (p-1)\}$. From (10.14) of Chap. 10, the SVD of $\mathcal{C}$ yields

$$\mathcal{C} = \Gamma \Lambda \Delta^\top, \tag{15.9}$$

where Γ contains the eigenvectors of $\mathcal{C}\mathcal{C}^\top$, Δ the eigenvectors of $\mathcal{C}^\top \mathcal{C}$ and $\Lambda = \text{diag}(\lambda_1^{1/2}, \ldots, \lambda_R^{1/2})$ with $\lambda_1 \geq \lambda_2 \geq \ldots \geq \lambda_R$ (the eigenvalues of $\mathcal{C}\mathcal{C}^\top$). Equation (15.9) implies that

$$c_{ij} = \sum_{k=1}^R \lambda_k^{1/2} \gamma_{ik} \delta_{jk}. \tag{15.10}$$

Note that (15.3) can be rewritten as

$$\text{tr}(\mathcal{C}\mathcal{C}^\top) = \sum_{k=1}^R \lambda_k = \sum_{i=1}^n \sum_{j=1}^p c_{ij}^2 = t. \tag{15.11}$$

This relation shows that the SVD of $\mathcal{C}$ decomposes the total χ^2 value rather than, as in Chap. 10, the total variance.

The duality relations between the row and the column space (10.11) are now, for $k = 1, \ldots, R$, given by

$$\begin{aligned} \delta_k &= \frac{1}{\sqrt{\lambda_k}} \mathcal{C}^\top \gamma_k, \\ \gamma_k &= \frac{1}{\sqrt{\lambda_k}} \mathcal{C} \delta_k. \end{aligned} \tag{15.12}$$

The projections of the rows and the columns of $\mathcal{C}$ are given by

$$\begin{aligned} \mathcal{C}\delta_k &= \sqrt{\lambda_k} \gamma_k, \\ \mathcal{C}^\top \gamma_k &= \sqrt{\lambda_k} \delta_k. \end{aligned} \tag{15.13}$$

Note that the eigenvectors satisfy

$$\delta_k^\top \sqrt{b} = 0, \quad \gamma_k^\top \sqrt{a} = 0. \tag{15.14}$$

From (15.10) we see that the eigenvectors δ_k and γ_k are the objects of interest when analyzing the correspondence between the rows and the columns. Suppose, that the first eigenvalue in (15.10) is dominant, so that

$$c_{ij} \approx \lambda_1^{1/2} \gamma_{i1} \delta_{j1}. \tag{15.15}$$

In this case, if the coordinates γ_{i1} and δ_{j1} are both large (with the same sign) relative to the other coordinates, then c_{ij} will be large as well, indicating a positive association between the i-th row and the j-th column category of the contingency table. If γ_{i1} and δ_{j1} were both large with opposite signs, then there would be a negative association between the i-th row and j-th column.

In many applications, the first two eigenvalues, λ_1 and λ_2, dominate and the percentage of the total χ^2 explained by the eigenvectors γ_1 and γ_2 and δ_1 and δ_2 is large. In this case, (15.13) and (γ_1, γ_2) can be used to obtain a graphical display of the n rows of the table $((\delta_1, \delta_2)$ play a similar role for the p columns of the table). The interpretation of the proximity between row and column points will be interpreted as above with respect to (15.10).

In correspondence analysis, we use the projections of weighted rows of $\mathcal{C}$ and the projections of weighted columns of $\mathcal{C}$ for graphical displays. Let r_k ($n \times 1$) be the projections of $\mathcal{A}^{-1/2}\mathcal{C}$ on δ_k and s_k ($p \times 1$) be the projections of $\mathcal{B}^{-1/2}\mathcal{C}^\top$ on γ_k ($k = 1, \ldots, R$):

$$\begin{aligned} r_k &= \mathcal{A}^{-1/2}\mathcal{C}\delta_k = \sqrt{\lambda_k}\mathcal{A}^{-1/2}\gamma_k, \\ s_k &= \mathcal{B}^{-1/2}\mathcal{C}^\top\gamma_k = \sqrt{\lambda_k}\mathcal{B}^{-1/2}\delta_k. \end{aligned} \tag{15.16}$$

These vectors have the property that

$$\begin{aligned} r_k^\top a &= 0, \\ s_k^\top b &= 0. \end{aligned} \tag{15.17}$$

The obtained projections on each axis $k = 1, \ldots, R$, are centered at zero with the natural weights given by a (the marginal frequencies of the rows of $\mathcal{X}$) for the row coordinates r_k and by b (the marginal frequencies of the columns of $\mathcal{X}$) for the column coordinates s_k (compare this to expression (15.14)). As a result, the origin is the center of gravity for all of the representations. We also know from (15.16) and the SVD of $\mathcal{C}$ that

$$\begin{aligned} r_k^\top \mathcal{A} r_k &= \lambda_k, \\ s_k^\top \mathcal{B} s_k &= \lambda_k. \end{aligned} \tag{15.18}$$

From the duality relation between δ_k and γ_k (see (15.12)) we obtain

$$r_k = \frac{1}{\sqrt{\lambda_k}} \mathcal{A}^{-1/2} \mathcal{C} \mathcal{B}^{1/2} s_k,$$
$$s_k = \frac{1}{\sqrt{\lambda_k}} \mathcal{B}^{-1/2} \mathcal{C}^\top \mathcal{A}^{1/2} r_k, \qquad (15.19)$$

which can be simplified to

$$r_k = \sqrt{\frac{x_{\bullet\bullet}}{\lambda_k}} \mathcal{A}^{-1} \mathcal{X} s_k,$$
$$s_k = \sqrt{\frac{x_{\bullet\bullet}}{\lambda_k}} \mathcal{B}^{-1} \mathcal{X}^\top r_k. \qquad (15.20)$$

These vectors satisfy the relations (15.1) and (15.2) for each $k = 1, \ldots, R$, simultaneously.

As in Chap. 10, the vectors r_k and s_k are referred to as factors (row factor and column factor, respectively). They have the following means and variances:

$$\bar{r}_k = \frac{1}{x_{\bullet\bullet}} r_k^\top a = 0,$$
$$\bar{s}_k = \frac{1}{x_{\bullet\bullet}} s_k^\top b = 0, \qquad (15.21)$$

and

$$\mathrm{Var}(r_k) = \frac{1}{x_{\bullet\bullet}} \sum_{i=1}^n x_{i\bullet} r_{ki}^2 = \frac{r_k^\top \mathcal{A} r_k}{x_{\bullet\bullet}} = \frac{\lambda_k}{x_{\bullet\bullet}},$$
$$\mathrm{Var}(s_k) = \frac{1}{x_{\bullet\bullet}} \sum_{j=1}^p x_{\bullet j} s_{kj}^2 = \frac{s_k^\top \mathcal{B} s_k}{x_{\bullet\bullet}} = \frac{\lambda_k}{x_{\bullet\bullet}}. \qquad (15.22)$$

Hence, $\lambda_k / \sum_{k=1}^R \lambda_j$, which is the part of the k-th factor in the decomposition of the χ^2 statistic t, may also be interpreted as the proportion of the variance explained by the factor k. The proportions

$$C_a(i, r_k) = \frac{x_{i\bullet} r_{ki}^2}{\lambda_k}, \quad \text{for } i = 1, \ldots, n, \ k = 1, \ldots, R, \qquad (15.23)$$

are called the absolute contributions of row i to the variance of the factor r_k. They show which row categories are most important in the dispersion of the k-th row factors. Similarly, the proportions

$$C_a(j, s_k) = \frac{x_{\bullet j} s_{kj}^2}{\lambda_k}, \quad \text{for } j = 1, \ldots, p, \ k = 1, \ldots, R, \qquad (15.24)$$

are called the absolute contributions of column j to the variance of the column factor s_k. These absolute contributions may help to interpret the graph obtained by correspondence analysis.

15.3   Correspondence Analysis in Practice

The graphical representations on the axes $k = 1, 2, \ldots, R$, of the n rows and of the p columns of $\mathcal{X}$ are provided by the elements of r_k and s_k. Typically, two-dimensional displays are often satisfactory if the cumulated percentage of variance explained by the first two factors, $\Psi_2 = \frac{\lambda_1 + \lambda_2}{\sum_{k=1}^{R} \lambda_k}$, is sufficiently large.

The interpretation of the graphs may be summarized as follows:

– The proximity of two rows (two columns) indicates a similar profile in these two rows (two columns), where "profile" refers to the conditional frequency distribution of a row (column); those two rows (columns) are almost proportional. The opposite interpretation applies when the two rows (two columns) are far apart.
– The proximity of a particular row to a particular column indicates that this row (column) has a particularly important weight in this column (row). In contrast to this, a row that is quite distant from a particular column indicates that there are almost no observations in this column for this row (and vice versa). Of course, as mentioned above, these conclusions are particularly true when the points are far away from 0.
– The origin is the average of the factors r_k and s_k. Hence, a particular point (row or column) projected close to the origin indicates an average profile.
– The absolute contributions are used to evaluate the weight of each row (column) in the variances of the factors.
– All the interpretations outlined above must be carried out in view of the quality of the graphical representation which is evaluated, as in PCA, using the cumulated percentage of variance.

Remark 15.1 Note that correspondence analysis can also be applied to more general $(n \times p)$ tables $\mathcal{X}$ which in a "strict sense" are not contingency tables.

As long as statistical (or natural) meaning can be given to sums over rows and columns, Remark 15.1 holds. This implies, in particular, that all of the variables are measured in the same units. In that case, $x_{\bullet\bullet}$ constitutes the total frequency of the observed phenomenon, and is shared between individuals (n rows) and between variables (p columns). Representations of the rows and columns of $\mathcal{X}$, r_k and s_k, have the basic property (15.19) and show which variables have important weights for each individual and vice versa. This type of analysis is used as an alternative to PCA. PCA is mainly concerned with covariances and correlations, whereas correspondence analysis analyses a more general kind of association. (see Exercises 15.3 and 15.10.)

Example 15.3 A survey of Belgium citizens who regularly read a newspaper was conducted in the 1980s. They were asked where they lived. The possible answers were 10 regions: 7 provinces (Antwerp, Western Flanders, Eastern Flanders, Hainant, Liège, Limbourg, and Luxembourg) and 3 regions around Brussels (Flemish Brabant, Walloon Brabant, and the city of Brussels). They were also asked what kind of newspapers they read on a regular basis. There were 15 possible answers split up into 3 classes: Flemish newspapers (label begins with the letter v), French newspapers

Table 15.1 Eigenvalues and percentages of the variance, Example 15.3

λ_j	Percentage of variance	Cumulated percentage
183.40	0.653	0.653
43.75	0.156	0.809
25.21	0.090	0.898
11.74	0.042	0.940
8.04	0.029	0.969
4.68	0.017	0.985
2.13	0.008	0.993
1.20	0.004	0.997
0.82	0.003	1.000
0.00	0.000	1.000

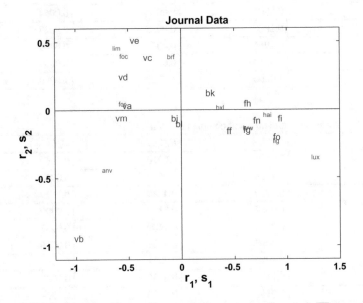

Fig. 15.1 Projection of rows (the 15 newspapers) and columns (the 10 regions) MVAcorrjourn

(label begins with f), and both languages together (label begins with b). The data set is given in Sect. B.11. The eigenvalues of the factorial correspondence analysis are given in Table 15.1.

Two-dimensional representations will be quite satisfactory since the first two eigenvalues account for 81 % of the variance. Figure 15.1 shows the projections of the rows (the 15 newspapers) and of the columns (the 10 regions).

As expected, there is a high association between the regions and the type of newspapers which is read. In particular, v_b (Gazet van Antwerp) is almost exclusively read in the province of Antwerp (this is an extreme point in the graph). The points

Table 15.2 Absolute contributions of row factors r_k

	$C_a(i, r_1)$	$C_a(i, r_2)$	$C_a(i, r_3)$
v_a	0.0563	0.0008	0.0036
v_b	0.1555	0.5567	0.0067
v_c	0.0244	0.1179	0.0266
v_d	0.1352	0.0952	0.0164
v_e	0.0253	0.1193	0.0013
f_f	0.0314	0.0183	0.0597
f_g	0.0585	0.0162	0.0122
f_h	0.1086	0.0024	0.0656
f_i	0.1001	0.0024	0.6376
b_j	0.0029	0.0055	0.0187
b_k	0.0236	0.0278	0.0237
b_l	0.0006	0.0090	0.0064
v_m	0.1000	0.0038	0.0047
f_n	0.0966	0.0059	0.0269
f_0	0.0810	0.0188	0.0899
Total	1.0000	1.0000	1.0000

Table 15.3 Absolute contributions of column factors s_k

	$C_a(j, s_1)$	$C_a(j, s_2)$	$C_a(j, s_3)$
brw	0.0887	0.0210	0.2860
bxl	0.1259	0.0010	0.0960
anv	0.2999	0.4349	0.0029
brf	0.0064	0.2370	0.0090
foc	0.0729	0.1409	0.0033
for	0.0998	0.0023	0.0079
hai	0.1046	0.0012	0.3141
lig	0.1168	0.0355	0.1025
lim	0.0562	0.1162	0.0027
lux	0.0288	0.0101	0.1761
Total	1.0000	1.0000	1.0000

on the left all belong to Flanders, whereas those on the right all belong to Wallonia.
Notice that the Walloon Brabant and the Flemish Brabant are not far from Brussels.
Brussels is close to the center (average) and also close to the bilingual newspapers.
It is shifted a little to the right of the origin due to the majority of French-speaking
people in the area.

The absolute contributions of the first 3 factors are listed in Tables 15.2 and 15.3.
The row factors r_k are in Table 15.2 and the column factors s_k are in Table 15.3.

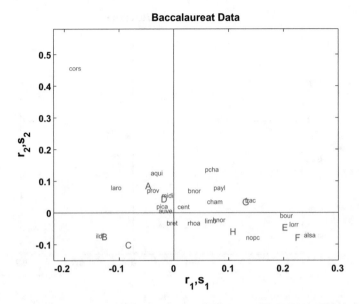

Fig. 15.2 Correspondence analysis including Corsica MVAcorrbac

Table 15.4 Eigenvalues and percentages of explained variance (including Corsica)

Eigenvalues λ	Percentage of variances	Cumulated percentage
2436.2	0.5605	0.561
1052.4	0.2421	0.803
341.8	0.0786	0.881
229.5	0.0528	0.934
152.2	0.0350	0.969
109.1	0.0251	0.994
25.0	0.0058	1.000
0.0	0.0000	1.000

They show, for instance, the important role of Antwerp and the newspaper v_b in determining the variance of both factors. Clearly, the first axis expresses linguistic differences between the three parts of Belgium. The second axis shows a larger dispersion between the Flemish region than the French-speaking regions. Note also that the third axis shows an important role of the category "f_i" (other French newspapers) with the Walloon Brabant "brw" and the Hainant "hai" showing the most important contributions. The coordinate of "f_i" on this axis is negative (not shown here) and so are the coordinates of "brw" and "hai". Apparently, these two regions also seem to feature a greater proportion of readers of more local newspapers.

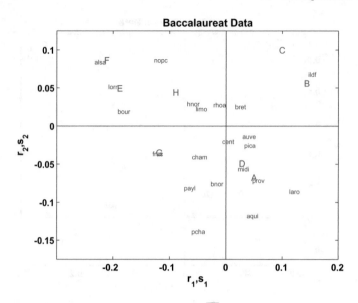

Fig. 15.3 Correspondence analysis excluding Corsica ⊙ MVAcorrbac

Table 15.5 Eigenvalues and percentages of explained variance (excluding Corsica)

Eigenvalues λ	Percentage of variances	Cumulated percentage
2408.6	0.5874	0.587
909.5	0.2218	0.809
318.5	0.0766	0.887
195.9	0.0478	0.935
149.3	0.0304	0.971
96.1	0.0234	0.994
22.8	0.0056	1.000
0.0	0.0000	1.000

Example 15.4 Applying correspondence analysis to the French baccalauréat data (Sect. B.14) leads to Fig. 15.2. Excluding Corsica we obtain Fig. 15.3. The different modalities are labeled A, ..., H and the regions are labeled ILDF, ..., CORS. The results of the correspondence analysis are given in Table 15.4 and Fig. 15.2.

The first two factors explain 80 % of the total variance. It is clear from Fig. 15.2 that Corsica (in the upper left) is an outlier. The analysis is therefore redone without Corsica and the results are given in Table 15.5 and Fig. 15.3. Since Corsica has such a small weight in the analysis, the results have not changed much.

The projections on the first three axes, along with their absolute contribution to the variance of the axis, are summarized in Table 15.6 for the regions and in Table 15.7 for baccalauréats.

Table 15.6 Coefficients and absolute contributions for regions, Example 15.4

Region	r_1	r_2	r_3	$C_a(i, r_1)$	$C_a(i, r_2)$	$C_a(i, r_3)$
ILDF	0.1464	0.0677	0.0157	0.3839	0.2175	0.0333
CHAM	−0.0603	−0.0410	−0.0187	0.0064	0.0078	0.0047
PICA	0.0323	−0.0258	−0.0318	0.0021	0.0036	0.0155
HNOR	−0.0692	0.0287	0.1156	0.0096	0.0044	0.2035
CENT	−0.0068	−0.0205	−0.0145	0.0001	0.0030	0.0043
BNOR	−0.0271	−0.0762	0.0061	0.0014	0.0284	0.0005
BOUR	−0.1921	0.0188	0.0578	0.0920	0.0023	0.0630
NOPC	−0.1278	0.0863	−0.0570	0.0871	0.1052	0.1311
LORR	−0.2084	0.0511	0.0467	0.1606	0.0256	0.0608
ALSA	−0.2331	0.0838	0.0655	0.1283	0.0439	0.0767
FRAC	−0.1304	−0.0368	−0.0444	0.0265	0.0056	0.0232
PAYL	−0.0743	−0.0816	−0.0341	0.0232	0.0743	0.0370
BRET	0.0158	0.0249	−0.0469	0.0011	0.0070	0.0708
PCHA	−0.0610	−0.1391	−0.0178	0.0085	0.1171	0.0054
AQUI	0.0368	−0.1183	0.0455	0.0055	0.1519	0.0643
MIDI	0.0208	−0.0567	0.0138	0.0018	0.0359	0.0061
LIMO	−0.0540	0.0221	−0.0427	0.0033	0.0014	0.0154
RHOA	−0.0225	0.0273	−0.0385	0.0042	0.0161	0.0918
AUVE	0.0290	−0.0139	−0.0554	0.0017	0.0010	0.0469
LARO	0.0290	−0.0862	−0.0177	0.0383	0.0595	0.0072
PROV	0.0469	−0.0717	0.0279	0.0142	0.0884	0.0383

Table 15.7 Coefficients and absolute contributions for baccalauréats, Example 15.4

Baccal	s_1	s_2	s_3	$C_a(j, s_1)$	$C_a(j, s_2)$	$C_a(j, s_3)$
A	0.0447	−0.0679	0.0367	0.0376	0.2292	0.1916
B	0.1389	0.0557	0.0011	0.1724	0.0735	0.0001
C	0.0940	0.0995	0.0079	0.1198	0.3556	0.0064
D	0.0227	−0.0495	−0.0530	0.0098	0.1237	0.4040
E	−0.1932	0.0492	−0.1317	0.0825	0.0141	0.2900
F	−0.2156	0.0862	0.0188	0.3793	0.1608	0.0219
G	−0.1244	−0.0353	0.0279	0.1969	0.0421	0.0749
H	−0.0945	0.0438	−0.0888	0.0017	0.0010	0.0112

Table 15.8 Eigenvalues and explained proportion of variance, Example 15.5

λ_j	Percentage of variance	Cumulated percentage
4399.0	0.4914	0.4914
2213.6	0.2473	0.7387
1382.4	0.1544	0.8932
870.7	0.0973	0.9904
51.0	0.0057	0.9961
34.8	0.0039	1.0000
0.0	0.0000	0.0000

The interpretation of the results may be summarized as follows. Table 15.7 shows that the baccalauréats B on one side and F on the other side are most strongly responsible for the variation on the first axis. The second axis mostly characterizes an opposition between baccalauréats A and C. Regarding the regions, Ile de France plays an important role in each axis. On the first axis, it is opposed to Lorraine and Alsace, whereas on the second axis, it is opposed to Poitou-Charentes and Aquitaine. All of this is confirmed in Fig. 15.3.

On the right side, there are the more classical baccalauréats and on the left, more technical ones. The regions on the left side have thus larger weights in the technical baccalauréats. Note also that most of the southern regions of France are concentrated in the lower part of the graph near the baccalauréat A.

Finally, looking at the 3-rd axis, we see that it is dominated by the baccalauréat E (negative sign) and to a lesser degree by H (negative) (as opposed to A (positive sign)). The dominating regions are HNOR (positive sign), opposed to NOPC and AUVE (negative sign). For instance, HNOR is particularly poor in baccalauréat D.

Example 15.5 The U.S. crime data set (Sect. B.8) gives the number of crimes in the 50 states of the U.S. classified in 1985 for each of the following seven categories: murder, rape, robbery, assault, burglary, larceny, and auto theft. The analysis of the contingency table, limited to the first two factors, provides the following results (see Table 15.8).

Looking at the absolute contributions (not reproduced here, see Exercise 15.6), it appears that the first axis is robbery ($+$) versus larceny ($-$) and auto-theft ($-$) axis and that the second factor contrasts assault ($-$) to auto-theft ($+$). The dominating states for the first axis are the North-Eastern States MA ($+$) and NY ($+$) contrasting the Western States WY ($-$) and ID ($-$). For the second axis, the differences are seen between the Northern States (MA ($+$) and RI ($+$)) and the Southern States AL ($-$), MS ($-$) and AR ($-$). These results can be clearly seen in Fig. 15.4 where all the states and crimes are reported. The figure also shows in which states the proportion of a particular crime category is higher or lower than the national average (the origin).

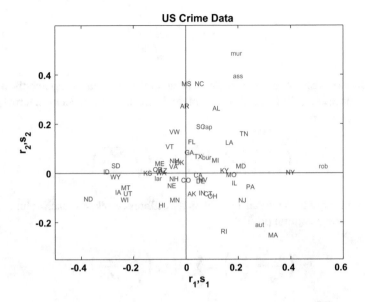

Fig. 15.4 Projection of rows (the 50 states) and columns (the 7 crime categories) **Q** MVAcorrcrime

Biplots

The biplot is a low-dimensional display of a data matrix $\mathcal{X}$ where the rows and columns are represented by points. The interpretation of a biplot is specifically directed toward the scalar products of lower dimensional factorial variables and is designed to approximately recover the individual elements of the data matrix in these scalar products. Suppose that we have a (10×5) data matrix with elements x_{ij}. The idea of the biplot is to find 10 row points $q_i \in \mathbb{R}^k$ ($k < p$, $i = 1, \ldots, 10$) and 5 column points $t_j \in \mathbb{R}^k$ ($j = 1, \ldots, 5$) such that the 50 scalar products between the row and the column vectors closely approximate the 50 corresponding elements of the data matrix $\mathcal{X}$. Usually, we choose $k = 2$. For example, the scalar product between q_7 and t_4 should approximate the data value x_{74} in the seventh row and the fourth column. In general, the biplot models the data x_{ij} as the sum of a scalar product in some low-dimensional subspace and a residual "error" term:

$$
\begin{aligned}
x_{ij} &= q_i^{\top} t_j + e_{ij} \\
&= \sum_k q_{ik} t_{jk} + e_{ij}.
\end{aligned}
\tag{15.25}
$$

To understand the link between correspondence analysis and the biplot, we need to introduce a formula which expresses x_{ij} from the original data matrix (see (15.3)) in terms of row and column frequencies. One such formula, known as the "reconstitution formula", is (15.10):

$$x_{ij} = E_{ij}\left(1 + \frac{\sum_{k=1}^{R} \lambda_k^{\frac{1}{2}} \gamma_{ik}\delta_{jk}}{\sqrt{\frac{x_{i\bullet}x_{\bullet j}}{x_{\bullet\bullet}}}}\right) \tag{15.26}$$

Consider now the row profiles $x_{ij}/x_{i\bullet}$ (the conditional frequencies) and the average row profile $x_{i\bullet}/x_{\bullet\bullet}$. From (15.26) we obtain the difference between each row profile and this average:

$$\left(\frac{x_{ij}}{x_{i\bullet}} - \frac{x_{i\bullet}}{x_{\bullet\bullet}}\right) = \sum_{k=1}^{R} \lambda_k^{\frac{1}{2}}\gamma_{ik}\left(\sqrt{\frac{x_{\bullet j}}{x_{i\bullet}x_{\bullet\bullet}}}\right)\delta_{jk}. \tag{15.27}$$

By the same argument we can also obtain the difference between each column profile and the average column profile:

$$\left(\frac{x_{ij}}{x_{\bullet j}} - \frac{x_{\bullet j}}{x_{\bullet\bullet}}\right) = \sum_{k=1}^{R} \lambda_k^{\frac{1}{2}}\gamma_{ik}\left(\sqrt{\frac{x_{i\bullet}}{x_{\bullet j}x_{\bullet\bullet}}}\right)\delta_{jk}. \tag{15.28}$$

Now, if $\lambda_1 \gg \lambda_2 \gg \lambda_3, \ldots$, we can approximate these sums by a finite number of K terms (usually $K = 2$) using (15.16) to obtain

$$\left(\frac{x_{ij}}{x_{\bullet j}} - \frac{x_{i\bullet}}{x_{\bullet\bullet}}\right) = \sum_{k=1}^{K} \left(\frac{x_{i\bullet}}{\sqrt{\lambda_k}x_{\bullet\bullet}}r_{ki}\right)s_{kj} + e_{ij}, \tag{15.29}$$

$$\left(\frac{x_{ij}}{x_{i\bullet}} - \frac{x_{\bullet j}}{x_{\bullet\bullet}}\right) = \sum_{k=1}^{K} \left(\frac{x_{\bullet j}}{\sqrt{\lambda_k}x_{\bullet\bullet}}s_{kj}\right)r_{ki} + e'_{ij}, \tag{15.30}$$

where e_{ij} and e'_{ij} are error terms. Equation (15.30) shows that if we consider displaying the differences between the row profiles and the average profile, then the projection of the row profile r_k and a rescaled version of the projections of the column profile s_k constitute a biplot of these differences. Equation (15.29) implies the same for the differences between the column profiles and this average.

Summary
↪ Correspondence analysis is a factorial decomposition of contingency tables. The p-dimensional individuals and the n-dimensional variables can be graphically represented by projecting onto spaces of smaller dimension.
↪ The practical computation consists of first computing a spectral decomposition of $\mathcal{A}^{-1}\mathcal{X}\mathcal{B}^{-1}\mathcal{X}^{\top}$ and $\mathcal{B}^{-1}\mathcal{X}^{\top}\mathcal{A}^{-1}\mathcal{X}$ which have the same first p eigenvalues. The graphical representation is obtained by plotting $\sqrt{\lambda_1}r_1$ vs. $\sqrt{\lambda_2}r_2$ and $\sqrt{\lambda_1}s_1$ vs. $\sqrt{\lambda_2}s_2$. Both plots might be displayed in the same graph taking into account the appropriate orientation of the eigenvectors r_i, s_j.

Continued on next page

Summary (continue)
$\hookrightarrow$ Correspondence analysis provides a graphical display of the association measure $c_{ij} = (x_{ij} - E_{ij})^2/E_{ij}$.
$\hookrightarrow$ Biplot is a low-dimensional display of a data matrix where the rows and columns are represented by points

15.4 Exercises

Exercise 15.1 Show that the matrices $A^{-1}\mathcal{X}B^{-1}\mathcal{X}^\top$ and $B^{-1}\mathcal{X}^\top A^{-1}\mathcal{X}$ have an eigenvalue equal to 1 and that the corresponding eigenvectors are proportional to $(1, \ldots, 1)^\top$.

Exercise 15.2 Verify the relations in (15.8), (15.14) and (15.17).

Exercise 15.3 Perform a correspondence analysis for the car marks data (Sect. 22.7). Explain how this table can be considered as a contingency table.

Exercise 15.4 Compute the χ^2-statistic of independence for the French baccalauréat data.

Exercise 15.5 Prove that $C = A^{-1/2}(\mathcal{X} - E)B^{-1/2}\sqrt{x_{\bullet\bullet}}$ and $E = \frac{ab^\top}{x_{\bullet\bullet}}$ and verify (15.20).

Exercise 15.6 Perform the full correspondence analysis of the U.S. crime data (Sect. B.8) and determine the absolute contributions for the first three axes. How can you interpret the third axis? Try to identify the state with one of the four regions to which it belongs. Do you think the four regions have a different behavior with respect to crime?

Exercise 15.7 Consider a $(n \times n)$ contingency table being a diagonal matrix $\mathcal{X}$. What do you expect the factors r_k, s_k to be like?

Exercise 15.8 Assume that after some reordering of the rows and the columns, the contingency table has the following structure:

$$\mathcal{X} = \begin{array}{c|c|c} & J_1 & J_2 \\ \hline I_1 & * & 0 \\ \hline I_2 & 0 & * \end{array}$$

That is, the rows I_i only have weights in the columns J_i, for $i = 1, 2$. What do you expect the graph of the first two factors to look like?

Exercise 15.9 Redo Exercise 15.8 using the following contingency table:

$$\mathcal{X} = \begin{array}{c|ccc} & J_1 & J_2 & J_3 \\ \hline I_1 & * & 0 & 0 \\ I_2 & 0 & * & 0 \\ I_3 & 0 & 0 & * \end{array}$$

Exercise 15.10 Consider the French food data (Sect. B.6). Given that all of the variables are measured in the same units (Francs), explain how this table can be considered as a contingency table. Perform a correspondence analysis and compare the results to those obtained in the NPCA analysis in Chap. 11.

Chapter 16
Canonical Correlation Analysis

Complex multivariate data structures are better understood by studying low-dimensional projections. For a joint study of two data sets, we may ask what type of low-dimensional projection helps in finding possible joint structures for the two samples. The canonical correlation analysis is a standard tool of multivariate statistical analysis for discovery and quantification of associations between two sets of variables.

The basic technique is based on projections. One defines an index (projected multivariate variable) that maximally correlates with the index of the other variable for each sample separately. The aim of canonical correlation analysis is to maximize the association (measured by correlation) between the low-dimensional projections of the two data sets. The canonical correlation vectors are found by a joint covariance analysis of the two variables. The technique is applied to a marketing example where the association of a price factor and other variables (like design, sportiness) is analyzed. Tests are given on how to evaluate the significance of the discovered association.

16.1 Most Interesting Linear Combination

The associations between two sets of variables may be identified and quantified by canonical correlation analysis. The technique was originally developed by Hotteling (Hotteling (1953)) who analyzed how arithmetic speed and arithmetic power are related to reading speed and reading power. Other examples are the relation between governmental policy variables and economic performance variables and the relation between job and company characteristics.

Suppose we are given two random variables $X \in \mathbb{R}^q$ and $Y \in \mathbb{R}^p$. The idea is to find an index describing a (possible) link between X and Y. Canonical correlation analysis (CCA) is based on linear indices, i.e., linear combinations

© Springer Nature Switzerland AG 2019
W. K. Härdle and L. Simar, *Applied Multivariate Statistical Analysis*,
https://doi.org/10.1007/978-3-030-26006-4_16

$$a^\top X \quad \text{and} \quad b^\top Y$$

of the random variables. Canonical correlation analysis searches for vectors a and b such that the relation of the two indices $a^\top x$ and $b^\top y$ is quantified in some interpretable way. More precisely, one is looking for the "most interesting" projections a and b in the sense that they maximize the correlation

$$\rho(a, b) = \rho_{a^\top X b^\top Y} \tag{16.1}$$

between the two indices.

Let us consider the correlation $\rho(a, b)$ between the two projections in more detail. Suppose that

$$\begin{pmatrix} X \\ Y \end{pmatrix} \sim \left(\begin{pmatrix} \mu \\ \nu \end{pmatrix}, \begin{pmatrix} \Sigma_{XX} & \Sigma_{XY} \\ \Sigma_{YX} & \Sigma_{YY} \end{pmatrix} \right)$$

where the sub-matrices of this covariance structure are given by

$$\mathsf{Var}(X) = \Sigma_{XX} \ (q \times q)$$
$$\mathsf{Var}(Y) = \Sigma_{YY} \ (p \times p)$$
$$\mathsf{Cov}(X, Y) = \mathsf{E}(X - \mu)(Y - \nu)^\top = \Sigma_{XY} = \Sigma_{YX}^\top \ (q \times p).$$

Using (3.7) and (4.26),

$$\rho(a, b) = \frac{a^\top \Sigma_{XY} b}{(a^\top \Sigma_{XX} a)^{1/2} (b^\top \Sigma_{YY} b)^{1/2}} . \tag{16.2}$$

Therefore, $\rho(ca, b) = \rho(a, b)$ for any $c \in \mathbb{R}^+$. Given the invariance of scale we may rescale projections a and b and thus we can equally solve

$$\max_{a,b} \quad a^\top \Sigma_{XY} b$$

under the constraints

$$a^\top \Sigma_{XX} a = 1$$
$$b^\top \Sigma_{YY} b = 1.$$

For this problem, define

$$\mathcal{K} = \Sigma_{XX}^{-1/2} \Sigma_{XY} \Sigma_{YY}^{-1/2}. \tag{16.3}$$

Recall the singular value decomposition of $\mathcal{K}(q \times p)$ from Theorem 2.2. The matrix $\mathcal{K}$ may be decomposed as

$$\mathcal{K} = \Gamma \Lambda \Delta^\top$$

with

$$\Gamma = (\gamma_1, \ldots, \gamma_k)$$
$$\Delta = (\delta_1, \ldots, \delta_k) \tag{16.4}$$
$$\Lambda = \mathrm{diag}(\lambda_1^{1/2}, \ldots, \lambda_k^{1/2})$$

where by (16.3) and (2.15),

$$k = \mathrm{rank}(\mathcal{K}) = \mathrm{rank}(\Sigma_{XY}) = \mathrm{rank}(\Sigma_{YX}) \, ,$$

and $\lambda_1 \geq \lambda_2 \geq \ldots \lambda_k$ are the nonzero eigenvalues of $\mathcal{N}_1 = \mathcal{K}\mathcal{K}^\top$ and $\mathcal{N}_2 = \mathcal{K}^\top\mathcal{K}$ and γ_i and δ_j are the standardized eigenvectors of $\mathcal{N}_1$ and $\mathcal{N}_2$ respectively.

Define now for $i = 1, \ldots, k$ the vectors

$$a_i = \Sigma_{XX}^{-1/2}\gamma_i, \tag{16.5}$$
$$b_i = \Sigma_{YY}^{-1/2}\delta_i, \tag{16.6}$$

which are called the *canonical correlation vectors*. Using these canonical correlation vectors we define the *canonical correlation variables*

$$\eta_i = a_i^\top X \tag{16.7}$$
$$\varphi_i = b_i^\top Y. \tag{16.8}$$

The quantities $\rho_i = \lambda_i^{1/2}$ for $i = 1, \ldots, k$ are called the *canonical correlation coefficients*.

From the properties of the singular value decomposition given in (16.4) we have

$$\mathrm{Cov}(\eta_i, \eta_j) = a_i^\top \Sigma_{XX} a_j = \gamma_i^\top \gamma_j = \begin{cases} 1 & i = j, \\ 0 & i \neq j. \end{cases} \tag{16.9}$$

The same is true for $\mathrm{Cov}(\varphi_i, \varphi_j)$. The following theorem tells us that the canonical correlation vectors are the solution to the maximization problem of (16.1).

Theorem 16.1 *For any given r, $1 \leq r \leq k$, the maximum*

$$C(r) = \max_{a,b} a^\top \Sigma_{XY} b \tag{16.10}$$

subject to

$$a^\top \Sigma_{XX} a = 1, \quad b^\top \Sigma_{YY} b = 1$$

and

$$a_i^\top \Sigma_{XX} a = 0 \, for \, i = 1, \ldots, r-1$$

is given by

$$C(r) = \rho_r = \lambda_r^{1/2}$$

and is attained when $a = a_r$ and $b = b_r$.

Proof The proof is given in three steps.
(i) Fix a and maximize over b, i.e., solve:

$$\max_b \left(a^\top \Sigma_{XY} b\right)^2 = \max_b \left(b^\top \Sigma_{YX} a\right)\left(a^\top \Sigma_{XY} b\right)$$

subject to $b^\top \Sigma_{YY} b = 1$. By Theorem 2.5 the maximum is given by the largest eigenvalue of the matrix

$$\Sigma_{YY}^{-1} \Sigma_{YX} a a^\top \Sigma_{XY}.$$

By Corollary 2.2, the only nonzero eigenvalue equals

$$a^\top \Sigma_{XY} \Sigma_{YY}^{-1} \Sigma_{YX} a. \tag{16.11}$$

(ii) Maximize (16.11) over a subject to the constraints of the Theorem. Put $\gamma = \Sigma_{XX}^{1/2} a$ and observe that (16.11) equals

$$\gamma^\top \Sigma_{XX}^{-1/2} \Sigma_{XY} \Sigma_{YY}^{-1} \Sigma_{YX} \Sigma_{XX}^{-1/2} \gamma = \gamma^\top \mathcal{K}^\top \mathcal{K} \gamma.$$

Thus, solve the equivalent problem

$$\max_\gamma \gamma^\top \mathcal{N}_1 \gamma \tag{16.12}$$

subject to $\gamma^\top \gamma = 1$, $\gamma_i^\top \gamma = 0$ for $i = 1, \ldots, r - 1$.

Note that the γ_i's are the eigenvectors of $\mathcal{N}_1$ corresponding to its first $r - 1$ largest eigenvalues. Thus, as in Theorem 11.3, the maximum in (16.12) is obtained by setting γ equal to the eigenvector corresponding to the r-th largest eigenvalue, i.e., $\gamma = \gamma_r$ or equivalently $a = a_r$. This yields

$$C^2(r) = \gamma_r^\top \mathcal{N}_1 \gamma_r = \lambda_r \gamma_r^\top \gamma = \lambda_r.$$

(iii) Show that the maximum is attained for $a = a_r$ and $b = b_r$. From the SVD of $\mathcal{K}$ we conclude that $\mathcal{K} \delta_r = \rho_r \gamma_r$ and hence

$$a_r^\top \Sigma_{XY} b_r = \gamma_r^\top \mathcal{K} \delta_r = \rho_r \gamma_r^\top \gamma_r = \rho_r.$$

Let

$$\begin{pmatrix} X \\ Y \end{pmatrix} \sim \left(\begin{pmatrix} \mu \\ \nu \end{pmatrix}, \begin{pmatrix} \Sigma_{XX} & \Sigma_{XY} \\ \Sigma_{YX} & \Sigma_{YY} \end{pmatrix} \right).$$

The canonical correlation vectors

$$a_1 = \Sigma_{XX}^{-1/2} \gamma_1,$$

$$b_1 = \Sigma_{YY}^{-1/2} \delta_1$$

maximize the correlation between the canonical variables

$$\eta_1 = a_1^\top X,$$

$$\varphi_1 = b_1^\top Y.$$

The covariance of the canonical variables η and φ is given in the next theorem.

Theorem 16.2 *Let η_i and φ_i be the i-th canonical correlation variables ($i = 1, \ldots, k$). Define $\eta = (\eta_1, \ldots, \eta_k)$ and $\varphi = (\varphi_1, \ldots, \varphi_k)$. Then*

$$\mathsf{Var}\begin{pmatrix} \eta \\ \varphi \end{pmatrix} = \begin{pmatrix} \mathcal{I}_k & \Lambda \\ \Lambda & \mathcal{I}_k \end{pmatrix}$$

with Λ given in (16.4).

This theorem shows that the canonical correlation coefficients, $\rho_i = \lambda_i^{1/2}$, are the covariances between the canonical variables η_i and φ_i and that the indices $\eta_1 = a_1^\top X$ and $\varphi_1 = b_1^\top Y$ have the maximum covariance $\sqrt{\lambda_1} = \rho_1$.

The following theorem shows that canonical correlations are invariant with respect to linear transformations of the original variables.

Theorem 16.3 *Let $X^* = \mathcal{U}^\top X + u$ and $Y^* = \mathcal{V}^\top Y + v$ where $\mathcal{U}$ and $\mathcal{V}$ are nonsingular matrices. Then the canonical correlations between X^* and Y^* are the same as those between X and Y. The canonical correlation vectors of X^* and Y^* are given by*

$$a_i^* = \mathcal{U}^{-1} a_i,$$
$$b_i^* = \mathcal{V}^{-1} b_i. \tag{16.13}$$

Summary
$\hookrightarrow$ Canonical correlation analysis aims to identify possible links between two (sub-)sets of variables $X \in \mathbb{R}^q$ and $Y \in \mathbb{R}^p$. The idea is to find indices $a^\top X$ and $b^\top Y$ such that the correlation $\rho(a, b) = \rho_{a^\top X b^\top Y}$ is maximal.
$\hookrightarrow$ The maximum correlation (under constraints) is attained by setting $a_i = \Sigma_{XX}^{-1/2} \gamma_i$ and $b_i = \Sigma_{YY}^{-1/2} \delta_i$, where γ_i and δ_i denote the eigenvectors of $\mathcal{K}\mathcal{K}^\top$ and $\mathcal{K}^\top\mathcal{K}$, $\mathcal{K} = \Sigma_{XX}^{-1/2} \Sigma_{XY} \Sigma_{YY}^{-1/2}$ respectively.
$\hookrightarrow$ The vectors a_i and b_i are called canonical correlation vectors.
$\hookrightarrow$ The indices $\eta_i = a_i^\top X$ and $\varphi_i = b_i^\top Y$ are called canonical correlation variables.

Continued on next page

Summary (continue)		
↪	The values $\rho_1 = \sqrt{\lambda_1}, \ldots, \rho_k = \sqrt{\lambda_k}$, which are the square roots of the nonzero eigenvalues of $\mathcal{K}\mathcal{K}^\top$ and $\mathcal{K}^\top\mathcal{K}$, are called the canonical correlation coefficients. The covariance between the canonical correlation variables is $\mathrm{Cov}(\eta_i, \varphi_i) = \sqrt{\lambda_i}, i = 1, \ldots, k$.	
↪	The first canonical variables, $\eta_1 = a_1^\top X$ and $\varphi_1 = b_1^\top Y$, have the maximum covariance $\sqrt{\lambda_1}$.	
↪	Canonical correlations are invariant with respect to linear transformations of the original variables X and Y.	

16.2 Canonical Correlation in Practice

In practice we have to estimate the covariance matrices Σ_{XX}, Σ_{XY} and Σ_{YY}. Let us apply the canonical correlation analysis to the car marks data (see Sect. 22.12). In the context of this data set one is interested in relating price variables with variables such as sportiness, safety. In particular, we would like to investigate the relation between the two variables *non-depreciation of value* and *price of the car* and all other variables.

Example 16.1 We perform the canonical correlation analysis on the data matrices $\mathcal{X}$ and $\mathcal{Y}$ that correspond to the set of values {Price, Value Stability} and {Economy, Service, Design, Sporty car, Safety, Easy handling}, respectively. The estimated covariance matrix S is given by

$$
\begin{array}{ccccccccc}
 & \text{Price} & \text{Value} & \text{Econ.} & \text{Serv.} & \text{Design} & \text{Sport.} & \text{Safety} & \text{Easy h.}
\end{array}
$$

$$
S = \left(\begin{array}{cc|cccccc}
1.41 & -1.11 & 0.78 & -0.71 & -0.90 & -1.04 & -0.95 & 0.18 \\
-1.11 & 1.19 & -0.42 & 0.82 & 0.77 & 0.90 & 1.12 & 0.11 \\
\hline
0.78 & -0.42 & 0.75 & -0.23 & -0.45 & -0.42 & -0.28 & 0.28 \\
-0.71 & 0.82 & -0.23 & 0.66 & 0.52 & 0.57 & 0.85 & 0.14 \\
-0.90 & 0.77 & -0.45 & 0.52 & 0.72 & 0.77 & 0.68 & -0.10 \\
-1.04 & 0.90 & -0.42 & 0.57 & 0.77 & 1.05 & 0.76 & -0.15 \\
-0.95 & 1.12 & -0.28 & 0.85 & 0.68 & 0.76 & 1.26 & 0.22 \\
0.18 & 0.11 & 0.28 & 0.14 & -0.10 & -0.15 & 0.22 & 0.32
\end{array}\right).
$$

Hence,

$$\mathcal{S}_{XX} = \begin{pmatrix} 1.41 & -1.11 \\ -1.11 & 1.19 \end{pmatrix}, \quad \mathcal{S}_{XY} = \begin{pmatrix} 0.78 & -0.71 & -0.90 & -1.04 & -0.95 & 0.18 \\ -0.42 & 0.82 & 0.77 & 0.90 & 1.12 & 0.11 \end{pmatrix},$$

$$\mathcal{S}_{YY} = \begin{pmatrix} 0.75 & -0.23 & -0.45 & -0.42 & -0.28 & 0.28 \\ -0.23 & 0.66 & 0.52 & 0.57 & 0.85 & 0.14 \\ -0.45 & 0.52 & 0.72 & 0.77 & 0.68 & -0.10 \\ -0.42 & 0.57 & 0.77 & 1.05 & 0.76 & -0.15 \\ -0.28 & 0.85 & 0.68 & 0.76 & 1.26 & 0.22 \\ 0.28 & 0.14 & -0.10 & -0.15 & 0.22 & 0.32 \end{pmatrix}.$$

It is interesting to see that value stability and price have a negative covariance. This makes sense since highly priced vehicles tend to loose their market value at a faster pace than medium priced vehicles.

Now we estimate $\mathcal{K} = \Sigma_{XX}^{-1/2}\,\Sigma_{XY}\,\Sigma_{YY}^{-1/2}$ by

$$\widehat{\mathcal{K}} = \mathcal{S}_{XX}^{-1/2}\,\mathcal{S}_{XY}\,\mathcal{S}_{YY}^{-1/2}$$

and perform a singular value decomposition of $\widehat{\mathcal{K}}$:

$$\widehat{\mathcal{K}} = \mathcal{G}\mathcal{L}\mathcal{D}^{\top} = (g_1, g_2)\ \mathrm{diag}(\ell_1^{1/2}, \ell_2^{1/2})\ (d_1, d_2)^{\top}$$

where the ℓ_i's are the eigenvalues of $\widehat{\mathcal{K}}\widehat{\mathcal{K}}^{\top}$ and $\widehat{\mathcal{K}}^{\top}\widehat{\mathcal{K}}$ with $\mathrm{rank}(\widehat{\mathcal{K}}) = 2$, and g_i and d_i are the eigenvectors of $\widehat{\mathcal{K}}\widehat{\mathcal{K}}^{\top}$ and $\widehat{\mathcal{K}}^{\top}\widehat{\mathcal{K}}$, respectively. The canonical correlation coefficients are

$$r_1 = \ell_1^{1/2} = 0.98, \quad r_2 = \ell_2^{1/2} = 0.89.$$

The high correlation of the second two canonical variables can be seen in Fig. 16.1. The second canonical variables are

$$\widehat{\eta}_1 = \widehat{a}_1^{\top}x = 1.602\,x_1 + 1.686\,x_2$$
$$\widehat{\varphi}_1 = \widehat{b}_1^{\top}y = 0.568\,y_1 + 0.544\,y_2 - 0.012\,y_3 - 0.096\,y_4 - 0.014\,y_5 + 0.915\,y_6.$$

Note that the variables y_1 (economy), y_2 (service) and y_6 (easy handling) have positive coefficients on $\widehat{\varphi}_1$. The variables y_3 (design), y_4 (sporty car) and y_5 (safety) have a negative influence on $\widehat{\varphi}_1$.

The canonical variable η_1 may be interpreted as a price and value index. The canonical variable φ_1 is mainly formed from the qualitative variables economy, service and handling with negative weights on design, safety and sportiness. These variables may therefore be interpreted as an appreciation of the value of the car. The sportiness has a negative effect on the price and value index, as do the design and the safety features.

Testing the Canonical Correlation Coefficients

The hypothesis that the two sets of variables $\mathcal{X}$ and $\mathcal{Y}$ are uncorrelated may be tested (under normality assumptions) with Wilks likelihood ratio statistic Gibbins (1985):

Fig. 16.1 The second
canonical variables for the
car marks data Q
MVAcancarm

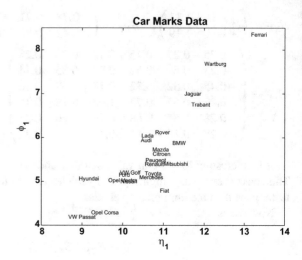

$$T^{2/n} = \left| \mathcal{I} - S_{YY}^{-1} S_{YX} S_{XX}^{-1} S_{XY} \right| = \prod_{i=1}^{k} (1 - \ell_i).$$

This statistic unfortunately has a rather complicated distribution. Bartlett (1939) provides an approximation for large n:

$$- \{n - (p + q + 3)/2\} \log \prod_{i=1}^{k} (1 - \ell_i) \sim \chi_{pq}^2. \qquad (16.14)$$

A test of the hypothesis that only s of the canonical correlation coefficients are nonzero may be based (asymptotically) on the statistic

$$- \{n - (p + q + 3)/2\} \log \prod_{i=s+1}^{k} (1 - \ell_i) \sim \chi_{(p-s)(q-s)}^2. \qquad (16.15)$$

Example 16.2 Consider Example 16.1 again. There are $n = 40$ persons that have rated the cars according to different categories with $p = 2$ and $q = 6$. The canonical correlation coefficients were found to be $r_1 = 0.98$ and $r_2 = 0.89$. Bartlett's statistic (16.14) is therefore

$$-\{40 - (2 + 6 + 3)/2\} \log\{(1 - 0.98^2)(1 - 0.89^2)\} = 165.59 \sim \chi_{12}^2$$

which is highly significant (the 99% quantile of the χ_{12}^2 is 26.23). The hypothesis of no correlation between the variables $\mathcal{X}$ and $\mathcal{Y}$ is therefore rejected.

Let us now test whether the second canonical correlation coefficient is different from zero. We use Bartlett's statistic (16.15) with $s = 1$ and obtain

$$-\{40 - (2 + 6 + 3)/2\} \log\{(1 - 0.89^2)\} = 54.19 \sim \chi_5^2$$

which is again highly significant with the χ_5^2 distribution.

Canonical Correlation Analysis with Qualitative Data

The canonical correlation technique may also be applied to qualitative data. Consider for example the contingency table $\mathcal{N}$ of the French baccalauréat data. The data set is given in Sect. B.14 in Appendix B. The CCA cannot be applied directly to this contingency table since the table does not correspond to the usual data matrix structure. We may wish, however, to explain the relationship between the row r and column c categories. It is possible to represent the data in a $(n \times (r + c))$ data matrix $\mathcal{Z} = (\mathcal{X}, \mathcal{Y})$ where n is the total number of frequencies in the contingency table $\mathcal{N}$ and $\mathcal{X}$ and $\mathcal{Y}$ are matrices of zero-one dummy variables. More precisely, let

$$x_{ki} = \begin{cases} 1 & \text{if the } k\text{-th individual belongs to the } i\text{-th row category} \\ 0 & \text{otherwise} \end{cases}$$

and

$$y_{kj} = \begin{cases} 1 & \text{if the } k\text{-th individual belongs to the } j\text{-th column category} \\ 0 & \text{otherwise} \end{cases}$$

where the indices range from $k = 1, \ldots, n$, $i = 1, \ldots, r$ and $j = 1, \ldots, c$. Denote the cell frequencies by n_{ij} so that $\mathcal{N} = (n_{ij})$ and note that

$$x_{(i)}^\top y_{(j)} = n_{ij},$$

where $x_{(i)}$ ($y_{(j)}$) denotes the i-th (j-th) column of $\mathcal{X}$ ($\mathcal{Y}$).

Example 16.3 Consider the following example where

$$\mathcal{N} = \begin{pmatrix} 3 & 2 \\ 1 & 4 \end{pmatrix}.$$

The matrices $\mathcal{X}$, $\mathcal{Y}$ and $\mathcal{Z}$ are therefore

$$\mathcal{X} = \begin{pmatrix} 1 & 0 \\ 1 & 0 \\ 1 & 0 \\ 1 & 0 \\ 1 & 0 \\ 0 & 1 \\ 0 & 1 \\ 0 & 1 \\ 0 & 1 \\ 0 & 1 \end{pmatrix}, \quad \mathcal{Y} = \begin{pmatrix} 1 & 0 \\ 1 & 0 \\ 1 & 0 \\ 0 & 1 \\ 0 & 1 \\ 1 & 0 \\ 0 & 1 \\ 0 & 1 \\ 0 & 1 \\ 0 & 1 \end{pmatrix}, \quad \mathcal{Z} = (\mathcal{X}, \mathcal{Y}) = \begin{pmatrix} 1 & 0 & 1 & 0 \\ 1 & 0 & 1 & 0 \\ 1 & 0 & 1 & 0 \\ 1 & 0 & 0 & 1 \\ 1 & 0 & 0 & 1 \\ 0 & 1 & 1 & 0 \\ 0 & 1 & 0 & 1 \\ 0 & 1 & 0 & 1 \\ 0 & 1 & 0 & 1 \\ 0 & 1 & 0 & 1 \end{pmatrix}.$$

The element n_{12} of $\mathcal{N}$ may be obtained by multiplying the first column of $\mathcal{X}$ with the second column of $\mathcal{Y}$ to yield

$$x_{(1)}^\top y_{(2)} = 2.$$

The purpose is to find the canonical variables $\eta = a^\top x$ and $\varphi = b^\top y$ that are maximally correlated. Note, however, that x has only one nonzero component and therefore an "individual" may be directly associated with its canonical variables or score (a_i, b_j). There will be n_{ij} points at each (a_i, b_j) and the correlation represented by these points may serve as a measure of dependence between the rows and columns of $\mathcal{N}$.

Let $\mathcal{Z} = (\mathcal{X}, \mathcal{Y})$ denote a data matrix constructed from a contingency table $\mathcal{N}$. Similar to Chap. 14 define

$$c = x_{i\bullet} = \sum_{j=1}^{c} n_{ij},$$

$$d = x_{\bullet j} = \sum_{i=1}^{r} n_{ij},$$

and define $C = \mathrm{diag}(c)$ and $D = \mathrm{diag}(d)$. Suppose that $x_{i\bullet} > 0$ and $x_{\bullet j} > 0$ for all i and j. It is not hard to see that

$$nS = \mathcal{Z}^\top \mathcal{H}\mathcal{Z} = \mathcal{Z}^\top \mathcal{Z} - n\bar{z}\bar{z}^\top = \begin{pmatrix} nS_{XX} & nS_{XY} \\ nS_{YX} & nS_{YY} \end{pmatrix}$$

$$= \left(\frac{n}{n-1}\right) \begin{pmatrix} C - n^{-1}cc^\top & \mathcal{N} - \widehat{\mathcal{N}} \\ \mathcal{N}^\top \widehat{\mathcal{N}}^\top & D - n^{-1}dd^\top \end{pmatrix}$$

where $\widehat{\mathcal{N}} = cd^\top/n$ is the estimated value of $\mathcal{N}$ under the assumption of independence of the row and column categories.

Note that

$$(n-1)S_{XX}1_r = C1_r - n^{-1}cc^\top 1_r = c - c(n^{-1}c^\top 1_r) = c - c(n^{-1}n) = 0$$

and therefore S_{XX}^{-1} does not exist. The same is true for S_{YY}^{-1}. One way out of this difficulty is to drop one column from both $\mathcal{X}$ and $\mathcal{Y}$, say the first column. Let $\bar{c}$ and $\bar{d}$ denote the vectors obtained by deleting the first component of c and d.

Define $\bar{C}$, $\bar{D}$ and $\bar{S}_{XX}$, $\bar{S}_{YY}$, $\bar{S}_{XY}$ accordingly and obtain

$$(n\bar{S}_{XX})^{-1} = \bar{C}^{-1} + n_{i\bullet}^{-1}1_r 1_r^\top$$

$$(n\bar{S}_{YY})^{-1} = \bar{D}^{-1} + n_{\bullet j}^{-1}1_c 1_c^\top$$

so that (16.3) exists. The score associated with an individual contained in the first row (column) category of $\mathcal{N}$ is 0.

The technique described here for purely qualitative data may also be used when the data is a mixture of qualitative and quantitative characteristics. One has to "blow up" the data matrix by dummy zero-one values for the qualitative data variables.

Summary
$\hookrightarrow$ In practice we estimate Σ_{XX}, Σ_{XY}, Σ_{YY} by the empirical covariances and use them to compute estimates ℓ_i, g_i, d_i for λ_i, γ_i, δ_i from the SVD of $\widehat{\mathcal{K}} = \mathcal{S}_{XX}^{-1/2} \mathcal{S}_{XY} \mathcal{S}_{YY}^{-1/2}$.
$\hookrightarrow$ The signs of the coefficients of the canonical variables tell us the direction of the influence of these variables.

16.3 Exercises

Exercise 16.1 Show that the eigenvalues of $\mathcal{K}\mathcal{K}^{\top}$ and $\mathcal{K}^{\top}\mathcal{K}$ are identical (Hint: Use Theorem 2.6).

Exercise 16.2 Perform the canonical correlation analysis for the following subsets of variables: $\mathcal{X}$ corresponding to {price} and $\mathcal{Y}$ corresponding to {economy, easy handling} from the car marks data (Sect. 22.12).

Exercise 16.3 Calculate the first canonical variables for Example 16.1. Interpret the coefficients.

Exercise 16.4 Use the SVD of matrix $\mathcal{K}$ to show that the canonical variables η_1 and η_2 are not correlated.

Exercise 16.5 Verify that the number of nonzero eigenvalues of matrix $\mathcal{K}$ is equal to rank(Σ_{XY}).

Exercise 16.6 Express the singular value decomposition of matrices $\mathcal{K}$ and $\mathcal{K}^{\top}$ using eigenvalues and eigenvectors of matrices $\mathcal{K}^{\top}\mathcal{K}$ and $\mathcal{K}\mathcal{K}^{\top}$.

Exercise 16.7 What will be the result of CCA for $Y = X$?

Exercise 16.8 What will be the results of CCA for $Y = 2X$ and for $Y = -X$?

Exercise 16.9 What results do you expect if you perform CCA for X and Y such that $\Sigma_{XY} = 0$? What if $\Sigma_{XY} = \mathcal{I}_p$?

References

M.S. Bartlett, A note on tests of significance in multivariate analysis. Proceedings of the Cambridge Philosophical Society **35**, 180–185 (1939)

R. Gibbins, *Canonical Analysis (A Review with Application in Ecology* (Springer, Berlin, 1985)

H. Hotelling, New light on the correlation coefficient and its transformation. J. R. Stat. Soc. Ser B **15**(1), 193–232 (1953)

Chapter 17
Multidimensional Scaling

One major aim of multivariate data analysis is dimension reduction. For data measured in Euclidean coordinates, Factor Analysis and Principal Component Analysis are dominantly used tools. In many applied sciences data is recorded as ranked information. For example, in marketing, one may record "product A is better than product B". High-dimensional observations therefore often have mixed data characteristics and contain relative information (w.r.t. a defined standard) rather than absolute coordinates that would enable us to employ one of the multivariate techniques presented so far.

Multidimensional scaling (MDS) is a method based on proximities between objects, subjects, or stimuli used to produce a spatial representation of these items. Proximities express the similarity or dissimilarity between data objects. It is a dimension reduction technique since the aim is to find a set of points in low dimension (typically two dimensions) that reflect the relative configuration of the high-dimensional data objects. The metric MDS is concerned with such a representation in Euclidean coordinates. The desired projections are found via an appropriate spectral decomposition of a distance matrix.

The metric MDS solution may result in projections of data objects that conflict with the ranking of the original observations. The nonmetric MDS solves this problem by iterating between a monotizing algorithmic step and a least squares projection step. The examples presented in this chapter are based on reconstructing a map from a distance matrix and on marketing concerns such as ranking of the outfit of cars.

17.1 The Problem

Multidimensional scaling (MDS) is a mathematical tool that uses proximities between objects, subjects or stimuli to produce a spatial representation of these items. The proximities are defined as any set of numbers that express the amount of

© Springer Nature Switzerland AG 2019
W. K. Härdle and L. Simar, *Applied Multivariate Statistical Analysis*,
https://doi.org/10.1007/978-3-030-26006-4_17

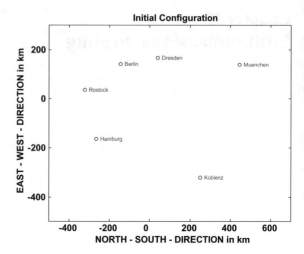

similarity or dissimilarity between pairs of objects, subjects, or stimuli. In contrast to the techniques considered so far, MDS does not start from the raw multivariate data matrix $\mathcal{X}$, but from a $(n \times n)$ dissimilarity or distance matrix, $\mathcal{D}$, with the elements δ_{ij} and d_{ij} respectively. Hence, the underlying dimensionality of the data under investigation is in general not known.

MDS is a data reduction technique because it is concerned with the problem of finding a set of points in low dimension that represents the "configuration" of data in high dimension. The "configuration" in high dimension is represented by the distance or dissimilarity matrix $\mathcal{D}$.

MDS techniques are often used to understand how people perceive and evaluate certain signals and information. For instance, political scientists use MDS techniques to understand why political candidates are perceived by voters as being similar or dissimilar. Psychologists use MDS to understand the perceptions and evaluations of speech, colors and personality traits, among other things. Last but not least, in marketing researchers use MDS techniques to shed light on the way consumers evaluate brands and to assess the relationship between product attributes.

In short, the primary purpose of all MDS techniques is to uncover structural relations or patterns in the data and to represent it in a simple geometrical model or picture. One of the aims is to determine the dimension of the model (the goal is a low-dimensional, easily interpretable model) by finding the d-dimensional space in which there is maximum correspondence between the observed proximities and the distances between points measured on a metric scale.

Multidimensional scaling based on proximities is usually referred to as metric MDS, whereas the more popular nonmetric MDS is used when the proximities are measured on an ordinal scale.

Example 17.1 A good example of how MDS works is given by Dillon and Goldstein (1984, p. 108). Suppose one is confronted with a map of Germany and asked to

Fig. 17.2 Metric MDS solution for the inter-city road distances after reflection and 90° rotation
Q MVAMDScity2

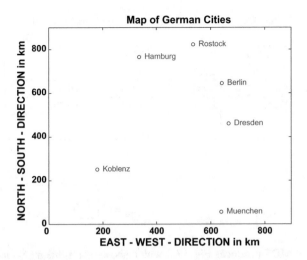

Table 17.1 Inter-city distances

	Berlin	Dresden	Hamburg	Koblenz	Munich	Rostock
Berlin	0	214	279	610	596	237
Dresden		0	492	533	496	444
Hamburg			0	520	772	140
Koblenz				0	521	687
Munich					0	771
Rostock						0

measure, with the use of a ruler and the scale of the map, some inter-city distances. Admittedly this is quite an easy exercise. However, let us now reverse the problem: One is given a set of distances, as in Table 17.1, and is asked to recreate the map itself. This is a far more difficult exercise, though it can be solved with a ruler and a compass in two dimensions. MDS is a method for solving this reverse problem in arbitrary dimensions. In Figs. 17.1 and 17.2 you can see the graphical representation of the metric MDS solution to Table 17.1 after rotating and reflecting the points representing the cities. Note that the distances given in Table 17.1 are road distances that in general do not correspond to Euclidean distances. In real-life applications, the problems are exceedingly more complex: there are usually errors in the data and the dimensionality is rarely known in advance.

Example 17.2 A further example is given in Table 17.2 where consumers noted their impressions of the dissimilarity of certain cars.

The dissimilarities in this table were in fact computed from Sect. 22.7 as Euclidean distances

Table 17.2 Dissimilarities for cars. (Numbering from Sect. 22.7)

	Audi 100	BMW 5	Citroen AX	Ferrari	...
Audi 100	0	2.232	3.451	3.689	...
BMW 5	2.232	0	5.513	3.167	...
Citroen AX	3.451	5.513	0	6.202	...
Ferrari	3.689	3.167	6.202	0	...
⋮	⋮	⋮	⋮	⋮	⋱

$$d_{ij} = \sqrt{\sum_{l=1}^{8}(x_{il} - x_{jl})^2}.$$

MDS produces Fig. 17.3 which shows a nonlinear relationship for all the cars in the projection. This enables us to build a nonlinear (quadratic) index with the Wartburg and the Trabant on the left and the Ferrari and the Jaguar on the right. We can construct an order or ranking of the cars based on the subjective impression of the consumers.

What does the ranking describe? The answer is given by Fig. 17.4 which shows the correlation between the MDS projection and the variables. Apparently, the first MDS direction is highly correlated with service $(-)$, value $(-)$, design $(-)$, sportiness $(-)$, safety $(-)$, and price $(+)$. We can interpret the first direction as the price direction since a bad mark in price ("high price") obviously corresponds with a good mark, say, in sportiness ("very sportive"). The second MDS direction is highly positively correlated with practicability. We observe from this data an almost orthogonal relationship between price and practicability.

In MDS a map is constructed in Euclidean space that corresponds to given distances. Which solution can we expect? The solution is determined only up to rotation, reflection, and shifts. In general, if $P_1, \ldots, P_n$ with coordinates $x_i = (x_{i1}, \ldots, x_{ip})^\top$ for $i = 1, \ldots, n$ represents a MDS solution in p dimensions, then $y_i = \mathcal{A}x_i + b$ with an orthogonal matrix $\mathcal{A}$ and a shift vector b also represents a MDS solution. A comparison of Figs. 17.1 and 17.2 illustrates this fact.

Solution methods that use only the rank order of the distances are termed *nonmetric methods* of MDS. Methods aimed at finding the points P_i directly from a distance matrix like the one in the Table 17.2 are called *metric methods*.

Summary

↪ MDS is a set of techniques which use distances or dissimilarities to project high-dimensional data into a low-dimensional space essential in understanding respondents perceptions and evaluations for all sorts of items.

↪ MDS starts with a $(n \times n)$ proximity matrix $\mathcal{D}$ consisting of dissimilarities $\delta_{i,j}$ or distances d_{ij}.

↪ MDS is an explorative technique and focuses on data reduction.

Continued on next page

Summary (continue)
$\hookrightarrow$ The MDS solution is indeterminate with respect to rotation, reflection and shifts.
$\hookrightarrow$ The MDS techniques are divided into metric MDS and nonmetric MDS.

17.2 Metric Multidimensional Scaling

Metric MDS begins with a $(n \times n)$ distance matrix $\mathcal{D}$ with elements d_{ij} where $i, j = 1, \ldots, n$. The objective of metric MDS is to find a configuration of points in p-dimensional space from the distances between the points such that the coordinates of the n points along the p dimensions yield a Euclidean distance matrix whose elements are as close as possible to the elements of the given distance matrix $\mathcal{D}$.

The Classical Solution

The classical solution is based on a distance matrix that is computed from a *Euclidean geometry*.

Definition 17.1 A $(n \times n)$ distance matrix $\mathcal{D} = (d_{ij})$ is Euclidean if for some points $x_1, \ldots, x_n \in \mathbb{R}^p$; $d_{ij}^2 = (x_i - x_j)^\top (x_i - x_j)$.

The following result tells us whether a distance matrix is Euclidean or not.

Theorem 17.1 *Define* $\mathcal{A} = (a_{ij})$, $a_{ij} = -\frac{1}{2} d_{ij}^2$ *and* $\mathcal{B} = \mathcal{HAH}$ *where* $\mathcal{H}$ *is the centering matrix.* $\mathcal{D}$ *is Euclidean if and only if* $\mathcal{B}$ *is positive semidefinite. If* $\mathcal{D}$ *is the distance matrix of a data matrix* $\mathcal{X}$, *then* $\mathcal{B} = \mathcal{H} \mathcal{X} \mathcal{X}^\top \mathcal{H}$. $\mathcal{B}$ *is called the inner product matrix.*

Recovery of coordinates

The task of MDS is to find the original Euclidean coordinates from a given distance matrix. Let the coordinates of n points in a p dimensional Euclidean space be given by x_i ($i = 1, \ldots, n$) where $x_i = (x_{i1}, \ldots, x_{ip})^\top$. Call $\mathcal{X} = (x_1, \ldots, x_n)^\top$ the coordinate matrix and assume $\bar{x} = 0$. The Euclidean distance between the i-th and j-th points is given by

$$d_{ij}^2 = \sum_{k=1}^{p} (x_{ik} - x_{jk})^2. \tag{17.1}$$

The general b_{ij} term of $\mathcal{B}$ is given by:

$$b_{ij} = \sum_{k=1}^{p} x_{ik} x_{jk} = x_i^\top x_j. \tag{17.2}$$

It is possible to derive $\mathcal{B}$ from the known squared distances d_{ij}, and then from $\mathcal{B}$ the unknown coordinates:

Fig. 17.3 MDS solution on
the car data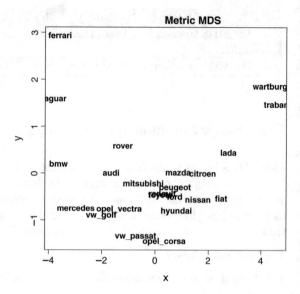
MVAmdscarm

Fig. 17.4 Correlation
between the MDS direction
and the variables
MVAmdscarm

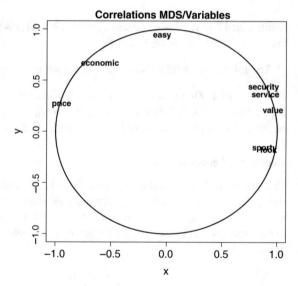

$$d_{ij}^2 = x_i^\top x_i + x_j^\top x_j - 2x_i^\top x_j$$
$$= b_{ii} + b_{jj} - 2b_{ij}. \qquad (17.3)$$

Centering of the coordinate matrix $\mathcal{X}$ implies that $\sum_{i=1}^n b_{ij} = 0$. Summing (17.3)
over i and j, we find:

$$\frac{1}{n}\sum_{i=1}^{n}d_{ij}^2 = \frac{1}{n}\sum_{i=1}^{n}b_{ii} + b_{jj}$$

$$\frac{1}{n}\sum_{j=1}^{n}d_{ij}^2 = b_{ii} + \frac{1}{n}\sum_{j=1}^{n}b_{jj}$$

$$\frac{1}{n^2}\sum_{i=1}^{n}\sum_{j=1}^{n}d_{ij}^2 = \frac{2}{n}\sum_{i=1}^{n}b_{ii}. \tag{17.4}$$

Solving (17.3) and (17.4) gives:

$$b_{ij} = -\frac{1}{2}(d_{ij}^2 - d_{i\bullet}^2 - d_{\bullet j}^2 + d_{\bullet\bullet}^2). \tag{17.5}$$

With $a_{ij} = -\frac{1}{2}d_{ij}^2$, and

$$a_{i\bullet} = \frac{1}{n}\sum_{j=1}^{n}a_{ij}$$

$$a_{\bullet j} = \frac{1}{n}\sum_{i=1}^{n}a_{ij}$$

$$a_{\bullet\bullet} = \frac{1}{n^2}\sum_{i=1}^{n}\sum_{j=1}^{n}a_{ij} \tag{17.6}$$

we get:

$$b_{ij} = a_{ij} - a_{i\bullet} - a_{\bullet j} + a_{\bullet\bullet}. \tag{17.7}$$

Define the matrix $\mathcal{A}$ as (a_{ij}), and observe that:

$$\mathcal{B} = \mathcal{H}\mathcal{A}\mathcal{H}. \tag{17.8}$$

The inner product matrix $\mathcal{B}$ can be expressed as

$$\mathcal{B} = \mathcal{X}\mathcal{X}^{\top}, \tag{17.9}$$

where $\mathcal{X} = (x_1, \ldots, x_n)^{\top}$ is the $(n \times p)$ matrix of coordinates. The rank of $\mathcal{B}$ is then

$$\mathrm{rank}(\mathcal{B}) = \mathrm{rank}(\mathcal{X}\mathcal{X}^{\top}) = \mathrm{rank}(\mathcal{X}) = p. \tag{17.10}$$

As required in Theorem 17.1 the matrix $\mathcal{B}$ is symmetric, positive semidefinite and of rank p, and hence it has p nonnegative eigenvalues and $n - p$ zero eigenvalues. $\mathcal{B}$ can now be written as

$$\mathcal{B} = \Gamma \Lambda \Gamma^{\top} \tag{17.11}$$

where $\Lambda = \mathrm{diag}(\lambda_1, \ldots, \lambda_p)$, the diagonal matrix of the eigenvalues of $\mathcal{B}$, and $\Gamma = (\gamma_1, \ldots, \gamma_p)$, the matrix of corresponding eigenvectors. Hence the coordinate matrix $\mathcal{X}$ containing the point configuration in $\mathbb{R}^p$ is given by

$$\mathcal{X} = \Gamma \Lambda^{\frac{1}{2}}. \tag{17.12}$$

How many dimensions?

The number of desired dimensions is small in order to provide practical interpretations, and is given by the rank of $\mathcal{B}$ or the number of nonzero eigenvalues λ_i. If $\mathcal{B}$ is positive semidefinite, then the number of nonzero eigenvalues gives the number of eigenvalues required for representing the distances d_{ij}.

The proportion of variation explained by p dimensions is given by

$$\frac{\sum_{i=1}^{p} \lambda_i}{\sum_{i=1}^{n-1} \lambda_i}. \tag{17.13}$$

It can be used for the choice of p. If $\mathcal{B}$ is not positive semidefinite we can modify (17.13) to

$$\frac{\sum_{i=1}^{p} \lambda_i}{\sum (\text{``positive eigenvalues''})}. \tag{17.14}$$

In practice the eigenvalues λ_i are almost always unequal to zero. To be able to represent the objects in a space with dimensions as small as possible we may modify the distance matrix to:

$$\mathcal{D}^* = d_{ij}^* \tag{17.15}$$

with

$$d_{ij}^* = \begin{cases} 0 & ; i = j \\ d_{ij} + e \geq 0 & ; i \neq j \end{cases} \tag{17.16}$$

where e is determined such that the inner product matrix $\mathcal{B}$ becomes positive semidefinite with a small rank.

Similarities

In some situations we do not start with distances but with similarities. The standard transformation (see Chap. 13) from a similarity matrix $\mathcal{C}$ to a distance matrix $\mathcal{D}$ is:

$$d_{ij} = (c_{ii} - 2c_{ij} + c_{jj})^{\frac{1}{2}}. \tag{17.17}$$

Theorem 17.2 _If $\mathcal{C} \leq 0$, then the distance matrix $\mathcal{D}$ defined by (17.17) is Euclidean with centered inner product matrix $\mathcal{B} = \mathcal{HCH}$._

Relation to Factorial Analysis

Suppose that the $(n \times p)$ data matrix $\mathcal{X}$ is centered so that $\mathcal{X}^{\top}\mathcal{X}$ equals a multiple of the covariance matrix $n\mathcal{S}$. Suppose that the p eigenvalues $\lambda_1, \ldots, \lambda_p$ of $n\mathcal{S}$ are distinct and non 0zero. Using the duality Theorem 10.4 of factorial analysis we see that $\lambda_1, \ldots, \lambda_p$ are also eigenvalues of $\mathcal{X}\mathcal{X}^{\top} = \mathcal{B}$ when $\mathcal{D}$ is the Euclidean distance matrix between the rows of $\mathcal{X}$. The k-dimensional solution to the metric MDS problem is thus given by the k first principal components of $\mathcal{X}$.

Optimality properties of the classical MDS solution

Let $\mathcal{X}$ be a $(n \times p)$ data matrix with some interpoint distance matrix $\mathcal{D}$. The objective of MDS is thus to find $\mathcal{X}_1$, a representation of $\mathcal{X}$ in a lower dimensional Euclidean space $\mathbb{R}^k$ whose interpoint distance matrix $\mathcal{D}_1$ is not far from $\mathcal{D}$. Let $\mathcal{L} = (\mathcal{L}_1, \mathcal{L}_2)$ be a $(p \times p)$ orthogonal matrix where $\mathcal{L}_1$ is $(p \times k)$. $\mathcal{X}_1 = \mathcal{X}\mathcal{L}_1$ represents a projection of $\mathcal{X}$ on the column space of $\mathcal{L}_1$; in other words, $\mathcal{X}_1$ may be viewed as a fitted configuration of $\mathcal{X}$ in $\mathbb{R}^k$. A measure of discrepancy between $\mathcal{D}$ and $\mathcal{D}_1 = (d_{ij}^{(1)})$ is given by

$$\phi = \sum_{i,j=1}^{n}(d_{ij} - d_{ij}^{(1)})^2. \tag{17.18}$$

Theorem 17.3 *Among all projections $\mathcal{X}\mathcal{L}_1$ of $\mathcal{X}$ onto k-dimensional subspaces of $\mathbb{R}^p$ the quantity ϕ in (17.18) is minimized when $\mathcal{X}$ is projected onto its first k principal factors.*

We see therefore that the metric MDS is identical to principal factor analysis as we have defined it in Chap. 10.

Summary
$\hookrightarrow$ Metric MDS starts with a distance matrix $\mathcal{D}$.
$\hookrightarrow$ The aim of metric MDS is to construct a map in Euclidean space that corresponds to the given distances.

$\hookrightarrow$ A practical algorithm is given as
 	
    1. start with distances d_{ij}
    2. define $\mathcal{A} = -\frac{1}{2}d_{ij}^2$
    3. put $\mathcal{B} = (a_{ij} - a_{i\bullet} - a_{\bullet j} + a_{\bullet\bullet})$
    4. find the eigenvalues $\lambda_1, \ldots, \lambda_p$ and the associated eigenvectors $\gamma_1, \ldots, \gamma_p$ where the eigenvectors are normalized so that $\gamma_i^{\top}\gamma_i = 1$.
    5. Choose an appropriate number of dimensions p (ideally $p = 2$)
    6. The coordinates of the n points in the Euclidean space are given by $x_{ij} = \gamma_{ij}\lambda_j^{1/2}$ for $i = 1, \ldots, n$ and $j = 1, \ldots, p$.

$\hookrightarrow$ Metric MDS is identical to principal components analysis.

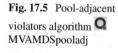

Fig. 17.5 Pool-adjacent
violators algorithm
MVAMDSpooladj

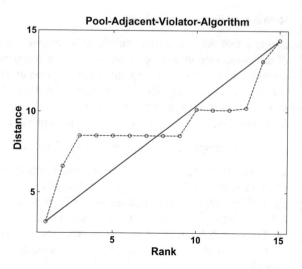

17.3 Nonmetric Multidimensional Scaling

The object of nonmetric MDS, as well as of metric MDS, is to find the coordinates
of the points in p-dimensional space, so that there is a good agreement between the
observed proximities and the interpoint distances. The development of nonmetric
MDS was motivated by two main weaknesses in the metric MDS (Fahrmeir and
Hamerle 1984, p. 679):

1. the definition of an explicit functional connection between dissimilarities and
   distances in order to derive distances out of given dissimilarities, and
2. the restriction to Euclidean geometry in order to determine the object configurations.

The idea of a nonmetric MDS is to demand a less rigid relationship between the
dissimilarities and the distances. Suppose that an unknown monotonic increasing
function f,

$$d_{ij} = f(\delta_{ij}), \tag{17.19}$$

is used to generate a set of distances d_{ij} as a function of given dissimilarities δ_{ij}. Here
f has the property that if $\delta_{ij} < \delta_{rs}$, then $f(\delta_{ij}) < f(\delta_{rs})$. The scaling is based on the
rank order of the dissimilarities. Nonmetric MDS is therefore ordinal in character.

The most common approach used to determine the elements d_{ij} and to obtain
the coordinates of the objects $x_1, x_2, \ldots, x_n$ given only rank order information is an
iterative process commonly referred to as the Shepard–Kruskal algorithm.

Shepard–Kruskal Algorithm

In the first step, called the initial phase, we calculate Euclidean distances $d_{ij}^{(0)}$ from an
arbitrarily chosen initial configuration $\mathcal{X}_0$ in dimension p^*, provided that all objects

Fig. 17.6 Ranks and distances 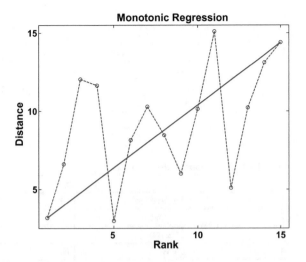 MVAMDSnonmstart

have different coordinates. One might use metric MDS to obtain these initial coordinates. The second step or nonmetric phase determines disparities $\widehat{d}_{ij}^{(0)}$ from the distances $d_{ij}^{(0)}$ by constructing a monotone regression relationship between the $d_{ij}^{(0)}$'s and δ_{ij}'s, under the requirement that if $\delta_{ij} < \delta_{rs}$, then $\widehat{d}_{ij}^{(0)} \leq \widehat{d}_{rs}^{(0)}$. This is called the weak monotonicity requirement. To obtain the disparities $\widehat{d}_{ij}^{(0)}$, a useful approximation method is the *pool-adjacent violators* (PAV) algorithm (see Fig. 17.5). Let

$$(i_1, j_1) > (i_2, j_2) > \ldots > (i_k, j_k) \qquad (17.20)$$

be the rank order of dissimilarities of the $k = n(n-1)/2$ pairs of objects. This corresponds to the points in Fig. 17.6. The PAV algorithm is described as follows: "beginning with the lowest ranked value of δ_{ij}, the adjacent $d_{ij}^{(0)}$ values are compared for each δ_{ij} to determine if they are monotonically related to the δ_{ij}'s. Whenever a block of consecutive values of $d_{ij}^{(0)}$ are encountered that violate the required monotonicity property the $d_{ij}^{(0)}$ values are averaged together with the most recent non-violator $d_{ij}^{(0)}$ value to obtain an estimator. Eventually this value is assigned to all points in the particular block".

In a third step, called the metric phase, the spatial configuration of $\mathcal{X}_0$ is altered to obtain $\mathcal{X}_1$. From $\mathcal{X}_1$ the new distances $d_{ij}^{(1)}$ can be obtained which are more closely related to the disparities $\widehat{d}_{ij}^{(0)}$ from step two.

Example 17.3 Consider a small example with four objects based on the car marks data set, see Table 17.3.

Our aim is to find a representation with $p^* = 2$ via MDS. Suppose that we choose as an initial configuration (Fig. 17.7) of $\mathcal{X}_0$ the coordinates given in Table 17.4.

Table 17.3 Dissimilarities δ_{ij} for car marks

j		1	2	3	4
i		Mercedes	Jaguar	Ferrari	VW
1	Mercedes	–			
2	Jaguar	3	–		
3	Ferrari	2	1	–	
4	VW	5	4	6	–

Table 17.4 Initial coordinates for MDS

i		x_{i1}	x_{i2}
1	Mercedes	3	2
2	Jaguar	2	7
3	Ferrari	1	3
4	VW	10	4

The corresponding distances $d_{ij} = \sqrt{(x_i - x_j)^\top (x_i - x_j)}$ are calculated in Table 17.5

A plot of the dissimilarities of Table 17.5 against the distance yields Fig. 17.8. This relation is not satisfactory since the ranking of the δ_{ij} did not result in a monotone relation of the corresponding distances d_{ij}. We apply therefore the PAV algorithm.

The first violator of monotonicity is the second point $(1, 3)$. Therefore we average the distances d_{13} and d_{23} to obtain the disparities

$$\widehat{d}_{13} = \widehat{d}_{23} = \frac{d_{13} + d_{23}}{2} = \frac{2.2 + 4.1}{2} = 3.17.$$

Applying the same procedure to $(2, 4)$ and $(1, 4)$ we obtain $\widehat{d}_{24} = \widehat{d}_{14} = 7.9$. The plot of δ_{ij} versus the disparities $\widehat{d}_{ij}$ represents a monotone regression relationship.

In the initial configuration (Fig. 17.7), the third point (Ferrari) could be moved so that the distance to object 2 (Jaguar) is reduced. This procedure however also alters the distance between objects 3 and 4. Care should be given when establishing a monotone relation between δ_{ij} and d_{ij}.

In order to assess how well the derived configuration fits the given dissimilarities Kruskal suggests a measure called STRESS1 that is given by

$$STRESS1 = \left(\frac{\sum_{i<j} (d_{ij} - \widehat{d}_{ij})^2}{\sum_{i<j} d_{ij}^2} \right)^{\frac{1}{2}}. \tag{17.21}$$

An alternative stress measure is given by

Table 17.5 Ranks and distances

i, j	d_{ij}	$rank(d_{ij})$	δ_{ij}
1, 2	5.1	3	3
1, 3	2.2	1	2
1, 4	7.3	4	5
2, 3	4.1	2	1
2, 4	8.5	5	4
3, 4	9.1	6	6

Fig. 17.7 Initial configuration of the MDS of the car data Q MVAnmdscar1

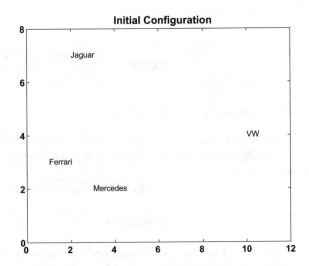

Fig. 17.8 Scatterplot of dissimilarities against distances Q MVAnmdscar2

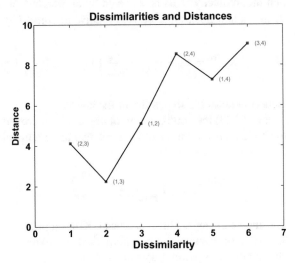

Table 17.6 STRESS calculations for car marks example

(i,j)	δ_{ij}	d_{ij}	$\widehat{d}_{ij}$	$(d_{ij} - \widehat{d}_{ij})^2$	d_{ij}^2	$(d_{ij} - \bar{d})^2$
$(2,3)$	1	4.1	3.15	0.9	16.8	3 .8
$(1,3)$	2	2.2	3.15	0.9	4.8	14.8
$(1,2)$	3	5.1	5.1	0	26.0	0.9
$(2,4)$	4	8.5	7.9	0.4	72.3	6.0
$(1,4)$	5	7.3	7.9	0.4	53.3	1.6
$(3,4)$	6	9.1	9.1	0	82.8	9.3
Σ		36.3		2.6	256.0	36.4

$$STRESS2 = \left(\frac{\sum_{i<j}(d_{ij} - \widehat{d}_{ij})^2}{\sum_{i<j}(d_{ij} - \bar{d})^2} \right)^{\frac{1}{2}}, \tag{17.22}$$

where $\bar{d}$ denotes the average distance.

Example 17.4 The Table 17.6 presents the STRESS calculations for the car example. The average distance is $\bar{d} = 36.4/6 = 6.1$. The corresponding STRESS measures are

$$STRESS1 = \sqrt{2.6/256} = 0.1$$

$$STRESS2 = \sqrt{2.6/36.4} = 0.27.$$

The goal is to find a point configuration that balances the effects STRESS and non monotonicity. This is achieved by an iterative procedure. More precisely, one defines a new position of object i relative to object j by

$$x_{il}^{NEW} = x_{il} + \alpha \left(1 - \frac{\widehat{d}_{ij}}{d_{ij}} \right)(x_{jl} - x_{il}), \quad l = 1, \dots, p^*. \tag{17.23}$$

Here α denotes the step width of the iteration.

By (17.23) the configuration of object i is improved relative to object j. In order to obtain an overall improvement relative to all remaining points one uses:

$$x_{il}^{NEW} = x_{il} + \frac{\alpha}{n-1} \sum_{j=1, j\neq i}^{n} \left(1 - \frac{\widehat{d}_{ij}}{d_{ij}} \right)(x_{jl} - x_{il}), \quad l = 1, \dots, p^*. \tag{17.24}$$

The choice of step width α is crucial. Kruskal proposes a starting value of $\alpha = 0.2$. The iteration is continued by a numerical approximation procedure, such as steepest descent or the Newton–Raphson procedure.

In a fourth step, the evaluation phase, the STRESS measure is used to evaluate whether or not its change as a result of the last iteration is sufficiently small that the procedure is terminated. At this stage the optimal fit has been obtained for a given dimension. Hence, the whole procedure needs to be carried out for several dimensions.

Example 17.5 Let us compute the new point configuration for $i = 3$ (Ferrari) (Fig. 17.9). The initial coordinates from Table 17.4 are

$$x_{31} = 1 \text{ and } x_{32} = 3.$$

Applying (17.24) yields (for $\alpha = 3$):

$$x_{31}^{NEW} = 1 + \frac{3}{4-1} \sum_{j=1, j\neq 3}^{4} \left(1 - \frac{\widehat{d_{3j}}}{d_{3j}}\right)(x_{j1} - 1)$$

$$= 1 + \left(1 - \frac{3.15}{2.2}\right)(3 - 1) + \left(1 - \frac{3.15}{4.1}\right)(2 - 1) + \left(1 - \frac{9.1}{9.1}\right)(10 - 1)$$

$$= 1 - 0.86 + 0.23 + 0$$

$$= 0.37.$$

Similarly we obtain $x_{32}^{NEW} = 4.36$.

To find the appropriate number of dimensions, p^*, a plot of the minimum STRESS value as a function of the dimensionality is made. One possible criterion in selecting the appropriate dimensionality is to look for an elbow in the plot. A rule of thumb that can be used to decide if a STRESS value is sufficiently small or not is provided by Kruskal:

$$S > 20\%, \text{ poor}; \quad S = 10\%, \text{ fair}; \quad S < 5\%, \text{ good}; \quad S = 0, \text{ perfect}. \qquad (17.25)$$

Summary
↪ Nonmetric MDS is only based on the rank order of dissimilarities.
↪ The object of nonmetric MDS is to create a spatial representation of the objects with low dimensionality.
↪ A practical algorithm is given as

1. Choose an initial configuration.
2. Find d_{ij} from the configuration.
3. Fit $\widehat{d_{ij}}$, the disparities, by the PAV algorithm.
4. Find a new configuration $\mathcal{X}_{n+1}$ by using the steepest descent.
5. Go to 2.

Fig. 17.9 First iteration for Ferrari using Shepard–Kruskal algorithm

⊙ MVAnmdscar3

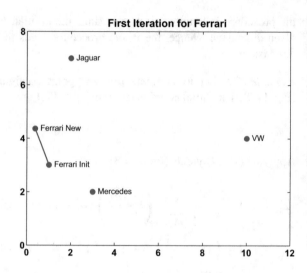

17.4 Exercises

Exercise 17.1 Apply the MDS method to the Swiss bank note data. What do you expect to see ?

Exercise 17.2 Using (17.6), show that (17.7) can be written in the form (17.2).

Exercise 17.3 Show that

1. $b_{ii} = a_{\bullet\bullet} - 2a_{i\bullet}; \; b_{ij} = a_{ij} - a_{i\bullet} - a_{\bullet j} + a_{\bullet\bullet}; \; i \neq j$
2. $\mathcal{B} = \sum_{i=1}^{p} x_i x_i^{\top}$
3. $\sum_{i=1}^{n} \lambda_i = \sum_{i=1}^{n} b_{ii} = \frac{1}{2n} \sum_{i,j=1}^{n} d_{ij}^2.$

Exercise 17.4 Redo a careful analysis of the car marks data based on the following dissimilarity matrix:

j i		1 Nissan	2 Kia	3 BMW	4 Audi
1	Nissan	–			
2	Kia	2	–		
3	BMW	4	6	–	
4	Audi	3	5	1	–

Exercise 17.5 Redo Exercise 17.4 with the U.S. crime data.

Exercise 17.6 Perform the MDS analysis on the Athletic Records data in Chap. 22. Can you see which countries are "close to each other"?

References

W.R. Dillon, M. Goldstein, *Multivariate Analysis* (Wiley, New York, 1984)
L. Fahrmeir, A. Hamerle, *Multivariate Statistische Verfahren* (De Gruyter, Berlin, 1984)

Chapter 18
Conjoint Measurement Analysis

Conjoint Measurement Analysis plays an important role in marketing. In the design of new products, it is valuable to know which components carry what kind of utility for the customer. Marketing and advertisement strategies are based on the perception of the new product's overall utility. It can be valuable information for a car producer to know whether a change in sportiness or a change in safety or comfort equipment is perceived as a higher increase in overall utility. The Conjoint Measurement Analysis is a method for attributing utilities to the components (part worths) on the basis of ranks given to different outcomes (stimuli) of the product. An important assumption is that the overall utility is decomposed as a sum of the utilities of the components.

In Sect. 18.1, we introduce the idea of Conjoint Measurement Analysis. We give two examples from the food and car industries. In Sect. 18.2, we shed light on the problem of designing questionnaires for ranking different product outcomes. In Sect. 18.3, we see that the metric solution of estimating the part-worths is given by solving a least squares problem. The estimated preference ordering may be nonmonotone. The nonmetric solution strategy takes care of this inconsistency by iterating between a least squares solution and the pool-adjacent violators algorithm.

18.1 Introduction

In the design and perception of new products, it is important to specify the contributions made by different facets or elements. The overall utility and acceptance of such a new product can then be estimated and understood as a possibly additive function of the elementary utilities. Examples are the design of cars, a food article, or the program of a political party. For a new type of margarine, one may ask whether a change in taste or presentation will enhance the overall perception of the product. The elementary utilities are here the presentation style and the taste (e.g., calory content). For a party program, one may want to investigate whether a stronger ecological or a

© Springer Nature Switzerland AG 2019
W. K. Härdle and L. Simar, *Applied Multivariate Statistical Analysis*,
https://doi.org/10.1007/978-3-030-26006-4_18

Table 18.1   Tester's ranking of cars

Car	1	2	3	4
Ranking	1	2	4	3

stronger social orientation gives a better overall profile of the party. For the marketing of a new car one may be interested in whether this new car should have a stronger active safety or comfort equipment or a more sporty note or combinations of both.

In Conjoint Measurement Analysis one assumes that the overall utility can be explained as an additive decomposition of the utilities of different elements. In a sample of questionnaires, people ranked the product types and thus revealed their preference orderings. The aim is to find the decomposition of the overall utility on the basis of observed data and to interpret the elementary or marginal utilities.

Example 18.1   A car producer plans to introduce a new car with features that appeal to the customer and that may help in promoting future sales. The new elements that are considered are comfort/safety components (e.g., active steering or GPS) and a sporty look (leather steering wheel and additional kW of the engine). The car producer has thus 4 lines of cars:

car 1: basic safety equipment and low sportiness
car 2: basic safety equipment and high sportiness
car 3: high safety equipment and low sportiness
car 4: high safety equipment and high sportiness

For the car producer it is important to rank these cars and to find out customers' attitudes toward a certain product line in order to develop a suitable marketing scheme. A tester may rank the cars as described in Table 18.1.

The elementary utilities here are the comfort equipment and the level of sportiness. Conjoint Measurement Analysis aims at explaining the rank order given by the test person as a function of these elementary utilities.

Example 18.2   A food producer plans to create a new margarine and varies the product characteristics "calories" (low vs. high) and "presentation" (a plastic pot vs. paper package) (Backhaus et al. 1996). We can view this in fact as ranking four products.

product 1: low calories and plastic pot
product 2: low calories and paper package
product 3: high calories and plastic pot
product 4: high calories and paper package

These four fictive products may now be ordered by a set of sample testers as described in Table 18.2.

The Conjoint Measurement Analysis aims to explain such a preference ranking by attributing *part-worths* to the different elements of the product. The part-worths are the utilities of the elementary components of the product.

Table 18.2 Tester's ranking of margarine

Product	1	2	3	4
Ranking	3	4	1	2

In interpreting the part-worths one may find that for a test person one of the elements has a higher value or utility. This may lead to a new design or to the decision that this utility should be emphasized in advertisement schemes.

Summary
↪ Conjoint Measurement Analysis is used in the design of new products.
↪ Conjoint Measurement Analysis tries to identify part-worth utilities that contribute to an overall utility.
↪ The part-worths enter additively into an overall utility.
↪ The interpretation of the part-worths gives insight into the perception and acceptance of the product.

18.2 Design of Data Generation

The product is defined through the properties of the components. A *stimulus* is defined as a combination of the different components. Examples 18.1 and 18.2 had four stimuli each. In the margarine example they were the possible combinations of the factors X_1 (calories) and X_2 (presentation). If a product property such as

$$X_3(\text{usage}) = \begin{cases} 1 \text{ bread} \\ 2 \text{ cooking} \\ 3 \text{ universal} \end{cases}$$

is added, then there are $3 \cdot 2 \cdot 2 = 12$ stimuli.

For the automobile Example 18.1 additional characteristics may be engine power and the number of doors. Suppose that the engines offered for the new car have $50, 70$, and 90 kW and that the car may be produced in 2-, 4-, or 5-door versions. These categories may be coded as

$$X_3(\text{power of engine}) = \begin{cases} 1 \text{ 50 kW} \\ 2 \text{ 70 kW} \\ 3 \text{ 90 kW} \end{cases}$$

and

$$X_4(\text{doors}) = \begin{cases} 1 \text{ 2 doors} \\ 2 \text{ 4 doors} \\ 3 \text{ 5 doors} \end{cases}.$$

Table 18.3 Trade-off matrices for margarine

X_3	X_1		X_3	X_2		X_1	X_2
1	1 2		1	1 2		1	1 2
2	1 2		2	1 2		2	1 2
3	1 2		3	1 2			

Table 18.4 Trade-off matrices for car design

X_4	X_3		X_4	X_2		X_4	X_1
1	1 2 3		1	1 2		1	1 2
2	1 2 3		2	1 2		2	1 2
3	1 2 3		3	1 2		3	1 2

X_3	X_2		X_3	X_1		X_2	X_1
1	1 2		1	1 2		1	1 2
2	1 2		2	1 2		2	1 2
3	1 2		3	1 2			

Both X_3 and X_4 have three factor levels each, whereas the first two factors X_1 (safety) and X_2 (sportiness) have only two levels. Altogether $2 \cdot 2 \cdot 3 \cdot 3 = 36$ stimuli are possible. In a questionnaire a tester would have to rank all 36 different products.

The *profile method* asks for the utility of each stimulus. This may be time consuming and tiring for a test person if there are too many factors and factor levels. Suppose that there are six properties of components with three levels each. This results in $3^6 = 729$ stimuli (i.e., 729 different products) that a tester would have to rank.

The *two-factor method* is a simplification and considers only two factors simultaneously. It is also called trade-off analysis. The idea is to present just two stimuli at a time and then to recombine the information. Trade-off analysis is performed by defining the trade-off matrices corresponding to stimuli of two factors only.

The trade-off matrices for the levels X_1, X_2, and X_3 from the margarine Example 18.2 are given in Table 18.3. The trade-off matrices for the new car outfit are described in Tabel 18.4.

The choice between the profile method and the trade-off analysis should be guided by consideration of the following aspects:

1. requirements on the test person,
2. time consumption, and
3. product perception.

The first aspect relates to the ability of the test person to judge the different stimuli. It is certainly an advantage of the trade-off analysis that one only has to consider two factors simultaneously. The two-factor method can be carried out more easily in a questionnaire without an interview.

The profile method incorporates the possibility of a complete product perception since the test person is not confronted with an isolated aspect (two factors) of the product. The stimuli may be presented visually in its final form (e.g., as a picture). With the number of levels and properties the number of stimuli rise exponentially with the profile method. The time to complete a questionnaire is therefore a factor in the choice of method.

In general, the product perception is the most important aspect and is therefore the profile method that is used the most. The time consumption aspect speaks for the trade-off analysis. There exist, however, clever strategies on selecting representation subsets of all profiles that bound the time investment. We therefore concentrate on the profile method in the following.

Summary
↪ A stimulus is a combination of different properties of a product.
↪ Conjoint measurement analysis is based either on a list of all factors (profile method) or on trade-off matrices (two-factor method).
↪ Trade-off matrices are used if there are too many factor levels.
↪ Presentation of trade-off matrices makes it easier for testers since only two stimuli have to be ranked at a time.

18.3 Estimation of Preference Orderings

On the basis of the reported preference values for each stimulus conjoint analysis determines the part-worths. Conjoint analysis uses an additive model of the form

$$Y_k = \sum_{j=1}^{J} \sum_{l=1}^{L_j} \beta_{jl}\, I(X_j = x_{jl}) + \mu, \text{ for } k = 1, \ldots, K \text{ and } \forall\, j \sum_{l=1}^{L_j} \beta_{jl} = 0. \quad (18.1)$$

X_j $(j = 1, \ldots, J)$ denote the factors, x_{jl} $(l = 1, \ldots, L_j)$ are the levels of each factor X_j and the coefficients β_{jl} are the part-worths. The constant μ denotes an overall level and Y_k is the observed preference for each stimulus and the total number of stimuli are

$$K = \prod_{j=1}^{J} L_j.$$

Equation (18.1) is without an error term for the moment. In order to explain how (18.1) may be written in the standard linear model form we first concentrate on $J = 2$ factors. Suppose that the factors engine power and airbag safety equipment have been ranked as follows: There are $K = 6$ preferences altogether. Suppose that

			Airbag	
			1	2
Engine	50 kW	1	1	3
	70 kW	2	2	6
	90 kW	3	4	5

the stimuli have been sorted so that Y_1 corresponds to engine level 1 and airbag level 1, Y_2 corresponds to engine level 1 and airbag level 2, and so on. Then model (18.1) reads:

$$Y_1 = \beta_{11} + \beta_{21} + \mu$$
$$Y_2 = \beta_{11} + \beta_{22} + \mu$$
$$Y_3 = \beta_{12} + \beta_{21} + \mu$$
$$Y_4 = \beta_{12} + \beta_{22} + \mu$$
$$Y_5 = \beta_{13} + \beta_{21} + \mu$$
$$Y_6 = \beta_{13} + \beta_{22} + \mu.$$

Now we would like to estimate the part-worths β_{jl}.

Example 18.3 In the margarine example let us consider the part-worths of $X_1 =$ usage and $X_2 =$ calories. We have $x_{11} = 1$, $x_{12} = 2$, $x_{13} = 3$, $x_{21} = 1$ and $x_{22} = 2$. (We momentarily relabeled the factors: X_3 became X_1). Hence $L_1 = 3$ and $L_2 = 2$. Suppose that a person has ranked the six different products as in Table 18.5.

If we order the stimuli as follows:

$$Y_1 = \text{Utility } (X_1 = 1 \wedge X_2 = 1)$$
$$Y_2 = \text{Utility } (X_1 = 1 \wedge X_2 = 2)$$
$$Y_3 = \text{Utility } (X_1 = 2 \wedge X_2 = 1)$$
$$Y_4 = \text{Utility } (X_1 = 2 \wedge X_2 = 2)$$
$$Y_5 = \text{Utility } (X_1 = 3 \wedge X_2 = 1)$$
$$Y_6 = \text{Utility } (X_1 = 3 \wedge X_2 = 2),$$

Table 18.5 Ranked products

			X_2 (calories)	
			Low	High
			1	2
X_1 (usage)	Bread	1	2	1
	Cooking	2	3	4
	Universal	3	6	5

Table 18.6 Metric solution for car example

	X_2 (airbags) 1	X_2 (airbags) 2	$\bar{p}_{x_1 \bullet}$	β_{1l}
50 kW 1	1	3	2	−1.5
X_1 (engine) 70 kW 2	2	6	4	0.5
90 kW 3	4	5	4.5	1
$\bar{p}_{x_2 \bullet}$	2.33	4.66	3.5	
β_{2l}	−1.16	1.16		

we obtain from Eq. (18.1) the same decomposition as above:

$$Y_1 = \beta_{11} + \beta_{21} + \mu$$
$$Y_2 = \beta_{11} + \beta_{22} + \mu$$
$$Y_3 = \beta_{12} + \beta_{21} + \mu$$
$$Y_4 = \beta_{12} + \beta_{22} + \mu$$
$$Y_5 = \beta_{13} + \beta_{21} + \mu$$
$$Y_6 = \beta_{13} + \beta_{22} + \mu.$$

Our aim is to estimate the part-worths β_{jl} as well as possible from a collection of tables like Table 18.5 that have been generated by a sample of test persons. First, the so-called metric solution to this problem is discussed and then a nonmetric solution.

Metric Solution

The problem of conjoint measurement analysis can be solved by the technique of analysis of variance (ANOVA). An important assumption underlying this technique is that the "distance" between any two adjacent preference orderings corresponds to the same difference in utility. That is, the difference in utility between the products ranked first and second is the same as the difference in utility between the products ranked fourth and fifth. Put differently, we treat the ranking of the products—which is a cardinal variable—as if it were a metric variable.

Introducing a mean utility μ Eq. (18.1) can be rewritten. The mean utility in the above Example 18.3 is $\mu = (1 + 2 + 3 + 4 + 5 + 6)/6 = 21/6 = 3.5$. In order to check the deviations of the utilities from this mean, we enlarge Table 18.5 by the mean utility $\bar{p}_{x_j \bullet}$, given a certain level of the other factor. The metric solution for the car example is given in Table 18.6:

Example 18.4 In the margarine example the resulting part-worths for $\mu = 3.5$ are

$$\beta_{11} = -2 \quad \beta_{21} = 0.16$$
$$\beta_{12} = 0 \quad \beta_{22} = -0.16.$$
$$\beta_{13} = 2$$

Table 18.7 Metric solution for Table 18.5

		X_2 (calories)			
		low	high		
		1	2	$\bar{p}_{x_{1\bullet}}$	β_{1l}
bread	1	2	1	1.5	-2
X_1 (usage) cooking	2	3	4	3.5	0
universal	3	6	5	5.5	2
$\bar{p}_{x_{2\bullet}}$		3.66	3.33	3.5	
β_{2l}		0.16	-0.16		

Table 18.8 Deviations between model and data

Stimulus	Y_k	$\hat{Y}_k$	$Y_k - \hat{Y}_k$	$(Y_k - \hat{Y}_k)^2$
1	2	1.66	0.33	0.11
2	1	1.33	-0.33	0.11
3	3	3.66	-0.66	0.44
4	4	3.33	0.66	0.44
5	6	5.66	0.33	0.11
6	5	5.33	-0.33	0.11
Σ	21	21	0	1.33

Note that $\sum_{l=1}^{L_j} \beta_{jl} = 0 \ (j = 1, \ldots, J)$. The estimated utility $\hat{Y}_1$ for the product with low calories and usage of bread, for example, is:

$$\hat{Y}_1 = \beta_{11} + \beta_{21} + \mu = -2 + 0.16 + 3.5 = 1.66.$$

The estimated utility $\hat{Y}_4$ for product 4 (cooking ($X_1 = 2$) and high calories ($X_2 = 2$)) is:

$$\hat{Y}_4 = \beta_{12} + \beta_{22} + \mu = 0 - 0.16 + 3.5 = 3.33.$$

The coefficients β_{jl} are computed as $\bar{p}_{x_{jl}} - \mu$, where $\bar{p}_{x_{jl}}$ is the average preference ordering for each factor level. For instance, $\bar{p}_{x_{11}} = 1/2 * (2 + 1) = 1.5$.

The fit can be evaluated by calculating the deviations of the fitted values to the observed preference orderings. In the rightmost column of Table 18.8 the quadratic deviations between the observed rankings (utilities) Y_k and the estimated utilities $\hat{Y}_k$ are listed.

The technique described that generated Table 18.7 is in fact the solution to a least squares problem. The conjoint measurement problem (18.1) may be rewritten as a linear regression model (with error $\varepsilon = 0$):

$$Y = \mathcal{X}\beta + \varepsilon \tag{18.2}$$

with $\mathcal{X}$ being a design matrix with dummy variables. $\mathcal{X}$ has the row dimension $K = \prod_{j=1}^{J} L_j$ (the number of stimuli) and the column dimension $D = \sum_{j=1}^{J} L_j - J$.

The reason for the reduced column number is that per factor only $(L_j - 1)$ vectors are linearly independent. Without loss of generality we may standardize the problem so that the last coefficient of each factor is omitted. The error term ε is introduced since even for one person the preference orderings may not fit the model (18.1).

Example 18.5 If we rewrite the β coefficients in the form

$$\begin{pmatrix} \beta_1 \\ \beta_2 \\ \beta_3 \\ \beta_4 \end{pmatrix} = \begin{pmatrix} \mu + \beta_{13} + \beta_{22} \\ \beta_{11} - \beta_{13} \\ \beta_{12} - \beta_{13} \\ \beta_{21} - \beta_{22} \end{pmatrix} \tag{18.3}$$

and define the design matrix $\mathcal{X}$ as

$$\mathcal{X} = \begin{pmatrix} 1 & 1 & 0 & 1 \\ 1 & 1 & 0 & 0 \\ 1 & 0 & 1 & 1 \\ 1 & 0 & 1 & 0 \\ 1 & 0 & 0 & 1 \\ 1 & 0 & 0 & 0 \end{pmatrix}, \tag{18.4}$$

then Eq. (18.1) leads to the linear model (with error $\varepsilon = 0$):

$$Y = \mathcal{X}\beta + \varepsilon. \tag{18.5}$$

The least squares solution to this problem is the technique used for Table 18.7.

In practice we have more than one person to answer the utility rank question for the different factor levels. The design matrix is then obtained by stacking the above design matrix n times. Hence, for n persons we have as a final design matrix:

$$\mathcal{X}^* = 1_n \otimes \mathcal{X} = \left. \begin{pmatrix} \mathcal{X} \\ \vdots \\ \mathcal{X} \end{pmatrix} \right\} n - \text{times}$$

which has dimension $(nK)(L - J)$ (where $L = \sum_{j=1}^{J} L_j$) and $Y^* = (Y_1^\top, ..., Y_n^\top)^\top$.

The linear model (18.5) can now be written as

$$Y^* = \mathcal{X}^*\beta + \varepsilon^*. \tag{18.6}$$

Given that the test people assign different rankings, the error term ε^* is a necessary part of the model.

Example 18.6 If we take the β vector as defined in (18.3) and the design matrix $\mathcal{X}$ from (18.4), we obtain the coefficients:

$$
\begin{aligned}
\hat{\beta}_1 &= 5.33 = \hat{\mu} + \hat{\beta}_{13} + \hat{\beta}_{22} \\
\hat{\beta}_2 &= -4 = \hat{\beta}_{11} - \hat{\beta}_{13} \\
\hat{\beta}_3 &= -2 = \hat{\beta}_{12} - \hat{\beta}_{13} \\
\hat{\beta}_4 &= 0.33 = \hat{\beta}_{21} - \hat{\beta}_{22} \\
\sum_{l=1}^{L_j} \hat{\beta}_{jl} &= 0.
\end{aligned}
\tag{18.7}
$$

Solving (18.7) we have:

$$
\begin{aligned}
\hat{\beta}_{11} &= \hat{\beta}_2 - \tfrac{1}{3}\left(\hat{\beta}_2 + \hat{\beta}_3\right) & = -2 \\
\hat{\beta}_{12} &= \hat{\beta}_3 - \tfrac{1}{3}\left(\hat{\beta}_2 + \hat{\beta}_3\right) & = 0 \\
\hat{\beta}_{13} &= -\tfrac{1}{3}\left(\hat{\beta}_2 + \hat{\beta}_3\right) & = 2 \\
\hat{\beta}_{21} &= \hat{\beta}_4 - \tfrac{1}{2}\hat{\beta}_4 = \tfrac{1}{2}\hat{\beta}_4 & = 0.16 \\
\hat{\beta}_{31} &= -\tfrac{1}{2}\hat{\beta}_4 & = -0.16 \\
\hat{\mu} &= \hat{\beta}_1 + \tfrac{1}{3}\left(\hat{\beta}_2 + \hat{\beta}_3\right) + \tfrac{1}{2}(\hat{\beta}_4) = 3.5.
\end{aligned}
\tag{18.8}
$$

In fact, we obtain the same estimated part-worths as in Table 18.7. The stimulus $k = 2$ corresponds to adding up β_{11}, β_{22}, and μ (see 18.3). Adding $\hat{\beta}_1$ and $\hat{\beta}_2$ gives:

$$
\hat{Y}_2 = 5.33 - 4 = 1.33.
$$

Nonmetric solution

If we drop the assumption that utilities are measured on a metric scale, we have to use (18.1) to estimate the coefficients from an adjusted set of estimated utilities. More precisely, we may use the monotone ANOVA as developed by (Kruskal, 1965). The procedure works as follows. First, one estimates model (18.1) with the ANOVA technique described above. Then one applies a monotone transformation $\hat{Z} = f(\hat{Y})$ to the estimated stimulus utilities. The monotone transformation f is used because the fitted values $\hat{Y}_k$ from (18.2) of the reported preference orderings Y_k may not be monotone. The transformation $\hat{Z}_k = f(\hat{Y}_k)$ is introduced to guarantee monotonicity of preference orderings. For the car example the reported Y_k values were $Y = (1,\ 3,\ 2,\ 6,\ 4,\ 5)^{\top}$. The estimated values are computed as

Fig. 18.1 Plot of estimated preference orderings versus revealed rankings and PAV fit 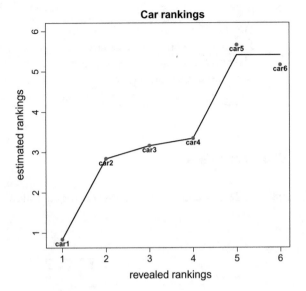 MVAcarrankings

$$\hat{Y}_1 = -1.5 - 1.16 + 3.5 = 0.84$$
$$\hat{Y}_2 = -1.5 + 1.16 + 3.5 = 3.16$$
$$\hat{Y}_3 = -0.5 - 1.16 + 3.5 = 2.84$$
$$\hat{Y}_4 = -0.5 + 1.16 + 3.5 = 5.16$$
$$\hat{Y}_5 = 1.5 - 1.16 + 3.5 \quad\, = 3.34$$
$$\hat{Y}_6 = 1.5 + 1.16 + 3.5 \quad\, = 5.66.$$

If we make a plot of the estimated preference orderings against the revealed ones, we obtain Fig. 18.1.

We see that the estimated $\hat{Y}_6 = 5.16$ is below the estimated $\hat{Y}_5 = 5.66$ and thus an inconsistency in ranking the utilities occurs. The monotone transformation $\hat{Z}_k = f(\hat{Y}_k)$ is introduced to make the relationship in Fig. 18.1 monotone. A very simple procedure consists of averaging the "violators" $\hat{Y}_6$ and $\hat{Y}_5$ to obtain 5.41. The relationship is then monotone but the model (18.1) may now be violated. The idea is therefore to iterate these two steps. This procedure is iterated until the stress measure (see Chap. 17)

$$\text{STRESS} = \frac{\sum\limits_{k=1}^{K}(\hat{Z}_k - \hat{Y}_k)^2}{\sum\limits_{k=1}^{K}(\hat{Y}_k - \bar{\hat{Y}})^2} \tag{18.9}$$

is minimized over β and the monotone transformation f. The monotone transformation can be computed by the so-called pool-adjacent violators (PAV) algorithm.

Summary
↪ The part-worths are estimated via the least squares method.
↪ The metric solution corresponds to analysis of variance in a linear model.
↪ The nonmetric solution iterates between a monotone regression curve fitting and determining the part-worths by ANOVA methodology.
↪ The fitting of data to a monotone function is done via the PAV algorithm.

18.4 Exercises

Exercise 18.1 Compute the part-worths for the following table of rankings:

	X_2		
	1	2	
X_1	1	1	2
	2	4	3
	3	6	5

Exercise 18.2 Consider again Example 18.5. Rewrite the design matrix $\mathcal{X}$ and the parameter vector β so that the overall mean effect μ is part of $\mathcal{X}$ and β, i.e., find the matrix $\mathcal{X}'$ and β' such that $Y = \mathcal{X}'\beta'$.

Exercise 18.3 Compute the design matrix for Example 18.5 for $n = 3$ persons ranking the margarine with X_1 and X_2.

Exercise 18.4 Construct an analog for Table 18.8 for the car example.

Exercise 18.5 Compute the part-worths on the basis of the following tables of rankings observed on $n = 3$ persons.

	X_2				X_2				X_2	
	1	1 2			1 3				3 1	
X_1	2	4 3		X_1	4 2		X_1	5 2	
	3	6 5			5 6				6 4	

Exercise 18.6 Suppose that in the car example a person has ranked cars by the profile method on the following characteristics:

There are $k = 18$ stimuli.

$$X_1 = \text{motor}$$
$$X_2 = \text{safety}$$
$$X_3 = \text{doors}$$

X_1	X_2	X_3	Preference
1	1	1	1
1	1	2	3
1	1	3	2
1	2	1	5
1	2	2	4
1	2	3	6

X_1	X_2	X_3	Preference
2	1	1	7
2	1	2	8
2	1	3	9
2	2	1	10
2	2	2	12
2	2	3	11

X_1	X_2	X_3	Preference
3	1	1	13
3	1	2	15
3	1	3	14
3	2	1	16
3	2	2	17
3	2	3	18

Estimate and analyze the part-worths.

References

K. Backhaus, B. Erichson, W. Plinke, R. Weiber, *Multivariate Analysemethoden* (Springer, Berlin, 1996)

J.B. Kruskal, Analysis of factorial experiments by estimating a monotone transformation of data. J. R. Stat. Soc. Ser B **27**, 251–263 (1965)

Chapter 19
Applications in Finance

A portfolio is a linear combination of assets. Each asset contributes to a weight c_j to the portfolio. The performance of such a portfolio is a function of the various returns of the assets and of the weights $c = (c_1, \ldots, c_p)^\top$. In this chapter, we investigate the "optimal choice" of the portfolio weights c. The optimality criterion is the mean–variance efficiency of the portfolio. Usually investors are risk-averse, therefore, we can define a mean–variance efficient portfolio to be a portfolio that has a minimal variance for a given desired mean return. Equivalently, we could try to optimize the weights for the portfolios with maximal mean return for a given variance (risk structure). We develop this methodology in the situations of (non)existence of riskless assets and discuss relations with the capital assets pricing model (CAPM).

19.1 Portfolio Choice

Suppose that one has a portfolio of p assets. The price of asset j at time i is denoted as p_{ij}. The return from asset j in a single time period (day, month, year, etc.) is

$$x_{ij} = \frac{p_{ij} - p_{i-1,j}}{p_{i-1,j}}.$$

We observe the vectors $x_i = (x_{i1}, \ldots, x_{ip})^\top$ (i.e., the returns of the assets which are contained in the portfolio) over several time periods. We stack these observations into a data matrix $\mathcal{X} = (x_{ij})$ consisting of observations of a random variable

$$X \sim (\mu, \Sigma).$$

The return of the portfolio is the weighted sum of the returns of the p assets:

© Springer Nature Switzerland AG 2019
W. K. Härdle and L. Simar, *Applied Multivariate Statistical Analysis*,
https://doi.org/10.1007/978-3-030-26006-4_19

$$Q = c^\top X, \tag{19.1}$$

where $c = (c_1, \ldots, c_p)^\top$ (with $\sum_{j=1}^p c_j = 1$) denotes the proportions of the assets in the portfolio. The mean return of the portfolio is given by the expected value of Q, which is $c^\top \mu$. The *risk* or *variance (squared volatility)* of the portfolio is given by the variance of Q (Theorem 4.6), which is equal to two times

$$\frac{1}{2} c^\top \Sigma c. \tag{19.2}$$

The reason for taking *half of* the variance of Q is merely technical. The optimization of (19.2) with respect to c is of course equivalent to minimizing $c^\top \Sigma c$. Our aim is to maximize the portfolio returns (19.1) given a bound on the volatility (19.2) or vice versa to minimize risk given a (desired) mean return of the portfolio.

Summary
$\hookrightarrow$ Given a matrix of returns $\mathcal{X}$ from p assets in n time periods, and that the underlying distribution is stationary, i.e., $X \sim (\mu, \Sigma)$, then the (theoretical) return of the portfolio is a weighted sum of the returns of the p assets, namely $Q = c^\top X$.
$\hookrightarrow$ The expected value of Q is $c^\top \mu$. For technical reasons one considers optimizing $\frac{1}{2} c^\top \Sigma c$. The risk or squared volatility is $c^\top \Sigma c = \mathsf{Var}(c^\top X)$.
$\hookrightarrow$ The portfolio choice, i.e., the selection of c, is such that the return is maximized for a given risk bound.

19.2   Efficient Portfolio

A variance efficient portfolio is one that keeps the risk (19.2) minimal under the constraint that the weights sum to 1, i.e., $c^\top 1_p = 1$. For a variance efficient portfolio, we therefore try to find the value of c that minimizes the Lagrangian

$$\mathcal{L} = \frac{1}{2} c^\top \Sigma c - \lambda(c^\top 1_p - 1). \tag{19.3}$$

A mean–variance efficient portfolio is defined as one that has minimal variance among all portfolios with the same mean. More formally, we have to find a vector of weights c such that the variance of the portfolio is minimal subject to two constraints:

1. a certain, prespecified mean return $\bar{\mu}$ has to be achieved and
2. the weights have to sum to one.

Mathematically speaking, we are dealing with an optimization problem under two constraints.

The Lagrangian function for this problem is given by

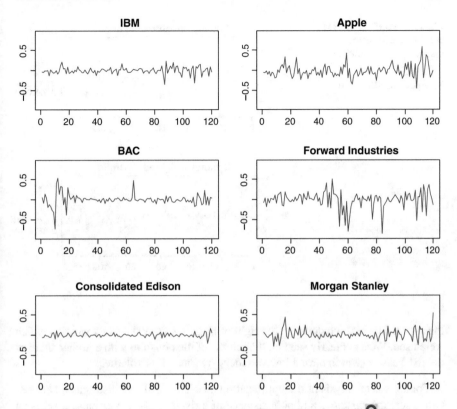

Fig. 19.1 Monthly Returns of six firms from January 2000 to December 2009 ⚙ MVAreturns

$$\mathcal{L} = c^\top \Sigma c + \lambda_1 (\bar{\mu} - c^\top \mu) + \lambda_2 (1 - c^\top 1_p).$$

With tools presented in Sect. 2.4 we can calculate the first-order condition for a minimum:

$$\frac{\partial \mathcal{L}}{\partial c} = 2\Sigma c - \lambda_1 \mu - \lambda_2 1_p = 0. \tag{19.4}$$

Example 19.1 Figure 19.1 shows the monthly returns from January 2000 to December 2009 of six stocks. The data is from Yahoo Finance.

For each stock we have chosen the same scale on the vertical axis (which gives the return of the stock). Note how the return of some stocks, such as Forward Industries and Apple, are much more volatile than the returns of other stocks, such as IBM or Consolidated Edison (Electric utilities).

As a very simple example consider two differently weighted portfolios containing only two assets, IBM and Forward Industries.

Figure 19.2 displays the monthly returns of the two portfolios. The portfolio in the upper panel consists of approximately 10% Forward Industries assets and 90% IBM assets. The portfolio in the lower panel contains an equal proportion of each of

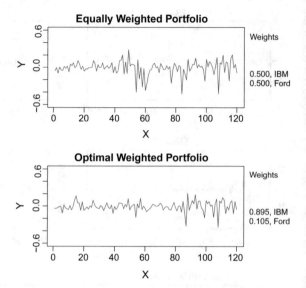

Fig. 19.2 Portfolio of IBM and Forward Industries assets, equal and efficient weights Q MVAportfol_IBM_Ford

the assets. The text windows on the right of Fig. 19.2 show the exact weights which were used. We can clearly see that the returns of the portfolio with a higher share of the IBM assets (which have a low variance) are much less volatile.

For an exact analysis of the optimization problem (19.4) we distinguish between two cases: the existence and nonexistence of a riskless asset. A riskless asset is an asset such as a zero bond, i.e., a financial instrument with a fixed nonrandom return (Franke et al. 2011).

Nonexistence of a Riskless Asset

Assume that the covariance matrix Σ is invertible (which implies positive definiteness). This is equivalent to the nonexistence of a portfolio c with variance $c^\top \Sigma c = 0$. If all assets are uncorrelated, Σ is invertible if all of the asset returns have positive variances. A riskless asset (uncorrelated with all other assets) would have zero variance since it has fixed, nonrandom returns. In this case Σ would not be positive definite.

The optimal weights can be derived from the first order condition (19.4) as

$$c = \frac{1}{2}\Sigma^{-1}(\lambda_1\mu + \lambda_2 1_p). \tag{19.5}$$

Multiplying this by a $(p \times 1)$ vector 1_p of ones, we obtain

$$1 = 1_p^\top c = \frac{1}{2}1_p^\top \Sigma^{-1}(\lambda_1\mu + \lambda_2 1_p),$$

which can be solved for λ_2 to get:

$$\lambda_2 = \frac{2 - \lambda_1 1_p^\top \Sigma^{-1} \mu}{1_p^\top \Sigma^{-1} 1_p}.$$

Plugging this expression into (19.5) yields

$$c = \frac{1}{2}\lambda_1 \left(\Sigma^{-1}\mu - \frac{1_p^\top \Sigma^{-1} \mu}{1_p^\top \Sigma^{-1} 1_p} \Sigma^{-1} 1_p \right) + \frac{\Sigma^{-1} 1_p}{1_p^\top \Sigma^{-1} 1_p}. \qquad (19.6)$$

For the case of a variance efficient portfolio there is no restriction on the mean of the portfolio ($\lambda_1 = 0$). The optimal weights are therefore:

$$c = \frac{\Sigma^{-1} 1_p}{1_p^\top \Sigma^{-1} 1_p}. \qquad (19.7)$$

This formula is identical to the solution of (19.3). Indeed, differentiation with respect to c gives

$$\Sigma c = \lambda 1_p$$

$$c = \lambda \Sigma^{-1} 1_p.$$

If we plug this into (19.3), we obtain

$$\mathcal{L} = \frac{1}{2}\lambda^2 1_p^\top \Sigma^{-1} 1_p - \lambda(\lambda 1_p^\top \Sigma^{-1} 1_p - 1)$$

$$= \lambda - \frac{1}{2}\lambda^2 1_p^\top \Sigma^{-1} 1_p.$$

This quantity is a function of λ and is minimal for

$$\lambda = (1_p^\top \Sigma^{-1} 1_p)^{-1}$$

since

$$\frac{\partial^2 \mathcal{L}}{\partial c^\top \partial c} = \Sigma > 0.$$

Theorem 19.1 *The variance efficient portfolio weights for returns* $X \sim (\mu, \Sigma)$ *are*

$$c_{opt} = \frac{\Sigma^{-1} 1_p}{1_p^\top \Sigma^{-1} 1_p}. \qquad (19.8)$$

Existence of a Riskless Asset

If an asset exists with variance equal to zero, then the covariance matrix Σ is not invertible. The notation can be adjusted for this case as follows: denote the return of the riskless asset by r (under the absence of arbitrage this is the interest rate), and partition the vector and the covariance matrix of returns such that the last component is the riskless asset. Thus, the last equation of the system (19.4) becomes

$$2\operatorname{Cov}(r, X) - \lambda_1 r - \lambda_2 = 0,$$

and, because the covariance of the riskless asset with any portfolio is zero, we have

$$\lambda_2 = -r\lambda_1. \qquad (19.9)$$

Let us for a moment modify the notation in such a way that in each vector and matrix the components corresponding to the riskless asset are excluded. For example, c is the weight vector of the *risky* assets (i.e., assets with positive variance), and c_0 denotes the proportion invested in the riskless asset. Obviously, $c_0 = 1 - 1_p^\top c$, and Σ the covariance matrix of the *risky* assets, is assumed to be invertible. Solving (19.4) using (19.9) gives

$$c = \frac{\lambda_1}{2}\Sigma^{-1}(\mu - r1_p). \qquad (19.10)$$

This equation may be solved for λ_1 by plugging it into the condition $\mu^\top c = \bar{\mu}$. This is the mean–variance efficient weight vector of the risky assets if a riskless asset exists. The final solution is

$$c = \frac{\bar{\mu}\Sigma^{-1}(\mu - r1_p)}{\mu^\top \Sigma^{-1}(\mu - r1_p)}. \qquad (19.11)$$

The variance optimal weighting of the assets in the portfolio depends on the structure of the covariance matrix as the following corollaries show.

Corollary 19.1 *A portfolio of uncorrelated assets whose returns have equal variances $(\Sigma = \sigma^2 \mathcal{I}_p)$ needs to be weighted equally:*

$$c_{opt} = p^{-1}1_p.$$

Proof Here we obtain $1_p^\top \Sigma^{-1} 1_p = \sigma^{-2} 1_p^\top 1_p = \sigma^{-2} p$ and therefore $c = \frac{\sigma^{-2}1_p}{\sigma^{-2}p} = p^{-1}1_p.$

Corollary 19.2 *A portfolio of correlated assets whose returns have equal variances, i.e.,*

$$\Sigma = \sigma^2 \begin{pmatrix} 1 & \rho & \cdots & \rho \\ \rho & 1 & \cdots & \rho \\ \vdots & \vdots & \ddots & \vdots \\ \rho & \rho & \cdots & 1 \end{pmatrix}, \qquad -\frac{1}{p-1} < \rho < 1$$

needs to be weighted equally:

$$c_{opt} = p^{-1}1_p.$$

Proof Σ can be rewritten as $\Sigma = \sigma^2 \left\{ (1 - \rho)\mathcal{I}_p + \rho 1_p 1_p^\top \right\}$. The inverse is

$$\Sigma^{-1} = \frac{\mathcal{I}_p}{\sigma^2(1 - \rho)} - \frac{\rho 1_p 1_p^\top}{\sigma^2(1 - \rho)\{1 + (p - 1)\rho\}}$$

since for a $(p \times p)$ matrix $\mathcal{A}$ of the form $\mathcal{A} = (a - b)\mathcal{I}_p + b 1_p 1_p^\top$ the inverse is generally given by

$$\mathcal{A}^{-1} = \frac{\mathcal{I}_p}{(a - b)} - \frac{b \, 1_p 1_p^\top}{(a - b)\{a + (p - 1)b\}}.$$

Hence

$$\begin{aligned}
\Sigma^{-1}1_p &= \frac{1_p}{\sigma^2(1 - \rho)} - \frac{\rho 1_p 1_p^\top 1_p}{\sigma^2(1 - \rho)\{1 + (p - 1)\rho\}} \\
&= \frac{[\{1 + (p - 1)\rho\} - \rho p]1_p}{\sigma^2(1 - \rho)\{1 + (p - 1)\rho\}} = \frac{\{1 - \rho\}1_p}{\sigma^2(1 - \rho)\{1 + (p - 1)\rho\}} \\
&= \frac{1_p}{\sigma^2\{1 + (p - 1)\rho\}}
\end{aligned}$$

which yields

$$1_p^\top \Sigma^{-1} 1_p = \frac{p}{\sigma^2\{1 + (p - 1)\rho\}}$$

and thus $c = p^{-1}1_p$.

Let us now consider assets with different variances. We will see that in this case the weights are adjusted to the risk.

Corollary 19.3 *A portfolio of uncorrelated assets with returns of different variances, i.e., $\Sigma = \mathrm{diag}(\sigma_1^2, \ldots, \sigma_p^2)$, has the following optimal weights*

$$c_{j,opt} = \frac{\sigma_j^{-2}}{\sum\limits_{l=1}^{p} \sigma_l^{-2}}, \quad j = 1, \ldots, p.$$

Proof From $\Sigma^{-1} = \mathrm{diag}(\sigma_1^{-2}, \ldots, \sigma_p^{-2})$ we have $1_p^\top \Sigma^{-1} 1_p = \sum_{l=1}^{p} \sigma_l^{-2}$ and therefore the optimal weights are $c_j = \sigma_j^{-2} / \sum\limits_{l=1}^{p} \sigma_l^{-2}$.

This result can be generalized for covariance matrices with block structures.

Corollary 19.4 *A portfolio of assets with returns* $X \sim (\mu, \Sigma)$, *where the covariance matrix has the form:*

$$\Sigma = \begin{pmatrix} \Sigma_1 & 0 & \cdots & 0 \\ 0 & \Sigma_2 & \ddots & \vdots \\ \vdots & \ddots & \ddots & \vdots \\ 0 & \cdots & 0 & \Sigma_r \end{pmatrix}$$

has optimal weights $c = (c_1, \dots, c_r)^\top$ *given by*

$$c_{j,opt} = \frac{\Sigma_j^{-1} 1}{1^\top \Sigma_j^{-1} 1}, \quad j = 1, \dots, r.$$

Summary

↪   An efficient portfolio is one that keeps the risk minimal under the constraint that a given mean return is achieved and that the weights sum to 1, i.e., that minimizes $\mathcal{L} = c^\top \Sigma c + \lambda_1(\bar{\mu} - c^\top \mu) + \lambda_2(1 - c^\top 1_p)$.

↪   If a riskless asset does not exist, the variance efficient portfolio weights are given by

$$c = \frac{\Sigma^{-1} 1_p}{1_p^\top \Sigma^{-1} 1_p}.$$

↪   If a riskless asset exists, the mean–variance efficient portfolio weights are given by

$$c = \frac{\bar{\mu} \Sigma^{-1}(\mu - r 1_p)}{\mu^\top \Sigma^{-1}(\mu - r 1_p)}.$$

↪   The efficient weighting depends on the structure of the covariance matrix Σ. Equal variances of the assets in the portfolio lead to equal weights, different variances lead to weightings proportional to these variances:

$$c_{j,opt} = \frac{\sigma_j^{-2}}{\sum\limits_{l=1}^{p} \sigma_l^{-2}}, \quad j = 1, \dots, p.$$

19.3 Efficient Portfolios in Practice

We can now demonstrate the usefulness of this technique by applying our method to the monthly market returns computed on the basis of transactions at the New York stock market and the NASDAQ stock market between January 2000 and December 2009.

Example 19.2 Recall that we had shown the portfolio returns with uniform and optimal weights in Fig. 19.2. The covariance matrix of the returns of IBM and Forward Industries is

$$S = \begin{pmatrix} 0.0073 & 0.0023 \\ 0.0023 & 0.0454 \end{pmatrix}.$$

Hence by (19.7) the optimal weighting is

$$\widehat{c} = \frac{S^{-1}1_2}{1_2^\top S^{-1}1_2} = (0.8952, 0.1048)^\top.$$

The effect of efficient weighting becomes even clearer when we expand the portfolio to six assets. The covariance matrix for the returns of all six firms introduced in Example 19.1 is

$$S = 10^{-3} \begin{pmatrix} 7.3 & 6.2 & 3.1 & 2.3 & -0.1 & 5.2 \\ 6.2 & 23.9 & 4.3 & 2.1 & 0.4 & 6.4 \\ 3.1 & 4.3 & 19.5 & -0.9 & 1.1 & 3.7 \\ 2.3 & 2.1 & -0.9 & 45.4 & -2.1 & 0.8 \\ -0.1 & 0.4 & 1.1 & -2.1 & 2.4 & -0.1 \\ 5.2 & 6.4 & 3.7 & 0.8 & -0.1 & 14.7 \end{pmatrix}.$$

Hence the optimal weighting is

$$\widehat{c} = \frac{S^{-1}1_6}{1_6^\top S^{-1}1_6} = (0.1894, -0.0139, 0.0094, 0.0580, 0.7112, 0.0458)^\top.$$

As we can clearly see, the optimal weights are quite different from the equal weights ($c_j = 1/6$). The weights which were used are shown in text windows on the right-hand side of Fig. 19.3.

This efficient weighting assumes stable covariances between the assets over time. Changing covariance structure over time implies weights that depend on time as well. This is part of a large body of literature on multivariate volatility models. For a review refer to Franke et al. (2011).

Fig. 19.3 Portfolio of all six assets, equal and efficient weights ⬤MVAportfol

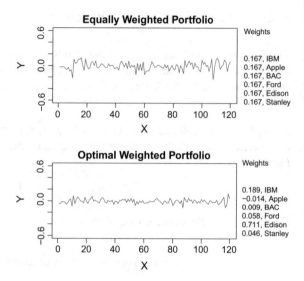

Summary
↪   Efficient portfolio weighting in practice consists of estimating the covariances of the assets in the portfolio and then computing efficient weights from this empirical covariance matrix.
↪   Note that this efficient weighting assumes stable covariances between the assets over time.

19.4   The Capital Asset Pricing Model (CAPM)

The CAPM considers the relation between a mean–variance efficient portfolio and an asset uncorrelated with this portfolio. Let us denote this specific asset return by y_0. The riskless asset with constant return $y_0 = r$ may be such an asset. Recall from (19.4) the condition for a mean–variance efficient portfolio:

$$2 \Sigma c - \lambda_1^- - \lambda_2 1_p = 0.$$

In order to eliminate λ_2, we can multiply (19.4) by $c^\top$ to get:

$$2 c^\top \Sigma c - \lambda_1 \bar{\mu} = \lambda_2.$$

Plugging this into (19.4), we obtain:

$$2 \Sigma c - \lambda_1^- = 2 c^\top \Sigma c 1_p - \lambda_1 \bar{\mu} 1_p$$

$$^- = \bar{\mu} 1_p + \frac{2}{\lambda_1} (\Sigma c - c^\top \Sigma c 1_p). \tag{19.12}$$

For the asset that is uncorrelated with the portfolio, Eq. (19.12) can be written as:

$$y_0 = \bar{\mu} - \frac{2}{\lambda_1} c^\top \Sigma c$$

since $y_0 = r$ is the mean return of this asset and is otherwise uncorrelated with the risky assets. This yields:

$$\lambda_1 = 2 \frac{c^\top \Sigma c}{\bar{\mu} - y_0} \tag{19.13}$$

and if (19.13) is plugged into (19.12):

$$\mu = \bar{\mu} 1_p + \frac{\bar{\mu} - y_0}{c^\top \Sigma c} (\Sigma c - c^\top \Sigma c 1_p)$$

$$\mu = y_0 1_p + \frac{\Sigma c}{c^\top \Sigma c} (\bar{\mu} - y_0)$$

$$\mu = y_0 1_p + \beta(\bar{\mu} - y_0) \tag{19.14}$$

with

$$\beta \stackrel{\text{def}}{=} \frac{\Sigma c}{c^\top \Sigma c}.$$

The relation (19.14) holds if there exists any asset that is uncorrelated with the mean–variance efficient portfolio c. The existence of a riskless asset is not a necessary condition for deriving (19.14). However, for this special case we arrive at the well-known expression

$$\mu = r 1_p + \beta(\bar{\mu} - r), \tag{19.15}$$

which is known as the *Capital Asset Pricing Model* (CAPM), see Franke et al. (2011). The *beta factor* β measures the relative performance with respect to riskless assets or an index. It reflects the sensitivity of an asset with respect to the whole market. The beta factor is close to 1 for most assets. A factor of 1.16, for example, means that the asset reacts in relation to movements of the whole market (expressed through an index like DAX or DOW JONES) 16 percent stronger than the index. This is of course true for both positive and negative fluctuations of the whole market.

Summary
↪   The weights of the mean–variance efficient portfolio satisfy $2\Sigma c - \lambda_1 \mu - \lambda_2 1_p = 0$.
↪   In the CAPM the mean of X depends on the riskless asset and the prespecified mean $\bar{\mu}$ as follows $\mu = r 1_p + \beta(\bar{\mu} - r)$.
↪   The beta factor β measures the relative performance with respect to riskless assets or an index and reflects the sensitivity of an asset with respect to the whole market.

19.5   Exercises

Exercise 19.1 Prove that the inverse of $\mathcal{A} = (a - b)\mathcal{I}_p + b1_p1_p^\top$ is given by

$$\mathcal{A}^{-1} = \frac{\mathcal{I}_p}{(a - b)} - \frac{b\,1_p1_p^\top}{(a - b)\{a + (p - 1)b\}}.$$

Exercise 19.2 The empirical covariance between the 120 returns of IBM and Forward Industries is 0.0023 (see Example 19.2). Test if the true covariance is zero. Hint: Use Fisher's Z-transform.

Exercise 19.3 Explain why in both Figs. 19.2 and 19.3 the portfolios have negative returns just before the end of the series, regardless of whether they are optimally weighted or not! (What happened in the mid-2007?)

Exercise 19.4 Apply the method used in Example 19.2 on the same data (Sect. 22.5) including also the Digital Equipment company. Obviously one of the weights is negative. Is this an efficient weighting?

Exercise 19.5 In the CAPM the β value tells us about the performance of the portfolio relative to the riskless asset. Calculate the β value for each single stock price series relative to the "riskless" asset IBM.

Reference

J. Franke, W. Härdle, C. Hafner, *Introduction to Statistics of Financial Markets*, 3rd edn. (Springer, Heidelberg, 2011)

Chapter 20
Computationally Intensive Techniques

It is generally accepted that training in statistics must include some exposure to the mechanics of computational statistics. This exposure to computational methods is of an essential nature when we consider extremely high-dimensional data. Computer-aided techniques can help us to discover dependencies in high dimensions without complicated mathematical tools. A draftman's plot (i.e., a matrix of pairwise scatter-plots like in Fig. 1.14) may lead us immediately to a theoretical hypothesis (on a lower dimensional space) on the relationship of the variables. Computer-aided techniques are therefore at the heart of multivariate statistical analysis.

With the rapidly increasing amount of data statistics faces a new challenge. While in the twentieth century the focus was on the mathematical precision of statistical modeling, the twenty-first century relies more and more on data analytic procedures that provide information (even for extremely large databases) on the fingertip. This demand on fast availability of condensed statistical information has changed the statistical paradigm and has shifted energy from mathematical analysis to computational analysis of course without loosing sight of the statistical core questions.

In this chapter we first present the concept of Simplicial Depth—a multivariate extension of the data depth concept of Sect. 1.1. We then present Projection Pursuit—a semiparametric technique which is based on a one-dimensional, flexible regression or on the idea of density smoothing applied to PCA-type projections. A similar model is underlying the sliced inverse regression (SIR) technique which we discuss in Sect. 20.3.

The next technique is called support vector machines and is motivated by non-linear classification (discrimination) problems. support vector machines (SVM) are classification methods based on statistical learning theory. A quadratic optimization problem determines so-called support vectors with high margin that guarantee maximal separability. Nonlinear classification is achieved by mapping the data into a feature space and finding a linear separating hyperplane in this feature space. Another advanced technique is CART—Classification and Regression Trees, a decision tree procedure developed by Breiman et al. (1984).

© Springer Nature Switzerland AG 2019

W. K. Härdle and L. Simar, *Applied Multivariate Statistical Analysis*,
https://doi.org/10.1007/978-3-030-26006-4_20

20.1   Simplicial Depth

Simplicial depth generalizes the notion of data depth as introduced in Sect. 1.1. This general definition allows us to define a multivariate median and to visually present high-dimensional data in low dimension. For univariate data we have well-known parameters of location which describe the center of a distribution of a random variable X. These parameters are, for example, the *mean*

$$\bar{x} = \frac{1}{n} \sum_{i=1}^{n} x_i, \tag{20.1}$$

or the *mode*

$$x_{mod} = \arg \max_{x} \hat{f}(x),$$

where $\hat{f}$ is the estimated density function of X (see Sect. 1.3). The *median*

$$x_{med} = \begin{cases} x_{\{(n+1)/2\}} & \text{if } n \text{ odd} \\ \frac{x_{(n/2)} + x_{(n/2+1)}}{2} & \text{otherwise,} \end{cases}$$

where $x_{(i)}$ is the order statistics of the n observations x_i, is yet another measure of location.

The first two parameters can be easily extended to multivariate random variables. The mean in higher dimensions is defined as in (20.1) and the mode accordingly,

$$x_{mod} = \arg \max_{x} \hat{f}(x)$$

with $\hat{f}$ the estimated multidimensional density function of X (see Sect. 1.3). The median poses a problem though since in a multivariate sense we cannot interpret the element-wise median

$$x_{med,j} = \begin{cases} x_{\{(n+1)/2\},j} & \text{if } n \text{ odd} \\ \frac{x_{(n/2),j} + x_{(n/2+1),j}}{2} & \text{otherwise} \end{cases} \tag{20.2}$$

as a point that is "most central". The same argument applies to other observations of a sample that have a certain "depth" as defined in Sect. 1.1. The "fourths" or the "extremes" are not defined in a straightforward way in higher (not even for two) dimensions.

An equivalent definition of the median in one dimension is given by the *simplicial depth*. It is defined as follows: For each pair of datapoints x_i and x_j we generate a closed interval, a one-dimensional simplex, which contains x_i and x_j as border points. Redefine the median as the datapoint x_{med}, which is enclosed in the maximum number of intervals:

$$x_{med} = \arg \max_{i} \#\{k, l; x_i \in [x_k, x_l]\}. \tag{20.3}$$

Fig. 20.1 Construction of
simplicial depth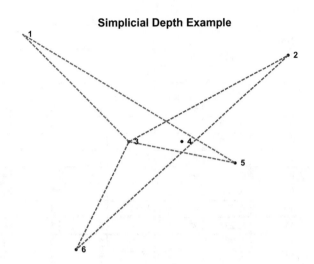
MVAsimdep1

With this definition of the median, the median is the "deepest" and "most central" point in a data set as discussed in Sect. 1.1. This definition involves a computationally intensive operation since we generate $n(n-1)/2$ intervals for n observations.

In two dimensions, the computation is even more intensive since the interval $[x_k, x_l]$ is replaced by a triangle constructed from three different datapoints. The median as the deepest point is then defined by that datapoint that is covered by the maximum number of triangles. In three dimensions, triangles become pyramids formed from four points and the median is that datapoint that lies in the maximum number of pyramids.

An example of the depth in two dimensions is given by the constellation of points given in Fig. 20.1. If we build, for example, the triangle of the points 1, 3, 5 (denoted as $\triangle\,135$ in Table 20.1), it contains the point 4. From Table 20.1 we count the number of coverages to obtain the simplicial depth values of Table 20.2.

In arbitrary dimension p, we look for datapoints that lie inside a simplex (or convex $hull$) formed from $p+1$ points. We therefore extend the definition of the median to the multivariate case as follows

$$x_{med} = \arg\max_i \#\{k_0, \ldots, k_p; x_i \in hull(x_{k_0}, \ldots, x_{k_p})\}. \qquad (20.4)$$

Here $k_0, \ldots, k_p$ denote the indices of $p+1$ datapoints. Thus for each datapoint we have a multivariate data depth. If we compute all the necessary simplices $hull(x_{k_0}, \ldots, x_{k_p})$, the computing time will unfortunately be exponential as the dimension increases.

In Fig. 20.2, we calculate the simplicial depth for a two-dimensional, 10-point distribution according to depth. It contains 100 data points with corresponding parameters controlling its spread. The deepest point, the two-dimensional median, is indicated as a big star in the center. The points with less depth are indicated via gray shades.

Table 20.1 Coverages for artificial configuration of points

Triangle			Coverages					
1	△ 123	∋	1	2	3			
2	△ 124	∋	1	2		4		
3	△ 125	∋	1	2			5	
4	△ 126	∋	1	2	3	4		6
5	△ 134	∋	1		3	4		
6	△ 135	∋	1		3	4	5	
7	△ 136	∋	1		3			6
8	△ 145	∋	1			4	5	
9	△ 146	∋	1		3	4		6
10	△ 156	∋	1		3	4	5	6
11	△ 234	∋		2	3	4		
12	△ 235	∋		2	3	4	5	
13	△ 236	∋		2	3	4		6
14	△ 245	∋		2		4	5	
15	△ 246	∋		2		4		6
16	△ 256	∋		2			5	6
17	△ 345	∋			3	4	5	
18	△ 346	∋			3	4		6
19	△ 356	∋			3		5	6
20	△ 456	∋				4	5	6

Table 20.2 Simplicial depths for artificial configuration of points

Point	1	2	3	4	5	6
Depth	10	10	12	14	8	8

Summary

↪ The "depth" of a datapoint in one dimension can be computed by counting all (closed) intervals of two datapoints which contain the datapoint.

↪ The "deepest" datapoint is the central point of the distribution, the median.

↪ The "depth" of a datapoint in arbitrary dimension p is defined as the number of simplices (constructed from $p + 1$ points) covering this point. It is called simplicial depth.

↪ A multivariate extension of the median is to take the "deepest" datapoint of the distribution.

↪ In the bivariate case we count all triangles of datapoints which contain the datapoint to compute its depth.

20.2 Projection Pursuit

"Projection Pursuit" stands for a class of exploratory projection techniques. This class contains statistical methods designed for analyzing high-dimensional data using low-dimensional projections. The aim of projection pursuit is to reveal possible nonlinear and therefore interesting structures hidden in the high-dimensional data. To what extent these structures are "interesting" is measured by an index. Exploratory Projection Pursuit (EPP) goes back to Kruskal (1969, 1972). The approach was successfully implemented for exploratory purposes by various other authors. The idea has been applied to regression analysis, density estimation, classification, and discriminant analysis.

Exploratory Projection Pursuit

In EPP, we try to find "interesting" low-dimensional projections of the data. For this purpose, a suitable index function $I(\alpha)$, depending on a normalized projection vector α, is used. This function will be defined such that "interesting" views correspond to local and global maxima of the function. This approach naturally accompanies the technique of principal component analysis (PCA) of the covariance structure of a random vector X. In PCA we are interested in finding the axes of the covariance ellipsoid. The index function $I(\alpha)$ is in this case the variance of a linear combination $\alpha^\top X$ subject to the normalizing constraint $\alpha^\top \alpha = 1$ (see Theorem 11.2). If we analyze a sample with a p-dimensional normal distribution, the "interesting" high-dimensional structure we find by maximizing this index is of course linear.

There are many possible projection indices, for simplicity the kernel based and polynomial based indices are reported. Assume that the p-dimensional random variable X is sphered and centered, that is, $\mathsf{E}(X) = 0$ and $\mathsf{Var}(X) = \mathcal{I}_p$. This will remove the effect of location, scale, and correlation structure. This covariance structure can be achieved easily by the Mahalanobis transformation (3.26).

Friedman and Tukey (1974) proposed to investigate the high-dimensional distribution of X by considering the index

$$I_{\mathrm{FT},h}(\alpha) = n^{-1} \sum_{i=1}^{n} \hat{f}_{h,\alpha}(\alpha^\top X_i) \qquad (20.5)$$

where $\hat{f}_{h,\alpha}$ denotes the kernel estimator (see Sect. 1.3)

$$\hat{f}_{h,\alpha}(z) = n^{-1} \sum_{j=1}^{n} K_h(z - \alpha^\top X_j) \qquad (20.6)$$

of the projected data. Note that (20.5) is an estimate of $\int f^2(z)dz$ where $z = \alpha^\top X$ is a one-dimensional random variable with mean zero and unit variance. If the high-dimensional distribution of X is normal, then each projection $z = \alpha^\top X$ is standard

Fig. 20.2 10-point
distribution according to
depth with the median shown
as a big star in the center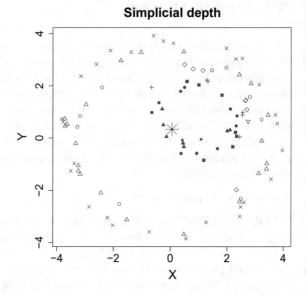
MVAsimdepex

is the standard normal density, a far more plausible candidate than the parabolic
density as a norm from which departure is to be regarded as "interesting". Thus

normal since $||\alpha|| = 1$ and since X has been centered and sphered by, e.g., the
Mahalanobis transformation.

The index should therefore be stable as a function of α if the high-dimensional
data is in fact normal. Changes in $I_{\mathrm{FT},h}(\alpha)$ with respect to α therefore indicate
deviations from normality. Hodges and Lehman (1956) showed that, given a mean
of zero and unit variance, the (compact support) density which minimizes $\int f^2$ is
uniquely given by

$$f(z) = \max\{0, c(b^2 - z^2)\},$$

where $c = 3/(20\sqrt{5})$ and $b = \sqrt{5}$. This is a parabolic density function, which is
equal to zero outside the interval $(-\sqrt{5}, \sqrt{5})$. A high value of the Friedman–Tukey
index indicates a larger departure from the parabolic form.

An alternative index is based on the negative of the entropy measure, i.e.,
$\int -f \log f$. The density for zero mean and unit variance which minimizes the index

$$\int f \log f$$

is the standard normal density, a far more plausible candidate than the parabolic
density as a norm from which departure is to be regarded as "interesting". Thus
in using $\int f \log f$ as a projection index we are really implementing the viewpoint
of seeing "interesting" projections as departures from normality. Yet another index
could be based on the Fisher information (see Sect. 6.2)

$$\int (f')^2/f.$$

To optimize the entropy index, it is necessary to recalculate it at each step of the numerical procedure. There is no method of obtaining the index via summary statistics of the multivariate data set, so the workload of the calculation at each iteration is determined by the number of observations. It is therefore interesting to look for approximations to the entropy index. Jones and Sibson (1987) suggested that deviations from the normal density should be considered as

$$f(x) = \varphi(x)\{1 + \varepsilon(x)\} \tag{20.7}$$

where the function ε satisfies

$$\int \varphi(u)\varepsilon(u)u^{-r}du = 0, \text{ for } r = 0, 1, 2. \tag{20.8}$$

In order to develop the Jones and Sibson (1987) index it is convenient to think in terms of cumulants $\kappa_3 = \mu_3 = \mathsf{E}(X^3), \kappa_4 = \mu_4 = \mathsf{E}(X^4) - 3$ (see Sect. 1.3). The standard normal density satisfies $\kappa_3 = \kappa_4 = 0$, an index with any hope of tracking the entropy index must at least incorporate information up to the level of symmetric departures (κ_3 or κ_4 not zero) from normality. The simplest of such indices is a positive definite quadratic form in κ_3 and κ_4. It must be invariant under sign-reversal of the data since both $\alpha^\top X$ and $-\alpha^\top X$ should show the same kind of departure from normality. Note that κ_3 is odd under sign-reversal, i.e., $\kappa_3(\alpha^\top X) = -\kappa_3(-\alpha^\top X)$. The cumulant κ_4 is even under sign-reversal, i.e., $\kappa_4(\alpha^\top X) = \kappa_4(-\alpha^\top X)$. The quadratic form in κ_3 and κ_4 measuring departure from normality cannot include a mixed $\kappa_3\kappa_4$ term.

For the density (20.7) one may conclude with (20.8) that

$$\int f(u) \log(u)du \approx \frac{1}{2} \int \varphi(u)\varepsilon(u)du.$$

Now if f is expressed as a Gram–Charliér expansion

$$f(x)\varphi(x) = \{1 + \kappa_3 H_3(x)/6 + \kappa_4 H_4(x)/24 + ...\} \tag{20.9}$$

Kendall and Stuart (1977, p. 169) where H_r is the r-th Hermite polynomial, then the truncation of (20.9) and use of orthogonality and normalization properties of Hermite polynomials with respect to φ yields

$$\frac{1}{2} \int \varphi(x)\varepsilon^2(x)dx = \left(\kappa_3^2 + \kappa_4^2/4\right)/12.$$

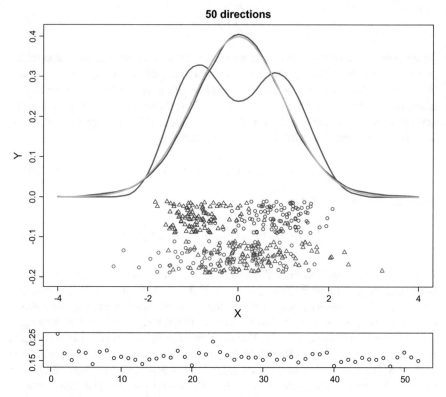

Fig. 20.3 Exploratory Projection Pursuit for the Swiss bank notes data (green = standard normal, red = best, blue = worst) 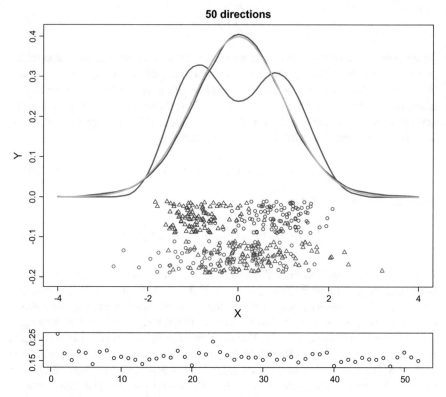 MVAppexample

The index proposed by Jones and Sibson (1987) is therefore

$$I_{JS}(\alpha) = \{\kappa_3^2(\alpha^\top X) + \kappa_4^2(\alpha^\top X)/4\}/12.$$

This index measures in fact the negative entropy difference $\int f \log f - \int \varphi \log \varphi$.

Example 20.1 The exploratory Projection Pursuit is used on the Swiss bank note data. For 50 randomly chosen one-dimensional projections of this six-dimensional data set we calculate the Friedman–Tukey index to evaluate how "interesting" their structures are.

Figure 20.3 shows the density for the standard, normally distributed data (green) and the estimated densities for the best (red) and the worst (blue) projections found. A dotplot of the projections is also presented. In the lower part of the figure we see the estimated value of the Friedman–Tukey index for each computed projection. From this information we can judge the nonnormality of the bank note data set since there is a lot of variation across the 50 random projections.

Projection Pursuit Regression

The problem in projection pursuit regression is to estimate a response surface

$$f(x) = \mathsf{E}(Y \mid x)$$

via approximating functions of the form

$$\hat{f}(x) = \sum_{k=1}^{M} g_k(\Lambda_k^\top x)$$

with nonparametric regression functions g_k and projection indices Λ_k. Given observations $\{(x_1, y_1), \dots, (x_n, y_n)\}$ with $x_i \in \mathbb{R}^p$ and $y_i \in \mathbb{R}$ the basic algorithm works as follows.

1. Set $r_i^{(0)} = y_i$ and $k = 1$.
2. Minimize

$$E_k = \sum_{i=1}^{n} \left\{ r_i^{(k-1)} - g_k(\Lambda_k^\top x_i) \right\}^2$$

   where Λ_k is an orthogonal projection matrix and g_k is a nonparametric regression estimator.
3. Compute new residuals

$$r_i^{(k)} = r_i^{(k-1)} - g_k(\Lambda_k^\top x_i).$$

4. Increase k and repeat the last two steps until E_k becomes small.

Although this approach seems to be simple, we encounter some problems. One of the most serious is that the decomposition of a function into sums of functions of projections may not be unique. An example is

$$z_1 z_2 = \frac{1}{4ab}\{(az_1 + bz_2)^2 - (az_1 - bz_2)^2\}.$$

Numerical improvements of this algorithm were suggested by Friedman and Stuetzle (1981).

Summary
↪ Exploratory Projection Pursuit is a technique used to find interesting structures in high-dimensional data via low-dimensional projections. Since the Gaussian distribution represents a standard situation, we define the Gaussian distribution as the most uninteresting.
↪ The search for interesting structures is done via a projection score like the Friedman–Tukey index $I_{FT}(\alpha) = \int f^2$. The parabolic distribution has the minimal score. We maximize this score over all projections.
↪ The Jones–Sibson index maximizes $$I_{JS}(\alpha) = \{\kappa_3(\alpha^\top X) + \kappa_4^2(\alpha^\top X)/4\}/12$$ as a function of α.
↪ The entropy index maximizes $$I_E(\alpha) = \int f(\alpha^\top X) \log f(\alpha^\top X)$$ where f is the density of $\alpha^\top X$.
↪ In Projection Pursuit Regression, the idea is to represent the unknown function by a sum of nonparametric regression functions on projections. The key problem is in choosing the number of terms and often the interpretability.

20.3   Sliced Inverse Regression

Sliced inverse regression (SIR) is a dimension reduction method proposed by Duan and Li (1991). The idea is to find a smooth regression function that operates on a variable set of projections. Given a response variable Y and a (random) vector $X \in \mathbb{R}^p$ of explanatory variables, SIR is based on the model:

$$Y = m(\beta_1^\top X, \ldots, \beta_k^\top X, \varepsilon), \qquad (20.10)$$

where $\beta_1, \ldots, \beta_k$ are unknown projection vectors, k is unknown and assumed to be less than p, $m : \mathbb{R}^{k+1} \to \mathbb{R}$ is an unknown function, and ε is the noise random variable with $\mathsf{E}(\varepsilon \,|\, X) = 0$.

Model (20.10) describes the situation where the response variable Y depends on the p-dimensional variable X only through a k-dimensional subspace. The unknown β_i's, which span this space, are called *effective dimension reduction directions* (EDR-directions). The span is denoted as *effective dimension reduction space* (EDR-space). The aim is to estimate the base vectors of this space, for which neither the length nor the direction can be identified. Only the space in which they lie is identifiable.

SIR tries to find this k-dimensional subspace of $\mathbb{R}^p$ which under the model (20.10) carries the essential information of the regression between X and Y. SIR also focuses on small k, so that nonparametric methods can be applied for the estimation of m. A direct application of nonparametric smoothing to X is for high dimension p generally not possible due to the sparseness of the observations. This fact is well known as the *curse of dimensionality*, see Huber (1985).

The name of SIR comes from computing the inverse regression (IR) curve. That means instead of looking for $\mathsf{E}\,(Y\,|\,X=x)$, we investigate $\mathsf{E}\,(X\,|\,Y=y)$, a curve in $\mathbb{R}^p$ consisting of p one-dimensional regressions. What is the connection between the IR and the SIR model (20.10)? The answer is given in the following theorem from Li (1991).

Theorem 20.1 *Given the model (20.10) and the assumption*

$$\forall b \in \mathbb{R}^p : \mathsf{E}\left(b^\top X\,|\,\beta_1^\top X = \beta_1^\top x, \ldots, \beta_k^\top X = \beta_k^\top x\right) = c_0 + \sum_{i=1}^{k} c_i \beta_i^\top x, \quad (20.11)$$

the centered IR curve $\mathsf{E}(X\,|\,Y=y) - \mathsf{E}(X)$ lies in the linear subspace spanned by the vectors $\Sigma\beta_i$, $i = 1, \ldots, k$, where $\Sigma = \mathsf{Cov}(X)$.

Assumption (20.11) is equivalent to the fact that X has an elliptically symmetric distribution, see Cook and Weisberg (1991). Hall and Li (1993) have shown that assumption (20.11) only needs to hold for the EDR-directions.

It is easy to see that for the standardized variable $Z = \Sigma^{-1/2}\{X - \mathsf{E}(X)\}$ the IR curve $m_1(y) = \mathsf{E}(Z\,|\,Y=y)$ lies in $\text{span}(\eta_1, \ldots, \eta_k)$, where $\eta_i = \Sigma^{1/2}\beta_i$. This means that the conditional expectation $m_1(y)$ is moving in $\text{span}(\eta_1, \ldots, \eta_k)$ depending on y. With b orthogonal to $\text{span}(\eta_1, \ldots, \eta_k)$, it follows that

$$b^\top m_1(y) = 0,$$

and further that

$$m_1(y)m_1(y)^\top b = \mathsf{Cov}\{m_1(y)\}b = 0.$$

As a consequence $\mathsf{Cov}\{\mathsf{E}(Z\,|\,y)\}$ is degenerated in each direction orthogonal to all EDR-directions η_i of Z. This suggests the following algorithm.

First, estimate $\mathsf{Cov}\{m_1(y)\}$ and then calculate the orthogonal directions of this matrix (for example, with eigenvalue/eigenvector decomposition). In general, the estimated covariance matrix will have full rank because of random variability, estimation errors and numerical imprecision. Therefore, we investigate the eigenvalues of the estimate and ignore eigenvectors having small eigenvalues. These eigenvectors $\hat{\eta}_i$ are estimates for the EDR-direction η_i of Z. We can easily rescale them to estimates $\hat{\beta}_i$ for the EDR-directions of X by multiplying by $\hat{\Sigma}^{-1/2}$, but then they are not necessarily orthogonal. SIR is strongly related to PCA. If all of the data falls into a single interval, which means that $\widehat{\mathsf{Cov}\{m_1(y)\}}$ is equal to $\widehat{\mathsf{Cov}(Z)}$, SIR coincides with PCA. Obviously, in this case any information about y is ignored.

The SIR Algorithm

The algorithm to estimate the EDR-directions via SIR is as follows:

1. Standardize x:

$$z_i = \hat{\Sigma}^{-1/2}(x_i - \bar{x}).$$

2. Divide the range of y_i into S nonoverlapping intervals *(slices)* $H_s, s = 1, \ldots, S$. n_s denotes the number of observations within slice H_s, and I_{H_s} the indicator function for this slice:

$$n_s = \sum_{i=1}^{n} I_{H_s}(y_i).$$

3. Compute the mean of z_i over all slices. This is a crude estimate $\widehat{m}_1$ for the *inverse regression curve* m_1:

$$\bar{z}_s = n_s^{-1} \sum_{i=1}^{n} z_i \; I_{H_s}(y_i).$$

4. Calculate the estimate for $\mathsf{Cov}\{m_1(y)\}$:

$$\widehat{V} = n^{-1} \sum_{s=1}^{S} n_s \bar{z}_s \bar{z}_s^{\top}.$$

5. Identify the eigenvalues $\hat{\lambda}_i$ and eigenvectors $\hat{\eta}_i$ of $\widehat{V}$.
6. Transform the standardized EDR-directions $\hat{\eta}_i$ back to the original scale. Now the estimates for the EDR-directions are given by

$$\hat{\beta}_i = \hat{\Sigma}^{-1/2} \hat{\eta}_i.$$

Remark 20.1 The number of different eigenvalues unequal to zero depends on the number of slices. The rank of $\widehat{V}$ cannot be greater than the number of slices-1 (the z_i sum up to zero). This is a problem for categorical response variables, especially for a binary response—where only one direction can be found.

SIR II

In the previous section we learned that it is interesting to consider the IR curve, that is, $\mathsf{E}(X \,|\, y)$. In some situations however SIR does not find the EDR-direction. We overcome this difficulty by considering the conditional covariance $\mathsf{Cov}(X \,|\, y)$ instead of the IR curve. An example where the EDR directions are not found via the SIR curve is given below.

Example 20.2 Suppose that $(X_1, X_2)^{\top} \sim N(0, \mathcal{I}_2)$ and $Y = X_1^2$. Then $\mathsf{E}(X_2 \,|\, y) = 0$ because of independence and $\mathsf{E}(X_1 \,|\, y) = 0$ because of symmetry. Hence, the EDR-direction $\beta = (1, 0)^{\top}$ is not found when the IR curve $\mathsf{E}(X \,|\, y) = 0$ is considered.

The conditional variance

$$\mathsf{Var}(X_1 \,|\, Y = y) = \mathsf{E}(X_1^2 \,|\, Y = y) = y,$$

offers an alternative way to find β. It is a function of y while $\mathsf{Var}(X_2 \,|\, y)$ is a constant.

The idea of SIR II is to consider the conditional covariances. The principle of SIR II is the same as before: investigation of the IR curve (here the conditional covariance instead of the conditional expectation). Unfortunately, the theory of SIR II is more complicated. The assumption of the elliptical symmetrical distribution of X has to be more restrictive, i.e., assuming the normality of X.

Given this assumption, one can show that the vectors with the largest distance to $\mathsf{Cov}(Z \,|\, Y = y) - \mathsf{E}\{\mathsf{Cov}(Z \,|\, Y = y)\}$ for all y are the most interesting for the EDR-space. An appropriate measure for the overall mean distance is, according to Li (1992),

$$\mathsf{E}\left(\|\, [\mathsf{Cov}(Z \,|\, Y = y) - \mathsf{E}\{\mathsf{Cov}(Z \,|\, Y = y)\}]\, b\|^2\right)$$
$$= b^\top \mathsf{E}\left(\|\mathsf{Cov}(Z \,|\, y) - \mathsf{E}\{\mathsf{Cov}(Z \,|\, y)\}\|^2\right) b. \qquad (20.12)$$

Equipped with this distance, we conduct again an eigensystem decomposition, this time for the above expectation $\mathsf{E}\left(\|\mathsf{Cov}(Z \,|\, y) - \mathsf{E}\{\mathsf{Cov}(Z \,|\, y)\}\|^2\right)$. Then we take the rescaled eigenvectors with the largest eigenvalues as estimates for the unknown EDR-directions.

The SIR II Algorithm

The algorithm of SIR II is very similar to the one for SIR, it differs in only two steps. Instead of merely computing the mean, the covariance of each slice has to be computed. The estimate for the above expectation (20.12) is calculated after computing all slice covariances. Finally, decomposition and rescaling are conducted, as before.

1. Do steps 1–3 of the SIR algorithm.
2. Compute the slice covariance matrix $\widehat{V}_s$:

$$\widehat{V}_s = (n_s - 1)^{-1} \sum_{i=1}^{n} I_{H_s}(y_i) z_i z_i^\top - n_s \bar{z}_s \bar{z}_s^\top.$$

3. Calculate the mean over all slice covariances:

$$\bar{V} = n^{-1} \sum_{s=1}^{S} n_s \widehat{V}_s.$$

4. Compute an estimate for (20.12):

Table 20.3 SIR: EDR-directions for simulated data

$\hat{\beta}_1$	$\hat{\beta}_2$	$\hat{\beta}_3$
0.452	0.881	0.040
0.571	−0.349	−0.787
0.684	−0.320	0.615

$$\widehat{V} = n^{-1} \sum_{s=1}^{S} n_s \left(\widehat{V}_s - \bar{V}\right)^2 = n^{-1} \sum_{s=1}^{S} n_s \widehat{V}_s^2 - \bar{V}^2.$$

5. Identify the eigenvectors and eigenvalues of $\widehat{V}$ and scale back the eigenvectors. This gives estimates for the SIR II EDR-directions:

$$\hat{\beta}_i = \hat{\Sigma}^{-1/2} \hat{\eta}_i.$$

Example 20.3 The result of SIR is visualized in four plots in Fig. 20.4: the left two show the response variable versus the first respectively second direction. The upper right plot consists of a three-dimensional plot of the first two directions and the response. The last picture shows $\hat{\Psi}_k$, the ratio of the sum of the first k eigenvalues and the sum of all eigenvalues, similar to principal component analysis.

The data are generated according to the following model:

$$y_i = \beta_1^\top x_i + (\beta_1^\top x_i)^3 + 4 \left(\beta_2^\top x_i\right)^2 + \varepsilon_i,$$

where the x_i's follow a three-dimensional normal distribution with zero mean, the covariance equal to the identity matrix, $\beta_2 = (1, -1, -1)^\top$, and $\beta_1 = (1, 1, 1)^\top$. ε_i is standard, normally distributed and $n = 300$. Corresponding to model (20.10), $m(u, v, \varepsilon) = u + u^3 + v^2 + \varepsilon$. The situation is depicted in Figs. 20.5 and 20.6.

Both algorithms were conducted using the slicing method with 20 elements in each slice. The goal was to find β_1 and β_2 with SIR. The data are designed such that SIR can detect β_1 because of the monotonic shape of $\{\beta_1^\top x_i + (\beta_1^\top x_i)^3\}$, while SIR II will search for β_2, as in this direction the conditional variance on y is varying.

If we normalize the eigenvalues for the EDR-directions in Table 20.3 such that they sum up to one, the resulting vector is $(0.852, 0.086, 0.062)$. As can be seen in the upper left plot of Fig. 20.4, there is a functional relationship found between the first index $\hat{\beta}_1^\top x$ and the response. Actually, β_1 and $\hat{\beta}_1$ are nearly parallel, that is, the normalized inner product $\hat{\beta}_1^\top \beta_1 / \{||\hat{\beta}_1|| ||\beta_1||\} = 0.9894$ is very close to one.

The second direction along β_2 is probably found due to the good approximation, but SIR does not provide it clearly, because it is "blind" with respect to the change of variance, as the second eigenvalue indicates.

For SIR II, the normalized eigenvalues are $(0.706, 0.185, 0.108)$, that is, about 69% of the variance is explained by the first EDR-direction (Table 20.4). Here, the

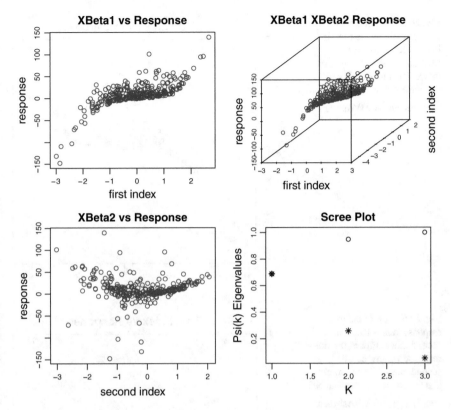

Fig. 20.4 SIR: The left plots show the response versus the estimated EDR-directions. The upper right plot is a three-dimensional plot of the first two directions and the response. SIR: The left plots show the response versus the estimated EDR-directions. The upper right plot is a three-dimensional plot of the first two directions and the response. The lower right plot shows the eigenvalues $\hat{\lambda}_i$ (∗) and the cumulative sum (○) ◘ MVAsirdata

Table 20.4 SIR II: EDR-directions for simulated data

$\hat{\beta}_1$	$\hat{\beta}_2$	$\hat{\beta}_3$
−0.272	0.964	−0.001
0.670	0.100	0.777
0.690	0.244	−0.630

normalized inner product of β_2 and $\hat{\beta}_1$ is 0.9992. The estimator $\hat{\beta}_1$ estimates in fact β_2 of the simulated model. In this case, SIR II found the direction where the second moment varies with respect to $\beta_2^\top x$ (Fig. 20.7).

In summary, SIR has found the direction which shows a strong relation regarding the conditional expectation between $\beta_1^\top x$ and y, and SIR II has found the direction where the conditional variance is varying, namely, $\beta_2^\top x$.

Fig. 20.5 Plot of the true
response versus the true first
index. Plot of the true
response versus the true first
index. The monotonic and
the convex shapes can be
clearly seen 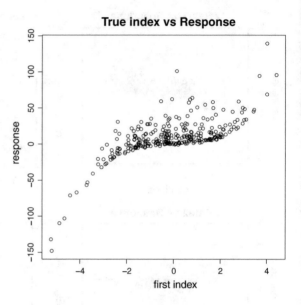 MVAsirdata

Fig. 20.6 Plot of the true
response versus the true
second index. Plot of the true
response versus the true
second index. The monotonic
and the convex shapes can be
clearly seen 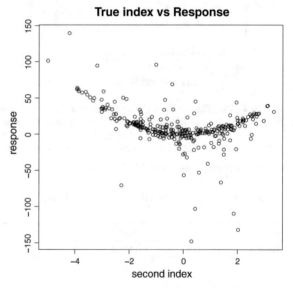 MVAsirdata

The behavior of the two SIR algorithms is as expected. In addition, we have seen that it is worthwhile to apply both versions of SIR. It is possible to combine SIR and SIR II (Cook and Weisberg 1991; Schott 1994; Li 1991), directly, or to investigate higher conditional moments. For the latter it seems to be difficult to obtain theoretical results.

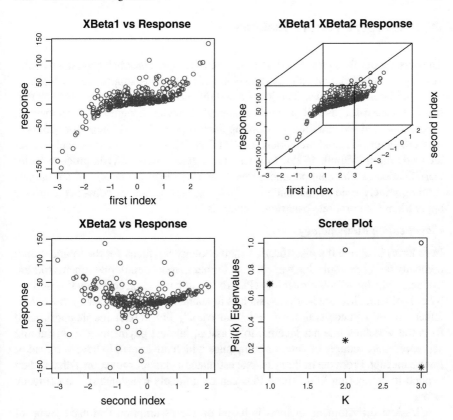

Fig. 20.7 SIR II mainly sees the direction β_2. SIR II mainly sees the direction β_2. The left plots show the response versus the estimated EDR-directions. The upper right plot is a three-dimensional plot of the first two directions and the response. The lower right plot shows the eigenvalues $\hat{\lambda}_i$ (∗) and the cumulative sum (∘) Q MVAsir2data

Summary
↪ SIR serves as a dimension reduction tool for regression problems.
↪ Inverse regression avoids the *curse of dimensionality*.
↪ The dimension reduction can be conducted without estimation of the regression function $y = m(x)$.
↪ SIR searches for the effective dimension reduction (EDR) by computing the inverse regression IR.
↪ SIR II uses the EDR on computing the inverse conditional variance.
↪ SIR might miss EDR directions that are found by SIR II.

20.4   Support Vector Machines

The purpose of this section is to introduce one of the most promising among recently developed multivariate nonlinear statistical techniques: the support vector machine (SVM). The SVM is a classification method that is based on statistical learning theory. It has been successfully applied to optical character recognition, early medical diagnostics, and text classification. One application where SVMs outperformed other methods is electric load prediction (EUNITE 2001), another one is optical character recognition (Vapnik 1995). In a variety of applications, SVMs produce better classification results than parametric methods (e.g., logit analysis) and are outperforming widely used nonparametric techniques, such as neural networks. Here we apply SVMs to corporate bankruptcy analysis.

Classification Methodology

In order to illustrate the classification methodology we focus for the moment on a company rating example that we will treat further in more detail. Investment risks are evaluated via the default probability (PD) for a company. Each company is described by a set of variables (predictors) x, such as financial ratios, and its class y that can be either $y = -1$ ("successful") or $y = 1$ ("bankrupt"). Financial ratios are constructed from the variables like net income, total assets, interest payments, etc. A training set represents a sample of data for companies which are known to have survived or gone bankrupt. From the training set one estimates a classifier function f that is then applied to computing PDs. These PDs can be uniquely translated into a company rating.

Classical discriminant analysis is based on the assumption that each group of observations is normally distributed with the same variance–covariance matrix but different means. Under such a formulation the discriminating function will be linear, see Theorem 14.2. Figure 20.8 displays this situation: if some linear combination of predictors (called Z-score in the context of bankruptcy analysis) is greater than a particular threshold value z_0 the observation under consideration is regarded as belonging to $y = 1$; if $Z < z_0$ the observation would belong to $y = -1$ (successful). One can change the labels "$-1, +1$" to the more standard notation "0, 1". The current labeling is done only for mathematical convenience.

The Z-score is

$$Z_i = a_1 x_{i1} + a_2 x_{i2} + \ldots + a_p x_{ip} = a^\top x_i,$$

where $x_i = (x_{i1}, \ldots, x_{ip})^\top \in \mathbb{R}^p$ are predictors for the i-th company. The classification based on the Z-score are necessarily linear and, therefore, may not handle more complex situations as in Fig. 20.9 when nonlinear classifiers, such as those generated by SVMs, can produce better results.

Expected versus Empirical Risk Minimization

A nonlinear classifier function f may be described by a function class $\mathcal{F}$. $\mathcal{F}$ is fixed a priori, e.g., it can be the class of linear classifiers (hyperplanes). A good classifier

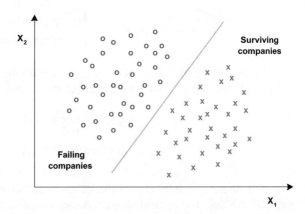

Fig. 20.8 A linear classification function in the case of linearly separable data

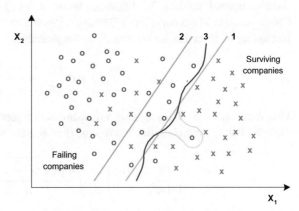

Fig. 20.9 Different linear classification functions (1) and (2) and a nonlinear one (3) in the linearly non-separable case

optimizes some criterion that tells us how well f separates the classes. As in (14.4) one considers the minimization of the expected risk:

$$R(f) = \int \frac{1}{2} |f(x) - y| \, dF(x, y). \tag{20.13}$$

The joint distribution $F(x, y)$, however, is never known in practical applications and must be estimated from the *training set* $\{x_i, y_i\}_{i=1}^{n}$. By replacing $F(x, y)$ with the empirical cdf $F_n(x, y)$ one obtains the empirical risk:

$$\hat{R}(f) = \frac{1}{n} \sum_{i=1}^{n} \frac{1}{2} |f(x_i) - y_i|. \tag{20.14}$$

The empirical risk is an average value of loss over the training set, while the expected risk is the expected value of loss under the true probability measure. The loss is given by

$$L(x, y) = \frac{1}{2}|f(x) - y| = \begin{cases} 0, & \text{if classification is correct,} \\ 1, & \text{if classification is wrong.} \end{cases}$$

One sees here that it is convenient to work with the labels "-1, 1" for y. The solutions to the problems of expected and empirical risk minimization:

$$f_{opt} = \arg\min_{f \in \mathcal{F}} R(f), \tag{20.15}$$

$$\hat{f}_n = \arg\min_{f \in \mathcal{F}} \widehat{R}(f), \tag{20.16}$$

generally do not coincide (Fig. 20.10), although converge as $n \to \infty$ if $\mathcal{F}$ is not too large. According to statistical learning theory (Vapnik 1995), it is possible to get a uniform upper bound on the difference between $R(f)$ and $\hat{R}(f)$ via the Vapnik–Chervonenkis (VC) theory. The VC bound states that there is a function ϕ (monotone increasing in h) so that for all $f \in \mathcal{F}$ with a probability $1 - \eta$:

$$R(f) \le \widehat{R}(f) + \phi\left\{ \frac{h}{n}, \frac{\log(\eta)}{n} \right\}. \tag{20.17}$$

Here h denotes the VC dimension, a measure of complexity of the involved function class $\mathcal{F}$. For a linear classification rule $g(x) = \text{sign}(x^\top w + b)$:

$$\phi\left\{ \frac{h}{n}, \frac{\log(\eta)}{n} \right\} = \sqrt{\frac{h\left(\log \frac{2n}{h}\right) - \log \frac{\eta}{4}}{n}}, \tag{20.18}$$

where h is the VC dimension. By plotting the function

$$\phi(u, v) = \{-u \cdot \log 2u + \log 4 - v\}^{-1/2}$$

for small u one sees the monotonicity of $\phi(u, v)$. In fact one can show that

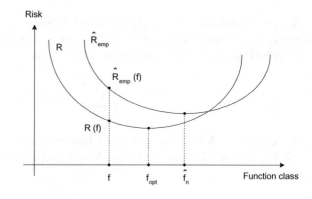

Fig. 20.10 The minima f_{opt} and $\hat{f}_n$ of the expected (R) and empirical ($\hat{R}$) risk functions generally do not coincide

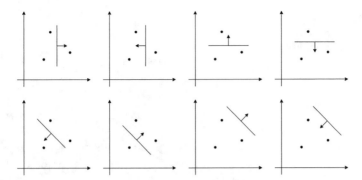

Fig. 20.11 Eight possible ways of shattering three points on the plane with a linear indicator function

$$\frac{\partial \phi\left(\frac{h}{n}, \frac{\log(\eta)}{n}\right)}{\partial h} \geqslant 0$$

if and only if $2n \geqslant h$. For a linear classifier with $h = p + 1$ this is an easy condition to meet.

The VC dimension of a set $\mathcal{F}$ of functions in a d-dimensional space is h if some function $f \in \mathcal{F}$ can shatter h objects $\{x_i \in \mathbb{R}^d, i = 1, ..., h\}$, in all 2^h possible configurations and no set $\{x_j \in \mathbb{R}^d, j = 1, ..., q\}$ with $q > h$, exists that satisfies this property. For example, three points on a plane ($d = 2$) can be shattered by linear indicator functions in $2^h = 2^3 = 8$ ways, whereas four points can not be shattered in $2^q = 2^4 = 16$ ways. Thus, the VC dimension of the set of linear indicator functions in a two-dimensional space is $h = 3$, see Fig. 20.11. The expression for the VC bound (20.17) involves the VC dimension h, a parameter controlling complexity of $\mathcal{F}$. The term $\phi\left\{\frac{h}{n}, \frac{\log(\eta)}{n}\right\}$ introduces a penalty for excessive complexity of a classifier function. The higher is the complexity of $f \in \mathcal{F}$ the higher are h and therefore ϕ. There is a trade-off between the number of classification errors on the training set and the complexity of the classifier function. If the complexity were not controlled for, it would be possible to construct a classifier function with no classification errors on the training set notwithstanding how low its generalization ability would be.

The SVM in the Linearly Separable Case

First we will describe the SVM in the linearly separable case. The family $\mathcal{F}$ of classification functions in the data space is given by

$$\mathcal{F} = \left\{x^\top w + b, w \in \mathbb{R}^p, b \in \mathbb{R}\right\} \tag{20.19}$$

In order to determine the support vectors we choose $f \in \mathcal{F}$ (or equivalently (w, b)) such that the so-called margin—the corridor between the separating hyperplanes— is maximal. This situation is illustrated in Fig. 20.12. The margin is equal to $d_- + d_+$. The classification function is a hyperplane plus the margin zone, where, in the

Fig. 20.12 The separating
hyperplane $x^\top w + b = 0$
and the margin in the linearly
separable case

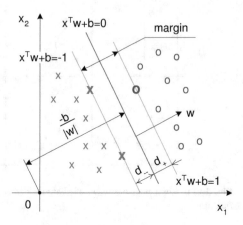

separable case, no observations can lie. It separates the points from both classes
with the highest "safest" distance (margin) between them. It can be shown that
margin maximization corresponds to the reduction of complexity as given by the VC
dimension of the SVM classifier. Apparently, the separating hyperplane is defined
only by the *support vectors* that hold the hyperplanes parallel to the separating one.
In Fig. 20.12 there are three support vectors that are marked with bold style: two
crosses and one circle. We come now to the description of the SVM selection.

Let $x^\top w + b = 0$ be a separating hyperplane. Then d_+ (d_-) will be the shortest
distance to the closest objects from the classes $+1$ (-1). Since the separation can be
done without errors, all observations $i = 1, 2, ..., n$ must satisfy:

$$x_i^\top w + b \geq +1 \quad \text{for} \quad y_i = +1$$
$$x_i^\top w + b \leq -1 \quad \text{for} \quad y_i = -1$$

We can combine both constraints into one:

$$y_i(x_i^\top w + b) - 1 \geq 0 \quad i = 1, 2, ..., n \tag{20.20}$$

The *canonical hyperplanes* $x_i^\top w + b = \pm 1$ are parallel and the distance between
each of them and the separating hyperplane is $d_+ = d_- = 1/\|w\|$. To maximize the
margin $d_+ + d_- = 2/\|w\|$ one therefore minimizes the Euclidean norm $\|w\|$ or its
square $\|w\|^2$.

The Lagrangian for the primal problem that corresponds to margin maximization
subject to constraint (20.20) is

$$L_P(w, b) = \frac{1}{2}\|w\|^2 - \sum_{i=1}^{n} \alpha_i \{y_i(x_i^\top w + b) - 1\} \tag{20.21}$$

The Karush–Kuhn–Tucker (KKT) (Gale et al. 1951) first-order optimality conditions are

$$\frac{\partial L_P}{\partial w} = 0: \quad w - \sum_{i=1}^{n} \alpha_i y_i x_i = 0$$

$$\frac{\partial L_P}{\partial b} = 0: \quad \sum_{i=1}^{n} \alpha_i y_i = 0$$

$$y_i(x_i^\top w + b) - 1 \geq 0, \quad i = 1, ..., n$$

$$\alpha_i \geq 0$$

$$\alpha_i \{ y_i(x_i^\top w + b) - 1 \} = 0$$

From these first-order condition, we can derive $w = \sum_{i=1}^{n} \alpha_i y_i x_i$ and therefore the summands in (20.21) read:

$$\frac{1}{2} \|w\|^2 = \frac{1}{2} \sum_{i=1}^{n} \sum_{j=1}^{n} \alpha_i \alpha_j y_i y_j x_i^\top x_j$$

$$-\sum_{i=1}^{n} \alpha_i \{ y_i(x_i^\top w + b) - 1 \} = -\sum_{i=1}^{n} \alpha_i y_i x_i^\top \sum_{j=1}^{n} \alpha_j y_j x_j + \sum_{i=1}^{n} \alpha_i$$

$$= -\sum_{i=1}^{n} \sum_{j=1}^{n} \alpha_i \alpha_j y_i y_j x_i^\top x_j + \sum_{i=1}^{n} \alpha_i$$

Substituting this into (20.21) we obtain the Lagrangian for the dual problem:

$$L_D(\alpha) = \sum_{i=1}^{n} \alpha_i - \frac{1}{2} \sum_{i=1}^{n} \sum_{j=1}^{n} \alpha_i \alpha_j y_i y_j x_i^\top x_j. \qquad (20.22)$$

The primal and dual problems are

$$\min_{w,b} L_P(w, b)$$

$$\max_{\alpha} L_D(\alpha) \quad \text{s.t.} \quad \alpha_i \geq 0, \quad \sum_{i=1}^{n} \alpha_i y_i = 0.$$

Since the optimization problem is convex the dual and primal formulations give the same solution.

Those points i for which the equation $y_i(x_i^\top w + b) = 1$ holds are called support vectors. After "training the support vector machine", i.e., solving the dual problem above and deriving Lagrange multipliers (they are equal to 0 for nonsupport vectors) one can classify a company. One uses the classification rule:

Fig. 20.13 The separating
hyperplane $x^\top w + b = 0$
and the margin in the linearly
non-separable case

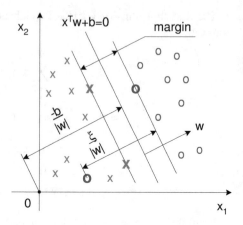

$$g(x) = \text{sign}\left(x^\top w + b\right), \tag{20.23}$$

where $w = \sum_{i=1}^{n} \alpha_i y_i x_i$ and $b = \frac{1}{2}(x_{+1} + x_{-1}) w$. x_{+1} and x_{-1} are two support vectors belonging to different classes for which $y(x^\top w + b) = 1$. The value of the classification function (the score of a company) can be computed as

$$f(x) = x^\top w + b. \tag{20.24}$$

Each score $f(x)$ uniquely corresponds to a default probability (PD). The higher $f(x)$ the higher the PD.

SVMs in the Linearly Non-separable Case

In the linearly non-separable case the situation is like in Fig. 20.13. The slack variables ξ_i represent the violation from strict separation. In this case the following inequalities can be induced from Fig. 20.13:

$$\begin{aligned} x_i^\top w + b &\geq 1 - \xi_i \text{ for } y_i = 1, \\ x_i^\top w + b &\leq -1 + \xi_i \text{ for } y_i = -1, \\ \xi_i &\geq 0. \end{aligned}$$

They can be combined into two constraints:

$$y_i(x_i^\top w + b) \geq 1 - \xi_i \tag{20.25}$$

$$\xi_i \geq 0. \tag{20.26}$$

SVM classification again maximizes the margin given a family of classification functions $\mathcal{F}$.

The penalty for misclassification, the classification error $\xi_i \geq 0$, is related to the distance from a misclassified point x_i to the canonical hyperplane bounding its class.

If $\xi_i > 0$, an error in separating the two sets occurs. The objective function corresponding to penalized margin maximization is then formulated as

$$\frac{1}{2} \|w\|^2 + C \sum_{i=1}^{n} \xi_i, \tag{20.27}$$

where the parameter C characterizes the weight given to the classification errors. The minimization of the objective function with constraint (20.25) and (20.26) provides the highest possible margin in the case when classification errors are inevitable due to the linearity of the separating hyperplane. Under such a formulation the problem is convex.

The Lagrange function for the primal problem is

$$L_P(w, b, \xi) = \frac{1}{2} \|w\|^2 + C \sum_{i=1}^{n} \xi_i - \sum_{i=1}^{n} \alpha_i \{ y_i (x_i^\top w + b) - 1 + \xi_i \} - \sum_{i=1}^{n} \mu_i \xi_i, \tag{20.28}$$

where $\alpha_i \geq 0$ and $\mu_i \geq 0$ are Lagrange multipliers. The primal problem is formulated as

$$\min_{w, b, \xi} L_P(w, b, \xi).$$

The first-order conditions in this case are

$$\frac{\partial L_P}{\partial w} = 0 : \quad w - \sum_{i=1}^{n} \alpha_i y_i x_i = 0$$

$$\frac{\partial L_P}{\partial b} = 0 : \quad \sum_{i=1}^{n} \alpha_i y_i = 0$$

$$\frac{\partial L_P}{\partial \xi_i} = 0 : \quad C - \alpha_i - \mu_i = 0$$

With the conditions for the Lagrange multipliers:

$$\alpha_i \geq 0$$
$$\mu_i \geq 0$$
$$\alpha_i \{ y_i (x_i^\top w + b) - 1 + \xi_i \} = 0$$
$$\mu_i \xi_i = 0$$

Note that $\sum_{i=1}^{n} \alpha_i y_i b = 0$ therefore similar to the linear separable case the primal problem translates into:

$$L_D(\alpha) = \frac{1}{2} \sum_{i=1}^{n} \sum_{j=1}^{n} \alpha_i \alpha_j y_i y_j x_i^\top x_j - \sum_{i=1}^{n} \alpha_i y_i x_i^\top \sum_{j=1}^{n} \alpha_j y_j x_j$$

$$+ C \sum_{i=1}^{n} \xi_i + \sum_{i=1}^{n} \alpha_i - \sum_{i=1}^{n} \alpha_i \xi_i - \sum_{i=1}^{n} \mu_i \xi_i$$

$$= \sum_{i=1}^{n} \alpha_i - \frac{1}{2} \sum_{i=1}^{n} \sum_{j=1}^{n} \alpha_i \alpha_j y_i y_j x_i^\top x_j + \sum_{i=1}^{n} \xi_i (C - \alpha_i - \mu_i)$$

Since the last term is 0 we derive the dual problem as

$$L_D(\alpha) = \sum_{i=1}^{n} \alpha_i - \frac{1}{2} \sum_{i=1}^{n} \sum_{j=1}^{n} \alpha_i \alpha_j y_i y_j x_i^\top x_j, \qquad (20.29)$$

and the dual problem is posed as

$$\max_{\alpha} L_D(\alpha),$$

subject to:

$$0 \leq \alpha_i \leq C,$$
$$\sum_{i=1}^{n} \alpha_i y_i = 0.$$

Nonlinear Classification

The SVMs can also be generalized to the nonlinear case. In order to obtain nonlinear classifiers as in Fig. 20.14 one maps the data with a nonlinear structure via a function $\Psi : \mathbb{R}^p \mapsto \mathbb{H}$ into a very large dimensional space $\mathbb{H}$ where the classification rule is (almost) linear. Note that all the training vectors x_i appear in L_D (20.29) only as scalar products of the form $x_i^\top x_j$. In the nonlinear SVM situations this transforms to $\psi(x_i)^\top \psi(x_j)$.

The so-called *kernel trick* is to compute this scalar product via a kernel function. These kernel functions are actually related to those we presented in Sect. 1.3. If a kernel function K exists such that $K(x_i, x_j) = \Psi(x_i)^\top \Psi(x_j)$, then it can be used without knowing the transformation Ψ explicitly. A necessary and sufficient condition for a symmetric function $K(x_i, x_j)$ to be a kernel is given by Mercer's theorem (Mercer 1909). It requires positive definiteness, i.e., for any data set $x_1, \ldots, x_n$ and any real numbers $\lambda_1, \ldots, \lambda_n$ the function K must satisfy

$$\sum_{i=1}^{n} \sum_{j=1}^{n} \lambda_i \lambda_j K(x_i, x_j) \geq 0. \qquad (20.30)$$

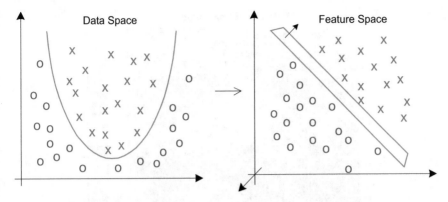

Fig. 20.14 Mapping into a three-dimensional feature space from a two-dimensional data space $\mathbb{R}^2 \mapsto \mathbb{R}^3$. The transformation $\Psi(x_1, x_2) = (x_1^2, \sqrt{2}x_1x_2, x_2^2)^\top$ corresponds to the kernel function $K(x_i, x_j) = (x_i^\top x_j)^2$

Some examples of kernel functions are

- $K(x_i, x_j) = e^{-\|x_i - x_j\|/2\sigma^2}$—the isotropic Gaussian kernel with constant σ
- $K(x_i, x_j) = e^{-(x_i - x_j)^\top r^{-2} \Sigma^{-1}(x_i - x_j)/2}$—the stationary Gaussian kernel with an anisotropic radial basis with constant r and variance–covariance matrix Σ from training set
- $K(x_i, x_j) = (x_i^\top x_j + 1)^p$—the polynomial kernel of degree p
- $K(x_i, x_j) = \tanh(kx_i^\top x_j - \delta)$—the hyperbolic tangent kernel with constant k and δ.

SVMs for Simulated Data

The basic parameters of SVMs are on the scaling r of the anisotropic radial basis functions (in the stationary Gaussian kernel) and the capacity C. The parameter r controls the local resolution of the SVM in the sense that smaller r create smaller curvature of the margin. The capacity C controls the amount of slack to allow for unclassified observations. A large C would create a very rough and curved margin where C close to zero makes the margin more smooth.

One of the guinea pig tests for a classification algorithm is the data described as "orange peel", i.e., when two groups of observations have similar means, their variance, however, being different. The classification results in this case are presented in Fig. 20.15. An SVM with a radial basis kernel is highly suitable for such a kind of data.

Another popular nonlinear test is the classification of "spiral data". We generated two spirals with the distance between them equal 1.0 that span over 3π radian. The SVM was chosen with $r = 0.1$ and $C = 10/n$. The SVM was able to separate the classes without an error if noise with parameters $\varepsilon_i \sim N(0, 0.1^2 \mathcal{I})$ was injected into the pure spiral data (Fig. 20.16). Obviously, both the "orange peel" and the "spiral data" are not linearly separable.

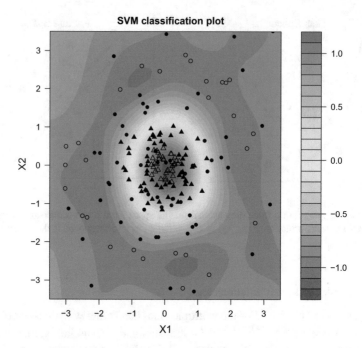

SVM classification plot

Fig. 20.15 SVM classification results for the "orange peel" data, $n = 200$, $d = 2$, $n_{-1} = n_{+1} = 100$, $x_{+1,i} \sim N((0, 0)^{\top}, 2^2 \mathcal{I})$, $x_{-1,i} \sim N((0, 0)^{\top}, 0.5^2 \mathcal{I})$ with SVM parameters $r = 0.5$ and $C = 20/200$ ⦿ MVAsvmOrangePeel

Solution of the SVM Classification Problem

The standard SVM optimization problem (20.29), which is a quadratic optimization problem, is usually solved by means of quadratic programming (QP). This technique, however, is notorious for (i) its bad scaling properties (the time required to solve the problem is proportional to n^3, where n is the number of observations), (ii) implementation difficulty, and (iii) enormous memory requirements. With the QP technique the whole kernel matrix of the size $n \times n$ has to be fit in the memory, which, assuming that each variable takes up 10 bytes of memory, will require $10 \times n \times n$ bytes. This means that 1 million observation (which is not unusual for practical applications such as credit scoring) will require 12000 TB (terabytes) or 10000000 MB of operating memory to store. With a typical size of the computer memory of 512 MB no more than around 5000 observations can be processed. Thus, the main emphasis in designing new algorithms was made on using special properties of SVMs to speed up the solution and reduce memory requirements.

Scoring Companies

For our illustration we selected the largest bankrupt companies with the capitalization of no less than 1 billion USD. The data set used in this work is from the Credit reform database provided by the Research Data Center (RDC) of the Humboldt Universität

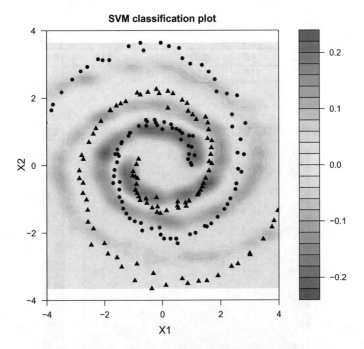

Fig. 20.16 SVM classification results for the noisy spiral data. The spirals spread over 3π radian; the distance between the spirals equals 1.0. $d = 2$, $n_{-1} = n_{+1} = 100$, $n = 200$. The noise was injected with the parameters $\varepsilon_i \sim N(0, 0.1^2\mathcal{I})$. The separation is perfect with SVM parameters $r = 0.1$ and $C = 10/200$ ☒ MVAsvmSpiral

zu Berlin. It contains financial information from about 20000 solvent and 1000 insolvent German companies. The period spans from 1996 to 2002 and in the case of the insolvent companies the information is gathered 2 years before the insolvency took place. The last annual report of a company before it goes bankrupt receives the indicator $y = 1$ and for the rest (solvent) companies $y = -1$.

We are given 28 variables, i.e., cash, inventories, equity, EBIT, number of employees, and branch code. From the original data, we create common financial indicators which are denoted as $x1, \cdots, x25$. These ratios can be grouped into four categories such as profitability, leverage, liquidity, and activity.

Obviously, data for the year 1996 are missing and we will exclude them for further calculations. In order to reduce the effect of the outliers on the results, all observations that exceeded the upper limit of IQ (Inter-quartile range) or the lower limit of IQ were replaced with these values. To demonstrate how performance changes, we will use the Accounts Payable (AP) turnover (named $X24$) and ratio of Operating Income (OI) and Total Asset (TA) (named $X3$). We choose randomly 50 solvent and 50 insolvent companies. The statistical description of financial ratios is summarized in Table 20.5.

Table 20.5 Descriptive statistics for financial ratios

Ratio	$q_{0.05}$		Med.		$q_{0.95}$		IQR	
OI/TA	−0	22	0	00	0	10	0	06
AP/Sales	0	03	0	14	0	36	0	10

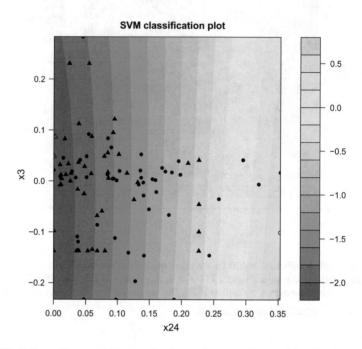

Fig. 20.17 Ratings of companies in two dimensions. Low complexity of classifier functions with $\sigma = 100$ and $C = 1$. Percentage of misclassification is 0.43. The separation is perfect with SVM parameters $r = 0.1$ and $C = 10/200$ 🔍 MVAsvmSig100C1

Keep in mind that different kernels will influence performance. We will use one of the most common ones, the isotropic Gaussian kernel. Triangles and circles in Fig. 20.17 represent successful and failing companies from the training set, respectively. The colored background corresponds to different score values f. The more blue the area, the higher the score and the greater the probability of default. Most successful companies lying in the red area have positive profitability and a reasonable activity.

Figure 20.17 presents the classification results for an SVM using isotropic Gaussian kernel with $\sigma = 100$ and the fixed capacity $C = 1$. With given priors, the SVM has trouble classifying between solvent and insolvent company. The radial base σ, which determines the minimum radius of a group, is too large. Notice that SVM do a poor job of distinguishing between groups even though most observations are used as support vector.

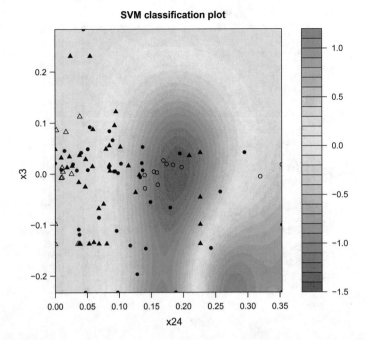

Fig. 20.18 Ratings of companies in two dimensions. The case of an average complexity of classifier functions with $\sigma = 2$ and capacity is fixed at $C = 1$. Percentage of misclassification is reduced to 0.27 🔍 MVAsvmSig2C1

The applied SVMs differed in two aspects: (i) their capacity that is controlled by the coefficient C in (20.28) and (ii) the complexity of classifier functions controlled in our case by the isotropic radial basis in the Gaussian kernel. In Fig. 20.18 The value σ is reduced to 2 while C remains the same. SVM start recognizing the difference between solvent and insolvent companies resulting in sharper cluster. Figure 20.19 demonstrate the effect of the changing capacity to the classification result. The optimization of SVM parameters (C and σ) can be done by using grid search method or an other advance algorithm so-called Genetic Algorithm.

Figure 20.20 shows a cumulative accuracy profile (CAP) curve which is particularly useful in that it simultaneously measures Type I and Type II errors. In statistical terms, the CAP curve represents the cumulative probability of default events for different percentiles of the risk score scale. Now, we introduce accuracy ratio (AR) derived from CAP curve for measuring and comparing the performance of credit risk model. Therefore, AR is defined as the ratio of the area between a model CAP curve and the random curve to the area between the perfect CAP curve and the random CAP curve (see Fig. 20.20). Perfect classification is attained if the value of AR is equal to one.

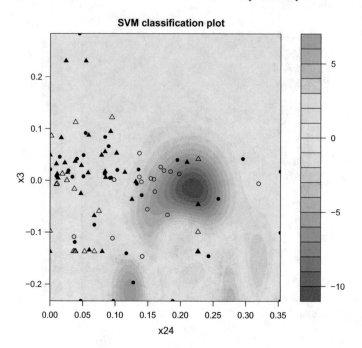

Fig. 20.19 Ratings of companies in two dimensions. High capacity ($C = 200$) with radial basis is fixed at $\sigma = 0.5$. Percentage of misclassification is 0.10 🔍 MVAsvmSig05C200

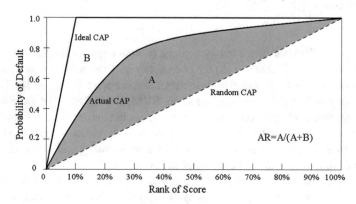

Fig. 20.20 Cumulative accuracy profile (CAP) curve

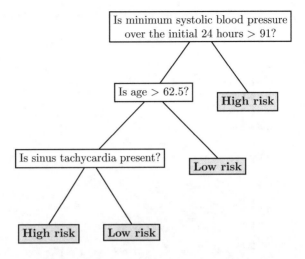

Fig. 20.21 Decision tree for low/high patients

Summary
↪ SVM classification is done by mapping the data into feature space and finding a separating hyperplane there.
↪ The support vectors are determined via a quadratic optimization problem.
↪ SVM produces highly nonlinear classification boundaries.

20.5   Classification and Regression Trees

Classification and regression trees (CART) is a method of data analysis developed by a group of American statisticians Breiman et al. (1984). The aim of CART is to classify observations into a subset of *known classes* or to predict levels of regression functions. CART is a nonparametric tool which is designed to represent decision rules in a form of so-called *binary trees*. Binary trees split a learning sample parallel to the coordinate axis and represent the resulting data clusters hierarchically starting from a *root node* for the whole learning sample itself and ending with relatively homogenous buckets of observations.

Regression trees are constructed in a similar way but the final buckets do not represent classes but rather approximations to an unknown regression functions at a particular point of the independent variable. In this sense regression trees are estimates via a nonparametric regression model. Here we provide an outlook of how decision trees are created, what challenges arise during practical applications and, of course, a number of examples will illustrate the power of CART.

How Does CART Work ?

Consider the example of how high risk patients (those who will not survive at least 30 days after a heart attack is admitted) were identified at San Diego Medical Center,

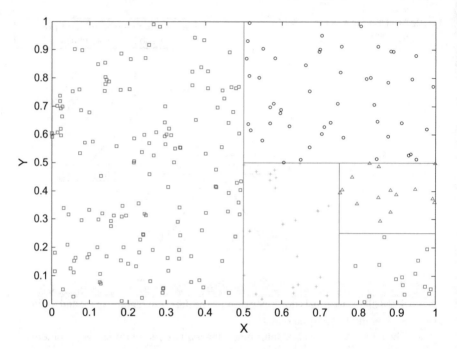

Fig. 20.22 CART orthogonal splitting example where each color corresponds to one cluster

University of California on the basis of initial 24-hour data. A classification rule using at most three decisions (questions) is presented in Fig. 20.21. Left branches of the tree represent cases of positive answers, right branches—negative ones so that, e.g., if minimum systolic blood pressure over the last 24 h is less or equal 91, then the patient belongs to *the high risk* group. In this example the dependant variable is binary: low risk (0) and high risk (1).

A different situation occurs when we are interested in the expected *amount* of days the patient will be able to survive. The decision tree will probably change and the *terminal nodes* will now indicate a mean expected number of days the patient will survive. This situation describes a regression tree rather than a classification tree.

In a more formal setup let Y be a dependent variable—binary or continuous and $X \in \mathbb{R}^d$. We are interested in approximating

$$f(x) = \mathrm{E}(Y|X = x)$$

For the definition of conditional expectations we refer to Sect. 4.2. CART estimates this function f by a step function that is constructed via splits along the coordinate axis. An illustration is given in Fig. 20.22. The regression function $f(x)$ is approximated by the values of the step function. The splits along the coordinate axes are to be determined from the data.

The following simple one dimensional example shows that the choice of splits points involves some decisions. Suppose that $f(x) = I(x \in [0, 1]) + 2 I(x \in [1, 2])$ is a simple step function with a step at $x = 1$. Assume now that one observes $Y_i = f(x_i) + \varepsilon_i$, $X_i \sim U[0, 2]$, $\varepsilon_i \sim N(0, 1)$. By going through the X data points as possible split points one sees that in the neighborhood of $x = 1$ one has two possibilities: one simply takes the X_i left to 1 or the observation right to 1. In order to make such splits unique one averages these neighboring points.

Impurity Measures

A more formal framework on how to split and where to split needs to be developed. Suppose there are n observations in the learning sample and n_j is the overall number of observations belonging to class j, $j = 1, \ldots, J$. The *class probabilities* are

$$\pi(j) = \frac{n_j}{n}, j = 1, \ldots, J \tag{20.31}$$

$\pi(j)$ is the proportion of observations belonging to a particular class. Let $n(t)$ be the number of observations at node t and $n_j(t)$—the number of observations belonging to the j-th class at t. The frequency of the event that an observation of the j-th class falls into node t is

$$p(j, t) = \pi(j) \frac{n_j(t)}{n_j} \tag{20.32}$$

The proportion of observations at t are $p(t) = \sum_{j=1}^{J} p(j, t)$ the *conditional probability* of an observation to belong to class j given that it is at node t is

$$p(j|t) = \frac{p(j, t)}{p(t)} = \frac{n_j(t)}{n(t)} \tag{20.33}$$

Define now a degree of class homogeneity in a given node. This characteristic—an *impurity measure* $i(t)$—will represent a class homogeneity indicator for a given tree node and hence will help to find optimal splits. Define an *impurity function* $\iota(t)$ which is determined on $(p_1, \ldots, p_J) \in [0, 1]^J$ with $\sum_{j=1}^{J} p_j = 1$ so that

1. ι has a unique maximum at point $\left(\frac{1}{J}, \frac{1}{J}, \ldots, \frac{1}{J}\right)$;
2. ι has a unique minimum at points $(1, 0, 0, \ldots, 0), (0, 1, 0, \ldots, 0), \ldots, (0, 0, 0, \ldots, 1)$;
3. ι is a symmetric function of $p_1, \ldots, p_J$

Each function satisfying these conditions is called an impurity function. Given ι, define the *impurity measure* $i(t)$ for a node t as

$$i(t) = \iota\{p(1|t), p(2|t), \ldots, p(J|t)\} \tag{20.34}$$

Fig. 20.23 Parent and child nodes hierarchy

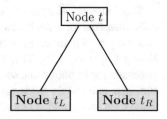

Denote an arbitrary data split by s, then for a given node t which we will call a *parent node* two *child nodes* described in Fig. 20.23 arise: t_L and t_R representing observations meeting and not meeting the split criterion s. A fraction p_L of data from t falls to the left child node and $p_R = 1 - p_L$ is the share of data in t_R.

A *quality measure* of how well split s works is

$$\Delta i(s, t) = i(t) - p_L i(t_L) - p_R i(t_R) \tag{20.35}$$

The higher the value of $\Delta i(s, t)$ the better split we have since data impurity is reduced. In order to find an optimal split s it is natural to maximize $\Delta i(s, t)$. Note that in (20.35) for different splits s, the value of $i(t)$ remains constant, hence it is equivalent to find

$$
\begin{aligned}
s^* &= \operatorname*{argmax}_{s} \Delta i\,(s, t) \\
&= \operatorname*{argmax}_{s} \{-p_L i\,(t_L) - p_R i\,(t_R)\} \\
&= \operatorname*{argmax}_{s} \{p_L i\,(t_L) + p_R i\,(t_R)\}
\end{aligned}
$$

where t_L and t_R are implicit functions of s. This splitting procedure is repeated until one arrives at a minimal bucket size. Classes are then assigned to terminal nodes using the following rule:

$$\text{If } p(j|t) = \max_{i} p(i|t), \text{ then } j^*(t) = j \tag{20.36}$$

If the maximum is not unique, then $j^*(t)$ is assigned randomly to those classes for which $p(i|t)$ takes its maximum value. The crucial question is of course to define an impurity function $i\,(t)$. A natural definition of impurity is via a *variance* measure: Assign 1 to all observations at node t belonging to class j and 0 to others. A sample variance estimate for node t observations is $p(j|t)\{1 - p(j|t)\}$.

Summing over all J classes we obtain the *Gini index*:

$$i\,(t) = \sum_{j=1}^{J} p(j|t)\{1 - p(j|t)\} = 1 - \sum_{j=1}^{J} p^2(j|t) \tag{20.37}$$

The Gini index is an impurity function $\iota(p_1, \ldots, p_J)$, $p_j = p(j|t)$. It is not hard to see that the Gini index is a convex function. Since $p_L + p_R = 1$, we get

$$
\begin{aligned}
i(t_L)p_L + i(t_R)p_R &= \iota\{p(1|t_L), \ldots, p(J|t_L)\} p_L + \iota\{p(1|t_R), \ldots, p(J|t_R)\} p_R \\
&\leq \iota\{p_L p(1|t_L) + p_R p(1|t_R), \ldots, p_L p(J|t_L) + p_R p(J|t_R)\}
\end{aligned}
$$

where inequality becomes an equality in case $p(j|t_L) = p(j|t_R)$, $j = 1, \ldots, J$.
Recall that

$$
\frac{p(j, t_L)}{p(t)} = \frac{p(t_L)}{p(t)} \cdot \frac{p(j, t_L)}{p(t_L)} = p_L p(j|t_L)
$$

and since

$$
p(j|t) = \frac{p(j, t_L) + p(j, t_R)}{p(t)} = p_L p(j|t_L) + p_R p(j|t_R)
$$

we can conclude that

$$
i(t_L)p_L + i(t_R)p_R \leq i(t) \tag{20.38}
$$

Hence each variant of data split leads to $\Delta i(s, t) > 0$ unless $p(j|t_R) = p(j|t_L) = p(j|t)$, i.e., when no split decreases class heterogeneity.

Impurity measures can be defined in a number of different ways, for practical applications the so-called *twoing rule* can be considered. Instead of maximizing impurity change at a particular node, the twoing rule tries to balance as if the learning sample had only two classes. The reason for such an algorithm is that such a decision rule is able to distinguish observations between general factors on top levels of the tree and take into account specific data characteristics at lower levels.

If $S = \{1, \ldots, J\}$ is the set of learning sample classes, divide it into two subsets

$$
S_1 = \{j_1, \ldots, j_n\}, \text{ and } S_2 = S \backslash S_1
$$

All observations belonging to S_1 get dummy class 1, and the rest dummy class 2. The next step is to calculate $\Delta i(s, t)$ for different s *as if there were only two (dummy) classes*. Since actually $\Delta i(s, t)$ depends on S_1, the value $\Delta i(s, t, S_1)$ is maximized. Now apply a *two-step procedure*: first, find $s^*(S_1)$ maximizing $\Delta i(s, t, S_1)$ and second, find a *superclass* S_1^* maximizing $\Delta i\{s^*(S_1), t, S_1\}$. In other words the idea of twoing is to find a combination of superclasses at each node that maximizes the impurity increment for two classes.

This method provides one big advantage: it finds so-called *strategic nodes*, i.e., nodes filtering observations in the way that they are different to the maximum feasible extent. Although applying the twoing rule may seem to be desirable especially for data with a big number of classes, another challenge arises: computational speed. Let's assume that the learning sample has J classes, then a set S can be split into S_1 and S_2 by 2^{J-1} ways. For 11 classes data this will create more than 1000 com-

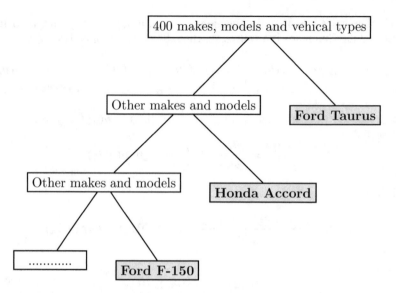

Fig. 20.24 Classification tree constructed by Gini index

binations. Fortunately the following result helps to reduce drastically the amount of computations.

It can be proven Breiman et al. (1984) that in a classification task with two classes and impurity measure $p(1|t)p(2|t)$ for an arbitrary split s a superclass $S_1(s)$ is determined by

$$S_1(s) = \{j : p(j|t_L) \geq p(j|t_R)\},$$

$$\max_{S_1} \Delta i(s, t, S_1) = \frac{p_L p_R}{4} \left\{ \sum_{j=1}^{J} |p(j|t_L) - p(j|t_R)| \right\}^2 \tag{20.39}$$

Hence the twoing rule can be applied in practice as well as Gini index, although the first criterion works a bit slower.

Gini Index and Twoing Rule in Practice

In this section we look at practical issues of using these two rules. Consider a learning data set from Salford Systems with 400 observations characterizing automobiles: their make, type, color, technical parameters, age, etc. The aim is to build a decision tree splitting different cars by their characteristics based on feasible relevant parameters. The classification tree constructed using the Gini index is given in Fig. 20.24.

A particular feature here is that at each node observations belonging to one make are filtered out, i.e., observations with most striking characteristics are separated. As a result a decision tree is able to pick out automobile makes quite easily.

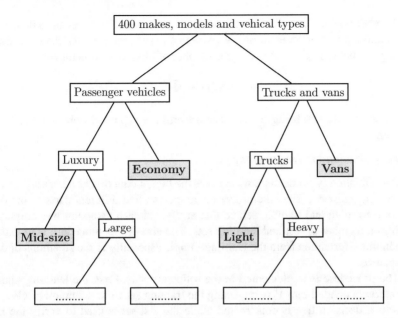

Fig. 20.25 Classification tree constructed by twoing index

The twoing rule based tree Fig. 20.25 for the same data is different. Instead of specifying particular car makes at each node, application of the twoing rule results in strategic nodes, i.e., questions which distinguish between different car classes to the maximum extent. This feature can be vital when high-dimensional data sets with a big number of classes are processed.

Optimal Size of a Decision Tree

Up to now we were interested in determining the best split s^* at a particular node. The next and perhaps more important question is how to determine the optimal tree size, i.e., when to *stop splitting*. If each terminal node has only class homogenous data set, then every point of the learning sample can be flawlessly classified using this *maximum tree*. But can be such an approach fruitful?

The maximum tree is a case of overspecification. Some criterion is required to stop data splitting. Since tree building is dependent on $\Delta i(s, t)$, a criterion is to stop data splitting if

$$\Delta i(s, t) < \bar{\beta}, \tag{20.40}$$

where $\bar{\beta}$ is some threshold value.

The value of $\bar{\beta}$ is to be chosen in a subjective way and this is unfortunately a drawback. Empirical simulations show that the impurity increment is frequently non-monotone, that is why even for small $\bar{\beta}$ the tree may be underparametrized. Setting even smaller values for $\bar{\beta}$ will probably remedy the situation but at the cost of tree overparametrization.

Another way to determine the adequate shape of a decision tree is to demand a minimum number of observations $\overline{N}$ (bucked size) at each terminal node. A disadvantage is that if at terminal node t the number of observations is higher

$$N(t) > \overline{N} \tag{20.41}$$

then this node is also being split as data are still not supposed to be clustered well enough.

Cross-Validation for Tree Pruning

Cross-validation is a procedure which uses the bigger data part as a *training set* and the rest as a *test set*. Then the process is looped so that different parts of the data become learning and training set, so that at the end each datapoint was employed both as a member of test and learning sets. The aim of this procedure is to extract maximum information from the learning sample especially in the situations of data scarceness.

The procedure is implemented in the following way. First, the learning sample is *randomly* divided into V parts. Using the training set from the union of $(V - 1)$ subsets a decision tree is constructed while the test set is used to verify the tree quality. This procedure is looped over all possible subsets.

Unfortunately for small values of V cross-validation estimates can be *unstable* since each iteration a cluster of data is selected *randomly* and the number of iterations itself is relatively small, thus the overall estimation result is somewhat random. Nowadays cross-validation with $V = 10$ is an industry standard and for many applications a good balance between computational complexity and statistical precision.

Cost-Complexity Function and Cross-Validation

Another method taken into account is *tree complexity*, i.e., the *number of terminal nodes*. The maximum tree will get a penalty for its big size, on the other hand, it will be able to make perfect in-sample predictions. Small trees will, of course, get lower penalty for their size but their prediction abilities are limited. Optimization procedure based on such a trade-off criterion could determine a good decision tree.

Define *the internal misclassification error* of an arbitrary observation at node t as $e(t) = 1 - \max_j p(j \mid t)$, define also $E(t) = e(t)p(t)$. Then *internal misclassification tree error* is $E(T) = \sum_{t \in \tilde{T}} E(t)$ where $\tilde{T}$ is a set of terminal nodes. The estimates are called *internal* because they are based solely on the learning sample. It may seem that $E(T)$ as a tree quality measure is sufficient but unfortunately it is not so. Consider the case of the maximum tree, here $E(T_{MAX}) = 0$, i.e., the tree is of best configuration.

For any subtree $T (\leq T_{MAX})$ define the number of terminal nodes $|\tilde{T}|$ as a measure of its complexity. The following cost-complexity function can be used:

$$E_\alpha(T) = E(T) + \alpha |\tilde{T}| \tag{20.42}$$

where $\alpha \geq 0$ is a complexity parameter and $\alpha \left|\widetilde{T}\right|$ is a cost component. The more complex the tree (high number of terminal nodes) the lower is $E(T)$ but at the same time the higher is the penalty $\alpha \left|\widetilde{T}\right|$ and vice versa.

The number of subtrees of T_{MAX} is finite. Hence pruning of T_{MAX} leads to creation of a subtree sequence $T_1, T_2, T_3, \ldots$ with a decreasing number of terminal nodes.

An important question is if a subtree $T \leq T_{MAX}$ for a given α minimizing $E_\alpha(T)$ always exists and whether it is unique?

In Breiman et al. (1984) it is shown that for $\forall \alpha \geq 0$ there exists an optimal tree $T(\alpha)$ in the sense that

1. $E_\alpha \{T(\alpha)\} = \min_{T \leq T_{MAX}} E_\alpha(T) = \min_{T \leq T_{MAX}} \left\{E(T) + \alpha \left|\widetilde{T}\right|\right\}$
2. if $E_\alpha(T) = E_\alpha \{T(\alpha)\}$ then $T(\alpha) \leq T$.

This result is a proof of existance, but also a proof of uniqueness: consider another subtree T' so that T and T' both minimize E_α and are not nested, then $T(\alpha)$ does not exist in accordance with second condition.

The idea of introducing cost-complexity function at this stage is to check only a subset of different subtrees of T_{MAX}: optimal subtrees for different values of α. The starting point is to define the first optimal subtree in the sequence so that $E(T_1) = E(T_{MAX})$ and the size of T_1 is minimum among other subtrees with the same cost level. To get T_1 out of T_{MAX} for each terminal node of T_{MAX} it is necessary to verify the condition $E(t) = E(t_L) + E(t_R)$ and if it is fulfilled—node t is pruned. The process is looped until no extra pruning is available—the resulting tree $T(0)$ becomes T_1.

Define a node t as an *ancestor* of t' and t' as *descendant* of t if there is a connected path down the tree leading from t to t'. Consider Fig. 20.26 where nodes t_4, t_5, t_8, t_9, t_{10}, and t_{11} are descendants of t_2 while nodes t_6 and t_7 are not descendants of t_2 although they are positioned lower since it is not possible to connect them with a path from t_2 to these nodes without engaging t_1. Nodes t_4, t_2 and t_1 are ancestors of t_9 and t_3 is not ancestor of t_9.

Define the *branch* T_t of the tree T as a subtree based on node t and all its descendants. An example is given in Fig. 20.27. Pruning a branch T_t from a tree T means deleting all descendant nodes of t. Denote the transformed tree as $T - T_t$. Pruning the branch T_{t_2} results in the tree described in Fig. 20.28.

For any branch T_t define the *internal misclassification estimate* as

$$E(T_t) = \sum_{t' \in \widetilde{T}_t} E(t'), \qquad (20.43)$$

where $\widetilde{T}_t$ is the set of terminal nodes of T_t. Hence for an arbitrary node t of T_t:

$$E(t) > E(T_t) \qquad (20.44)$$

Consider now the *cost-complexity misclassification estimate* for branches or single nodes. Define for a single node $\{t\}$:

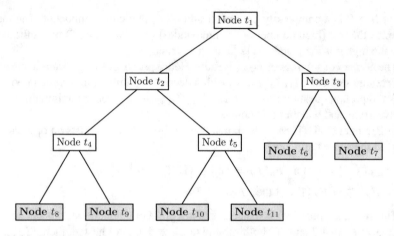

Fig. 20.26 Decision tree hierarchy

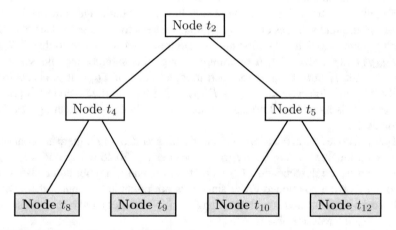

Fig. 20.27 The branch T_{t_2} of the original tree T

Fig. 20.28 $T - T_{t_2}$ the
pruned tree T

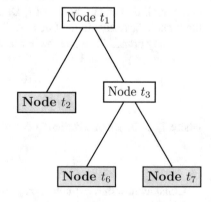

$$E(\{t\}) = E(t) + \alpha \tag{20.45}$$

and for a branch:

$$E_\alpha(T_t) = E(T_t) + \alpha \left|\widetilde{T}_t\right| \tag{20.46}$$

When $E_\alpha(T_t) < E_\alpha(\{t\})$ the branch T_t is preferred to a single node $\{t\}$ according to cost-complexity. For some α both (20.45), (20.46) will become equal. This critical value of α can be determined from

$$E_\alpha(T_t) < E_\alpha(\{t\}) \tag{20.47}$$

which is equivalent to

$$\alpha < \frac{E(t) - E(T_t)}{\left|\widetilde{T}_t\right| - 1}, \tag{20.48}$$

where $\alpha > 0$ since $E(t) > E(T_t)$.

To obtain the next member of the subtrees sequence, i.e., T_2 out of T_1 a special node called *weak link* is determined. For this purpose a function $g_1(t)$, $t \in T_1$ is defined as

$$g_1(t) = \begin{cases} \frac{E(t) - E(T_t)}{\left|\widetilde{T}_t\right| - 1}, & t \notin \widetilde{T}_1 \\ +\infty, & t \in \widetilde{T}_1 \end{cases} \tag{20.49}$$

Node $\bar{t}_1$ is a weak link in T_1 if

$$g_1(\bar{t}_1) = \min_{t \in T_1} g_1(t) \tag{20.50}$$

and a new value for α_2 is defined as

$$\alpha_2 = g_1(\bar{t}_1) \tag{20.51}$$

A new tree $T_2 \prec T_1$ in the sequence is obviously defined by pruning the branch $T_{\bar{t}_1}$, i.e.,

$$T_2 = T_1 - T_{\bar{t}_1} \tag{20.52}$$

The process is looped until root node $\{t_0\}$—the final member of sequence—is reached. When there are multiple weak links detected, for instance, $g_k(\bar{t}_k) = g_k(\bar{t}'_k)$, then both branches are pruned, i.e., $T_{k+1} = T_k - T_{\bar{t}_k} - T_{\bar{t}'_k}$.

In this way it is possible to get the sequence of optimal subtrees $T_{MAX} \succ T_1 \succ T_2 \succ T_3 \succ \ldots \succ \{t_0\}$ for which it is possible to prove that the sequence $\{\alpha_k\}$ is increasing, i.e., $\alpha_k < \alpha_{k+1}$, $k \geq 1$ and $\alpha_1 = 0$. For $k \geq 1$: $\alpha_k \leq \alpha < \alpha_{k+1}$ and $T(\alpha) = T(\alpha_k) = T_k$.

Practically this tells us how to implement the search algorithm. First, the maximum tree T_{MAX} is taken, then T_1 is found and a weak link $\bar{t}_1$ is detected and branch $T_{\bar{t}_1}$ is pruned off, α_2 is calculated and the process is continued.

Table 20.6 Typical pruning speed

Tree	T_1	T_2	T_3	T_4	T_5	T_6	T_7	T_8	T_9	T_{10}	T_{11}	T_{12}	T_{13}
$|\widetilde{T}_k|$	71	63	58	40	34	19	10	9	7	6	5	2	1

When the algorithm is applied to T_1, the number of pruned nodes is usually quite significant. For instance, consider the following typical empirical evidence (see Table 20.6). When the trees become smaller, the difference in the number of terminal nodes also gets smaller.

Finally, it is worth mentioning that the sequence of optimally pruned subtrees is a subset of trees which might be constructed using direct method of internal misclassification estimator minimization given a fixed number of terminal nodes. Consider an example of tree $T(\alpha)$ with seven terminal nodes, then there is no other subtree T with seven terminal nodes having lower $E(T)$. Otherwise

$$E_\alpha(T) = E(T) + 7\alpha < E_\alpha\{T(\alpha)\} = \min_{T \le T_{MAX}} E_\alpha(T)$$

which is impossible by definition.

Applying the method of V-fold cross-validation to the sequence $T_{MAX} \succ T_1 \succ T_2 \succ T_3 \succ \ldots \succ \{t_0\}$, an *optimal tree* is determined. On the other hand it is frequently pointed out that choice of tree with minimum value of $E^{CV}(T)$ is not always adequate since $E^{CV}(T)$ is not too robust, i.e., there is a whole range of values $E^{CV}(T)$ satisfying $E^{CV}(T) < E^{CV}_{MIN}(T) + \varepsilon$ for small $\varepsilon > 0$. Moreover, when $V < N$ a simple change of random generator seed will definitely result in changed values of $|\widetilde{T}_k|$ minimizing $\widehat{E}(T_K)$. Hence a so-called *one standard error* empirical rule is applied which states that if T_{k_0} is the tree minimizing $E^{CV}(T_{k_0})$ from the sequence $T_{MAX} \succ T_1 \succ T_2 \succ T_3 \succ \ldots \succ \{t_0\}$, then a value k_1 and a correspondent tree T_{k_1} are selected so that

$$\underset{k_1}{\arg\max}\ \widehat{E}(T_{k_1}) \le \widehat{E}(T_{k_0}) + \sigma\left\{\widehat{E}(T_{k_0}),\right\} \tag{20.53}$$

where $\sigma(\cdot)$ denotes sample estimate of standard error and $\widehat{E}(\cdot)$—the relevant sample estimators.

The dotted line on Fig. 20.29 shows the area where the values of $\widehat{E}(T_k)$ only slightly differ from $\min_{|\widetilde{T}_k|} \widehat{E}(T_k)$. The left edge which is roughly equivalent to 16 terminal nodes shows the application of one standard error rule. The use of one standard error rule allows not only to achieve more robust results but also to get trees of lower complexity given the error comparable with $\min_{|\widetilde{T}_k|} \widehat{E}(T_k)$.

Regression Trees

Up to now we concentrate on classification trees. Although *regression trees* share a similar logical framework, there are some differences which need to be addressed.

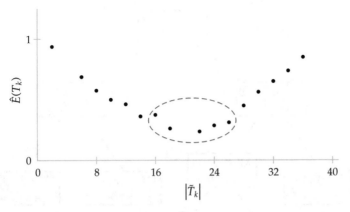

Fig. 20.29 The example of relationship between $\widehat{E}(T_k)$ and number of terminal nodes

The important difference between classification and regression trees is the type of dependent variable Y. When Y is discrete, a decision tree is called a classification tree, a regression tree is a decision tree with a *continuous* dependent variable.

Gini index and twoing rule discussed in previous sections assume that the number of classes is finite and hence introduce some measures based mainly on $p(j|t)$ for arbitrary class j and node t. But since in case of continuous dependent variable there are no more classes, this approach cannot be used anymore unless groups of continuous values are effectively substituted with artificial classes. Since there are no classes anymore—how can be the maximum regression tree determined? Analogously with discrete case, absolute homogeneity can be then described only after some adequate impurity measure for regression trees is introduced.

Recall the idea of *Gini index*, then it becomes quite natural to use the *variance* as impurity indicator. Since for each node data variance can be easily computed, then splitting criterion for an arbitrary node t can be written as

$$s^* = \underset{s}{\operatorname{argmax}}\ [p_L \operatorname{var}\{t_L(s)\} + p_R \operatorname{var}\{t_R(s)\}], \qquad (20.54)$$

where t_L and t_R are emerging child nodes which are, of course, directly dependent on the choice of s^*.

Hence the maximum regression tree can be easily defined as a structure where each node has only the same predicted values. It is important to point out that since continuous data have much higher chances to take different values comparing with discrete ones, the size of maximum regression tree is usually very big.

When the maximum regression tree is properly defined, it is then of no problem to get an optimal size tree. Like with classification trees, maximum regression tree is usually supposed to be upwardly pruned with the help of cost-complexity function and cross-validation. That is why the majority of results presented above is applied to regression trees as well.

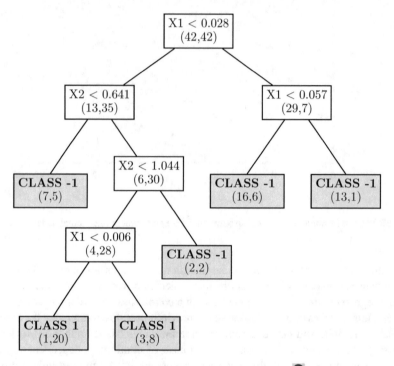

Fig. 20.30 Decision tree for bankruptcy data set: Gini index, $\bar{N} = 30$ 🔍 MVACARTBan1

Bankruptcy Analysis

This section provides a practical study on bankruptcy data involving decision trees. A data set with 84 observations representing different companies is constituted by three variables:

- net income to total assets ratio
- total liabilities to total assets ratio
- company status (-1 if bankrupt and 1 if not)

The data is described in Sect. 22.9.

The goal is to predict and describe the company status given the two primary financial ratios. Since no additional information like the functional form of possible relationship is available, the use of a *classification tree* is an active alternative.

The tree given in Fig. 20.30 was constructed using the Gini index and a $\bar{N} = 30$ constraint, i.e., the number of points in each of the terminal nodes can not be more than 30. Numbers in parentheses displayed on terminal nodes are observation quantities belonging to Class 1 and Class -1.

If we loose the constraint to $\bar{N} = 10$, the decision rule changes, see Fig. 20.31. How exactly did the situation change? Consider the Class 1 terminal nodes of the tree in Fig. 20.30. The first one contains 21 observations and thus was split for $\bar{N} = 10$.

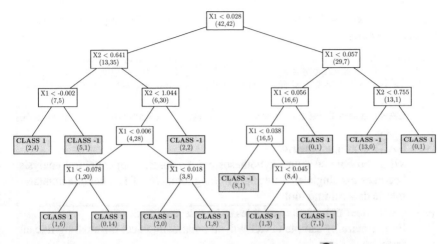

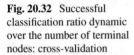

Fig. 20.31 Decision tree for bankruptcy data set: Gini index, $\bar{N} = 10$ MVACARTBan2

Fig. 20.32 Successful
classification ratio dynamic
over the number of terminal
nodes: cross-validation

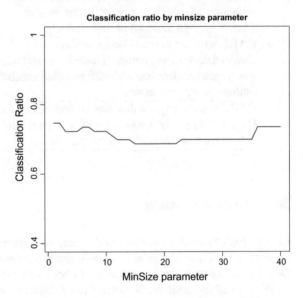

When it was split two new nodes of *different classes* emerged and for both of them
the impurity measure has decreased.

We may conclude that $\bar{N} \approx 10$ is a good choice and analyzing the tree produced
we can state that for this particular example the net income to total assets (X_1)
ratio appears to be an important class indicator. The successful classification ratio
dynamic over the number of terminal nodes is shown in Fig. 20.32. It is chosen by
cross-validation method.

For this example with relatively small sample size we construct two maximum
trees—using the Gini and twoing rules, see Figs. 20.33 and 20.34. Looking at both

decision trees we see that the choice of impurity measure is not so important as the right choice of tree size.

Summary
↪ CART is a tree based method splitting the data sequentially into a binary tree.
↪ CART determined the nodes by minimizing an impurity measure at each mode.
↪ CART is nonparametric: When no data structure hypotheses are available, nonparametric analysis becomes the single effective data mining tool. CART is a flexible nonparametric data mining tool.
↪ CART does not require variables to be selected in advance: From a learning sample CART will automatically select the most significant ones.
↪ CART is very efficient in computational terms: Although all possible data splits are analyzed, the CART architecture is flexible enough to do all of them quickly.
↪ CART is robust to the effect of outliers: Due to data-splitting nature of decision rules creation it is possible to distinguish between data sets with different characteristics and hence to neutralize outliers in separate nodes.
↪ CART can use any combination of continuous and categorical data: Researchers are no longer limited to a particular class of data and will be able to capture more real-life examples.

20.6 Boston Housing

Coming back to the Boston housing data set, we compare the results of exploratory projection pursuit on the original data $\mathcal{X}$ and the transformed data $\widehat{\mathcal{X}}$ motivated in Sect. 1.9. So we exclude X_4 (indicator of Charles River) from the present analysis.

The aim of this analysis is to see from a different angle whether our proposed transformations yield more normal distributions and whether it will yield data with less outliers. Both effects will be visible in our projection pursuit analysis.

We first apply the Jones and Sibson index to the non-transformed data with 50 randomly chosen 13-dimensional directions. Figure 20.35 displays the results in the following form.

In the lower part, we see the values of the Jones and Sibson index. It should be constant for 13-dimensional normal data. We observe that this is clearly not the case. In the upper part of Fig. 20.35 we show the standard normal density as a green curve and two densities corresponding to two extreme index values. The red, slim curve corresponds to the maximal value of the index among the 50 projections. The blue

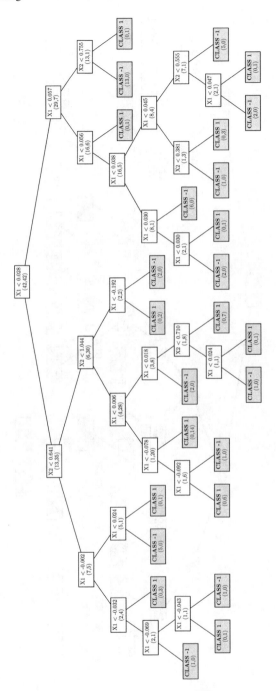

Fig. 20.33 Maximum tree constructed employing Gini index MVACARTGiniTree1

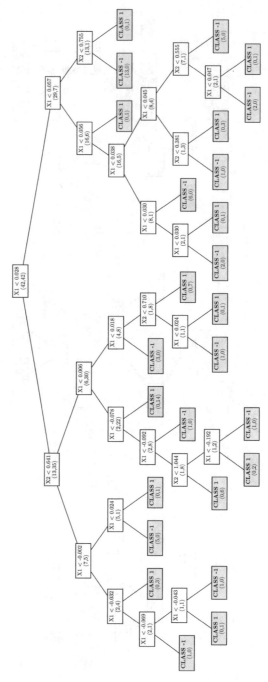

Fig. 20.34 Maximum tree constructed employing twoing rule 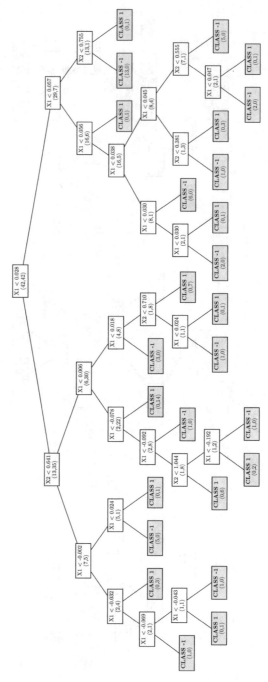 MVACARTTwoingTree1

Fig. 20.35 Projection
Pursuit with the
Sibson–Jones index with 13
original variables

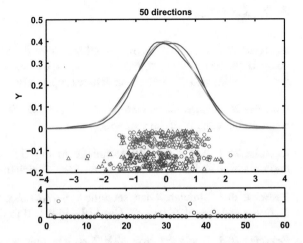

Q MVAppsib

Fig. 20.36 Projection
Pursuit with the
Sibson–Jones index with 13
transformed variables

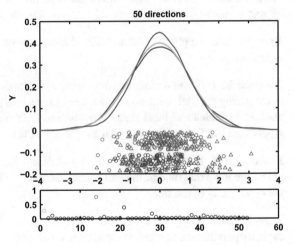

Q MVAppsib

curve, which is close to the normal, corresponds to the minimal value of the Jones and Sibson index. The corresponding values of the indices have the same color in the lower part of Fig. 20.35. Below the densities, a jitter plot shows the distribution of the projected points $\alpha^\top x_i$ $(i = 1, \ldots, 506)$. We conclude from the outlying projection in the red distribution that several points are in conflict with the normality assumption.

Figure 20.36 presents an analysis with the same design for the transformed data. We observe in the lower part of the figure values that are much lower for the Jones and Sibson index (by a factor of 10) with lower variability which suggests that the transformed data is closer to the normal. ("Closeness" is interpreted here in the sense of the Jones and Sibson index.) This is confirmed by looking to the upper part of Fig. 20.36 which has a significantly less outlying structure than in Fig. 20.35.

20.7 Exercises

Exercise 20.1 Calculate the Simplicial Depth for the Swiss bank notes data set and compare the results to the univariate medians. Calculate the Simplicial Depth again for the genuine and counterfeit bank notes separately.

Exercise 20.2 Construct a configuration of points in $\mathbb{R}^2$ such that $x_{med,j}$ from (20.2) is not in the "center" of the scatterplot.

Exercise 20.3 Apply the SIR technique to the U.S. companies data with $Y =$ *market value* and $X =$ *all other variables*. Which directions do you find?

Exercise 20.4 Simulate a data set with $X \sim N_4(0, I_4)$, $Y = (X_1 + 3X_2)^2 + (X_3 - X_4)^4 + \varepsilon$ and $\varepsilon \sim N(0, (0.1)^2)$. Use SIR and SIR II to find the EDR directions.

Exercise 20.5 Apply the Projection Pursuit technique on the Swiss bank notes data set and compare the results to the PC analysis and the Fisher discriminant rule.

Exercise 20.6 Apply the SIR and SIR II technique on the car data set in Sect. 22.3 with $Y = price$.

Exercise 20.7 Generate four regions on the two-dimensional unit square by sequentially cutting parallel to the coordinate axes. Generate 100 two-dimensional Uniform random variables and label them according to their presence in the above regions. Apply the CART algorithm to find the regions bound and to classify the observations.

Exercise 20.8 Modify Exercise 20.7 by defining the regions as lying above and below the main diagonal of the unit square. Make a CART analysis and comment on the complexity of the tree.

Exercise 20.9 Apply the SVM with different radial basis parameter r and different capacity parameter c in order to separate two circular data sets. This example is often called the Orange Peel exercise and involves two Normal distributions $N(\mu, \Sigma_i)$, $i = 1, 2$, with covariance matrices $\Sigma_1 = 2I_2$ and $\Sigma_2 = 0.5I_2$.

Exercise 20.10 The noisy spiral data set consists of two intertwining spirals that need to be separated by a nonlinear classification method. Apply the SVM with different radial basis parameter r and capacity parameter c in order to separate the two spiral data sets.

Exercise 20.11 Apply the SVM to separate the bankrupt from the surviving (profitable) companies using the profitability and leverage ratios given in the Bankruptcy data set in Sect. 22.9.

References

L. Breiman, J.H. Friedman, R. Olshen, C.J. Stone, *Classification and Regression Trees* (Wadsworth, Belmont, 1984)

R.D. Cook, S. Weisberg, Comment on "Sliced inverse regression for dimension reduction". J. Am. Stat. Assoc. **86**(414), 328–332 (1991)

N. Duan, K.-C. Li, Slicing regression: a link-free regression method. Ann. Stat. **19**(2), 505–530 (1991)

EUNITE. Electricity load forecast competition of the European Network on Intelligent Technologies for Smart Adaptive Systems (2001). http://neuron.tuke.sk/competition/

J.H. Friedman, W. Stuetzle, Projection Pursuit Classification (1981). (Unpublished manuscript)

J.H. Friedman, J.W. Tukey, A projection pursuit algorithm for exploratory data analysis. IEEE Trans. Comput. C **23**, 881–890 (1974)

P. Hall, K.-C. Li, On almost linearity of low dimensional projections from high dimensional data. Ann. Stat. **21**(2), 867–889 (1993)

J.L. Hodges, E.L. Lehman, The efficiency of some non-parametric competitors of the t-test. Ann. Math. Stat. **27**, 324–335 (1956)

P. Huber, Projection pursuit. Ann. Stat. **13**(2), 435–475 (1985)

M.C. Jones, R. Sibson, What is projection pursuit? (With discussion). J. R. Stat. Soc. Ser. A **150**(1), 1–36 (1987)

K. Kendall, S. Stuart, *Distribution Theory*, vol. 1, The Advanced Theory of Statistics (Griffin, London, 1977)

J.B. Kruskal, Toward a practical method which helps uncover the structure of a set of observations by finding the line tranformation which optimizes a new "index of condensation", in *Statistical Computation*, ed. by R.C. Milton, J.A. Nelder (Academic Press, New York, 1969), pp. 427–440

J.B. Kruskal, Linear transformation of multivariate data to reveal clustering, in *Multidimensional Scaling: Theory and Applications in the Behavioural Sciences*, volume 1, ed. by R.N. Shepard, A.K. Romney, S.B. Nerlove (Seminar Press, London, 1972), pp. 179–191

K.-C. Li, Sliced inverse regression for dimension reduction (with discussion). J. Am. Stat. Assoc. **86**(414), 316–342 (1991)

K.-C. Li, On principal Hessian directions for data visualization and dimension reduction: another application of Stein's lemma. J. Am. Stat. Assoc. **87**, 1025–1039 (1992)

J. Mercer, Functions of positive and negative type and their connection with the theory of integral equations. Philos. Trans. R. Soc. Lond. **209**, 415–446 (1909)

J.R. Schott, Determining the dimensionality in sliced inverse regression. J. Am. Stat. Assoc. **89**(425), 141–148 (1994)

V. Vapnik, *The Nature of Statistical Learning Theory* (Springer, New York, 1995)

Part IV
Appendix

Chapter 21
Symbols and Notations

21.1 Basics

X, Y	random variables or vectors
$X_1, X_2, \ldots, X_p$	random variables
$X = (X_1, \ldots, X_p)^\top$	random vector
$X \sim \cdot$	X has distribution $\cdot$
$\mathcal{A}, \mathcal{B}$	matrices in Sect. 2.1
Γ, Δ	matrices in Sect. 2.2
$\mathcal{X}, \mathcal{Y}$	data matrices in Sect. 3.1
Σ	covariance matrix in Sect. 3.1
1_n	vector of ones $(\underbrace{1, \ldots, 1}_{n\text{-times}})^\top$ in Sect. 2.1
0_n	vector of zeros $(\underbrace{0, \ldots, 0}_{n\text{-times}})^\top$ in Sect. 2.1
$I(.)$	indicator function, i.e., for a set M is $I = 1$ on M, $I = 0$ otherwise
i	$\sqrt{-1}$
$\Rightarrow$	implication
$\Leftrightarrow$	equivalence
$\approx$	approximately equal
$\otimes$	Kronecker product
iff	if and only if, equivalence.

21.2 Mathematical Abbreviations

$\operatorname{tr}(\mathcal{A})$	trace of matrix $\mathcal{A}$
$\operatorname{hull}(x_1, \ldots, x_k)$	convex hull of points $\{x_1, \ldots, x_k\}$
$\operatorname{diag}(\mathcal{A})$	diagonal of matrix $\mathcal{A}$

© Springer Nature Switzerland AG 2019
W. K. Härdle and L. Simar, *Applied Multivariate Statistical Analysis*,
https://doi.org/10.1007/978-3-030-26006-4_21

rank$(\mathcal{A})$ rank of matrix $\mathcal{A}$
det$(\mathcal{A})$ determinant of matrix $\mathcal{A}$
$C(\mathcal{A})$ column space of matrix $\mathcal{A}$.

21.3 Samples

x, y observations of X and Y
$x_1, \ldots, x_n = \{x_i\}_{i=1}^n$ sample of n observations of X
$\mathcal{X} = \{x_{ij}\}_{i=1,\ldots,n;\, j=1,\ldots,p}$ $(n \times p)$ data matrix of observations of $X_1, \ldots, X_p$
                                 or of $X = (X_1, \ldots, X_p)^T$ in Sect. 3.1
$x_{(1)}, \ldots, x_{(n)}$ the order statistic of $x_1, \ldots, x_n$ in Sect. 1.1
$\mathcal{H}$ centering matrix, $\mathcal{H} = \mathcal{I}_n - n^{-1} 1_n 1_n^{\top}$ in Sect. 3.3

21.4 Densities and Distribution Functions

$f(x)$ density of X
$f(x, y)$ joint density of X and Y
$f_X(x), f_Y(y)$ marginal densities of X and Y
$f_{X_1}(x_1), \ldots, f_{X_p}(x_2)$ marginal densities of $X_1, \ldots, X_p$
$\hat{f}_h(x)$ histogram or kernel estimator of $f(x)$ in Sect. 1.2
$F(x)$ distribution function of X
$F(x, y)$ joint distribution function of X and Y
$F_X(x), F_Y(y)$ marginal distribution functions of X and Y
$\varphi(x)$ density of the standard normal distribution
$\Phi(x)$ standard normal distribution function
$\varphi_X(t)$ characteristic function of X
m_k k-th moment of X
κ_j cumulants or semi-invariants of X.

21.5 Moments

$\mathsf{E}X, \mathsf{E}Y$ mean values of random variables or vectors X and Y
                                 in Sect. 3.1
$\sigma_{XY} = \mathsf{Cov}(X, Y)$ covariance between random variables X and Y in
                                 Sect. 3.1
$\sigma_{XX} = \mathsf{Var}(X)$ variance of random variable X in Sect. 3.1

$$\rho_{XY} = \frac{\mathrm{Cov}(X, Y)}{\sqrt{\mathrm{Var}(X)\mathrm{Var}(Y)}}$$
correlation between random variables X and Y in Sect. 3.2

$\Sigma_{XY} = \mathrm{Cov}(X, Y)$ covariance between r. vectors X and Y, i.e., $\mathrm{Cov}(X, Y) = E(X - EX)(Y - EY)^{\top}$

$\Sigma_{XX} = \mathrm{Var}(X)$ covariance matrix of the random vector X.

21.6 Empirical Moments

$$\bar{x} = \frac{1}{n} \sum_{i=1}^{n} x_i$$
average of X sampled by $\{x_i\}_{i=1,\dots,n}$ in Sect. 1.1

$$s_{XY} = \frac{1}{n} \sum_{i=1}^{n} (x_i - \bar{x})(y_i - \bar{y})$$
empirical covariance of random variables X and Y sampled by $\{x_i\}_{i=1,\dots,n}$ and $\{y_i\}_{i=1,\dots,n}$ in Sect. 3.1

$$s_{XX} = \frac{1}{n} \sum_{i=1}^{n} (x_i - \bar{x})^2$$
empirical variance of random variable X sampled by $\{x_i\}_{i=1,\dots,n}$ in Sect. 3.1

$$r_{XY} = \frac{s_{XY}}{\sqrt{s_{XX} s_{YY}}}$$
empirical correlation of X and Y in Sect. 3.2

$\mathcal{S} = \{s_{X_i X_j}\} = x^{\top}\mathcal{H}x$ emp. cov. matrix of $X_1, \dots, X_p$ or of the random vector $X = (X_1, \dots, X_p)^{\top}$ in Sects. 3.1 and 3.3

$\mathcal{R} = \mathcal{D}^{-1/2}\mathcal{S}\mathcal{D}^{-1/2}$ emp. corr. matrix of $X_1, \dots, X_p$ or of the random vector $X = (X_1, \dots, X_p)^{\top}$ in Sects. 3.2 and 3.3.

21.7 Distributions

$\varphi(x)$ density of the standard normal distribution

$\Phi(x)$ distribution function of the standard normal distribution

$N(0, 1)$ standard normal or Gaussian distribution

$N(\mu, \sigma^2)$ normal distribution with mean μ and variance σ^2

$N_p(\mu, \Sigma)$ p-dimensional normal distribution with mean μ and covariance matrix Σ

$\xrightarrow{\mathcal{L}}$ convergence in distribution   in Sect. 4.5

CLT   Central Limit Theorem   in in Sect. 4.5

χ_p^2 χ^2 distribution with p degrees of freedom

$\chi_{1-\alpha;p}^2$ $1 - \alpha$ quantile of the χ^2 distribution with p degrees of freedom

t_n t-distribution with n degrees of freedom

$t_{1-\alpha/2;n}$ $1 - \alpha/2$ quantile of the t-distribution with n d.f.

$F_{n,m}$ F-distribution with n and m degrees of freedom

$F_{1-\alpha;n,m}$ $1 - \alpha$ quantile of the F-distribution with n and m degrees of freedom

$T^2_{p,n}$ Hotelling T^2-distribution with p and n degrees of freedom.

Chapter 22
Data

22.1 Boston Housing Data

The Boston housing data set was collected by Harrison and Rubinfeld (1978). It comprise 506 observations for each census district of the Boston metropolitan area. The data set was analyzed in Belsley et al. (1980).

X_1: per capita crime rate,
X_2: proportion of residential land zoned for large lots,
X_3: proportion of nonretail business acres,
X_4: Charles River (1 if tract bounds river, 0 otherwise),
X_5: nitric oxides concentration,
X_6: average number of rooms per dwelling,
X_7: proportion of owner-occupied units built prior to 1940,
X_8: weighted distances to five Boston employment centers,
X_9: index of accessibility to radial highways,
X_{10}: full-value property tax rate per \$10,000,
X_{11}: pupil/teacher ratio,
X_{12}: $1000(B - 0.63)^2 I(B < 0.63)$ where B is the proportion of African American,
X_{13}: % lower status of the population,
X_{14}: median value of owner-occupied homes in \$1000.

22.2 Swiss Bank Notes

Six variables measured on 100 genuine and 100 counterfeit old Swiss 1000-franc bank notes. The data stem from Flury and Riedwyl (1988). The columns correspond to the following six variables:

X_1: Length of the bank note,
X_2: Height of the bank note, measured on the left,

© Springer Nature Switzerland AG 2019
W. K. Härdle and L. Simar, *Applied Multivariate Statistical Analysis*,
https://doi.org/10.1007/978-3-030-26006-4_22

X_3: Height of the bank note, measured on the right,
X_4: Distance of inner frame to the lower border,
X_5: Distance of inner frame to the upper border, and
X_6: Length of the diagonal.

Observations 1–100 are the genuine bank notes and the other 100 observations are the counterfeit bank notes.

22.3   Car Data

The car data set Chambers et al. (1983) consists of 13 variables measured for 74 car types. The abbreviations in this section are as follows:

X_1: P   Price,
X_2: M   Mileage (in miles per gallone),
X_3: R78   Repair record 1978 (5-point scale; 5 best, 1 worst),
X_4: R77   Repair record 1977 (scale as before),
X_5: H   Headroom (in inches),
X_6: R   Rear seat clearance (distance from front seat back to rear seat, in inches),
X_7: Tr   Trunk space (in cubic feet),
X_8: W   Weight (in pound),
X_9: L   Length (in inches),
X_{10}: T   Turning diameter (required to make a U-turn, in feet),
X_{11}: D   Displacement (in cubic inches),
X_{12}: G   Gear ratio for high gear,
X_{13}: C   Company headquarter (1—U.S., 2—Japan, 3—Europe).

22.4   Classic Blue Pullovers Data

This is a synthetic data set consisting of 10 measurements of 4 variables. The story: A textile shop manager is studying the sales of "classic blue" pullovers over 10 periods. He uses three different marketing methods and hopes to understand his sales as a fit of these variables using statistics. The variables measured are the following:

X_1: Numbers of sold pullovers,
X_2: Price (in EUR),
X_3: Advertisement costs in local newspapers (in EUR), and
X_4: Presence of a sales assistant (in hours per period).

22.5 U.S. Companies Data

The data set consists of measurements for 79 U.S. companies. The abbreviations in this section are as follows:

X_1: A   Assets (USD),
X_2: S   Sales (USD),
X_3: MV   Market Value (USD),
X_4: P   Profits (USD),
X_5: CF   Cash Flow (USD), and
X_6: E   Employees.

22.6 French Food Data

The data set consists of the average expenditures on food for several different types of families in France (manual workers = MA, employees = EM, managers = CA) with different numbers of children (2, 3, 4, or 5 children). The data is taken from Lebart et al. (1982).

22.7 Car Marks

The synthetic data represents averaged marks for 23 car types from a sample of 40 persons. The marks range from 1 (very good) to 6 (very bad) like German school marks. The variables are the following:

X_1: A   Economy,
X_2: B   Service,
X_3: C   Non-depreciation of value,
X_4: D   Price, Mark 1 for very cheap cars
X_5: E   Design,
X_6: F   Sporty car,
X_7: G   Safety, and
X_8: H   Easy handling.

22.8 U.S. Crime Data

This is a open data set by Uniform Crime Reporting Statistics consisting of 50 measurements of 7 variables. It states for 1 year (1985) the reported number of crimes in the 50 states of the U.S. classified according to 7 categories (X_3–X_9).

X_1: land area (land),
X_2: population 1985 (popu 1985),
X_3: murder (murd),
X_4: rape,
X_5: robbery (robb),
X_6: assault (assa),
X_7: burglary (burg),
X_8: larcery (larc),
X_9: autothieft (auto),
X_{10}: U.S. states region number (reg), and
X_{11}: U.S. states division number (div).

Division numbers		Region numbers	
New England	1	Northeast	1
Mid Atlantic	2	Midwest	2
E N Central	3	South	3
W N Central	4	West	4
S Atlantic	5		
E S Central	6		
W S Central	7		
Mountain	8		
Pacific	9		

22.9   Bankruptcy Data I

The data are the profitability, leverage, and bankruptcy indicators for 84 companies.

The data set contains information on 42 of the largest companies that filed for protection against creditors under Chap. 11 of the U.S. Bankruptcy Code in 2001–2002 after the stock market crash of 2000. The bankrupt companies were matched with 42 surviving companies with the closest capitalizations and the same U.S. industry classification codes available through the Division of Corporate Finance of the Securities and Exchange Commission (SEC 2004).

The information for each company was collected from the annual reports for 1998–1999 (SEC 2004), i.e., 3 years prior to the defaults of the bankrupt companies. The following data set contains profitability and leverage ratios calculated, respectively, as the ratio of net income (NI) and total assets (TA) and the ratio of total liabilities (TL) and total assets (TA).

22.10 Bankruptcy Data II

Altman (1968), quoted by Morrison (1990b), reports financial data on 66 banks.
X1 = (working capital)/(total assets),
X2 = (retained earnings)/(total assets),
X3 = (earnings before interest and taxes)/(total assets),
X4 = (market value equity)/(book value of total liabilities),
X5 = (sales)/(total assets).

The first 33 observations correspond to bankrupt banks and the last 33 for solvent banks as indicated by the last columns: values of y.

22.11 Journaux Data

This is a data set that was created from a survey completed in the 1980s in Belgium questioning people's reading habits. They were asked where they live (10 regions comprised of 7 provinces and 3 regions around Brussels) and what kind of newspaper they read on a regular basis. The 15 possible answers belong to 3 classes: Flemish newspapers (first letter v), French newspapers (first letter f), and both languages (first letter b).

X_1: WaBr   Walloon Brabant.
X_2: Brar   Brussels area.
X_3: Antw   Antwerp.
X_4: FlBr   Flemish Brabant.
X_5: OcFl   Occidental Flanders.
X_6: OrFl   Oriental Flanders.
X_7: Hain   Hainaut.
X_8: Lièg   Liège.
X_9: Limb   Limburg.
X_{10}: Luxe   Luxembourg.

22.12 Timebudget Data

In Volle (1985), we can find data on 28 individuals identified according to sex, country where they live, professional activity and matrimonial status, which indicates the amount of time each person spent on ten categories of activities over 100 days (100·24 h = 2400 h total in each row) in the year 1976.

X_1: prof :   professional activity,
X_2: tran :   transportation linked to professional activity,

X_3: hous :   household occupation,
X_4: kids :   occupation linked to children,
X_5: shop :   shopping,
X_6: pers :   time spent for personal care,
X_7: eat :   eating,
X_8: slee :   sleeping,
X_9: tele :   watching television, and
X_{10}: leis :   other leisures.

maus: active men in the U.S.,
waus: active women in the U.S.,
wnus: nonactive women in the U.S.,
mmus: married men in U.S.,
wmus: married women in U.S.,
msus: single men in U.S.,
wsus: single women in U.S.,
mawe: active men from Western countries,
wawe: active women from Western countries,
wnwe: nonactive women from Western countries,
mmwe: married men from Western countries,
wmwe: married women from Western countries,
mswe: single men from Western countries,
wswe: single women from Western countries,
mayo: active men from Yugoslavia,
wayo: active women from Yugoslavia,
wnyo: nonactive women from Yugoslavia,
mmyo: married men from Yugoslavia,
wmyo: married women from Yugoslavia,
msyo: single men from Yugoslavia,
wsyo: single women from Yugoslavia,
maes: active men from eastern countries,
waes: active women from eastern countries,
wnes: nonactive women from eastern countries,
mmes: married men from eastern countries,
wmes: married women from eastern countries,
mses: single men from eastern countries,
wses: single women from eastern countries.

22.13   Vocabulary Data

This example of the evolution of the vocabulary of children can be found in Bock
(1975). Data are drawn from test results on file in the Records Office of the Laboratory
School of the University of Chicago. They consist of scores, obtained from a cohort

of pupils from the eighth through eleventh grade levels, on alternative forms of the vocabulary section of the Coorperative Reading Test. It provides the following scaled scores shown for the sample of 64 subjects (the origin and units are fixed arbitrarily).

22.14 French Baccalauréat Frequencies

The data consist of observations of 202100 baccalauréats from France in 1976 and give the frequencies for different sets of modalities classified into regions. For a reference, see Bourouche and Saporta (1980). The variables (modalities) are the following:

X_1:A   Philosophy-Letters,
X_2:B   Economics and Social Sciences,
X_3:C   Mathematics and Physics,
X_4:D   Mathematics and Natural Sciences,
X_5:E   Mathematics and Techniques,
X_6:F   Industrial Techniques,
X_7:G   Economic Techniques, and
X_8:H   Computer Techniques.

References

D. Harrison, D.L. Rubinfeld, Hedonic prices and the demand for clean air. J. Environ. Econ. Manag. **5**, 81–102 (1978)

D.A. Belsley, E. Kuh, R.E. Welsch, *Regression Diagnostics* (Wiley, 1980)

J.M. Chambers, W.S. Cleveland, B. Kleiner, P.A. Tukey, *Graphical Methods for Data Analysis* (Duxbury Press, Boston, 1983)

L. Lebart, A. Morineau, J.P. Fénelon, *Traitement des données statistiques* (Dunod, Paris, 1982)

SEC. Securities and Exchange Commission: Archive of historical documents (2004). http://www.sec.gov/cgi-bin/srch-edgar

E.I. Altman, Financial ratios, discriminant analysis and the prediction of corporate bankruptcy. J. Finan. **23**, 589–609 (1968)

D.F. Morrison, *Multivariate Statistical Methods*, 3rd edn. (McGraw-Hill Publishing Company, New York, 1990b)

V. Volle, *Analyse des Données* (Economica, Paris, 1985)

R.D. Bock, *Multivariate Statistical Methods In Behavioral Research* (Mc Graw-Hill, New York, 1975)

Index

A

Actual Error Rate (AER), 404
Adaptive weights clustering, 381
Admissible, 401
Agglomerative techniques, 374
Allocation rules, 395
Analysis of Variance (ANOVA), 91
Andrews' Curves, 23
Angle between two vectors, 64
Apparent Error Rate (APER), 404

B

Bayes discriminant rule, 400
Bernoulli distribution, 130, 131
Best line, 287
Binary structure, 365
Binomial sampling, 253
Biplots, 427
Bootstrap, 161
Bootstrap sample, 163
Boston Housing, 35, 100, 222, 242, 323, 355, 388, 534
Boxplot, 4, 5
    construction, 7

C

Canonical correlation, 431
Canonical correlation analysis, 431
Canonical correlation coefficient, 433
Canonical correlation variable, 433
Canonical correlation vector, 433
Capital Assets Pricing Model (CAPM), 484
Cauchy distribution, 140
Centering matrix, 81
Central Limit Theorem (CLT), 130, 132, 133

Centroid, 377
Characteristic functions, 113, 118
Classic blue pullovers, 74
Cluster algorithms, 370
Cluster analysis, 363
Cobb–Douglas production function, 234
Cochran theorem, 175
Coefficient of determination, 87, 97
    adjusted, 97
Column space, 66, 286
Common factors, 339
Common principal components, 320
Common Principle Components Analysis (CPCA), 320
Communality, 340
Complete linkage, 377
Computationally intensive techniques, 487
Concentration ellipsoid, 126
Conditional approximations, 172
Conditional covariance, 498, 499
Conditional density, 109
Conditional distribution, 170
Conditional expectation, 115, 497, 499
Conditional pdf, 108
Confidence interval, 133
Confusion matrix, 403
Contingency table, 244, 253, 413
Contrast, 208
Convex hull, 489
Copula, 110, 153
Copulae, 151
Correlation, 76
    multiple, 172
Correspondence analysis, 413
Covariance, 72
Covariance matrix
    decomposition, 300

© Springer Nature Switzerland AG 2019
W. K. Härdle and L. Simar, *Applied Multivariate Statistical Analysis*,
https://doi.org/10.1007/978-3-030-26006-4

properties, 114
Cramer-Rao, 189
Cramer–Rao-lower bound, 188
Cramer–Wold, 120
Cumulant, 120
Cumulative distribution function (cdf), 108
Curse of dimensionality, 497

D
Data depth, 489
Degrees of freedom, 93
Dendrogram, 376
Density functions, 108
Determinant, 48
Deviance, 248
Diagonal matrix, 49
Dice, 366
Discriminant analysis, 395
Discriminant rule, 396
Discrimination rules in practice, 402
Dissimilarity of cars, 445
Distance
    d, 61
    Euclidean, 61
    iso-distance curves, 61
Distance matrix, 447
Distance measures, 367
Distribution, 108
Draftsman's plot, 19
Duality relations, 292
Duality theorem, 451

E
Effective dimension reduction directions,
    496, 498
Effective dimension reduction space, 496
Efficient portfolio, 476
Eigenvalues, 51
Eigenvectors, 51
Elastic net, 276
Elliptical distribution, 179
Elliptically symmetric distribution, 497
Existence of a riskless asset, 479
Expected cost of misclassification, 397
Explained variation, 87
Exploratory projection pursuit, 491
Extremes, 7

F
Faces, 21
Factor analysis, 337

Factor analysis model, 338
Factorial axis, 288, 289
Factorial method, 315
Factorial representation, 294, 296
Factorial variable, 288, 295
Factor model, 345
Factors, 286
Factor scores, 353
Farthest Neighbor, 377
Fisher information, 190
Fisher information matrix, 188, 190
Fisher's linear discrimination function, 405
Five-number summary, 5
Flury faces, 23
Fourths, 5
French food expenditure, 318
F-spread, 6
F-test, 94
Full model, 93
Fuzzy clustering, 373

G
General multinormal distribution, 176
G-inverse, 51
    nonuniqueness, 54
Gradient, 58
Group-building algorithm, 364

H
Heavy-tailed distributions, 135
Hessian, 58
Hexagon, 32
Hexagon binning, 32
Hexagon plot, 32
Hierarchical algorithm, 374
Histograms, 11
Hotelling T^2-distribution, 176
Hyperbolic, 138

I
Idempotent matrix, 49
Identity matrix, 49
Independence copula, 111
Independent, 77, 109
Inertia, 294, 296
Information matrix, 189
Interpretation of the factors, 341
Interpretation of the principal components,
    307
Invariance of scale, 341
Inverse, 50

Inverse regression, 497, 498

J
Jaccard, 366
Jacobian, 123
Jordan decomposition, 53, 54

K
Kernel densities, 13
Kernel estimator, 14
K-means clustering, 371
K-median clustering, 373
K-medoids clustering, 372
K-mode clustering, 373
Kulczynski, 366
Kullback–Leibler divergence, 383

L
Laplace distribution, 139
Lasso, 262
Likelihood function, 184
Likelihood ratio test, 196
Limit theorems, 129
Linear discriminant analysis, 399
Linear regression, 84
Linear transformation, 83
Link function, 496
Loadings, 339, 340
   non-uniqueness, 342
Logit models, 251
Log-likelihood function, 184
Log-linear, 244

M
Mahalanobis distance, 399
Mahalanobis transformation, 84, 125
Marginal densities, 109
Marketing strategies, 92
Maximum likelihood discriminant rule, 396
Maximum likelihood estimator, 184
MDS direction, 446
Mean–variance, 475, 476
Median, 5, 488
Metric methods, 446
Mixture model, 142
Model with interactions, 238
Moments, 113
Multidimentional scaling, 443
Multinormal, 127, 167
Multinormal distribution, 125

Multivariate Generalized Hyperbolic Distribution, 145
Multivariate Laplace distribution, 148
Multivariate median, 489
Multivariate t-distribution, 148, 179

N
Nearest Neighbor, 377
Nonexistence of a riskless asset, 478
Nonhomogeneous, 83
Nonmetric methods of MDS, 446
Nonmetric solution, 467
Normal distribution, 185
Normal-inverse Gaussian, 138
Normalized Principal Components (NPCs), 313
Norm of a vector, 64
Null space, 66

O
Odds, 253
Order statistics, 5
Orthogonal complement, 67
Orthogonal matrix, 49
Orthonormed, 289
Outliers, 3
Outside bars, 6

P
Parallel profiles, 219
Partitioned covariance matrix, 168
Partitioned matrices, 59
Partitioning algorithms, 371
Pearson chi-square, 248
Pearson chi-square test for independence, 249
Pool-Adjacent Violators (PAV) algorithm, 453, 471
Portfolio analysis, 475
Portfolio choice, 475
Positive-definite, 55
Positive definiteness, 58
Positive or negative dependence, 21
Positive semidefinite, 55, 82
Principal axes, 62
Principal component method, 349
Principal components, 303
Principal Components Analysis (PCA), 497, 500
Principal components in practice, 303
Principal components technique, 304

Principal components transformation, 300, 303
Principal factors, 347
Profile analysis, 218
Profile method, 464
Projection matrix, 67
Projection pursuit, 491
Projection pursuit regression, 495
Projection vector, 496
Proximity between objects, 365
Proximity measure, 364
p-value, 249

Q
Quadratic discriminant analysis, 401
Quadratic forms, 55
Quadratic response model, 233
Quality of the representations, 317

R
Randomized discriminant rule, 400
Rank, 48
Reduced model, 93
Rotation, 66, 351
Row space, 286
Russel and Rao (RR), 366

S
Sampling distributions, 129
Scatterplot matrix, 19
Separation line, 19
Similarity of objects, 365
Simple Analysis of Variance (ANOVA), 91
Simple Matching, 366
Single linkage, 377
Single matching, 367
Singular normal distribution, 127
Singular Value Decomposition (SVD), 54, 293
Sliced inverse regression, 496, 500
    algorithm, 498
Sliced inverse regression II, 498–500
    algorithm, 499
Solution
    nonmetric, 470
Specific factors, 339
Specific variance, 340

Spectral clustering, 385
Spectral decompositions, 53
Spherical distribution, 178
Standardized Linear Combinations (SLC), 300
Statistics, 129
Stimulus, 463
Student's t-distribution, 85
Student's t with n, 138
Summary statistics, 81
Sum of squares, 93
Supervised learning, 363, 364
Support vector machines, 504
Swiss bank data, 4
Symmetric matrix, 49

T
Tanimoto, 366
Three-way tables, 246
Total variation, 87
Trace, 48
Trade-off analysis, 464
Transformations, 123
Transpose, 50
T-test, 85
Two-factor method, 464

U
Unbiased estimator, 189
Uncorrelated factors, 339
Unexplained variation, 87
Unit vector, 64
Unsupervised learning, 363
Upper triangular matrix, 49

V
Variance explained by PCs, 311
Varimax criterion, 351
Varimax method, 351
Varimax rotation method, 351

W
Ward clustering, 378
Wishart density, 176
Wishart distribution, 174, 176

Printed in the United States
By Bookmasters